Lecture Notes in Computer Science 3091

Springer
Berlin
Heidelberg
New York
Hong Kong
London
Milan
Paris
Tokyo

Vincent van Oostrom (Ed.)

Rewriting Techniques and Applications

15th International Conference, RTA 2004
Aachen, Germany, June 3-5, 2004
Proceedings

Springer

Volume Editor

Vincent van Oostrom
Universiteit Utrecht, Department of Philosophy
Heidelberglaan 8, 3584 CS Utrecht, The Netherlands
E-mail: Vincent.vanOostrom@phil.uu.nl

Library of Congress Control Number: Applied for

CR Subject Classification (1998): F.4, F.3.2, D.3, I.2.2-3, I.1

ISSN 0302-9743
ISBN 3-540-22153-0 Springer-Verlag Berlin Heidelberg New York

Springer-Verlag is a part of Springer Science+Business Media

springeronline.com

Printed in Germany

Typesetting: Camera-ready by author, data conversion by Olgun Computergrafik
Printed on acid-free paper SPIN: 11011743 06/3142 5 4 3 2 1 0

Preface

This volume contains the proceedings of the *15th International Conference on Rewriting Techniques and Applications* (RTA 2004), which was held June 2–5, 2004, at the RWTH Aachen in Germany. RTA is the major forum for the presentation of research on all aspects of rewriting. Previous RTA conferences took place in Dijon (1985), Bordeaux (1987), Chapel Hill (1989), Como (1991), Montreal (1993), Kaiserslautern (1995), Rutgers (1996), Sitges (1997), Tsukuba (1998), Trento (1999), Norwich (2000), Utrecht (2001), Copenhagen (2002), and Valencia (2003).

The program committee selected 19 papers for presentation, including five system descriptions, from a total of 43 submissions. In addition, there were invited talks by Neil Jones, Aart Middeldorp, and Robin Milner.

Many people helped to make RTA 2004 a success. I am grateful to the members of the program committee and the external referees for reviewing the submissions and maintaining the high standards of the RTA conferences. It is a great pleasure to thank the conference chair Jürgen Giesl and the other members of the local organizing committee. They were in charge of the local organization of all events partaking in the Federated Conference on Rewriting, Deduction, and Programming (RDP). Apart from RTA 2004, these events were:

- 2nd International Workshop on Higher-Order Rewriting
  (Delia Kesner, Femke van Raamsdonk, and Joe Wells),
- 5th International Workshop on Rule-Based Programming
  (Slim Abdennadher and Christophe Ringeissen),
- 13th International Workshop on Functional and (Constraint) Logic Programming (Herbert Kuchen),
- IFIP Working Group 1.6 on Term Rewriting (Claude Kirchner),
- 4th International Workshop on Reduction Strategies in Rewriting and Programming (Sergio Antoy and Yoshihito Toyama),
- 7th International Workshop on Termination
  (Michael Codish and Aart Middeldorp).

I thank the organizers of all these events for making RTA 2004 more attractive by collocating with it. Finally, I gratefully acknowledge the financial support of the Deutsche Forschungsgemeinschaft (DFG).

April 2004 Vincent van Oostrom

Preface

Conference Organization

Program Chair

Vincent van Oostrom	*Universiteit Utrecht*

Conference Chair

Jürgen Giesl	*RWTH Aachen*

Program Committee

Zena Ariola	*Oregon*
Jürgen Giesl	*Aachen*
Masahito Hasegawa	*Kyoto*
Hélène Kirchner	*Nancy*
Pierre Lescanne	*Lyon*
Klaus Madlener	*Kaiserslautern*
Narciso Martí-Oliet	*Madrid*
Paul-André Melliès	*Paris*
Oege de Moor	*Oxford*
Vincent van Oostrom	*Utrecht*
Frank Pfenning	*Carnegie Mellon*
Ashish Tiwari	*SRI*
Ralf Treinen	*ENS Cachan*
Roel de Vrijer	*Amsterdam*

Organizing Committee

Jürgen Giesl
Elke Ohlenforst
Arnd Gehrmann
Peter Schneider-Kamp
René Thiemann

RTA Steering Committee

Franz Baader	*Dresden*	
Pierre Lescanne	*Lyon*	
Aart Middeldorp	*Innsbruck*	(chair)
Femke van Raamsdonk	*Amsterdam*	(publicity chair)
Robert Nieuwenhuis	*Barcelona*	
Ralf Treinen	*Cachan*	

List of Referees

Andreas Abel
Jürgen Avenhaus
Franco Barbanera
Alessandro Berarducci
Stefan Blom
Olivier Bournez
Patricia Bouyer
Mark van den Brand
Thierry Cachat
Adam Cichon
Horatiu Cirstea
Dan Dougherty
Steven Eker
Olivier Fissore
David de Frutos-Escrig
Thomas Genet
Alfons Geser
Silvia Ghilezan
Isabelle Gnaedig
Guillem Godoy Balil
Gunter Grieser
Makoto Hamana
Thomas Hillenbrand
Daniel Hirschkoff
Petra Hofstedt
Florent Jacquemard
Patricia Johann
Thierry Joly
Neil Jones
Deepak Kapur
Delia Kesner
Jeroen Ketema
Sava Krstic
Frédéric Lang
Sébastien Limet
Denis Lugiez
Claude Marché
Maurice Margenstern
Ralph Matthes
Aart Middeldorp
Pierre-Etienne Moreau
Paliath Narendran
Monica Nesi
Joachim Niehren
Tobias Nipkow
Hitoshi Ohsaki
Nicolas Ollinger
Peter Ölveczky
Friedrich Otto
Miguel Palomino
Ricardo Peña Marí
Brigitte Pientka
Leonor Prensa Nieto
Maurizio Proietti
Femke van Raamsdonk
Christophe Ringeissen
Mario Rodríguez-Artalejo
Andrea Sattler-Klein
Manfred Schmidt-Schauss
Klaus Schneider
Peter Schneider-Kamp
Helmut Seidl
Damien Sereni
Natarajan Shankar
Ganesh Sittampalam
Mark-Oliver Stehr
Toshinori Takai
Bas Terwijn
René Thiemann
Yoshihito Toyama
Akihiko Tozawa
Xavier Urbain
Rakesh Verma
Laurent Vigneron
Eelco Visser
Fer-Jan de Vries
Clemens Grabmayer
Benjamin Wack
Martijn Warnier
Rolf Wiehagen
Hans Zantema

Table of Contents

Termination Analysis of the Untyped λ-Calculus

Neil D. Jones[1] and Nina Bohr[2]

[1] DIKU, University of Copenhagen
[2] IT University of Copenhagen

Abstract. An algorithm is developed that, given an untyped λ-expression, can certify that its call-by-value evaluation will terminate. It works by an extension of the "size-change principle" earlier applied to first-order programs. The algorithm is sound (and proven so in this paper) but not complete: some λ-expressions may in fact terminate under call-by-value evaluation, but not be recognised as terminating.
The *intensional* power of size-change termination is reasonably high: It certifies as terminating all primitive recursive programs, and many interesting and useful general recursive *algorithms* including programs with mutual recursion and parameter exchanges, and Colson's "minimum" algorithm. Further, the approach allows free use of the Y combinator, and so can identify as terminating a substantial subset of PCF.
The *extensional power* of size-change termination is the set of functions computable by size-change terminating programs. This lies somewhere between Péter's multiple recursive functions and the class of ϵ_0-recursive functions.

1 Introduction

The *size-change* analysis of [5] can show termination of first-order functional programs whose parameter values have a well-founded size order. The method is reasonably general, easily automated, and does not require human invention of lexical or other parameter orders. This paper applies similar ideas to establish termination of *higher-order* programs. For simplicity and generality we focus on the simplest such language, the λ-calculus. We expect that the framework can be naturally extended to higher-order functional programs, e.g., functional subsets of Scheme or ML.

1.1 An Example of Size-Change Analysis

Example 1. A motivating example is the size-change termination analysis of a first-order program, using the framework of [5]. Consider a program with functions f and g defined by mutual recursion:

```
f(x,y)   = if x=0 then y else 1: g(x,y,y)
g(u,v,w) = if w=0 then 3:f(u-1,w) else 2:g(u,v-1,w+2)
```

The three function calls have been labeled 1, 2 and 3. The "control flow graph" in Figure 1 shows the calling function and called function of each call, e.g.,

V. van Oostrom (Ed.): RTA 2004, LNCS 3091, pp. 1–23, 2004.

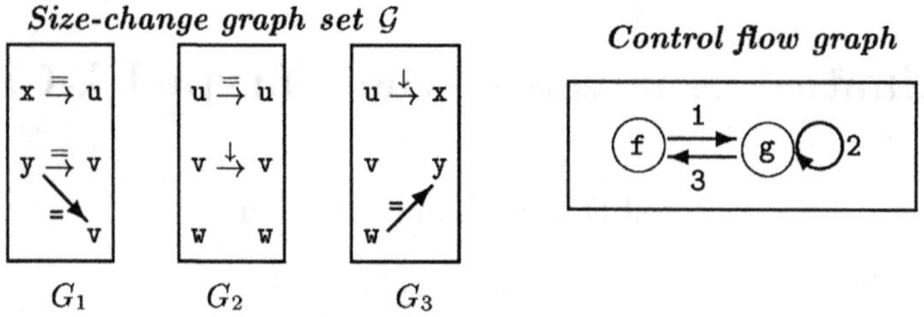

Fig. 1. Call graph and size-change graphs for the example first-order program.

$3 : \mathtt{g} \to \mathtt{f}$. Each call is associated with a "size-change graph", e.g., G_3 for call 3, that describes the data flow from the calling function's parameters to the called function's parameters.

Termination reasoning: Consider (in order to prove it impossible) any *infinite size-change graph sequence* $\mathcal{M} = g_1 g_2 \ldots \in \{G_1, G_2, G_3\}^\omega$ that follows the program's control flow:

Case 1: $\mathcal{M} = \ldots (G_2)^\omega$ ends in infinitely many G_2's: In this case, *variable* v *descends infinitely.*

Case 2: $\mathcal{M} = \ldots (G_1 G_2^* G_3)^\omega$. In this case, *variable* u *descends infinitely.*

Both cases are impossible (assuming, as we do, that the data value set is well-founded). Therefore a call of *any program function with any data will terminate.*

End of example.

1.2 Definitions and Terminology

We describe the structure of size-change graphs and state the size-change termination condition.

Definition 1.

1. *A* size-change graph $A \xrightarrow{G} B$ *consists of a* source *set A; a* target *set B; and a set of labeled arcs $G \subseteq A \times \{=, \downarrow\} \times B$. If A, B are clear from context, we identify $A \xrightarrow{G} B$ with its arc set G.*
2. *The* identity *size-change graph for A is $A \xrightarrow{id_A} A$ where $id_A = \{\mathtt{x} \xrightarrow{=} \mathtt{x} \mid \mathtt{x} \in A\}$.*
3. *Size-change graphs $A \xrightarrow{G_1} B$ and $C \xrightarrow{G_2} D$ are* composible *if $B = C$.*
4. *The* sequential composition *of size-change graphs $A \xrightarrow{G_1} B$ and $B \xrightarrow{G_2} C$ is $A \xrightarrow{G_1;G_2} C$ where*

$$G_1; G_2 = \left\{\mathtt{x} \xrightarrow{\downarrow} \mathtt{z} \mid \ \downarrow \ \in \{r, s \mid \mathtt{x} \xrightarrow{r} \mathtt{y} \in G_1 \text{ and } \mathtt{y} \xrightarrow{s} \mathtt{z} \in G_2 \text{ for some } \mathtt{y} \in B\} \right\}$$
$$\cup \left\{\mathtt{x} \xrightarrow{=} \mathtt{z} \mid \{=\} = \{r, s \mid \mathtt{x} \xrightarrow{r} \mathtt{y} \in G_1 \text{ and } \mathtt{y} \xrightarrow{s} \mathtt{z} \in G_2 \text{ for some } \mathtt{y} \in B\} \right\}$$

Lemma 1. *Sequential composition is associative.* $A \xrightarrow{G} B$ *implies* $id_A; G = G; id_B = G$.

Definition 2. *A* multipath $\mathcal{M}$ *over a set* $\mathcal{G}$ *of size-change graphs is a finite or infinite composible sequence of graphs in* $\mathcal{G}$. *Define*

$$\mathcal{G}^\omega = \{\mathcal{M} = G_0, G_1, \ldots \mid \textit{graphs } G_i, G_{i+1} \textit{ are composible for } i = 0, 1, 2, \ldots\}$$

Definition 3.

1. *A* thread *in a multipath* $\mathcal{M} = G_0, G_1, G_2, \ldots$ *is a sequence* $t = a_j \xrightarrow{r_j} a_{j+1} \xrightarrow{r_{j+1}} \ldots$ *such that* $a_k \xrightarrow{r_k} a_{k+1} \in G_k$ *for every* $k \geq j$ *(and each* r_k *is* $=$ *or* $\downarrow$*.)*
2. *Thread* t *is of* infinite descent *if* $r_k = \downarrow$ *for infinitely many* $k \geq j$.
3. *A set* $\mathcal{G}$ *of size-change graphs satisfies the* size-change condition *if every* $\mathcal{M} \in \mathcal{G}^\omega$ *contains at least one thread of infinite descent.*

The size-change condition is decidable; its worst-case complexity, as a function of the size of the program being analysed, is shown complete for PSPACE in [5].

The example revisited: The program of Figure 1 has three size-change graphs, one for each of the calls `1 : f → g`, `2 : g → g`, `3 : g → f`, so $\mathcal{G} = \{A \xrightarrow{G_1} B, B \xrightarrow{G_2} B, B \xrightarrow{G_3} A\}$ where $A = \{\mathtt{x}, \mathtt{y}\}$ and $B = \{\mathtt{u}, \mathtt{v}, \mathtt{w}\}$. (Note: the vertical layout of the size-change graphs in Figure 1 is inessential, though intuitively appealing. One could simply write, for instance, $G_3 = \{\mathtt{u} \xrightarrow{\downarrow} \mathtt{x}, \mathtt{w} \xrightarrow{=} \mathtt{y}\}$.)

The termination reasoning shows that $\mathcal{G}$ satisfies the size-change condition: Every infinite multipath has either a thread that decreases `u` infinitely, or a thread that decreases `v` infinitely.

2 The Call-by-Value λ-Calculus

We first summarise some standard definitions and results.

Definition 4. *Exp is the set of all* λ*-expressions that can be formed by these syntax rules, where* `@` *is the* application operator *(sometimes omitted). We use the* `teletype` *font for* λ*-expressions.*

```
e, P  ::=  x | e @ e | λx.e
x     ::=  Variable name
```

- *The set of* free variables $fv(\mathtt{e})$ *is defined in the usual way:* $fv(\mathtt{x}) = \{\mathtt{x}\}$, $fv(\mathtt{e@e'}) = fv(\mathtt{e}) \cup fv(\mathtt{e'})$ *and* $fv(\lambda\mathtt{x.e}) = fv(\mathtt{e}) \setminus \{\mathtt{x}\}$. *A closed* λ*-expression* `e` *satisfies* $fv(\mathtt{e}) = \emptyset$.
- *A* program, *usually denoted by* `P`, *is any closed* λ*-expression.*
- *The set of* subxpressions *of a* λ*-expression* `e` *is denoted by* $subexp(\mathtt{e})$.

The following is standard, e.g., [9]. Notation: $\mathsf{e}[v/\mathtt{x}]$ (β-reduction) is the result of substituting v for all free occurrences of $\mathtt{x}$ in e and renaming λ-bound variables if needed to avoid capture.

Definition 5. (Call-by-value semantics) *The* call-by-value evaluation relation *is defined by the following inference rules, with judgement form* $\mathsf{e} \Downarrow v$ *where* e *is a* λ*-expression and* $v \in \mathit{ValueS}$*. ValueS (for "standard value") is the set of all abstractions* $\lambda\mathtt{x}.\mathsf{e}$.

$$\text{(ValueS)}\ \frac{}{v \Downarrow v}\ \text{(If } v \in \mathit{ValueS}\text{)}$$

$$\text{(ApplyS)}\ \frac{\mathsf{e}_1 \Downarrow \lambda\mathtt{x}.\mathsf{e}_0 \qquad \mathsf{e}_2 \Downarrow v_2 \qquad \mathsf{e}_0[v_2/\mathtt{x}] \Downarrow v}{\mathsf{e}_1 @ \mathsf{e}_2 \Downarrow v}$$

Lemma 2. (Determinism) *If* $\mathsf{e} \Downarrow v$ *and* $\mathsf{e} \Downarrow w$ *then* $v = w$.

Lemma 3. *If* e *is closed and its* λ*-variables are all distinct, then a deduction of* $\mathsf{e} \Downarrow v$ *involves no renaming.*

3 Challenges in Termination Analysis of the λ-Calculus

The size-change termination analysis of [5] is based on several concepts including

1. Identifying nontermination as being caused by *infinite sequences of state transitions.*
2. A fixed set of *program control points.*
3. *Observable decreases* in data value sizes.
4. *Construction* of one size-change graph for each function call.
5. Finding the program's entire *control flow graph*, and the call sequences that follow it.

At first sight *all* these concepts seem to be absent from the λ-calculus, except that an application must be a call; and even then, it is not a priori clear *which* function is being called. We will show, one step at a time, that all the concepts do in fact exist in call-by-value λ-calculus evaluation.

3.1 Identifying Nontermination (Challenge 1)

We will sometimes write $\mathsf{e} \Downarrow$ to mean $\mathsf{e} \Downarrow v$ for some $v \in \mathit{ValueS}$, and write $\mathsf{e} \not\Downarrow$ to mean there is no $v \in \mathit{ValueS}$ such that $\mathsf{e} \Downarrow v$, i.e., if evaluation of e does not terminate.

A proof of $\mathsf{e} \Downarrow v$ is a finite object, and no such proof exists if the computation of e fails to terminate. In order to trace an arbitrary computation, terminating or not, we introduce the "calls" relation $\mathsf{e} \to \mathsf{e}'$. The rationale is straightforward: $\mathsf{e} \to \mathsf{e}'$ if in order to deduce $\mathsf{e} \Downarrow v$ for some value v, it is necessary first to deduce $\mathsf{e}' \Downarrow u$ for some u, i.e., some inference rule has form $\frac{\ldots\ \mathsf{e}' \Downarrow ?\ \ldots}{\mathsf{e} \Downarrow ?}$. Applying this to Definition 5 gives the following.

Definition 6. *The* call relation $\to \subseteq Exp \times Exp$ *is* $\to = \underset{r}{\to} \cup \underset{d}{\to} \cup \underset{c}{\to}$ *where* $\underset{r}{\to}, \underset{d}{\to}, \underset{c}{\to}$ *are defined by the following inference rules*[1].

$$\text{(OperatorS)} \quad \frac{}{\mathsf{e}_1@\mathsf{e}_2 \underset{r}{\to} \mathsf{e}_1}$$

$$\text{(OperandS)} \quad \frac{\mathsf{e}_1 \Downarrow v_1}{\mathsf{e}_1@\mathsf{e}_2 \underset{d}{\to} \mathsf{e}_2}$$

$$\text{(CallS)} \quad \frac{\mathsf{e}_1 \Downarrow \lambda\mathsf{x}.\mathsf{e}_0 \qquad \mathsf{e}_2 \Downarrow v_2}{\mathsf{e}_1@\mathsf{e}_2 \underset{c}{\to} \mathsf{e}_0[v_2/\mathsf{x}]}$$

As usual, we write $\to^+$ *for the transitive closure of* $\to$, *and* $\to^*$ *for its reflexive transitive closure.*

A familiar example: Call-by-value reduction of the combinator $\Omega = (\lambda\mathsf{x}.\mathsf{x}@\mathsf{x})@(\lambda\mathsf{y}.\mathsf{y}@\mathsf{y})$ yields an infinite call chain:

$$\Omega = (\lambda\mathsf{x}.\mathsf{x}@\mathsf{x})@(\lambda\mathsf{y}.\mathsf{y}@\mathsf{y}) \to (\lambda\mathsf{y}.\mathsf{y}@\mathsf{y})@(\lambda\mathsf{y}.\mathsf{y}@\mathsf{y}) \to (\lambda\mathsf{y}.\mathsf{y}@\mathsf{y})@(\lambda\mathsf{y}.\mathsf{y}@\mathsf{y}) \to \ldots$$

In fact, this linear-call-sequence behaviour is typical of nonterminating computations:

Lemma 4. (*NIS, or* <u>**N**</u>*ontermination* <u>**I**</u>*s* <u>**S**</u>*equential*) *Let* P *be a program. Then* $\mathsf{P} \not\Downarrow$ *if and only if there exists an infinite call chain*

$$\mathsf{P} = \mathsf{e}_0 \to \mathsf{e}_1 \to \mathsf{e}_2 \to \ldots$$

Proof See the Appendix. □

Rules (CallS) and (ApplyS) have a certain overlap: $\mathsf{e}_2 \Downarrow v_2$ appears in both, as does $\mathsf{e}_0[v_2/\mathsf{x}]$. Thus the (Call) rule can be used as an intermediate step to simplify the (Apply) rule. Variations on the following combined set will be used in the rest of the paper:

Definition 7. *(Combined evaluate and call rules, standard semantics)*

$$\text{(ValueS)} \quad \frac{}{v \Downarrow v} \quad (\text{If } v \in \mathit{Value})$$

$$\text{(OperatorS)} \quad \frac{}{\mathsf{e}_1@\mathsf{e}_2 \underset{r}{\to} \mathsf{e}_1} \qquad \text{(OperandS)} \quad \frac{\mathsf{e}_1 \Downarrow v_1}{\mathsf{e}_1@\mathsf{e}_2 \underset{d}{\to} \mathsf{e}_2}$$

$$\text{(CallS)} \quad \frac{\mathsf{e}_1 \Downarrow \lambda\mathsf{x}.\mathsf{e}_0 \qquad \mathsf{e}_2 \Downarrow v_2}{\mathsf{e}_1@\mathsf{e}_2 \underset{c}{\to} \mathsf{e}_0[v_2/\mathsf{x}]} \qquad \text{(ApplyS)} \quad \frac{\mathsf{e}_1@\mathsf{e}_2 \underset{c}{\to} \mathsf{e}' \qquad \mathsf{e}' \Downarrow v}{\mathsf{e}_1@\mathsf{e}_2 \Downarrow v}$$

[1] Naming: r, d in $\underset{r}{\to}, \underset{d}{\to}$ are the last letters of *operato***r** and *operan***d**, and c in $\underset{c}{\to}$ stands for “call”.

3.2 Finitely Describing the Computation Space (Challenge 2)

An equivalent *environment-based* semantics addresses the lack of program control points.

Definition 8. (States, values and environments) *State, Value, Env are the smallest sets such that*

$$\begin{array}{lcll} State & = & \{ \quad \mathtt{e} : \rho & \mid \mathtt{e} \in Exp, \rho \in Env \text{ and } fv(\mathtt{e}) \subseteq dom(\rho) \} \\ Value & = & \{ \quad \lambda\mathtt{x}.\mathtt{e} : \rho & \mid \lambda\mathtt{x}.\mathtt{e} : \rho \in State \} \\ Env & = & \{ \rho : X \to Value & \mid X \text{ is a finite set of variables} \} \end{array}$$

The empty environment with domain $X = \emptyset$ is written $[]$. The evaluation judgement form is $s \Downarrow v$ where $s \in State, v \in Value$.

We follow the pattern of Definition 7, except that substitution (β-reduction) $\mathtt{e}_0[v_2/\mathtt{x}]$ of the (ApplyS) rule is replaced by a "lazy substitution" that just updates the environment in the (Call) rule.

Definition 9. (Environment-based evaluation and call semantics) *The evaluation and call relations $\Downarrow, \to$ are defined by the following inference rules, where* $\to \; = \; \underset{r}{\to} \cup \underset{d}{\to} \cup \underset{c}{\to}$.

(Value) $\dfrac{}{v \Downarrow v}$ (If $v \in Value$) (Var) $\dfrac{}{\mathtt{x}\colon \rho \Downarrow \rho(\mathtt{x})}$

(Operator) $\dfrac{}{\mathtt{e}_1@\mathtt{e}_2 : \rho \underset{r}{\to} \mathtt{e}_1 : \rho}$ (Operand) $\dfrac{\mathtt{e}_1 : \rho \Downarrow v_1}{\mathtt{e}_1@\mathtt{e}_2 : \rho \underset{d}{\to} \mathtt{e}_2 : \rho}$

(Call) $\dfrac{\mathtt{e}_1 : \rho \Downarrow \lambda\mathtt{x}.\mathtt{e}_0 : \rho_0 \qquad \mathtt{e}_2 : \rho \Downarrow v_2}{\mathtt{e}_1@\mathtt{e}_2 : \rho \underset{c}{\to} \mathtt{e}_0 : \rho_0[\mathtt{x} \mapsto v_2]}$ (Apply) $\dfrac{\mathtt{e}_1@\mathtt{e}_2 : \rho \underset{c}{\to} \mathtt{e}' : \rho' \qquad \mathtt{e}' : \rho' \Downarrow v}{\mathtt{e}_1@\mathtt{e}_2 : \rho \Downarrow v}$

Remark: *A tighter version of these rules would "shrink-wrap" the environment ρ in a state $\mathtt{e} : \rho$, so that $dom(\rho) = fv(\mathtt{e})$ (rather than $\supseteq$). For instance, the conclusion of* (Operator) *would be*[2] $\mathtt{e}_1@\mathtt{e}_2 : \rho \underset{r}{\to} \mathtt{e}_1 : \rho_{|fv(\mathtt{e}_1)}$. *This has no significant effect on computations.*

Following the lines of [9], the environment-based semantics is shown equivalent to the standard semantics in the sense that they have the same termination behaviour, and when evaluation terminates the computed values are related by function $F : Exp \times Env \to Exp$ defined as

$$F(\mathtt{e} : \rho) = \mathtt{e}[F(\rho(\mathtt{x}_1))/\mathtt{x}_1, ..., F(\rho(\mathtt{x}_k))/\mathtt{x}_k] \text{ where } \{\mathtt{x}_1, .., \mathtt{x}_k\} = dom(\rho) \cap fv(e)$$

Lemma 5. $\mathtt{P} : [] \Downarrow v$ *(by Definition 9) if and only if* $\mathtt{P} \Downarrow F(v)$ *(by Definition 5).*

Proof is in the Appendix. The following is proven in the same way as Lemma 4.

[2] The *restriction* of ρ to a finite set A of variables is the environment $\rho_{|A}$ with domain $A \cap dom(\rho)$ that agrees with ρ on this domain.

Lemma 6. *(NIS, or* **N***ontermination* **I***s* **S***equential) Let* P *be a program. Then* $P : [] \not\Downarrow$ *if and only if there exists an infinite call chain*

$$P : [] = e_0 : \rho_0 \to e_1 : \rho_1 \to e_2 : \rho_2 \to \dots$$

3.3 The Subexpression Property

Definition 10. *Given a state s, we define its* expression support *exp_sup(s) by*

$$exp\_sup(e : \rho) = subexp(e) \cup \bigcup_{x \in fv(e)} exp\_sup(\rho(x))$$

Lemma 7. (Subexpression property) *If* $s \Downarrow s'$ *or* $s \to s'$ *then* $exp\_sup(s) \supseteq exp\_sup(s')$.

Corollary 1. *If* $P : [] \Downarrow \lambda x.e : \rho$ *then* $\lambda x.e \in subexp(P)$.

Proof of Lemma 7 is by induction on the proof of $s \Downarrow v$ or $s \to s'$. Base cases: $s = x : \rho$ and $s = \lambda x.e : \rho$ are immediate. For rule (Call) suppose $e_1 : \rho \Downarrow \lambda x.e_0 : \rho_0$ and $e_2 : \rho \Downarrow v_2$. By induction

$$exp\_sup(e_1 : \rho) \supseteq exp\_sup(\lambda x.e_0 : \rho_0) \quad \text{and} \quad exp\_sup(e_2 : \rho) \supseteq exp\_sup(v_2)$$

Thus

$$\begin{aligned} exp\_sup(e_1@e_2 : \rho) &= exp\_sup(e_1 : \rho) \cup exp\_sup(e_2 : \rho) \supseteq \\ exp\_sup(\lambda x.e_0 : \rho_0) \cup exp\_sup(v_2) &\supseteq exp\_sup(e_0 : \rho_0[x \mapsto v_2]) \end{aligned}$$

For rule (Apply) we have $exp\_sup(e_1@e_2 : \rho) \supseteq exp\_sup(e' : \rho') \supseteq exp\_sup(v)$. Cases (Operator), (Operand) are immediate. □

3.4 A Control Point Is a Subexpression of a λ-Program(Challenge 2)

The subexpression property does not hold for the standard rewriting semantics, but it is the starting point for our program analysis: A *control point* will be a subexpression of the program P being analysed, and our analyses will bind program flow information to subexpressions of P.

By the NIS Lemma 6, if $P \not\Downarrow$ then there exists an infinite call chain

$$P : [] = e_0 : \rho_0 \to e_1 : \rho_1 \to e_2 : \rho_2 \to \dots$$

By Lemma 7, $e_i \in subexp(P)$ for each i. Our termination-detecting algorithm will focus on the *size relations between consecutive environments* ρ_i and ρ_{i+1} in this chain. Since *subexp*(P) is a finite set, at least one subexpression e occurs infinitely often, so "self-loops" will be of particular interest.

Example 2. Figure 2 shows the combinator $\Omega = (\lambda x.x@x)@(\lambda y.y@y)$ as a tree whose subexpressions are labeled by numbers. To its right is the "calls" relation $\rightarrow$. It has an infinite call chain:

$$\Omega : [] \rightarrow x@x : \rho_1 \rightarrow y@y : \rho_2 \rightarrow y@y : \rho_2 \rightarrow y@y : \rho_2 \rightarrow \ldots$$

or $1 : [] \rightarrow 3 : \rho_1 \rightarrow 7 : \rho_2 \rightarrow 7 : \rho_2 \rightarrow \ldots$ where $\rho_1 = [x \mapsto \lambda y.y@y : []]$ and $\rho_2 = [y \mapsto \lambda y.y@y : []]$

The set of states reachable from $P : []$ is finite, so this computation enters a "repetitive loop."

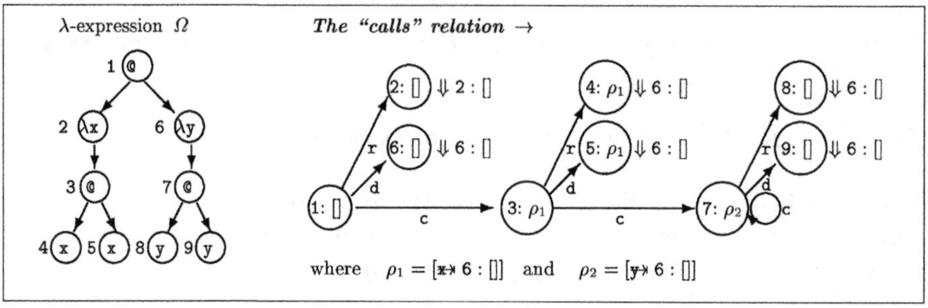

Fig. 2. A λ-expression and its call relation.

It is also possible that a computation will reach infinitely many states that are all different. Since all states have an expression component lying in a set of fixed size, and each expression in the environment also lies in this finite set, in an infinite state set $\mathcal{S}$ there will be states whose *environment depths* are arbitrarily large.

4 Size-Change Analysis of the Untyped λ-Calculus

Our end goal, given program P, is correctly to assert the nonexistence of infinite call chains starting at $P : []$. Our means are size-change graphs G that "safely" describe the relation between states s_1 and s_2 in a call $s_1 \rightarrow s_2$ or an evaluation $s_1 \Downarrow s_2$. First, we develop a "value decrease" notion suitable for the untyped λ-calculus. This is essentially the "height" of a closure value $e : \rho$.

4.1 Size Changes in a Computation (Challenge 3)

The *support* of a state $s = e : \rho$ is given by

$$support(e : \rho) = \{e : \rho\} \cup \bigcup_{x \in fv(e)} support(\rho(x))$$

Definition 11. *Relations* $s_1 \succeq s_2$ *and* $s_1 \succ s_2$.

- $s_1 \succeq s_2$ *holds if* $support(s_1) \ni s_2$;
- $s_1 \succeq s_2$ *holds if* $s_1 = \mathtt{e}_1 : \rho_1$ *and* $s_2 = \mathtt{e}_2 : \rho_2$, *where* $subexp(\mathtt{e}_1) \ni \mathtt{e}_2$ *and* $\rho_1(\mathtt{x}) = \rho_2(\mathtt{x})$ *for all* $\mathtt{x} \in fv(\mathtt{e}_2)$; *and*
- $s_1 \succ s_2$ *holds if* $s_1 \succeq s_2$ *and* $s_1 \neq s_2$.

Sets $support(\mathtt{e} : \rho)$ and $subexp(\mathtt{e})$ are clearly finite, yielding:

Lemma 8. *The relation* $\succ \subseteq State \times State$ *is well-founded.*

Definition 12. *We now relate states to the components of a size-change graph:*

1. *The* graph basis *of a state* $s = \mathtt{e} : \rho$ *is* $gb(s) = fv(\mathtt{e}) \cup \{\bullet\}$.
2. *Suppose* $s_1 = \mathtt{e}_1 : \rho_1$ *and* $s_2 = \mathtt{e}_2 : \rho_2$. *A size-change graph* G *relating the pair* (s_1, s_2) *has* source $gb(s_1)$ *and* target $gb(s_2)$.
3. *We abbreviate* $(fv(\mathtt{e}_1) \cup \{\bullet\}) \xrightarrow{G} (fv(\mathtt{e}_2) \cup \{\bullet\})$ *to* $\mathtt{e}_1 \xrightarrow{G} \mathtt{e}_2$.
4. *The* valuation function $\overline{s} : gb(s) \to Value$ *of a state* s *is defined by:*

$$\overline{s}(\bullet) = s \quad \text{and} \quad \overline{\mathtt{e} : \rho}(\mathtt{x}) = \rho(\mathtt{x})$$

Definition 13. *Let* $s_1 = \mathtt{e}_1 : \rho_1$ *and* $s_2 = \mathtt{e}_2 : \rho_2$. *Size-change graph* $\mathtt{e}_1 \xrightarrow{G} \mathtt{e}_2$ *is* safe[3] *for* (s_1, s_2) *if*

$$a_1 \xrightarrow{=} a_2 \in G \;\text{ implies }\; \overline{s_1}(a_1) = \overline{s_2}(a_2) \;\text{ and}$$

$$a_1 \xrightarrow{\downarrow} a_2 \in G \;\text{ implies }\; \overline{s_1}(a_1) \succ \overline{s_2}(a_2)$$

Explanation: an arc $\bullet \xrightarrow{r} \bullet$ in G r-relates the two states $s_1 = \mathtt{e}_1 : \rho_1$ and $s_2 = \mathtt{e}_2 : \rho_2$. Relation $r \in \{=, \downarrow\}$ corresponds to $s_1 = s_2$ or $s_1 \succ s_2$ as in Definition 11. Further, an arc $\mathtt{x} \xrightarrow{r} \mathtt{y}$ similarly relates the value of $\mathtt{x}$ in environment ρ_1 to the value of $\mathtt{y}$ in environment ρ_2. An arc $\bullet \xrightarrow{r} \mathtt{y}$ in G r-relates the state $s_1 = \mathtt{e}_1 : \rho_1$ to the value of $\mathtt{y}$ in environment ρ_2, and an arc $\mathtt{x} \xrightarrow{r} \bullet$ similarly relates the value of $\mathtt{x}$ in environment ρ_1 to the state $s_2 = \mathtt{e}_2 : \rho_2$.

Definition 14. *A set* $\mathcal{G}$ *of size-change graphs is* safe for program $\mathtt{P}$ *if* $\mathtt{P} : [] \to^* s_1 \to s_2$ *implies some* $G \in \mathcal{G}$ *is safe for the pair* (s_1, s_2).

Example 3. Figure 3 shows a graph set $\mathcal{G}$ that is safe for program $\Omega = (\lambda\mathtt{x}.\mathtt{x@x})(\lambda\mathtt{y}.\mathtt{y@y})$. For brevity, each subexpression of Ω is referred to by number in the diagram of $\mathcal{G}$. Subexpression $\mathtt{1} = \Omega$ has no free variables, so arcs from node $\mathtt{1}$ are labeled with size-change graphs $G_0 = \emptyset$.

Theorem 1. *If* $\mathcal{G}$ *is safe for program* $\mathtt{P}$ *and satisfies the size-change condition, then call-by-value evaluation of* $\mathtt{P}$ *terminates.*

[3] The term "safe" comes from abstract interpretation [3]. An alternative would be "sound."

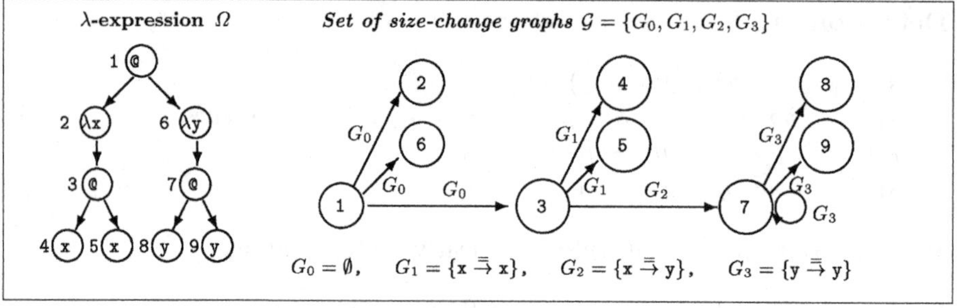

Fig. 3. A set of size-change graphs that safely describes Ω's nonterminating computation.

Proof Suppose call-by-value-evaluation of P does not terminate. Then by Lemma 6 there is an infinite call chain

$$\mathtt{P} : [] = \mathtt{e}_0 : \rho_0 \to \mathtt{e}_1 : \rho_1 \to \mathtt{e}_2 : \rho_2 \to \dots$$

Letting $s_i = \mathtt{e}_i : \rho_i$, by safety of $\mathcal{G}$ (Definition 14), there is a size-change graph $G_i \in \mathcal{G}$ that safely describes each pair (s_i, s_{i+1}). By the size-change condition (Definition 3) the multipath $\mathcal{M} = G_0, G_1, \dots$ has an infinite thread $t = a_j \overset{r_j}{\to} a_{j+1} \overset{r_{j+1}}{\to} \dots$ such that $k \geq j$ implies $a_k \overset{r_k}{\to} a_{k+1} \in G_k$, and each r_k is $\downarrow$ or $=$, and there are infinitely many $r_k = \downarrow$. Consider the value sequence $\overline{s_j}(a_j), \overline{s_{j+1}}(a_{j+1}), \dots$. By safety of G_k (Definition 13) we have $\overline{s_k}(a_k) \succeq \overline{s_{k+1}}(a_{k+1})$ for every $k \geq j$, and infinitely many proper decreases $\overline{s_k}(a_k) \succ \overline{s_{k+1}}(a_{k+1})$. However this is impossible since by Lemma 8 the relation $\succ$ on *Value* is well-founded.

Conclusion: call-by-value-evaluation of P terminates. □

The goal is partly achieved: We have found a sufficient condition on a set of size-change graphs to guarantee program termination. What we have not yet done is to find an algorithm to *construct* a size-change graph set $\mathcal{G}$ that is safe for P (The safety condition of Definition 14 is in general undecidable, so enumeration of all graphs won't work.) Our graph construction algorithm is developed in two steps:

- First, the exact evaluation and call relations are "instrumented" so as to produce safe size-change graphs during evaluation.
- Second, an *abstract interpretation* of these rules yields a computable over-approximation $\mathcal{G}$ that contains all graphs that can be built during exact evaluation.

4.2 Safely Describing Value Flows in a Single Computation (Challenge 4)

First, the exact evaluation and call relations are "instrumented" so as to produce safe size-change graphs during evaluation.

Definition 15. (Evaluation and call with graph generation) *The extended evaluation and call judgement forms are* $\mathsf{e} : \rho \to \mathsf{e}' : \rho', G$ *and* $\mathsf{e} : \rho \Downarrow \mathsf{e}' : \rho', G$*. The inference rules are:*

(ValueG) $$\frac{}{\lambda\mathsf{x.e} : \rho \Downarrow \lambda\mathsf{x.e} : \rho,\, id^{=}_{\lambda\mathsf{x.e}}}$$

(VarG) $$\frac{}{\mathsf{x} : \rho \Downarrow \rho(\mathsf{x}), \{\mathsf{x} \xrightarrow{=} \bullet\} \cup \{\mathsf{x} \xrightarrow{\downarrow} \mathsf{y} \mid \mathsf{y} \in fv(\mathsf{e}')\}} \quad (\text{If } \rho(\mathsf{x}) = \mathsf{e}' : \rho')$$

(OperatorG) $$\frac{}{\mathsf{e}_1@\mathsf{e}_2 : \rho \underset{r}{\to} \mathsf{e}_1 : \rho,\, id^{\downarrow}_{\mathsf{e}_1}}$$ (OperandG) $$\frac{\mathsf{e}_1 : \rho \Downarrow v_1}{\mathsf{e}_1@\mathsf{e}_2 : \rho \underset{d}{\to} \mathsf{e}_2 : \rho,\, id^{\downarrow}_{\mathsf{e}_2}}$$

(CallG) $$\frac{\mathsf{e}_1 : \rho \Downarrow \lambda\mathsf{x.e}_0 : \rho_0, G_1 \qquad \mathsf{e}_2 : \rho \Downarrow v_2, G_2}{\mathsf{e}_1@\mathsf{e}_2 : \rho \underset{c}{\to} \mathsf{e}_0 : \rho_0[\mathsf{x} \mapsto v_2],\, G_1^{-\bullet} \cup G_2^{\bullet \mapsto \mathsf{x}}}$$

(ApplyG) $$\frac{\mathsf{e}_1@\mathsf{e}_2 : \rho \underset{c}{\to} \mathsf{e}' : \rho', G' \qquad \mathsf{e}' : \rho' \Downarrow v, G}{\mathsf{e}_1@\mathsf{e}_2 : \rho \Downarrow v, (G'; G)}$$

Notations used in the rules (x, y, z *are variables and not* $\bullet$*):*

$$\begin{array}{rl}
id^{=}_{\mathsf{e}} & \textit{stands for } \{\bullet \xrightarrow{=} \bullet\} \cup \{\mathsf{x} \xrightarrow{=} \mathsf{x} \mid \mathsf{x} \in fv(\mathsf{e})\} \\
id^{\downarrow}_{\mathsf{e}} & \textit{stands for } \{\bullet \xrightarrow{\downarrow} \bullet\} \cup \{\mathsf{x} \xrightarrow{=} \mathsf{x} \mid \mathsf{x} \in fv(\mathsf{e})\} \\
G_1^{-\bullet} & \textit{stands for } \{\, \mathsf{y} \xrightarrow{r} \mathsf{z} \mid\ \mathsf{y} \xrightarrow{r} \mathsf{z} \in G_1\} \cup \{\, \bullet \xrightarrow{\downarrow} \mathsf{z} \mid\ \bullet \xrightarrow{r} \mathsf{z} \in G_1\} \\
G_2^{\bullet \mapsto \mathsf{x}} & \textit{stands for } \{\, \mathsf{y} \xrightarrow{r} \mathsf{x} \mid\ \mathsf{y} \xrightarrow{r} \bullet \in G_2\,\} \cup \{\, \bullet \xrightarrow{\downarrow} \mathsf{x} \mid\ \bullet \xrightarrow{r} \bullet \in G_2\,\}
\end{array}$$

The diagram of Figure 4 may be of some use in visualising data-flow during evaluation of $\mathsf{e}_1@\mathsf{e}_2$. States are in ovals and triangles represent environments. In the application $\mathsf{e}_1@\mathsf{e}_2 : \rho$ on the left, operator $\mathsf{e}_1 : \rho$ evaluates to $\lambda\mathsf{x.e}_0 : \rho_0, G_1$ and operand $\mathsf{e}_2 : \rho$ evaluates to $\mathsf{e}' : \rho', G_2$. The size change graphs G_1 and G_2 show relations between variables bound in their environments. There is a call from the application $\mathsf{e}_1@\mathsf{e}_2 : \rho$ to $\mathsf{e}_0 : \rho_0[\mathsf{x} \mapsto \mathsf{e}' : \rho']$ the body of the operator-value with the environment extended with a binding of x to the operand-value $\mathsf{e}' : \rho'$.

Lemma 9. $s \to s', G$ (by Definition 15) *iff* $s \to s'$ (by Definition 7)*. Further,* $s \Downarrow s', G$ *iff* $s \Downarrow s'$*.*

Theorem 2. (The extracted graphs are safe) $s \to s', G$ *or* $s \Downarrow s', G$ *implies* G *is safe for* (s, s')*.*

Proof The Lemma is immediate since the new rules extend the old, without any restriction on their applicability. Proof of "safety" is by a case analysis deferred to the Appendix. □

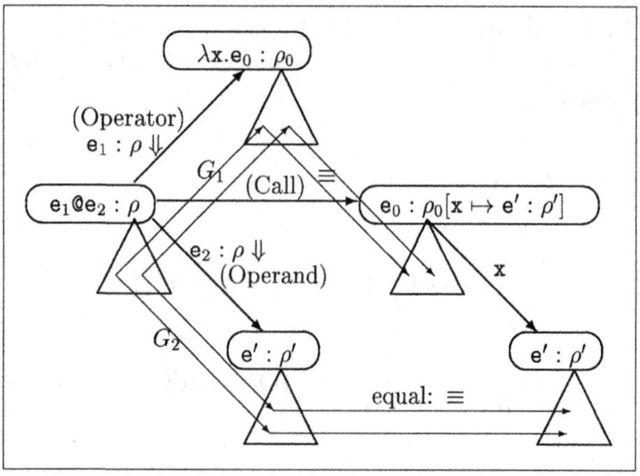

Fig. 4. Data-flow in an application.

4.3 Abstract Interpretation of the "Calls" and "Evaluates-to" Relations (Challenge 5)

A coarse approximation may be obtained by removing all environment components. To deal with the absence of environments the variable lookup rule is modified: If $\mathtt{e}_1@\mathtt{e}_2$ is *any* application in P such that $\mathtt{e}_1$ can evaluate to a value of form $\lambda\mathtt{x}.\mathtt{e}$ and $\mathtt{e}_2$ can evaluate to value v_2, then v_2 is regarded as a possible value of x.

The main virtue of these rules is that there are only finitely many possible judgements $\mathtt{e} \to \mathtt{e}'$ and $\mathtt{e} \Downarrow \mathtt{e}'$. Consequently, the runtime behavior of program P may be (approximately) analysed by exhaustively applying these inference rules. The next section will extend the rules so they also generate size-change graphs.

Definition 16. (Approximate evaluation and call rules) *The new judgement forms are* $\mathtt{e} \Downarrow \mathtt{e}'$ *and* $\mathtt{e} \to \mathtt{e}'$*. The inference rules are:*

$$\text{(ValueA)}\ \frac{}{\lambda\mathtt{x}.\mathtt{e} \Downarrow \lambda\mathtt{x}.\mathtt{e}} \qquad \text{(VarA)}\ \frac{\mathtt{e}_1@\mathtt{e}_2 \in \mathit{subexp}(\mathtt{P}) \quad \mathtt{e}_1 \Downarrow \lambda\mathtt{x}.\mathtt{e}_0 \quad \mathtt{e}_2 \Downarrow v_2}{\mathtt{x} \Downarrow v_2}$$

$$\text{(OperatorA)}\ \frac{}{\mathtt{e}_1@\mathtt{e}_2 \underset{r}{\to} \mathtt{e}_1} \qquad \text{(OperandA)}\ \frac{}{\mathtt{e}_1@\mathtt{e}_2 \underset{d}{\to} \mathtt{e}_2}$$

$$\text{(CallA)}\ \frac{\mathtt{e}_1 \Downarrow \lambda\mathtt{x}.\mathtt{e}_0 \quad \mathtt{e}_2 \Downarrow v_2}{\mathtt{e}_1@\mathtt{e}_2 \underset{c}{\to} \mathtt{e}_0} \qquad \text{(ApplyA)}\ \frac{\mathtt{e}_1@\mathtt{e}_2 \underset{c}{\to} \mathtt{e}' \quad \mathtt{e}' \Downarrow v}{\mathtt{e}_1@\mathtt{e}_2 \Downarrow v}$$

Remark: the (VarA) rule refers globally to P, the program being analysed. Notice that the approximate evaluation is nondeterministic: An expression may evaluate to more than one value.

Lemma 10. *Suppose* $\mathtt{P} : [] \to^* \mathtt{e} : \rho$. *If* $\mathtt{e} : \rho \Downarrow \mathtt{e}' : \rho'$ *then* $\mathtt{e} \Downarrow \mathtt{e}'$. *Further, if* $\mathtt{e} : \rho \to \mathtt{e}' : \rho'$ *then* $\mathtt{e} \to \mathtt{e}'$.

Proof Straightforward; see the Appendix. □

4.4 Construction of Size-Change Graphs by Abstract Interpretation

We now extend the coarse approximation to construct size-change graphs.

Definition 17. (Approximate evaluation and call with graph generation) *The judgement forms are now* $\mathtt{e} \to \mathtt{e}', G$ *and* $\mathtt{e} \Downarrow \mathtt{e}', G$.

$$\text{(ValueAG)} \quad \frac{}{\lambda \mathtt{x}.\mathtt{e} \Downarrow \lambda \mathtt{x}.\mathtt{e},\ id^{=}_{\lambda \mathtt{x}.\mathtt{e}}}$$

$$\text{(VarAG)} \quad \frac{\mathtt{e}_1 @ \mathtt{e}_2 \in \mathit{subexp}(\mathtt{P}) \quad \mathtt{e}_1 \Downarrow \lambda \mathtt{x}.\mathtt{e}_0, G_1 \qquad \mathtt{e}_2 \Downarrow v_2, G_2}{\mathtt{x} \Downarrow v_2, \{\mathtt{x} \overset{=}{\rightarrow} \bullet\} \cup \{\mathtt{x} \overset{\downarrow}{\rightarrow} \mathtt{y} \mid \mathtt{y} \in \mathit{fv}(v_2)\}}$$

$$\text{(OperatorAG)} \quad \frac{}{\mathtt{e}_1 @ \mathtt{e}_2 \underset{r}{\rightarrow} \mathtt{e}_1, id^{\downarrow}_{\mathtt{e}_1}} \qquad \text{(OperandAG)} \quad \frac{}{\mathtt{e}_1 @ \mathtt{e}_2 \underset{d}{\rightarrow} \mathtt{e}_2, id^{\downarrow}_{\mathtt{e}_2}}$$

$$\text{(CallAG)} \quad \frac{\mathtt{e}_1 \Downarrow \lambda \mathtt{x}.\mathtt{e}_0, G_1 \qquad \mathtt{e}_2 \Downarrow v_2, G_2}{\mathtt{e}_1 @ \mathtt{e}_2 \underset{c}{\rightarrow} \mathtt{e}_0, G_1^{-\bullet} \cup G_2^{\bullet \mapsto \mathtt{x}}} \qquad \text{(ApplyAG)} \quad \frac{\mathtt{e}_1 @ \mathtt{e}_2 \underset{c}{\rightarrow} \mathtt{e}', G' \qquad \mathtt{e}' \Downarrow v, G}{\mathtt{e}_1 @ \mathtt{e}_2 \Downarrow v, G'; G}$$

Lemma 11. *Suppose* $\mathtt{P} : [] \to^* \mathtt{e} : \rho$. *If* $\mathtt{e} : \rho \to \mathtt{e}' : \rho', G$ *then* $\mathtt{e} \to \mathtt{e}', G$. *Further, if* $\mathtt{e} : \rho \Downarrow \mathtt{e}' : \rho', G$ *then* $\mathtt{e} \Downarrow \mathtt{e}', G$.

Proof Straightforward; see the Appendix. □

Definition 18.

$$\begin{aligned} \mathit{absint}(\mathtt{P}) = \Big\{ & G_j \mid j > 0 \land \exists \mathtt{e}_i, G_i (0 \le i \le j) : \\ & \mathtt{P} = \mathtt{e}_0 \land (\mathtt{e}_0 \to \mathtt{e}_1, G_1) \land \ldots \land (\mathtt{e}_{j-1} \to \mathtt{e}_j, G_j) \Big\} \end{aligned}$$

Theorem 3.

1. *The set* $\mathit{absint}(\mathtt{P})$ *is safe for* $\mathtt{P}$.
2. *The set* $\mathit{absint}(\mathtt{P})$ *can be effectively computed from* $\mathtt{P}$.

Proof Part 1: Suppose $\mathtt{P} : [] = s_0 \to s_1 \to \ldots \to s_j$. Theorem 2 implies $s_i \to s_{i+1}, G_i$ where each G_i is safe for the pair (s_i, s_{i+1}). Let $s_i = \mathtt{e}_i : \rho_i$. By Lemma 11, $\mathtt{e}_i \to \mathtt{e}_{i+1}, G_{i+1}$. By the definition of $\mathit{absint}(\mathtt{P})$, $G_j \in \mathit{absint}(\mathtt{P})$.

Part 2: There is only a fixed number of subexpressions of $\mathtt{P}$, or of possible size-change graphs. Thus $\mathit{absint}(\mathtt{P})$ can be computed by applying Definition 17 exhaustively, starting with $\mathtt{P}$, until no new graphs or subexpressions are obtained.

□

5 Examples and Experimental Results

5.1 Simple Example

Using Church numerals ($\mathtt{n} = \lambda\mathtt{s}\lambda\mathtt{z}.\mathtt{s}^n(\mathtt{z})$), we expect `2 succ 0` to reduce to `succ(succ 0)`. However this contains unreduced redexes because call-by-value does not reduce under a λ, so we force the computation to carry on through by applying `2 succ 0` to the identity (twice). This gives:

```
2 succ 0 id1 id2 where
  succ = λm.λs.λz. m s (s z)
  id1  = λx.x
  id2  = λy.y
```

After writing this out in full as a λ-expression, our analyser yields (syntactically sugared):

```
    [λs2.λz2.(s2 @ (s2 @ z2))]          -- two --
  @ [λm.λs.λz. |15:|((m@s)@(s@z))]    -- succ --
  @ [λs1.λz1.z1]                        -- zero --
  @ [λx.x]                              -- id1 --
  @ [λy.y]                              -- id2 --
```

```
Output of loops from an analysis of this program:

15→* 15: [(m,>,m),(s,=,s),(z,=,z)], []

Size Change Termination: Yes
```

The loop occurs because application of 2 forces the code for the successor function to be executed twice, with decreasing argument values `m`. The notation for edges is a little different from previously, here `(m,>,m)` stands for $\mathtt{m} \xrightarrow{\downarrow} \mathtt{m}$.

5.2 $fnx = x + 2^n$ by Church Numerals

This more interesting program computes $fnx = x+2^n$ by higher-order primitive recursion. If `n` is a Church numeral then expression `n g x` reduces to $\mathtt{g}^n(\mathtt{x})$. Let `x` be the successor function, and `g` be a "double application" functional. Expressed in a readable named combinator form, we get:

```
f n x    where
f n    = if n=0 then succ else g(f(n-1))
g r a =  r(ra)
```

As a lambda-expression (applied to values $n = 3, x = 4$) this can be written:

```
[λn.λx. n                                          -- n --
        @ [λr.λa. 11: (r@ 13: (r@a))]              -- g --
        @ [λ k.λ s.λ z.(s@((k@s)@z))]              - succ-
        @ x ]                                      -- x --

@        [λs2.λz2. (s2@(s2@(s2@z2))) ]             -- 3 --
@        [λs1.λz1. (s1@(s1@(s1@(s1@z1))))]         -- 4 --
```

Following is the output from program analysis. The analysis found the following loops from a program point to itself with the associated size change graph and path. The first number refer to the program point, then comes a list of edges and last a list of numbers, the other program points that the loop passes through.

```
SELF Size Change Graphs, no repetition of graphs:

11 →* 11: [(r,>,r)]              []
11 →* 11: [(a,=,a),(r,>,r)]      [13]
13 →* 13: [(a,=,a),(r,>,r)]      [11]
13 →* 13: [(r,>,r)]              [11,11]

Size Change Termination: Yes
```

5.3 Ackermann's Function, Second-Order

This can be written without recursion using Church numerals as: `a m n` where `a = λm. m b succ` and `b = λg. λn. n g (g 1)`. Consequently `a m = b`^{m}`(succ)` and `b g n = g`$^{n+1}$`(1)`, which can be seen to agree with the usual first-order definition of Ackermann's function. Following is the same as a lambda-expression applied to argument values `m=2, n=3`, with numeric labels on some subexpressions.

```
(λm.m b succ) 2 3  =  (λm.m@b@succ)@2@3
(λm.m@(λg.λn.n@g@(g@1))@succ)@2@3
(λm.m@(λg.λn. 9: (n@g@ 13: (g@1)))@succ)@2@3
 where
 1    =  λs1.λz1.  17: (s1@z1)
succ  =  λk.λs.λz.  23: (s@ 25: (k@s@z))
 2    =  λs2.λz2.  s2@(s2@z2)
 3    =  λs3.λz3.  39: (s3@ 41: (s3@ 43: (s3@z3)))
```

Output from an analysis of this program is shown here.

```
SELF Size Change Graphs, no repetition of graphs:
(Because graphs are only taken once it is not always the case that
the
same loop is shown for all program points in its path)
```

```
 9 →*  9: [(•,>,n),(g,>,g)]                  [13]
 9 →*  9: [(g,>,g)]                          [17]
13 →* 13: [(g,>,g)]                          [9]
17 →* 17: [(s1,>,s1)]                        [9]
23 →* 23: [(k,>,k),(s,=,s),(z,=,z)]          [25]
23 →* 23: [(s,>,s)]                          [9]
23 →* 23: [(s,>,s),(z,>,k)]                  [25,17,9]
25 →* 25: [(k,>,k),(s,=,s),(z,=,z)]          [23]
25 →* 25: [(s,>,s),(z,>,k)]                  [17,9,23]
25 →* 25: [(s,>,s)]                          [23,9,23]
39 →* 39: [(s3,>,s3)]                        [9]
41 →* 41: [(s3,>,s3)]                        [9,39]
43 →* 43: [(s3,>,s3)]                        [9,39,41]

Size Change Termination: Yes
```

5.4 A Minimum Function, with General Recursion and Y-Combinator

We have another version of the termination analysis where programs can have as constants: natural numbers, predecessor, successor and zero-test, and also if-then-else expressions. In this setting it is possible to analyse the termination behaviour of a program with arbitrary natural number as input represented by •. This program computes the minimum of its two inputs using the call-by-value combinator $Y = \lambda$`p. [`λ`q.p@(`λ`s.q@q@s)] @ [`λ`t.p@(`λ`u.t@t@u)]`. The program, first as a first-order recursive definition.

```
    m x y = if x=0 then 0 else if y=0 then 0 else succ (m (pred
x) (pred y))
```

Now, in λ-expression form for analysis.

```
{λp. [λq.p@(λs.q@q@s)] @ [λt.p@(λu.t@t@u)]}    -- the Y combinator --
@
[ λm.λx.λy. [27:]if( (ztst @ x),
                     0,
                     [32:]    if( (ztst @ y),
                                  0,
                                  [37:]succ @ [39:] m @ (pred@x) @ (pred@y) ]
@ •
@ •
```

```
Output of loops from an analysis of this program:

27 →* 27: [(x,>,x),(y,>,y)]     [32,37,39]
32 →* 32: [(x,>,x),(y,>,y)]     [37,39,27]
37 →* 37: [(x,>,x),(y,>,y)]     [39,27,32]
39 →* 39: [(x,>,x),(y,>,y)]     [27,32,37]

Size Change Termination: Yes
```

5.5 Ackermann's Function, Second-Order with Constants and Y-Combinator

Ackermann's function can be written as: `a m n` where `a m = b`^{m}`(suc)` and `b g n = g`$^{n+1}$`(1)`. The following program expresses the computations of both a and b by loops, using the Y combinator (twice).

```
[λ y.λ y1.
(y1 @
λ a.λ m. 11: if( (ztst@m),
                λ v.(suc@v),
                19: ( (y @
                       λ b.λ f.λ n.
                       25: if( (ztst@n),
                               29: (f@1),
                               32: f@ 34: b @ f @ (pred@n))
                @ 41: a @ (pred@m)                      ]

@ {λp. [λq.p@(λs. q@q@s)] @ [λt.p@(λu. t@t@u)]}
@ {λp1. [λq1.p1@(λs. 72: q1@1q@s1)] @ [λt1.p1@(λu1. 81: t1@t1@u1)]}
@ •
@ •
```

```
Output of loops from an analysis of this program:

SELF Size Change Graphs no repetition of graphs:


11 →* 11: [(a,>,y),(m,>,m)]     [19,41,72]
11 →* 11: [(m,>,m)]             [19,41,72,11,19,41,72]
19 →* 19: [(a,>,y),(m,>,m)]     [41,72,11]
19 →* 19: [(m,>,m)]             [41,72,11,19,41,72,11]
25 →* 25: [(f,>,b),(f,>,f)]     [29]
25 →* 25: [(f,=,f),(n,>,n)]     [32,34]
25 →* 25: [(f,>,f)]             [29,25,32,34]
29 →* 29: [(f,>,f)]             [25]
32 →* 32: [(f,>,b),(f,>,f)]     [25]
32 →* 32: [(f,=,f),(n,>,n)]     [34,25]
32 →* 32: [(f,>,f)]             [25,32,34,25]
34 →* 34: [(f,=,f),(n,>,n)]     [25,32]
34 →* 34: [(f,>,b),(f,>,f)]     [25,29,25,32]
34 →* 34: [(f,>,f)]             [25,29,25,32,34,25,32]
41 →* 41: [(m,>,m)]             [72,11,19]
72 →* 72: [(s1,>,s1)]           [11,19,41]
81 →* 81: [(u1,>,u1)]           [11,19,41]

Size Change Termination: Yes
```

5.6 Imprecision of Abstract Interpretation

It is natural to wonder whether the gross approximation of Definition 16 comes at a cost. The (VarA) rule can in effect "mix up" different function applications, losing the coordination between operator and operand that is present in the exact semantics.

We have observed this in practice: The first time we had programmed Ackermann's using explicit recursion, we used the same instance of Y-combinator for both loops, so the single Y-combinator expression was "shared". The analysis did not discover that the program terminated.

However when this was replaced by the "unshared" version above, with two instances of the Y-combinator (`y` and `y1`) (one for each application), the problem disappeared and termination was correctly recognised.

6 Concluding Matters

Acknowledgements: The authors gratefully acknowledge detailed and constructive comments by Arne Glenstrup and Chin Soon Lee, and insightful comments by Luke Ong and David Wahlstedt.

Related work: Papers subsequent to [5] have used size-change graphs to find bounds on program running times [1]; solved related problems, e.g., to ensure that partial evaluation will terminate [4, 6]; and found more efficient (though less precise) algorithms [7]. Further, the thesis [8] extends the first-order size-change method [5] to handle higher-order named combinator programs. It uses a different approach than ours, and appears to be less general.

It is natural to compare the size-change condition with strongly normalising subsets of the λ-calculus, in particular subsets that are typable by various disciplines, ranging from simple types up to and including Girard's System F [2].

A notable difference is that general recursion better matches the habits of the working computer scientist: primitive recursion is unnatural for programming, and even more so when higher types are used. On the other hand, the class of functions computable by typable λ-expressions is enormous, and precisely characterised. Conclusion: more investigations need to be done.

Future work

1. Extend the analysis so it can recognise, given a program `P`, that `P @ v` will terminate for any choice of `v` from a given data set, e.g., Church or other numerals. This seems straightforward.
2. Prove or disprove **Conjecture:** The size-change method will recognise as terminating any simply typed λ-expression (sans types).
3. Prove or disprove **Conjecture:** The size-change method will recognise as terminating any System T expression (sans types, and coded using Church numerals).

Appendix

A Proof of Lemma 4

Proof *If:* Consider a proof tree of $\mathsf{P} \Downarrow v$. Each call rule of Definition 6 is associated with a use of rule (ApplyS) from Definition 5. There will be a call $\mathsf{e} \to \mathsf{e}'$ from each (ApplyS) conclusion $\mathsf{e} \Downarrow v$ to one of its three premises $\mathsf{e}' \Downarrow v'$ in the proof tree, and only these. Since the proof tree is finite, there will be no infinite call chains. (An inductive proof is omitted for brevity.)

Only if: Assume $\mathsf{P} \to^* \mathsf{e}$ and all call chains from e are finite. We prove by induction on the maximal length n of a call chain from e that $\mathsf{e} \Downarrow$.

$n = 0$: e is an abstraction that evaluates to itself.

$n > 0$: e must be an application $\mathsf{e} = \mathsf{e}_1@\mathsf{e}_2$. By rule (OperatorS) there is a call $\mathsf{e}_1@\mathsf{e}_2 \underset{d}{\to} \mathsf{e}_1$, and the maximal length of a call chain from e_1 is less than n. By induction there exists v_1 such that $\mathsf{e}_1 \Downarrow v_1$. We now conclude by rule (OperandS) that $\mathsf{e}_1@\mathsf{e}_2 \underset{r}{\to} \mathsf{e}_2$. By induction there exists v_2 such that $\mathsf{e}_2 \Downarrow v_2$.

All values are abstractions, so we can write $v_1 = \lambda \mathsf{x}.\mathsf{e}_0$. We now conclude by rule (CallS) that $\mathsf{e}_1@\mathsf{e}_2 \underset{c}{\to} \mathsf{e}_0[\mathsf{x} := v_2]$. By induction again, $\mathsf{e}_0[\mathsf{x} \mapsto v_2] \Downarrow v$ for some v. This gives us all premises for the (ApplyS) rule of Definition 5, so $\mathsf{e} = \mathsf{e}_1@\mathsf{e}_2 \Downarrow v$. □

B Proof of Lemma 5

Lemma 12. *$\mathsf{P} : [] \Downarrow v$ (by Definition 9) implies $\mathsf{P} \Downarrow F(v)$ (by Definition 5) for any program P.*

Proof Let $s \in State$, and assume $s \Downarrow v$ by Definition 9. This has a finite proof tree. We prove by induction on the height n of the proof tree that $s \Downarrow v$ implies $F(s) \Downarrow F(v)$ by Definition 5. It then follows that $\mathsf{P} : [] \Downarrow v$ implies $\mathsf{P} \Downarrow F(v)$.

$n = 0$: Two possibilities. First, $s \in Value$ implies there is an abstraction such that $s = \lambda \mathsf{x}.\mathsf{e} : \rho$ and $s = v$. $F(s) = F(v)$ is an abstraction $F(s) = F(\lambda \mathsf{x}.\mathsf{e} : \rho) = \lambda \mathsf{x}.F(\mathsf{e} : \rho_{|_{fv(\lambda x.e)}})$ hence $F(s) \Downarrow F(v)$. Second, $s = \mathsf{x} : \rho$ implies $s \Downarrow \rho(\mathsf{x}) = v = \lambda \mathsf{y}.\mathsf{e}' : \rho'$. By definition $F(s) = \mathsf{x}[F(\rho(\mathsf{x}))/\mathsf{x}] = F(\rho(\mathsf{x})) = F(v)$. Since v is an abstraction also $F(s) \Downarrow F(v)$ as before.

$n > 0$: Consider $s \in State$ such that $s \Downarrow v$ has evaluation tree of height $n > 0$. It must be an application $s = \mathsf{e}_1@\mathsf{e}_2 : \rho$, and the last rule applied must be the (Apply) rule. By induction we have $F(\mathsf{e}_1 : \rho) \Downarrow F(\lambda \mathsf{x}.\mathsf{e}_0 : \rho_0) = \lambda \mathsf{x}.F(\mathsf{e}_0 : \rho_{0|_{fv(\lambda x.e_0)}})$, and $F(\mathsf{e}_2 : \rho) \Downarrow F(v_2)$, and $F(\mathsf{e}_0 : \rho_0[\mathsf{x} \mapsto v_2]) \Downarrow F(v)$.

By definition of F we have $F(\mathsf{e}_0 : \rho_0[\mathsf{x} \mapsto v_2]) = F(\mathsf{e}_0 : \rho_{0|_{fv(\lambda x.e_0)}})[F(v_2)/x]$. All premises in the standard semantics (ApplyS) rule hold, so we conclude $F(s) = F(\mathsf{e}_1 : \rho)@F(\mathsf{e}_2 : \rho) \Downarrow F(v)$ as required. □

Lemma 13. *For any state $\mathsf{e} : \rho$ it holds that $\mathsf{e} : \rho \to \mathsf{e}' : \rho'$ implies $F(\mathsf{e} : \rho) \to F(\mathsf{e}' : \rho')$.*

Proof $\mathtt{e} : \rho \to \mathtt{e}' : \rho'$ implies $\mathtt{e} = \mathtt{e}_1@\mathtt{e}_2$. Clearly $F(\mathtt{e} : \rho) = F(\mathtt{e}_1@\mathtt{e}_2 : \rho) = F(\mathtt{e}_1 : \rho)@F(\mathtt{e}_2 : \rho)$. There are 3 possibilities for $\mathtt{e}' : \rho'$:

1. (Operator) rule has been applied $\mathtt{e}' : \rho' = \mathtt{e}_1 : \rho$. The (Operator) rule always applies to an application hence we also have $F(\mathtt{e}_1 : \rho)@F(\mathtt{e}_2 : \rho) \to F(\mathtt{e}_1 : \rho)$
2. (Operand) rule has been applied $\mathtt{e}' : \rho' = \mathtt{e}_2 : \rho$. Then it must hold for some v that $\mathtt{e}_1 : \rho \Downarrow v$, so by Lemma 12, $F(\mathtt{e}_1 : \rho) \Downarrow F(v)$. It follows that rule (Operand) can be applied and hence $F(\mathtt{e}_1 : \rho)@F(\mathtt{e}_2 : \rho) \to F(\mathtt{e}_2 : \rho)$
3. (Call) rule has been applied. For some $\lambda\mathtt{x}.\mathtt{e}_0 : \rho_0$ and v_2 we have $\mathtt{e}_1 : \rho \Downarrow \lambda\mathtt{x}.\mathtt{e}_0 : \rho_0$ and $\mathtt{e}_2 : \rho \Downarrow v_2$ and $\mathtt{e}' : \rho' = \mathtt{e}_0 : \rho_0[\mathtt{x} \mapsto v_2]$. By Lemma 12 we have $F(\mathtt{e}_2 : \rho) \Downarrow F(v_2)$ and $F(\mathtt{e}_1 : \rho) \Downarrow F(\lambda\mathtt{x}.\mathtt{e}_0 : \rho_0) = \lambda x.F(\mathtt{e}_0 : \rho_{0|fv(\lambda x.e_0)})$. Then by rule (Call) there is a call to $F(\mathtt{e}_0 : \rho_{0|fv(\lambda x.e_0)})[F(v_2)/\mathtt{x}]$. As before, $F(\mathtt{e}_0 : \rho_0[\mathtt{x} \mapsto v_2]) = F(\mathtt{e}_0 : \rho_{0|fv(\lambda\mathtt{x}.\mathtt{e}_0)})[F(v_2)/\mathtt{x}]$.

We conclude that in all cases when we have a call $\mathtt{e} : \rho \to \mathtt{e}' : \rho'$ then we also have a call $F(\mathtt{e} : \rho) \to F(\mathtt{e}' : \rho')$ □

Proof of Lemma 5 : "Only if" is Lemma 12. For "if" suppose $\mathtt{P} \Downarrow F(v)$. By Lemma 13 this implies $\mathtt{P} : [] \Downarrow v'$ for some v'. Lemma 2 (Determinism) implies $v = v'$. □

C Proof of Theorem 2

Proof For the "safety" theorem we use induction on proofs of $s \Downarrow s', G$ or $s \to s', G$. Safety of the constructed graphs for rules (ValueG), (OperatorG) and (OperandG) is immediate by Definitions 13 and 11.

The variable lookup rule **(VarG)** yields $\mathtt{x} : \rho \Downarrow \rho(\mathtt{x}), G$ with $G = \{\mathtt{x} \xrightarrow{=} \bullet\} \cup \{\mathtt{x} \xrightarrow{\downarrow} \mathtt{y} \mid \mathtt{y} \in \mathit{fv}(\mathtt{e}')\}$ and $\rho(\mathtt{x}) = \mathtt{e}' : \rho'$. By Definition 12, $\overline{\mathtt{x} : \rho}(\mathtt{x}) = \rho(\mathtt{x}) = \overline{\rho(\mathtt{x})}(\bullet)$, so arc $\mathtt{x} \xrightarrow{=} \bullet$ satisfies Definition 13. Further, if $\mathtt{x} \xrightarrow{\downarrow} \mathtt{y} \in G$ then $\mathtt{y} \in \mathit{fv}(\mathtt{e}')$. Thus $\rho'(\mathtt{y}) \in \mathit{support}(\mathtt{x} : \rho)$, so $\rho(\mathtt{x}) \succ \rho'(\mathtt{y})$ as required.

The rule **(CallG)** concludes $s \underset{c}{\to} s', G$, where $s = \mathtt{e}_1@\mathtt{e}_2 : \rho$ and $s' = \mathtt{e}_0 : \rho_0[\mathtt{x} \mapsto v_2]$ and $G = G_1^{-\bullet} \cup G_2^{\bullet \mapsto \mathtt{x}}$. Its premises are $\mathtt{e}_1 : \rho \Downarrow \lambda\mathtt{x}.\mathtt{e}_0 : \rho_0, G_1$ and $\mathtt{e}_2 : \rho \Downarrow v_2, G_2$. Let $v_2 = \mathtt{e}' : \rho'$. We assume inductively that G_1 is safe for $(\mathtt{e}_1 : \rho, \lambda\mathtt{x}.\mathtt{e}_0 : \rho_0)$ and that G_2 is safe for $(\mathtt{e}_2, v_2)$.

We wish to show safety: that $a \xrightarrow{=} a' \in G$ implies $\overline{s}(a) = \overline{s'}(a')$, and $a \xrightarrow{\downarrow} a' \in G$ implies $\overline{s}(a) \succ \overline{s'}(a')$. By definition of $G_1^{-\bullet}$ and $G_2^{\bullet \mapsto \mathtt{x}}$, $a \xrightarrow{r} a' \in G = G_1^{-\bullet} \cup G_2^{\bullet \mapsto \mathtt{x}}$ breaks into 6 cases:

Case 1: $\mathtt{y} \xrightarrow{\downarrow} \mathtt{z} \in G_1^{-\bullet}$ because $\mathtt{y} \xrightarrow{\downarrow} \mathtt{z} \in G_1$, where $\mathtt{y} \in \mathit{fv}(\mathtt{e}_1)$ and $\mathtt{z} \in \mathit{fv}(\lambda\mathtt{x}.\mathtt{e}_0)$. By safety of G_1, $\rho(\mathtt{y}) \succ \rho_0(\mathtt{z})$. Thus, as required,

$$\overline{s}(\mathtt{y}) = \overline{\mathtt{e}_1@\mathtt{e}_2 : \rho}(\mathtt{y}) = \rho(\mathtt{y}) \succ \rho_0(\mathtt{z}) = \rho_0[\mathtt{x} \mapsto v_2](\mathtt{z}) = \overline{s'}(\mathtt{z})$$

Case 2: $\mathtt{y} \xrightarrow{=} \mathtt{z} \in G_1^{-\bullet}$ because $\mathtt{y} \xrightarrow{=} \mathtt{z} \in G_1$. Like Case 1.

Case 3: $\bullet \xrightarrow{\downarrow} \mathtt{z} \in G_1^{-\bullet}$ because $\bullet \xrightarrow{r} \mathtt{z} \in G_1$, where $\mathtt{z} \in fv(\lambda\mathtt{x}.\mathtt{e}_0)$. Now $\bullet$ in G_1 refers to $\mathtt{e}_1 : \rho$, so $\mathtt{e}_1 : \rho \succeq \rho_0(\mathtt{z})$ by safety of G_1. Thus, as required,

$$\overline{s}(\bullet) = \mathtt{e}_1@\mathtt{e}_2 : \rho \succ \mathtt{e}_1 : \rho \succeq \rho_0(\mathtt{z}) = \rho_0[\mathtt{x} \mapsto v_2](\mathtt{z}) = \overline{s'}(\mathtt{z})$$

Case 4: $\mathtt{y} \xrightarrow{\downarrow} \mathtt{x} \in G_2^{\bullet \mapsto \mathtt{x}}$ because $\mathtt{y} \xrightarrow{\downarrow} \bullet \in G_2$, where $\mathtt{y} \in fv(\mathtt{e}_2)$. By safety of G_2, $\rho(\mathtt{y}) \succ v_2$. Thus, as required,

$$\overline{s}(\mathtt{y}) = \rho(\mathtt{y}) \succ v_2 = \rho_0[\mathtt{x} \mapsto v_2](\mathtt{x}) = \overline{s'}(\mathtt{x})$$

Case 5: $\mathtt{y} \xrightarrow{=} \mathtt{x} \in G_2^{\bullet \mapsto \mathtt{x}}$ because $\mathtt{y} \xrightarrow{=} \bullet \in G_2$. Like Case 4.

Case 6: $\bullet \xrightarrow{\downarrow} \mathtt{x} \in G_2^{\bullet \mapsto \mathtt{x}}$ because $\bullet \xrightarrow{r} \bullet \in G_2$. By safety of G_2, $\mathtt{e}_2 : \rho \succeq v_2$. Thus, as required,

$$\overline{s}(\bullet) = \mathtt{e}_1@\mathtt{e}_2 : \rho \succ \mathtt{e}_2 : \rho \succeq v_2 = \rho_0[\mathtt{x} \mapsto v_2](\mathtt{x}) = \overline{s'}(\mathtt{x})$$

The rule **(ApplyG)** concludes $s \Downarrow v, G'; G$ from premises $s \underset{c}{\rightarrow} s', G'$ and $s' \Downarrow v, G$, where $s = \mathtt{e}_1@\mathtt{e}_2 : \rho$ and $s' = \mathtt{e}' : \rho'$. We assume inductively that G' is safe for (s, s') and G is safe for (s', v). Let $G_0 = G'; G$.

We wish to show that G_0 is safe: that $a \xrightarrow{=} c \in G_0$ implies $\overline{s}(a) = \overline{v}(c)$, and $a \xrightarrow{\downarrow} c \in G_0$ implies $\overline{s}(a) \succ \overline{v}(c)$. First, consider the case $a \xrightarrow{=} c \in G_0$. Definition 1 implies $a \xrightarrow{=} b \in G'$ and $b \xrightarrow{=} c \in G$ for some b. Thus by the inductive assumptions we have $\overline{s}(a) = \overline{s'}(b) = \overline{v}(c)$, as required.

Second, consider the case $a \xrightarrow{\downarrow} c \in G_0$. Definition 1 implies $a \xrightarrow{r_1} b \in G'$ and $b \xrightarrow{r_2} c \in G$ for some b, where either or both of r_1, r_2 are $\downarrow$. By the inductive assumptions we have $\overline{s}(a) \succeq \overline{s'}(b)$ and $\overline{s'}(b) \succeq \overline{v}(c)$, and one or both of $\overline{s}(a) \succ \overline{s'}(b)$ and $\overline{s'}(b) \succ \overline{v}(c)$ hold. By Definition of $\succ$ and $\succeq$ this implies that $\overline{s}(a) \succ \overline{v}(c)$, as required. □

D Proof of Lemma 10

Proof The proof is by cases on which rule is applied to conclude $\mathtt{e} : \rho \Downarrow \mathtt{e}' : \rho'$ or $\mathtt{e} : \rho \rightarrow \mathtt{e}' : \rho'$. In all cases we show that some corresponding abstract interpretation rules can be applied to give the desired conclusion. The induction is on the total size of the proof[4] concluding that $\mathtt{e} : \rho \Downarrow \mathtt{e}' : \rho'$ or $\mathtt{e} : \rho \rightarrow \mathtt{e}' : \rho'$. The induction hypothesis is that the Lemma holds for all calls and evaluations performed in the computation before the last conclusion giving $\mathtt{e} : \rho \Downarrow \mathtt{e}' : \rho'$ or $\mathtt{e} : \rho \rightarrow \mathtt{e}' : \rho'$, i.e., we assume that the Lemma holds for premises of the rule last applied, and for any call and evaluation in the computation until then.

Base cases: Rule (Value), (Operator) and (Operand) in the exact semantics are modeled by axioms (ValueA), (OperatorA) and (OperandA) in the abstract

[4] This may be thought of as the number of steps in the computation of $\mathtt{e} : \rho \Downarrow \mathtt{e}' : \rho'$ or $\mathtt{e} : \rho \rightarrow \mathtt{e}' : \rho'$ starting from $P : []$.

semantics. These are the same as their exact-evaluation counterparts, after removal of environments for (ValueA) and (OperatorA), and a premise as well for (OperandA). Hence the Lemma holds if one of these rules were the last one applied.

The (Var) rule is, however, rather different from the (VarA) rule. If (Var) was applied to a variable $\mathtt{x}$ then the assumption is $\mathtt{P} : [] \to^* \mathtt{x} : \rho$ and $\mathtt{x} : \rho \Downarrow \mathtt{e}' : \rho'$. In this case $x \in dom(\rho)$ and $\mathtt{e}' : \rho' = \rho(x)$. Now $\mathtt{P} : [] \to^* \mathtt{x} : \rho$ begins from the empty environment, and we know all calls are from state to state. The only possible way $\mathtt{x}$ can have been bound is by a previous use of the (Call) rule, the only rule that extends an environment. The premises of this rule require that operator and operand in an application $e_1 @ e_2 : \rho''$ have previously been evaluated.

This requires that $e_1 @ e_2 \in subexp(P)$. By induction we can assume that the Lemma holds for $e_1 : \rho'' \Downarrow \lambda x.e_0 : \rho_0$ and $e_2 \Downarrow e' : \rho'$, so $e_1 \Downarrow \lambda x.e_0$ and $e_2 \Downarrow e'$ in the abstract semantics. Now we have all premises of rule (VarA), so we can conclude that $\mathtt{x} \Downarrow \mathtt{e}'$ as required.

For remaining rules (Apply) and (Call), when we assume that the Lemma holds for the premises in the rule applied to conclude $\mathtt{e} \Downarrow \mathtt{e}'$ or $\mathtt{e} : \rho \to \mathtt{e}' : \rho'$, then this gives us the premises for the corresponding rule for abstract interpretation. From this we can conclude the desired result. □

E Proof of Lemma 11

Proof The rules are the same as in Section 4.3, only extended with size-change graphs. We need to add to Lemma 10 that the size-change graphs generated for calls and evaluations can also be generated by the abstract interpretation. The proof is by cases on which rule is applied to conclude $\mathtt{e} \Downarrow \mathtt{e}', G$ or $\mathtt{e} : \rho \to \mathtt{e}' : \rho', G$.

We build on Lemma 10, and we saw in the proof of this that in abstract interpretation we can always use a rule corresponding to the one used in exact computation to prove corresponding steps. The induction hypothesis is that the Lemma holds for the premises of the rule in exact semantics.

Base case (VarAG): By Lemma 10 we have $\mathtt{x} : \rho \Downarrow \mathtt{e}' : \rho'$ implies $\mathtt{x} \Downarrow \mathtt{e}'$. The size-change graph built in (VarAG) is derived in the same way from $\mathtt{x}$ and $\mathtt{e}'$ as in rule (VarG), and they will therefore be identical.

For other call- and evaluation rules without premises, the abstract evaluation rule is as the exact-evaluation rule, only with environments removed, and the generated size-change graphs are not influenced by environments. Hence the Lemma will hold if these rules are applied.

For all other rules in a computation: When we know that Lemma 10 holds and assume that Lemma 11 hold for the premises, then we can conclude that if this rule is applied, then Lemma 11 holds by the corresponding rule from abstract interpretation. □

References

1. C.C. Frederiksen and N.D. Jones. Running-time Analysis and Implicit Complexity. Submitted to *Journal of Automated reasoning*Journal dadada.
2. J.Y. Girard, Y. Lafont, P. Taylor. *Proofs and Types.* Cambridge University Press, 1989.
3. N.D. Jones and F. Nielson. Abstract Interpretation: a Semantics-Based Tool for Program Analysis. In Handbook of Logic in Computer Science, pp. 527-629. Oxford University Press, 1994.
4. N.D. Jones and A. Glenstrup. Partial Evaluation Termination Analysis and Specialization-Point Insertion. ACM Transactions on Programming Languages and Systems, to appear in 2004.
5. C.S. Lee, N.D. Jones and A.M. Ben-Amram "The Size-Change Principle for Program Termination" POPL 2001: Proceedings 28th ACM SIGPLAN-SIGACT Symposium on Principles of Programming Languages, January 2001.
6. C.S. Lee. Finiteness analysis in polynomial time. In Static Analysis: 9th International Symposium, SAS 2002 (M Hermenegildo and G Puebla, eds.), pp. 493-508. Volume 2477 of Lecture Notes in Computer Science. Springer. September, 2002.
7. C.S. Lee. Program termination analysis in polynomial time. In Generative Programming and Component Engineering: ACM SIGPLAN/SIGSOFT Conference, GPCE 2002 (D Batory, C Consel, and W Taha, eds.), pp. 218-235. Volume 2487 of Lecture Notes in Computer Science.
8. C.S. Lee. "Program Termination Analysis and the Termination of Offline Partial Evaluation" Ph.D. thesis, University of Western Australia, March 2001
9. G.D. Plotkin. Call-by-name, call-by-value and the lambda-calculus. Theoretical Computer Science, 1, 1975.

A Type-Based Termination Criterion for Dependently-Typed Higher-Order Rewrite Systems

Frédéric Blanqui

LORIA & INRIA
615 rue du Jardin Botanique, BP 101, 54602 Villers-lès-Nancy, France
blanqui@loria.fr
http://www.loria.fr/~blanqui/

Abstract. Several authors devised type-based termination criteria for ML-like languages allowing non-structural recursive calls. We extend these works to general rewriting and dependent types, hence providing a powerful termination criterion for the combination of rewriting and β-reduction in the Calculus of Constructions.

1 Introduction

The Calculus of Constructions (CC) [13] is a powerful type system allowing polymorphic and dependent types. It is the basis of several proof assistants (Coq, Lego, Agda, ...) since it allows one to formalize the proofs of higher-order logic. In this context, it is essential to allow users to define functions and predicates in the most convenient way and to be able to decide whether a term is a proof of some proposition, and whether two terms/propositions are equivalent w.r.t. user definitions. As exemplified in [16, 10], a promising approach is rewriting. To this end, we need powerful criteria to check the termination of higher-order rewrite-based definitions combined with β-reduction.

In [10], we proved that such a combination is strongly normalizing if, on the one hand, first-order rewrite rules are strongly normalizing and non-duplicating and, on the other hand, higher-order rewrite rules satisfy a termination criterion based on the notion of computability closure and similar to primitive recursion. However, many rewrite systems do not satisfy these conditions, as division[1] on natural numbers *nat* for instance:

$$\begin{array}{lrcl}
(1) & -\;x\;0 & \rightarrow & x \\
(2) & -\;0\;x & \rightarrow & 0 \\
(3) & -\;(sx)\;(sy) & \rightarrow & -\;x\;y \\
(4) & /\;0\;x & \rightarrow & 0 \\
(5) & /\;(sx)\;y & \rightarrow & s\;(/\;(-\;x\;y)\;y)
\end{array}$$

[1] $(/\;x\;y)$ is the lower integer part of $\frac{x}{y+1}$.

V. van Oostrom (Ed.): RTA 2004, LNCS 3091, pp. 24–39, 2004.

Hughes *et al* [20], Xi [26], Giménez *et al* [18, 5] and Abel [2] devised termination criteria able to treat such examples by exploiting the way inductive types are usually interpreted [23]. Take for instance the addition on Brouwer's ordinals ord whose constructors are $0 : ord$, $s : ord \Rightarrow ord$ and $lim : (nat \Rightarrow ord) \Rightarrow ord$:

$$\begin{array}{lrcl}
(1) & +\ 0\ x & \rightarrow & x \\
(2) & +\ (sx)\ y & \rightarrow & s\ (+\ x\ y) \\
(3) & +\ (lim\ f)\ y & \rightarrow & lim\ ([x : nat](+\ (f\ x)\ y))
\end{array}$$

The usual computability-based technique for proving the termination of this function is to interpret ord as the fixpoint of the following monotone function φ on the powerset of the set of strongly normalizing terms $\mathcal{SN}$ ordered by inclusion:

$$\varphi(X) = \{t \in \mathcal{SN} \mid t \rightarrow^* su \Rightarrow u \in X; t \rightarrow^* limf \Rightarrow \forall u \in \mathcal{SN}, fu \in X\}$$

The fixpoint of φ, $[\![ord]\!]$, can be reached by transfinite iteration and every $t \in [\![ord]\!]$ is obtained after a smallest ordinal $o(t)$ of iterations, the order of t. This naturally defines an ordering: $t > u$ iff $o(t) > o(u)$, with which $lim\ f > fu$ for all $u \in \mathcal{SN}$.

Now, applying this technique to nat, we can easily check that $o(-tu) \leq o(t)$ and thus allow the recursive call with $-xy$ in the definition of $/$ above. We proceed by induction on $o(t)$, knowing that $-tu$ is computable (*i.e.* belongs to $[\![nat]\!]$) iff all its reducts are computable:

- If $-tu$ matches rule (1) then $o(-tu) = o(t)$.
- If $-tu$ matches rule (2) then $o(-tu) = 0 \leq o(t)$.
- If $-tu$ matches rule (3) then $t = st'$ and $u = su'$. By induction hypothesis, $o(-t'u') \leq o(t')$. Thus, $o(-tu) = 1 + o(-t'u') \leq 1 + o(t') = o(t)$.
- If $-tu$ matches no rule then $o(-tu) = 0 \leq o(t)$.

The idea of the previously cited authors is to add that size/index/stage information to the syntax in order to prove this automatically. Instead of a single type nat, they consider a family of types $\{nat^{\mathfrak{a}}\}_{\mathfrak{a}\in\omega}$ (higher-order types require ordinals bigger than ω), each type $nat^{\mathfrak{a}}$ being interpreted by the set obtained after $\mathfrak{a}$ iterations of the function φ for nat. For first-order data types, $\mathfrak{a}$ can be seen as the maximal number of constructors at the top of a term. Finally, they define a decidable type system in which $-$ (defined by *fixpoint/cases* constructions in their work) can be typed by $nat^{\alpha} \Rightarrow nat^{\beta} \Rightarrow nat^{\alpha}$, where α and β are size variables, meaning that the order of $-tu$ is not greater than the order of t.

This can also be interpreted as a way to automatically prove theorems on the size of the result of a function w.r.t. the size of its arguments with applications to complexity and resource bound certification, and compilation optimization (*e.g.* array bound checks elimination and vector-based memoisation).

In this paper, we extend this technique to the full Calculus of Algebraic Constructions [10] whose type conversion rule depends on user definitions, and to general rewrite-based definitions (including rewriting modulo equational theories treated elsewhere [7]) instead of definitions only based on *fixpoint/cases* constructions. However, several questions remain unanswered (*e.g.* subject reduction, matching on defined symbol, type inference) and left for future work.

We allow a richer size algebra than the one in [20, 5, 2] but we do not allow existential size variables and do not take into account conditionals as it can be done in Xi's work [26]. Note however that Xi is interested in the call-by-value normalization of closed simply-typed λ-terms with first-order data types, while we are interested in the strong normalization of the open terms of CAC.

The paper is organized as follows. Section 2 introduces the Calculus of Algebraic Constructions with Size Annotations (CACSA). Section 3 presents the termination criterion with some examples. Section 4 gives some elements of the termination proof (more details can be found in [9]). Finally, Section 5 proposes an extension (whose justification is ongoing) for capturing more definitions.

2 The Calculus of Algebraic Constructions with Size Annotations

CC is the full Pure Type System with set of *sorts* $\mathcal{S} = \{\star, \square\}$ and axiom $\star : \square$ [4]. The sort $\star$ is intended to be the universe of types and propositions, while $\square$ is intended to be the universe of predicate types. Let $\mathcal{X}$ be the set of variables.

The Calculus of Algebraic Constructions (CAC) [10] is an extension of CC with a set $\mathcal{F}$ of function or predicate *symbols* defined by a set $\mathcal{R}$ of (higher-order) rewrite rules [15, 22] whose left hand-sides are built from symbols and variables only. Every $x \in \mathcal{X} \cup \mathcal{F}$ is equipped with a sort s_x. We denote by $\mathcal{DF}$ the set of *defined* symbols, that is, the set of symbols f with a rule $f\boldsymbol{l} \to r \in \mathcal{R}$, and by $\mathcal{CF}$ the set $\mathcal{F} \setminus \mathcal{DF}$ of *constant* symbols. We add a superscript s to restrict these sets to objects of sort s.

Now, we assume given a first-order term algebra $\mathcal{A} = T(\mathcal{H}, \mathcal{Z})$, called the algebra of *size expressions*, built from a set $\mathcal{H}$ of *size symbols* of fixed arity and a set $\mathcal{Z}$ of *size variables*. Let $\mathcal{V}(t)$ be the set of size variables occurring in a term t. We assume that $\mathcal{H} \cap \mathcal{F} = \mathcal{Z} \cap \mathcal{X} = \emptyset$, $T(\mathcal{H}, \emptyset) \neq \emptyset$ and $\mathcal{A}$ is equipped with a quasi-ordering $\leq_{\mathcal{A}}$ stable by size substitution (if $a \leq_{\mathcal{A}} b$ then, for all size substitution φ, $a\varphi \leq_{\mathcal{A}} b\varphi$) such that $(\mathcal{A}, \leq_{\mathcal{A}})$ has a well-founded model $(\mathfrak{A}, \leq_{\mathfrak{A}})$:

Definition 1 (Size model). *A* pre-model *of $\mathcal{A}$ is given by a set $\mathfrak{A}$, an ordering $\leq_{\mathfrak{A}}$ on $\mathfrak{A}$ and a function $h_{\mathfrak{A}}$ from $\mathfrak{A}^n$ to $\mathfrak{A}$ for every n-ary size symbol $h \in \mathcal{H}$. A* size valuation *is a function ν from $\mathcal{Z}$ to $\mathfrak{A}$, naturally extended to a function on $\mathcal{A}$. A pre-model is a* model *if $a \leq_{\mathcal{A}} b$ implies $a\nu \leq_{\mathfrak{A}} b\nu$, for all size valuation ν. Such a model is* well-founded *if $>_{\mathfrak{A}}$ is well-founded.*

The Calculus of Algebraic Constructions with Size Annotations (CACSA) is an extension of CAC where constant predicate symbols are annotated by size expressions. The terms of CACSA are defined by the following grammar rule:

$$t ::= s \mid x \mid C^a \mid f \mid [x:t]t \mid (x:t)t \mid tt$$

where $C \in \mathcal{CF}^\square$, $f \in \mathcal{F} \setminus \mathcal{CF}^\square$ and $a \in \mathcal{A}$. We denote by $\mathcal{T}_{\mathcal{A}}(\mathcal{F}, \mathcal{X})$ the set of terms built from $\mathcal{F}$, $\mathcal{X}$ and $\mathcal{A}$. A product $(x:T)U$ with $x \notin \mathrm{FV}(U)$ is written $T \Rightarrow U$. We now assume that rewrite rules are built from annotated terms not containing size variables. Hence, if $t \to t'$ then, for all size substitution φ, $t\varphi \to t'\varphi$.

We also assume that every symbol f is equipped with a closed type $\tau_f = (\boldsymbol{x} : \boldsymbol{T})U$ with no size variable if $s_f = \Box$ (size variables are implicitly universally quantified otherwise), and $|\boldsymbol{l}| \le |\boldsymbol{x}|$ if $f\boldsymbol{l} \to r \in \mathcal{R}$, a set $\mathrm{Mon}^+(f) \subseteq A_f = \{1, \ldots, |\boldsymbol{x}|\}$ of *monotone arguments* and a disjoint set $\mathrm{Mon}^-(f) \subseteq A_f$ of *anti-monotone arguments.* For a size symbol h, $\mathrm{Mon}^+(h)$ (resp. $\mathrm{Mon}^-(h)$) is taken to be the arguments in which $h_{\mathfrak{A}}$ is monotone (resp. anti-monotone).

$$\text{(ax)} \quad \vdash \star : \Box$$

$$\text{(size)} \quad \frac{\vdash \tau_C : \Box}{\vdash C^a : \tau_C} \quad (C \in \mathcal{CF}^\Box,\ a \in \mathcal{A})$$

$$\text{(symb)} \quad \frac{\vdash \tau_f : s_f}{\vdash f : \tau_f \varphi} \quad (f \notin \mathcal{CF}^\Box)$$

$$\text{(var)} \quad \frac{\Gamma \vdash T : s_x}{\Gamma, x : T \vdash x : T} \quad (x \notin \mathrm{dom}(\Gamma))$$

$$\text{(weak)} \quad \frac{\Gamma \vdash t : T \quad \Gamma \vdash U : s_x}{\Gamma, x : U \vdash t : T} \quad (x \notin \mathrm{dom}(\Gamma))$$

$$\text{(prod)} \quad \frac{\Gamma \vdash U : s \quad \Gamma, x : U \vdash V : s'}{\Gamma \vdash (x : U)V : s'}$$

$$\text{(abs)} \quad \frac{\Gamma, x : U \vdash v : V \quad \Gamma \vdash (x : U)V : s}{\Gamma \vdash [x : U]v : (x : U)V}$$

$$\text{(app)} \quad \frac{\Gamma \vdash t : (x : U)V \quad \Gamma \vdash u : U}{\Gamma \vdash tu : V\{x \mapsto u\}}$$

$$\text{(sub)} \quad \frac{\Gamma \vdash t : T \quad \Gamma \vdash T' : s}{\Gamma \vdash t : T'} \quad (T \le T')$$

Fig. 1. Typing rules

An *environment* Γ is a sequence of variable-term pairs. Let $t \downarrow u$ iff there is v such that $t \to^* v \;{}^*\!\!\leftarrow u$, with $\to^*$ the reflexive and transitive closure of $\to = \to_\beta \cup \to_\mathcal{R}$. The typing rules of CACSA are given in Figure 1 and its subtyping rules in Figure 2. There are two differences with CAC. First, there is a new rule (size) for typing constant predicate symbols with size annotations, while the usual rule (symb) for typing symbols is restricted to the other symbols. Second, in CAC, the condition for (sub) is not $T \le T'$ but $T \downarrow T'$. Note that, if $\to$ is confluent then $\downarrow$ is equivalent to $\le$ without the subtyping rule (size).

Subtyping is necessary since size annotations are upper bounds. For instance, in an *if-then-else* expression, the *then*-branch does not need to exactly have the same type as the *else*-branch. Instead of subtyping, Xi uses singleton types, existential size variables and refinement types.

The way the subtyping relation is defined is due to Chen [12]. Replacing (red), (exp) and (refl) by $T \le U$ if $T \downarrow U$ would not allow us to prove that (trans) can be eliminated, which is essential for proving the compatibility of subtyping with the product construction (if $(x:U)V \le (x:U')V'$ then $U' \le U$ and $V \le V'$), which in turn enables one to prove that β preserves typing. Another consequence is that subtyping is decidable when applied on weakly normalizing terms. We refer the reader to [9] for more details on the meta-theory of our type system.

$$
\begin{array}{lcl}
\text{(refl)} & T \le T & \\[2ex]
\text{(size)} & C^a\boldsymbol{t} \le C^b\boldsymbol{t} & (C \in \mathcal{CF}^\square,\ a \le_{\mathcal{A}} b) \\[2ex]
\text{(prod)} & \dfrac{U' \le U \quad V \le V'}{(x:U)V \le (x:U')V'} & \\[2ex]
\text{(red)} & \dfrac{T' \le U'}{T \le U} & (T \rightarrow^* T',\ U' \ {}^*\!\!\leftarrow U) \\[2ex]
\text{(exp)} & \dfrac{T' \le U'}{T \le U} & (T \ {}^*\!\!\leftarrow T',\ U' \rightarrow^* U) \\[2ex]
\text{(trans)} & \dfrac{T \le U \quad U \le V}{T \le V} &
\end{array}
$$

Fig. 2. Subtyping rules

In this paper, we make two important assumptions:

(1) $\beta \cup \mathcal{R}$ is confluent. This is the case for instance if $\mathcal{R}$ is confluent and left-linear. Finding other sufficient conditions is an open problem.

(2) $\mathcal{R}$ preserves typing: if $l \rightarrow r \in \mathcal{R}$ and $\Gamma \vdash l\sigma : T$ then $\Gamma \vdash r\sigma : T$. Finding sufficient conditions with subtyping and dependent types does not seem easy. We leave the study of this problem for future work. With dependent or polymorphic symbols, requiring the existence of Γ and T such that $\Gamma \vdash l : T$ and $\Gamma \vdash r : T$ leads to non left-linear rules. In [10], we give general conditions avoiding the non-linearities implied by requiring l to be well-typed.

3 Constructor-Based Systems

We now study the case of CACSA's whose size algebra at least contains the following expressions:

$$a ::= \alpha \mid sa \mid \infty \mid \ldots$$

Following [5], when there is no other symbol, the ordering $\leq_{\mathcal{A}}$ on size expressions is defined as the smallest congruent quasi-ordering $\leq$ such that, for all a, $a < sa$ and $a \leq \infty$, and size expressions are interpreted in $\mathfrak{A} = \Omega + 1$, where Ω is the first uncountable ordinal, by taking $s_{\mathfrak{A}}(\mathfrak{a}) = \mathfrak{a} + 1$ if $\mathfrak{a} < \Omega$, $s_{\mathfrak{A}}(\Omega) = \Omega$ and $\infty_{\mathfrak{A}} = \Omega$.

One can easily imagine other size expressions like $a + b$, $max(a, b)$, ...

We now define the sets of positive and negative positions in a term, which will enforce monotonicity and anti-monotonicity properties respectively. Then, we define the set of admissible inductive types.

Definition 2 (Positive and negative positions). *The set of positions (words over $\{L, R, S\}$) in a term t is inductively defined as follows:*

- $\mathrm{Pos}(s) = \mathrm{Pos}(x) = \mathrm{Pos}(f) = \{\varepsilon\}$ *(empty word)*
- $\mathrm{Pos}((x : u)v) = \mathrm{Pos}([x : u]v) = \mathrm{Pos}(uv) = L.\mathrm{Pos}(u) \cup R.\mathrm{Pos}(v)$
- $\mathrm{Pos}(C^a) = \{\varepsilon\} \cup S.\mathrm{Pos}(a)$

Let $\mathrm{Pos}(x, t)$ *($x \in \mathcal{F} \cup \mathcal{X} \cup \mathcal{Z}$) be the set of positions of the free occurrences of x in t. The set of* positive positions *in t,* $\mathrm{Pos}^+(t)$*, and the set of* negative positions *in t,* $\mathrm{Pos}^-(t)$*, are simultaneously defined by induction on t:*

- $\mathrm{Pos}^\delta(s) = \mathrm{Pos}^\delta(x) = \{\varepsilon \mid \delta = +\}$
- $\mathrm{Pos}^\delta((x : U)V) = L.\mathrm{Pos}^{-\delta}(U) \cup R.\mathrm{Pos}^\delta(V)$
- $\mathrm{Pos}^\delta([x : U]v) = R.\mathrm{Pos}^\delta(v)$
- $\mathrm{Pos}^\delta(tu) = L.\mathrm{Pos}^\delta(t)$ *if* $t \neq f\boldsymbol{t}$
- $\mathrm{Pos}^\delta(f\boldsymbol{t}) = \{L^{|\boldsymbol{t}|} \mid \delta = +\} \cup \bigcup\{L^{|\boldsymbol{t}|-i}R.\mathrm{Pos}^{\varepsilon\delta}(t_i) \mid \varepsilon \in \{-, +\}, i \in \mathrm{Mon}^\varepsilon(f)\}$
- $\mathrm{Pos}^\delta(C^a\boldsymbol{t}) = \mathrm{Pos}^\delta(C\boldsymbol{t}) \cup \{L^{|\boldsymbol{t}|}S \mid \delta = +\}.\mathrm{Pos}^\delta(a)$.

where $\delta \in \{-, +\}$, $-+ = -$ and $-- = +$ (usual rules of signs).

Definition 3 (Constructor-based system). *We assume given a* precedence *$\leq_{\mathcal{F}}$ on $\mathcal{F}$ and that every $C \in \mathcal{CF}^\square$ with $C : (\boldsymbol{z} : \boldsymbol{V})\star$ is equipped with a set* $\mathrm{Cons}(C)$ *of* constructors, *that is, a set of constant symbols $f : (\boldsymbol{y} : \boldsymbol{U})C^a\boldsymbol{v}$ equipped with a set* $\mathrm{Acc}(f) \subseteq A_f$ *of* accessible *arguments such that:*

- *If there are $D \simeq_{\mathcal{F}} C$ such that* $\mathrm{Pos}(D, U_j) \neq \emptyset$ *then there is $\alpha \in \mathcal{Z}$ such that $\mathcal{V}(\tau_f) = \{\alpha\}$ and $a = s\alpha$.*
- *For all $j \in \mathrm{Acc}(c)$:*
  - *For all $D >_{\mathcal{F}} C$,* $\mathrm{Pos}(D, U_j) = \emptyset$.
  - *For all $D \simeq_{\mathcal{F}} C$ and $p \in \mathrm{Pos}(D, U_j)$, $p \in \mathrm{Pos}^+(U_j)$ and $U_j|_p = D^\alpha$.*
  - *For all $p \in \mathrm{Pos}(\alpha, U_j)$, $p = qS$, $U_j|_q = D^\alpha$ and $D \simeq_{\mathcal{F}} C$.*
  - *For all $x \in \mathrm{FV}^\square(U_j)$, there is ι_x with $v_{\iota_x} = x$ and $\mathrm{Pos}(x, U_j) \subseteq \mathrm{Pos}^+(U_j)$.*
- *For all $F \in \mathcal{DF}^\square$ and $F\boldsymbol{l} \rightarrow r \in \mathcal{R}$:*
  - *For all $G >_{\mathcal{F}} F$,* $\mathrm{Pos}(G, r) = \emptyset$.
  - *For all $i \in \mathrm{Mon}^\delta(F)$, $l_i \in \mathcal{X}^\square$ and $\mathrm{Pos}(l_i, r) \subseteq \mathrm{Pos}^\delta(r)$.*
  - *For all $x \in \mathrm{FV}^\square(r)$, there is κ_x with $l_{\kappa_x} = x$.*

The positivity conditions are usual. The restrictions on a and α are also present in [5, 2]. Section 5 proposes more general conditions. The conditions involving ι and κ mean that we restrict our attention to *small* inductive types for which predicate variables are parameters. See [6] for details about inductive types and weak/strong elimination.

An example is the inductive-recursive type $T : \star$ with constructors $v : nat \Rightarrow T$, $f : (list\ T) \Rightarrow T$ and $\mu : \neg\neg T \Rightarrow T$ (first-order terms with continuations), where $list : \star \Rightarrow \star$ is the type of polymorphic lists ($\mathrm{Mon}^+(list) = \{1\}$), $\neg : \star \Rightarrow \star$ ($\mathrm{Mon}^-(\neg) = \{1\}$) is defined by the rule $\neg\ A \rightarrow A \Rightarrow \bot$, and $\bot = (A : \star)A$.

We now give general conditions for rewrite rules to preserve strong normalization, based on the fundamental notion of *computability closure*. The computability closure of a term t is a set of terms that can be proved computable whenever t is computable. If, for every rule $f\boldsymbol{l} \rightarrow r$, r belongs to the computability closure of $\boldsymbol{l}$, then rules preserve computability, hence strong normalization.

In [10], the computability closure is inductively defined as a typing relation $\vdash_{\mathrm{c}}$ similar to $\vdash$ except for the (symb) case which is replaced by two new cases: (symb$^<$) for symbols strictly smaller than f, and (symb$^=$) for symbols equivalent to f whose arguments are structurally smaller than $\boldsymbol{l}$.

Here, (symb$^=$) is replaced by a new case for symbols equivalent to f whose arguments have, from typing, sizes smaller than those of $\boldsymbol{l}$. For comparing sizes, one can use metrics, similar to Dershowitz and Hoot's termination functions [14].

Definition 4 (Ordering on symbol arguments). *For every symbol $f : (\boldsymbol{x} : \boldsymbol{T})U$, we assume given two well-founded domains, $(D_f^{\mathcal{A}}, >_f^{\mathcal{A}})$ and $(D_f^{\mathfrak{A}}, >_f^{\mathfrak{A}})$, and two functions $\zeta_f^{\mathcal{A}} : \mathcal{A}^n \rightarrow D_f^{\mathcal{A}}$ and $\zeta_f^{\mathfrak{A}} : \mathfrak{A}^n \rightarrow D_f^{\mathfrak{A}}$ ($n = |\boldsymbol{x}|$) such that $(D_f^X, >_f^X) = (D_g^X, >_g^X)$ ($X \in \{\mathcal{A}, \mathfrak{A}\}$) whenever $f \simeq_{\mathcal{F}} g$, and we define:*

- $a_f^i = a$ *if* $T_i = C^a\boldsymbol{v}$, *and* $a_f^i = \infty$ *otherwise.*
- $(f, \varphi) >^{\mathcal{A}} (g, \psi)$ *iff* $f >_{\mathcal{F}} g$ *or* $f \simeq_{\mathcal{F}} g$ *and* $\zeta_f^{\mathcal{A}}(\boldsymbol{a}_f\varphi) >_f^{\mathcal{A}} \zeta_g^{\mathcal{A}}(\boldsymbol{a}_g\psi)$.
- $(f, \nu) >^{\mathfrak{A}} (g, \mu)$ *iff* $f >_{\mathcal{F}} g$ *or* $f \simeq_{\mathcal{F}} g$ *and* $\zeta_f^{\mathfrak{A}}(\boldsymbol{a}_f\nu) >_f^{\mathfrak{A}} \zeta_g^{\mathfrak{A}}(\boldsymbol{a}_g\mu)$.

Then, we assume that $>^{\mathcal{A}}$ is decidable and that $(f, \varphi) >^{\mathcal{A}} (g, \psi)$ implies $(f, \varphi\nu) >^{\mathfrak{A}} (g, \psi\nu)$ for all ν.

A simple metric is given by assigning a *status* to every symbol, that is, a non-empty sequence of multisets of positive integers, describing a simple combination of lexicographic and multiset comparisons. Given a set D and a status ζ of arity n (biggest integer occurring in it), we define $[\![\zeta]\!]_D$ on D^n as follows:

- $[\![M_1 \ldots M_k]\!]_D(\boldsymbol{x}) = ([\![M_1]\!]_D^m(\boldsymbol{x}), \ldots, [\![M_k]\!]_D^m(\boldsymbol{x}))$
- $[\![\{i_1, \ldots, i_p\}]\!]_D^m(\boldsymbol{x}) = \{x_{i_1}, \ldots, x_{i_p}\}$ (multiset)

Now, take $\zeta_f^X = [\![\zeta_f]\!]_X$, $D_f^X = \zeta_f^X(X^n)$ and $>_f^X = ((>_X)_{\mathrm{mul}})_{\mathrm{lex}}$.

For building the computability closure, one must start from the variables of the left hand-side. However, one cannot take any variable since, *a priori*, not every subterm of a computable term is computable. To this end, based on the interpretation of constant predicate symbols, we introduce the following notion:

Definition 5 (Accessibility). *We say that* $u : U$ *is* a-accessible *in* $t : T$, *written* $t : T \rhd_a u : U$, *iff* $t = f\boldsymbol{u}$, $f \in \mathrm{Cons}(C)$, $f : (\boldsymbol{y} : \boldsymbol{U})C^{s\alpha}\boldsymbol{v}$, $|\boldsymbol{u}| = |\boldsymbol{y}|$, $u = u_j$, $j \in \mathrm{Acc}(f)$, $T = C^{s\alpha\varphi}\boldsymbol{v}\gamma$, $U = U_j\gamma\varphi$, $\boldsymbol{y}\gamma = \boldsymbol{u}$, $\alpha\varphi = a$ *and* $\mathrm{Pos}(\alpha, \boldsymbol{u}) = \emptyset$.

A constructor $c : (\boldsymbol{y} : \boldsymbol{U})C^{a}\boldsymbol{v}$ *is* finitely branching[2] *iff, for all* $j \in \mathrm{Acc}(c)$, *either* $\mathrm{Pos}(\alpha, U_j) = \emptyset$ *or there exists* D *such that* $U_j = D^{\alpha}\boldsymbol{u}$. *We say that* $u : U$ *is* strongly a-accessible *in* $t : T$, *written* $t : T \blacktriangleright_a u : U$, *iff* $t : T \rhd_a u : U$, f *is a finitely branching constructor and* $\mathrm{Pos}(\alpha, U_j) \neq \emptyset$.

We say that $u : U$ *is* $*$-accessible modulo φ *in* $t : T$, *written* $t : T \gg_\varphi u : U$, *iff either* $t : T\varphi = u : U$ *and* $\varphi|_{\mathcal{V}(T)}$ *is a renaming*[3], *or* $t : T\varphi \blacktriangleright^* \rhd_\epsilon u : U$ *for some* $\epsilon \in \mathcal{Z}$.

This seems to restrict matching to constructors as in ML-like languages. However, one can prove that, for first-order data types, computability is equivalent to strong normalization [9]. Thus, every argument of a first-order symbol can be declared as accessible, and matching on defined first-order function symbols is possible. Meanwhile, it may be uneasy to find for these symbols output sizes and measures satisfying all the constraints required for subject reduction and recursive calls. More research has to be done on this subject.

Definition 6 (Termination criterion). *For every rule* $f\boldsymbol{l} \to r \in \mathcal{R}$ *with* $f : (\boldsymbol{x} : \boldsymbol{T})U$ *and* $\boldsymbol{x}\gamma = \boldsymbol{l}$, *we assume given a size substitution* φ. *The* computability closure *for this rule is given by the type system of Figure 3 on the set of terms* $\mathcal{T}_{\mathcal{A}}(\mathcal{F}', \mathcal{X}')$ *where* $\mathcal{F}' = \mathcal{F} \cup \mathrm{dom}(\Gamma)$, $\mathcal{X}' = \mathcal{X} \setminus \mathrm{dom}(\Gamma)$ *and, for all* $x \in \mathrm{dom}(\Gamma)$, $\tau_x = x\Gamma$ *and* $x <_{\mathcal{F}} f$. *The termination conditions are:*

- *Well-typedness: for all* $x \in \mathrm{dom}(\Gamma)$, $\vdash_{\mathrm{c}} l_i : T_i\varphi\gamma$.
- *Linearity:* Γ *is linear w.r.t. size variables.*
- *Accessibility: for all* $x \in \mathrm{dom}(\Gamma)$, *there are* i *and* β *such that* $l_i : T_i\gamma \gg_\varphi x : x\Gamma$, $T_i = C^{\beta}\boldsymbol{t}$ *and* $\mathcal{V}(\boldsymbol{t}) = \emptyset$.
- *Computability closure:* $\vdash_{\mathrm{c}} r : U\varphi\gamma$.
- *Positivity: for all* $\alpha \in \mathcal{V}(\boldsymbol{T})$, $\mathrm{Pos}(\alpha, U) \subseteq \mathrm{Pos}^+(U)$.
- *Safeness:* γ *is an injection from* $\mathrm{dom}^\square(\Gamma_f)$ *to* $\mathrm{dom}^\square(\Gamma)$.

The positivity condition on the output type of f w.r.t. size variables appears in the previous works on sized types too. It may be extended to more general continuity conditions [20, 1]. In [3], Abel gives an example of a function which is not terminating because it does not satisfy such a condition.

As for the safeness condition, it simply says that one cannot do matching or have non-linearities on predicate variables, which is known to lead to non-termination in some cases [19]. It is also part of other works on CC with inductive types [24] and rewriting [25].

The linearity, positivity, safeness and accessibility conditions are decidable. We think that the other conditions are decidable too, under the assumption that the satisfiability of inequality constraints in $\mathcal{A}$ is decidable. To this end, we

[2] Constructors of usual first-order data types are finitely branching.

[3] An injection from a finite subset of $\mathcal{Z}$ to $\mathcal{Z}$.

$$\text{(ax)}\quad \frac{}{\vdash_c \star : \square}$$

$$\text{(size)}\quad \frac{\vdash_c \tau_C : \square}{\vdash_c C^a : \tau_C} \qquad (C \in \mathcal{CF}^\square)$$

$$\text{(symb)}\quad \frac{\vdash_c \tau_g : s_g \quad (\forall i) \Delta \vdash_c y_i\delta : U_i\psi\delta}{\Delta \vdash_c g\boldsymbol{y}\delta : V\psi\delta} \qquad \begin{array}{c}(g \notin \mathcal{CF}^\square,\ g : (\boldsymbol{y} : \boldsymbol{U})V,\\ (g,\psi) <^A (f,\varphi))\end{array}$$

$$\text{(var)}\quad \frac{\Delta \vdash_c T : s_x}{\Delta, x : T \vdash_c x : T} \qquad (x \notin \mathrm{dom}(\Delta))$$

$$\text{(weak)}\quad \frac{\Delta \vdash_c t : T \quad \Delta \vdash_c U : s_x}{\Delta, x : U \vdash_c t : T} \qquad (x \notin \mathrm{dom}(\Delta))$$

$$\text{(prod)}\quad \frac{\Delta, x : U \vdash_c V : s}{\Delta \vdash_c (x : U)V : s}$$

$$\text{(abs)}\quad \frac{\Delta, x : U \vdash_c v : V \quad \Delta \vdash_c (x : U)V : s}{\Delta \vdash_c [x : U]v : (x : U)V}$$

$$\text{(app)}\quad \frac{\Delta \vdash_c t : (x : U)V \quad \Delta \vdash_c u : U}{\Delta \vdash_c tu : V\{x \mapsto u\}}$$

$$\text{(conv)}\quad \frac{\Delta \vdash_c t : T \quad \Delta \vdash_c T : s \quad \Delta \vdash_c T' : s}{\Delta \vdash_c t : T'} \qquad (T \leq T')$$

Fig. 3. Computability closure of $(f\boldsymbol{l} \rightarrow r, \Gamma, \varphi)$ with $f : (\boldsymbol{x} : \boldsymbol{T})U$ and $\boldsymbol{x}\gamma = \boldsymbol{l}$

prove the strong normalization of well-typed terms in Section 4, and describe a type inference algorithm in [9]. In practice, like Xi, we can restrict size expressions to linear arithmetic, for which the satisfiability of inequality constraints is decidable.

Note that, with polymorphic or dependent function symbols, the well-typedness condition makes the rules non left-linear. For instance, with concatenation on polymorphic list: $app\ A\ (cons\ A'\ x\ l)\ l' \rightarrow cons\ A\ x\ (app\ A\ l\ l')$, we need to take $A' = A$. In [10], we proved that, in CAC, this condition can be relaxed by relativizing the previous conditions with the substitution $\{A' \mapsto A\}$. The same technique should apply to CACSA.

We now give some examples satisfying these conditions:

Example 1 (Division on natural numbers). Take $nat : \star$, $0 : nat^0$, $s : nat^\alpha \Rightarrow nat^{s\alpha}$, $- : nat^\alpha \Rightarrow nat^\beta \Rightarrow nat^\alpha$ and $/ : nat^\alpha \Rightarrow nat^\beta \Rightarrow nat^\alpha$.

- For rule (3), take $\zeta_-(\alpha,\beta) = \alpha$, $\Gamma = x : nat^\delta, y : nat^\epsilon$, $\varphi = \{\alpha \mapsto s\delta, \beta \mapsto s\epsilon\}$ and $s <_\mathcal{F} -$. By (symb), $\vdash_c x : nat^\delta$ and $\vdash_c y : nat^\epsilon$. By (symb), $\vdash_c -xy : nat^\delta$ since $\zeta_-(\delta,\epsilon) = \delta < \zeta_-(s\delta, s\epsilon) = s\delta$. Thus, by (sub), $\vdash_c -xy : nat^{s\delta}$.

- For rule (5), take $\zeta_/(\alpha,\beta) = \alpha$, $\Gamma = x : nat^\delta, y : nat^\epsilon$, $\gamma = \{p \mapsto sx, q \mapsto y\}$, $\varphi = \{\alpha \mapsto s\delta, \beta \mapsto \epsilon\}$ and $- <_\mathcal{F} /$. By (symb), $\vdash_c x : nat^\delta$ and $\vdash_c y : nat^\epsilon$. By (symb), $\vdash_c -xy : nat^\delta$. By (symb), $\vdash_c /(-xy)y : nat^\delta$ since $\zeta_/(\delta,\epsilon) = \delta < \zeta_/(s\delta,\epsilon) = s\delta$. Thus, by (symb), $\vdash_c s(/(-xy)y) : nat^{s\delta}$.

Example 2 (Addition on Brouwer's ordinals). Take $ord : \star$, $0 : nat^0$, $s : nat^\alpha \Rightarrow nat^{s\alpha}$, $lim : (nat \Rightarrow ord^\alpha) \Rightarrow ord^{s\alpha}$ and $+ : nat^\alpha \Rightarrow nat^\beta \Rightarrow nat^\infty$. For rule (3), take $\zeta_+(\alpha,\beta) = \alpha$, $\Gamma = f : nat^\infty \Rightarrow ord^\delta, y : ord^\epsilon$, $\varphi = \{\alpha \mapsto s\delta, \beta \mapsto \epsilon\}$ and $s, lim <_\mathcal{F} +$. By (symb), $\vdash_c f : nat^\infty \Rightarrow ord^\delta$ and $\vdash_c y : ord^\epsilon$. Let $\Delta = x : nat^\infty$. By (var), $\Delta \vdash_c x : nat^\infty$. By (weak), $\Delta \vdash_c f : nat^\infty \Rightarrow ord^\delta$ and $\Delta \vdash_c y : ord^\epsilon$. By (app), $\Delta \vdash_c fx : ord^\delta$. By (symb), $\Delta \vdash_c +(fx)y : ord^\infty$ since $\zeta_+(\delta,\epsilon) = \delta < \zeta_+(s\delta,\epsilon) = s\delta$. By (abs), $\vdash_c [x : nat^\infty](+(fx)y) : (x : nat^\infty)ord^\delta$. Thus, by (symb), $\vdash_c lim([x : nat^\infty](+(fx)y)) : ord^{s\delta}$. This does not enter Xi's framework.

Example 3 (Huet and Hullot's reverse function). Take $list : \star$, $nil : list^0$, $cons : nat^\infty \Rightarrow list^\alpha \Rightarrow list^{s\alpha}$, $rev1 : nat^\infty \Rightarrow list^\infty \Rightarrow nat^\infty$, $rev2 : nat^\infty \Rightarrow list^\beta \Rightarrow list^\beta$ and $rev : list^\alpha \Rightarrow list^\alpha$.

$$
\begin{array}{lrcl}
(1) & rev1\ x\ nil & \rightarrow & x \\
(2) & rev1\ x\ (cons\ y\ l) & \rightarrow & rev1\ y\ l \\
(3) & rev2\ x\ nil & \rightarrow & nil \\
(4) & rev2\ x\ (cons\ y\ l) & \rightarrow & rev\ (cons\ x\ (rev\ (rev2\ y\ l))) \\
(5) & rev\ nil & \rightarrow & nil \\
(6) & rev\ (cons\ x\ l) & \rightarrow & cons\ (rev1\ x\ l)\ (rev2\ x\ l)
\end{array}
$$

For rule (4), take $\zeta_{rev}(\alpha) = 2\alpha$, $\zeta_{rev2}(\alpha,\beta) = 2\beta+1$, $\Gamma = x : nat^\infty, y : nat^\infty, l : list^\delta$, $\varphi = \{\beta \mapsto \delta+1\}$ and $rev \simeq_\mathcal{F} rev2 >_\mathcal{F} rev1 >_\mathcal{F} cons, nil$. Then, one can check that $\zeta_{rev2}(\infty,\delta+1) = 2\delta+3$ is strictly greater than $\zeta_{rev2}(\infty,\delta) = 2\delta+1$, $\zeta_{rev}(\delta) = 2\delta$ and $\zeta_{rev}(1+\delta) = 2\delta+2$.

4 Termination Proof

The termination proof follows the computability-based method of [10]. For lack of space, we just state the most important theorems. See [9] for details.

Let $\mathcal{R}_t$ be the set of possible interpretations for the terms of type t. $\mathcal{R}_s$ is made of sets of strongly normalizable terms. $\mathcal{R}_{(x:T)U}$ is made of the functions from $\mathcal{T} \times \mathcal{R}_T$ to $\mathcal{R}_U$ that are invariant by reduction or size substitution. $\mathcal{R}_t^m$ is the subset of $\mathcal{R}_t$ made of the functions that are monotone (resp. anti-monotone) in their monotone (resp. anti-monotone) arguments.

We first define the interpretation of types. Then, we prove monotonicity properties, the correctness of accessibility w.r.t. computability (accessible subterms of a computable term are computable), the correctness of the computability closure (every term of the computability closure is computable) and the computability of every symbol, hence the strong normalization of every well-typed term.

Definition 7 (Interpretation schema). *A candidate assignment is a function ξ from $\mathcal{X}$ to $\bigcup\{\mathcal{R}_t \mid t \in \mathcal{T}\}$. A candidate assignment ξ is a Γ-*assignment*, written $\xi \models \Gamma$, if, for all $x \in \mathrm{dom}(\Gamma)$, $x\xi \in \mathcal{R}_{x\Gamma}$.*

An interpretation *for a symbol $C \in \mathcal{CF}^\Box$ is a monotone function I from $\mathfrak{A}$ to $\mathcal{R}^m_{\tau_f}$. An* interpretation *for a symbol $f \notin \mathcal{CF}^\Box$ is an element of $\mathcal{R}^m_{\tau_f}$. An* interpretation *for a set $\mathcal{G}$ of predicate symbols is a function which, to every symbol $g \in \mathcal{G}$, associates an interpretation for g.*

The interpretation *of t w.r.t. a candidate assignment ξ, an interpretation I for $\mathcal{F}$, a substitution θ and a valuation ν, $[\![t]\!]^{I,\nu}_{\xi,\theta}$, is defined by induction on t:*

- $[\![t]\!]^{I,\nu}_{\xi,\theta} = \emptyset$ *if t is an object or a sort*
- $[\![F]\!]^{I,\nu}_{\xi,\theta} = I_F$ *if $F \in \mathcal{DF}^\Box$*
- $[\![C^a]\!]^{I,\nu}_{\xi,\theta} = I^{a\nu}_C$ *if $C \in \mathcal{CF}^\Box$*
- $[\![x]\!]^{I,\nu}_{\xi,\theta} = x\xi$
- $[\![(x:U)V]\!]^{I,\nu}_{\xi,\theta} = \{t \in \mathcal{T} \mid \forall u \in [\![U]\!]^{I,\nu}_{\xi,\theta}, \forall S \in \mathcal{R}_U, tu \in [\![V]\!]^{I,\nu}_{\xi^S_x,\theta^u_x}\}$
- $[\![[x:U]v]\!]^{I,\nu}_{\xi,\theta}(u,S) = [\![v]\!]^{I,\nu}_{\xi^S_x,\theta^u_x}$
- $[\![tu]\!]^{I,\nu}_{\xi,\theta} = [\![t]\!]^{I,\nu}_{\xi,\theta}(u\theta, [\![u]\!]^{I,\nu}_{\xi,\theta})$

where $\theta^u_x = \theta \cup \{x \mapsto u\}$ and $\xi^S_x = \xi \cup \{x \mapsto S\}$. A substitution θ is adapted *to a Γ-assignment ξ and a valuation ν, written $\xi, \theta \models_\nu \Gamma$, if $\mathrm{dom}(\theta) \subseteq \mathrm{dom}(\Gamma)$ and, for all $x \in \mathrm{dom}(\theta)$, $x\theta \in [\![x\Gamma]\!]^{I,\nu}_{\xi,\theta}$.*

We define the interpretation of predicate symbols by induction on $>_\mathcal{F}$. The definition of defined predicate symbols can be found in [10]. We now define the interpretation of constant predicate symbols by transfinite induction on $\mathfrak{a} \in \mathfrak{A}$.

Definition 8 (Interpretation of constant predicate symbols).

- $I^0_C(\boldsymbol{t},\boldsymbol{S})$ [4] *is the set of $u \in \mathcal{SN}$ that never reduces to a term of the form $f\boldsymbol{u}$ with $f \in \mathrm{Cons}(C)$, $f : (\boldsymbol{y}:\boldsymbol{U})C^a\boldsymbol{v}$, $|\boldsymbol{u}| = |\boldsymbol{y}|$ and $\mathrm{Acc}(f) \neq \emptyset$.*
- $I^{\mathfrak{a}+1}_C(\boldsymbol{t},\boldsymbol{S})$ *is the set of terms $u \in \mathcal{SN}$ such that, if u reduces to a constructor term $f\boldsymbol{u}$ with $f : (\boldsymbol{y}:\boldsymbol{U})C^{s\alpha}\boldsymbol{v}$ then, for all $j \in \mathrm{Acc}(f)$, $u_j \in [\![U_j]\!]^{I,\nu}_{\xi,\theta}$ with $\boldsymbol{y}\xi = \boldsymbol{S}_{\iota_y}$, $\boldsymbol{y}\theta = \boldsymbol{u}$ and $\alpha\nu = \mathfrak{a}$.*
- $I^{\mathfrak{b}}_C = \bigwedge_{\tau_C}(\{I^{\mathfrak{a}}_C \mid \mathfrak{a} < \mathfrak{b}\})$ *if $\mathfrak{b}$ is a limit ordinal.*

For $t \in I^\Omega_C(\boldsymbol{S})$, let $o_{C(\boldsymbol{S})}(t)$ be the smallest ordinal $\mathfrak{a}$ such that $t \in I^{\mathfrak{a}}_C(\boldsymbol{S})$.

The interpretation is well defined thanks to the assumptions made on constructors, and the following properties of the interpretation schema:

Lemma 1 (Monotonicity). *Let $\leq^+ = \leq$; $\leq^- = \geq$; $\xi \leq_x \xi'$ iff $x\xi \leq x\xi'$ and, for all $y \neq x$, $y\xi = y\xi'$; $I \leq_f I'$ iff $I_f \leq I'_f$ and, for all $g \neq f$, $I_g = I'_g$; $\nu \leq_\alpha \nu'$ iff $\alpha\nu \leq_\mathfrak{A} \alpha\nu'$ and, for all $\beta \neq \alpha$, $\beta\nu = \beta\nu'$. Assume that $\Gamma \vdash t : T$ and $\xi, \xi' \models \Gamma$.*

- *If $\xi \leq_x \xi'$ and $\mathrm{Pos}(x,t) \subseteq \mathrm{Pos}^\delta(t)$ then $[\![t]\!]^{I,\nu}_{\xi,\theta} \leq^\delta [\![t]\!]^{I,\nu}_{\xi',\theta}$.*
- *If $I \leq_f I'$ and $\mathrm{Pos}(f,t) \subseteq \mathrm{Pos}^\delta(t)$ then $[\![t]\!]^{I,\nu}_{\xi,\theta} \leq^\delta [\![t]\!]^{I',\nu}_{\xi,\theta}$.*

[4] In the following, we do not write $\boldsymbol{t}$ since the interpretation does not depend on it.

- *If* $\nu \le_\alpha \nu'$ *and* $\mathrm{Pos}(\alpha, t) \subseteq \mathrm{Pos}^\delta(t)$ *then* $[\![t]\!]^{I,\nu}_{\xi,\theta} \le^\delta [\![t]\!]^{I,\nu'}_{\xi,\theta}$.
- *If* $\Gamma \vdash T \le T' : s$, $T, T' \in \mathcal{WN}$, $[\![t]\!] = [\![t']\!]$ *whenever* $t \to t'$, *then* $[\![T]\!]^{I,\nu}_{\xi,\theta} \le [\![T']\!]^{I,\nu}_{\xi,\theta}$.

Theorem 1 (Accessibility correctness). *If* $t : T \gg_\varphi u : U$, $T = C^\beta \boldsymbol{t}$, $\mathcal{V}(\boldsymbol{t}) = \emptyset$ *and* $t\sigma \in [\![T]\!]^\mu_{\xi,\sigma}$ *then there is* ν *such that* $\beta\varphi\nu \le \beta\mu$ *and* $u\sigma \in [\![U]\!]^\nu_{\xi,\sigma}$.

Theorem 2 (Correctness of the computability closure). *Let* $(f\boldsymbol{l} \to r, \Gamma, \varphi) \in \mathcal{R}$, $f : (\boldsymbol{x} : \boldsymbol{T})U$ *and* $\boldsymbol{x}\gamma = \boldsymbol{l}$. *Assume that, for all* $(g, \mu) <^{\mathfrak{A}} (f, \varphi\nu)$, $g \in [\![\tau_g]\!]^\mu$. *If* $\Delta \vdash_{\mathrm{c}} t : T$ *and* $\xi, \sigma \models_\nu \Gamma, \Delta$ *then* $t\sigma \in [\![T]\!]^\nu_{\xi,\sigma}$.

Proof. By induction on $\Delta \vdash_{\mathrm{c}} t : T$. We only detail the case (symb). Since $(g, \psi) <^{\mathcal{A}} (f, \varphi)$, $(g, \psi\nu) <^{\mathfrak{A}} (f, \varphi\nu)$. Hence, by assumption, $g \in [\![\tau_g]\!]^{\psi\nu}$. Now, by induction hypothesis, $\boldsymbol{y}\delta\sigma \in [\![\boldsymbol{U}\psi\delta]\!]^\nu_{\xi,\sigma}$. By candidate substitution, there exists η such that $[\![\boldsymbol{U}\psi\delta]\!]^\nu_{\xi,\sigma} = [\![\boldsymbol{U}\psi]\!]^\nu_{\eta,\delta\sigma}$. By size substitution, $[\![\boldsymbol{U}\psi]\!]^\nu_{\eta,\delta\sigma} = [\![\boldsymbol{U}]\!]^{\psi\nu}_{\eta,\delta\sigma}$. Therefore, $g\boldsymbol{y}\delta\sigma \in [\![V]\!]^{\psi\nu}_{\eta,\delta\sigma} = [\![V\psi\delta]\!]^\nu_{\xi,\sigma}$. □

Lemma 2 (Computability of symbols). *For all* f *and* μ, $f \in [\![\tau_f]\!]^\mu$.

Proof. Assume that $\tau_f = (\boldsymbol{x} : \boldsymbol{T})U$ with U distinct from a product. $f \in [\![\tau_f]\!]^\mu$ iff, for all η, θ such that $\eta, \theta \models_\mu \Gamma_f$, $f\boldsymbol{x}\theta \in [\![U]\!]^\mu_{\eta,\theta}$. We prove it by induction on $((f, \mu), \theta)$ with $(>^{\mathfrak{A}}, \to)_{\mathrm{lex}}$ as well-founded ordering. □

Theorem 3 (Termination). $\beta \cup \mathcal{R}$ *is well-founded on well-typed terms.*

5 Towards Another Extension: Sized Constructors

By definition, constructors are restricted to types of the form $(\boldsymbol{y} : \boldsymbol{U})C^{s\alpha}\boldsymbol{v}$ with every occurrence of a type $D \simeq_{\mathcal{F}} C$ in $\boldsymbol{U}$ of the form D^α (this is so in [5, 2] too). However, some functions need more general size annotations [17]:

Example 4 (Paulson's normalization procedure of if-expressions). By taking the types $expr : \star$, $at : expr^0$, $if : expr^\alpha \Rightarrow expr^\beta \Rightarrow expr^\gamma \Rightarrow expr^{(\alpha+1)(\beta+\gamma+3)}$ and $nm : expr^\alpha \Rightarrow expr^\alpha$, one can prove the termination conditions for the rules:

$$\begin{array}{lrcl}
(1) & nm\ at & \to & at \\
(2) & nm\ (if\ at\ y\ z) & \to & if\ at\ (nm\ y)\ (nm\ z) \\
(3) & nm\ (if\ (if\ u\ v\ w)\ y\ z) & \to & nm\ (if\ u\ (nm\ (if\ v\ y\ z))\ (nm\ (if\ w\ y\ z)))
\end{array}$$

For rule (3), take $\zeta_{nm}(\alpha) = \alpha$, $\Gamma = u : expr^\alpha, v : expr^\beta, w : expr^\gamma, y : expr^\delta, z : expr^\epsilon$, $\upsilon = (\alpha+1)(\beta+\gamma+3)(\delta+\epsilon+3)$, $\varphi = \{\alpha \mapsto \upsilon\}$ and $nm >_{\mathcal{F}} at, if$. Then, one can check that υ is strictly greater than $(\beta+1)(\delta+\epsilon+3)$, $(\gamma+1)(\delta+\epsilon+3)$ and $(\alpha + 1)((\beta + 1)(\delta + \epsilon + 3) + (\gamma + 1)(\delta + \epsilon + 3) + 3)$.

The conditions on constructors imply also that non-recursive arguments are of size ∞ (*i.e.* undefined). So, there is no way to give different sizes to the terms of a non-recursive type. Yet, it may be very useful as shown by the type *blist* in the following example.

Example 5 (Quick sort). Take $bool : \star$, $true : bool^\infty$, $false : bool^\infty$, $blist : \star$, $pair : list^\alpha \Rightarrow list^\beta \Rightarrow blist^{max(\alpha,\beta)}$ or $pair : list^\alpha \Rightarrow list^\alpha \Rightarrow blist^\alpha$, $fst : blist^\alpha \Rightarrow list^\alpha$, $snd : blist^\alpha \Rightarrow list^\alpha$, $\leq : nat^\infty \Rightarrow nat^\infty \Rightarrow bool^\infty$, $pivot : nat^\infty \Rightarrow list^\alpha \Rightarrow blist^\alpha$, $qs : list^\infty \Rightarrow list^\infty \Rightarrow list^\infty$ and $qsort : list^\infty \Rightarrow list^\infty$.

$$\begin{array}{lll}
 & (3) \quad \leq\ 0\ x \rightarrow true & \\
(1)\ fst\ (pair\ x\ y) \rightarrow x & (4) \quad \leq\ (s\ x)\ 0 \rightarrow false & (6)\ if\ true\ x\ y \rightarrow x \\
(2)\ snd\ (pair\ x\ y) \rightarrow y & (5) \leq\ (s\ x)\ (s\ y) \rightarrow \leq\ x\ y & (7)\ if\ false\ x\ y \rightarrow y
\end{array}$$

$$\begin{array}{lrl}
(8) & pivot\ x\ nil \rightarrow & pair\ nil\ nil \\
(9) & pivot\ x\ (cons\ y\ l) \rightarrow & if\ (\leq\ y\ x)\ (pair\ (cons\ y\ u)\ v)\ (pair\ u\ (cons\ y\ v)) \\
 & & \text{where } u = fst\ (pivot\ x\ l) \text{ and } v = snd\ (pivot\ x\ l)
\end{array}$$

$$\begin{array}{lrl}
(10) & qs\ nil\ l \rightarrow & l \\
(11) & qs\ (cons\ x\ l)\ l' \rightarrow & qs\ u\ (cons\ x\ (qs\ v\ l')) \\
 & & \text{where } u = fst\ (pivot\ x\ l) \text{ and } v = snd\ (pivot\ x\ l)
\end{array}$$

$$\begin{array}{lrl}
(12) & qsort\ l \rightarrow & qs\ l\ nil
\end{array}$$

For rule (11), take $\zeta_{qs}(\alpha,\beta) = \alpha$, $\Gamma = x : nat^\infty, l : list^\delta, l' : list^\epsilon$, $\varphi = \{\alpha \mapsto s\delta, \beta \mapsto \epsilon\}$ and $qs >_\mathcal{F} pivot >_\mathcal{F} cons, pair, fst, snd$. By (symb), $\vdash_c x : nat^\infty$, $\vdash_c l : list^\delta$ and $\vdash_c l' : list^\epsilon$. By (symb), $\vdash_c pivot\ x\ l : blist^\delta$. By (symb), $\vdash_c u : list^\delta$ and $\vdash_c v : list^\delta$. By (symb), $\vdash_c qs\ v\ l' : list^\infty$. By (symb), $\vdash_c cons\ x\ (qs\ v\ l') : list^{s\infty}$. By (sub), $\vdash_c cons\ x\ (qs\ v\ l') : list^\infty$. Thus, by (symb), $\vdash_c qs\ u\ (cons\ x\ (qs\ v\ l')) : list^\infty$ since $\zeta_{qs}(\delta,\infty) = \delta < \zeta_{qs}(s\delta,\epsilon) = s\delta$.

Therefore, we naturally come to the following more general conditions, whose justification is ongoing.

Definition 9 (Sized constructors). *A type C is* non-recursive *if, for all constructor $f : (\boldsymbol{y} : \boldsymbol{U})C^a\boldsymbol{v}$ and $j \in \mathrm{Acc}(f)$, no $D \simeq C$ occurs in U_j. The first, third and fourth conditions of Definition 3 are replaced by the following ones:*

- *For all $j \in \mathrm{Acc}(f)$, $D \simeq_\mathcal{F} C$ and $p \in \mathrm{Pos}(D, U_j)$, $p \in \mathrm{Pos}^+(U_j)$ and $U_j|_p = D^\alpha$ for some $\alpha <_\mathcal{A} a$ ($\alpha \leq_\mathcal{A} a$ if C is non-recursive).*
- *For all $j \in \mathrm{Acc}(f)$, $\alpha \in \mathcal{V}(U_j)$ and $p \in \mathrm{Pos}(\alpha, U_j)$, there is $D \simeq_\mathcal{F} C$ and $q \in \mathrm{Pos}(D, U_j)$ such that $p = qS$.*

Note however that it still does not allow us to take $qs : list^\alpha \Rightarrow list^\beta \Rightarrow list^{\alpha+\beta}$ and thus $qsort : list^\alpha \Rightarrow list^\alpha$ since too much information is lost by taking $pair : list^\alpha \Rightarrow list^\beta \Rightarrow blist^{max(\alpha,\beta)}$. A solution would be to take $pair : list^\alpha \Rightarrow list^\beta \Rightarrow blist^{\langle\alpha,\beta\rangle}$ with $\langle\alpha,\beta\rangle$ interpreted as a pair of ordinals, and to say that $pivot$ has type $nat^\infty \Rightarrow list^\alpha \Rightarrow blist^{\langle\beta,\gamma\rangle}$ for some β and γ such that $\beta + \gamma = \alpha$, as it can be done in [26].

Another interest of Xi's framework is to take into account the semantics of conditional statements:

Example 6 (Mc Carthy's "91" function). Mc Carthy's "91" function f is defined by the following equations: $f(x) = f(f(x+11))$ if $x \leq 100$, and $f(x) = x - 10$ otherwise. In fact, f is equal to the function F such that $F(x) = 91$ if $x \leq 100$, and $F(x) = x - 10$ otherwise. A way to formalize this in CACSA would be to use conditional rewrite rules:

$$\begin{array}{llll} (1)\ f\ x & \rightarrow & f\ (f\ (+\ x\ 11)) & \text{if } \leq\ x\ 100 = true \\ (2)\ f\ x & \rightarrow & -\ x\ 10 & \text{if } \leq\ x\ 100 = false \end{array}$$

and take $f : nat^{\alpha} \Rightarrow nat^{F(\alpha)}$ and $\zeta_f^X(x) = max(0, 101 - x)$ as measure function, as it can be done in Xi's framework. Then, by taking into account the rewrite rule conditions, one could prove that, if $\Gamma = x : nat^{\delta}$ and $\leq\ x\ 100 = true$, then $\delta \leq 100$, $\zeta_f(\delta + 11) < \zeta_f(\delta)$ and $\zeta_f(F(\delta)) < \zeta_f(\delta)$.

6 Conclusion

The notion of computability closure, first introduced in [11] and further extended to higher-order pattern-matching [8], higher-order recursive path ordering [21, 25], type-level rewriting[10] and rewriting modulo equational theories [7], shows to be essential for extending to rewriting and dependent types type-based termination criteria for (polymorphic) λ-calculi with inductive types and case analysis [20, 26, 5, 2]. In contrast with what is suggested in [5], this notion, which is expressed as a sub-system of the whole type system (see Figure 3), allows pattern-matching and does not suffer from limitations one could find in systems relying on external guard predicates for recursive definitions.

We allow a richer size algebra than the one in [20, 5, 2] but do not allow existential size variables and conditional rewriting that are essential for capturing some size-preserving properties or some definitions as it can be done in [26]. Such extensions should allow us to subsume Xi's work completely.

Some questions also need further research. In particular, matching on defined symbols and decidability of type-checking. For type-checking, we believe that it is decidable if solving inequations in $\mathcal{A}$ is decidable. We already have preliminary results in this direction [9].

We made two important assumptions that also need further research. First, the confluence of $\beta \cup \mathcal{R}$, which is still an open problem when $\mathcal{R}$ is confluent, terminating and non left-linear. Second, the preservation of typing under rewriting for which we need to find decidable sufficient conditions.

We also assume that users provide appropriate sized types for function symbols and then check by our technique that the rewrite rules defining these function symbols are compatible with their types. An important extension would be to infer these types. Works in this direction already exist for ML-like languages.

Finally, by combining rewriting and subtyping in CC, this work may also be seen as an important step towards a better integration of membership equational logic and dependent type systems. Following [21, 25], we also think that it can serve as a basis for a higher-order extension of the General Path Ordering [14].

Acknowledgments

I thank Ralph Matthes for having invited me in Münich in February 2002, Andreas Abel for his useful comments on a previous version of this work (his technical report [2] and the discussions we had in Münich were the starting point of the present work), and the anonymous referees too.

References

1. A. Abel. Termination and productivity checking with continuous types. In *Proc. of TLCA'03*, LNCS 2701.
2. A. Abel. Termination checking with types. Technical Report 0201, Ludwig Maximilians Universität, München, Germany, 2002.
3. A. Abel. Termination checking with types, 2003. Submitted to ITA.
4. H. Barendregt. Lambda calculi with types. In S. Abramski, D. Gabbay, and T. Maibaum, editors, *Handbook of logic in computer science*, volume 2. Oxford University Press, 1992.
5. G. Barthe, M. J. Frade, E. Giménez, L. Pinto, and T. Uustalu. Type-based termination of recursive definitions. *Mathematical Structures in Computer Science*, 14(1):97–141, 2004.
6. F. Blanqui. Inductive types in the Calculus of Algebraic Constructions. In *Proc. of TLCA'03*, LNCS 2701.
7. F. Blanqui. Rewriting modulo in Deduction modulo. In *Proc. of RTA'03*, LNCS 2706.
8. F. Blanqui. Termination and confluence of higher-order rewrite systems. In *Proc. of RTA'00*, LNCS 1833.
9. F. Blanqui. A type-based termination criterion for dependently-typed higher-order rewrite systems. Draft. 38 pages. `http://www.loria.fr/~blanqui/`.
10. F. Blanqui. Definitions by rewriting in the Calculus of Constructions, 2003. To appear in Mathematical Structures in Computer Science.
11. F. Blanqui, J.-P. Jouannaud, and M. Okada. Inductive-data-type Systems. *Theoretical Computer Science*, 272:41–68, 2002.
12. G. Chen. *Subtyping, Type Conversion and Transitivity Elimination.* PhD thesis, Université Paris VII, France, 1998.
13. T. Coquand and G. Huet. The Calculus of Constructions. *Information and Computation*, 76(2–3):95–120, 1988.
14. N. Dershowitz and C. Hoot. Natural termination. *Theoretical Computer Science*, 142(2):179–207, 1995.
15. N. Dershowitz and J.-P. Jouannaud. Rewrite systems. In J. van Leeuwen, editor, *Handbook of Theoretical Computer Science*, vol. B, chap. 6. North-Holland, 1990.
16. G. Dowek and B. Werner. Proof normalization modulo. In *Proc. of TYPES'98*, LNCS 1657.
17. J. Giesl. Termination of nested and mutually recursive algorithms. *Journal of Automated Reasoning*, 19(1):1–29, 1997.
18. E. Giménez. Structural recursive definitions in type theory. In *Proc. of ICALP'98*, LNCS 1443.
19. R. Harper and J. Mitchell. Parametricity and variants of Girard's J operator. *Information Processing Letters*, 70:1–5, 1999.
20. J. Hughes, L. Pareto, and A. Sabry. Proving the correctness of reactive systems using sized types. In *Proc. of POPL'96*.
21. J.-P. Jouannaud and A. Rubio. The Higher-Order Recursive Path Ordering. In *Proc. of LICS'99*.

22. J. W. Klop, V. van Oostrom, and F. van Raamsdonk. Combinatory reduction systems: introduction and survey. *Theoretical Comp. Science*, 121:279–308, 1993.
23. N. P. Mendler. *Inductive Definition in Type Theory.* PhD thesis, Cornell University, United States, 1987.
24. M. Stefanova. *Properties of Typing Systems.* PhD thesis, Katholiecke Universiteit Nijmegen, The Netherlands, 1998.
25. D. Walukiewicz-Chrząszcz. Termination of rewriting in the Calculus of Constructions. *Journal of Functional Programming*, 13(2):339–414, 2003.
26. H. Xi. Dependent types for program termination verification. *Journal of Higher-Order and Symbolic Computation*, 15(1):91–131, 2002.

Termination of S-Expression Rewriting Systems: Lexicographic Path Ordering for Higher-Order Terms*

Yoshihito Toyama

RIEC, Tohoku University
Katahira 2-1-1, Aoba-ku, Sendai 980-8577, Japan
toyama@nue.riec.tohoku.ac.jp

Abstract. This paper expands the termination proof techniques based on the lexicographic path ordering to term rewriting systems over varyadic terms, in which each function symbol may have more than one arity. By removing the deletion property from the usual notion of the embedding relation, we adapt Kruskal's tree theorem to the lexicographic comparison over varyadic terms. The result presented is that finite term rewriting systems over varyadic terms are terminating whenever they are compatible with the lexicographic path order. The ordering is simple, but powerful enough to handle most of higher-order rewriting systems without λ-abstraction, expressed as S-expression rewriting systems.

1 Introduction

A term rewriting system [2, 15] is said to be terminating if all reduction sequences are finite. An important syntactical method to prove termination of a first-order term rewriting system is the one using the lexicographic path order (or the recursive path order with lexicographic status) relying on Kruskal's tree theorem [2, 4, 5, 7, 13, 15].

A higher-order rewriting system is a rewriting system to accommodate higher-order functions. Several syntactical methods for proving termination of them are presented [1, 6, 10–12, 14]. In the framework of the algebraic-functional systems, Jouannaud and Rubio [6] generalize the recursive path order (with lexicographic status) to higher-order rewrite rules, by adapting the notion of computability of typed λ calculus due to Tait and Girald. Concerning the termination method relying on Kruskal's tree theorem [9], Linfantsev and Bachmair [11] present the lexicographic path ordering for higher-order rewriting systems without λ-abstraction, in which higher-order terms are expressed by application and pairing. Aoto and Yamada [1] present a transformation method for applying the lexicographic path ordering to simply typed term rewriting systems. These syntactical methods have to use the types to guarantee well-foundedness of the

* This work was partially supported by grants from Japan Society for the Promotion of Science, No. 14580357, and from Ministry of Education, Culture, Sports, Science and Technology, No. 15017203.

V. van Oostrom (Ed.): RTA 2004, LNCS 3091, pp. 40–54, 2004.

presented orderings [1, 6, 11]. From this observation, a natural question arises: whether the syntactical methods relying entirely on Kruskal's tree theorem, without using the types, are really useful for proving termination of higher-order rewriting systems.

The main purpose of this paper is to give a positive answer to the above question; the lexicographic path ordering relying only on Kruskal's tree theorem is powerful enough to handle most of higher-order rewriting systems in the literature [1, 6, 10–12, 14]. The crucial point is the notion of simplification order [2, 4, 13, 15] over varyadic signatures, in which each function symbol may have more than one arity [4, 5]. Over varyadic signatures, the usual definition of simplification order requires the deletion property [4, 5], since it is tightly related to the use of Kruskal's tree theorem [9]. However, the lexicographic comparison is not compatible with this property. Hence, in the literature [2, 7, 15] the lexicographic path order (or the lexicographic comparison of the recursive path order with status) is restricted to fixed-arity (or bounded-varyadic [5]) signatures. In this paper, by removing the deletion property from the usual definition, we adapt the notion of simplification order to the lexicographic path order over varyadic signature.

The key result presented here is that finite term rewriting systems over varyadic signatures are terminating whenever they are compatible with the lexicographic path order. The result is simple but by no means trivial because we do not restrict to fixed-arity (or bounded-varyadic) signatures. To develop the termination method based on this result, we propose a new framework of the S-expressions built from constants and a varyadic function symbol of infinitely many arities, in which higher-order rewriting systems are expressed as S-expression rewriting systems. Termination of various examples is shown by the lexicographic path order on S-expressions, without using types. If we use the type information of higher-order rewrite rules, the power of our termination method can be strengthened more. For demonstrating this, we propose a non-termination preserving transformation of higher-order rewriting systems based on *currying* technique.

The remainder of this paper is organized as follows. After a preliminary section, in Section 3 we discuss the simplification ordering over varyadic signatures. In Section 4 the lexicographic path ordering over varyadic signatures is studied. In sections 5 and 6 we introduce S-expression rewriting systems and discuss termination of them. Section 7 shows that our ordering over S-expressions is a conservative extension of the usual ordering. Section 8 strengthens our ordering through transformational method based on the type information.

2 Preliminaries

We mainly follow the notation of [2, 15]. A signature is a set $\mathcal{F}$ of *function symbols* denoted by $f, g, h, \ldots$. Function symbols in $\mathcal{F}$ may be *varyadic*, i.e., have more than one arity. If at least one function symbol $f \in \mathcal{F}$ is varyadic, we say that $\mathcal{F}$ is varyadic; otherwise, $\mathcal{F}$ is fixed-arity. Elements of $\mathcal{F}$ of arity 0 are called constants. Let $\mathcal{V}$ be a countably infinite set of *variables* denoted by $x, y, z, \ldots, \alpha, \beta, \gamma, \ldots$ where $\mathcal{F} \cap \mathcal{V} = \phi$. The set of all *terms* built from $\mathcal{F}$ and $\mathcal{V}$ is denoted by $\mathcal{T}(\mathcal{F}, \mathcal{V})$, and terms are usually denoted by $s, t, r, \ldots$. The set of function symbols in a term t is denoted by $\mathcal{F}(t)$, and the set of variables is denoted by $\mathcal{V}(t)$.

A *substitution* σ is a mapping from $\mathcal{V}$ into $\mathcal{T}(\mathcal{F}, \mathcal{V})$. Substitutions are extended into homomorphisms from $\mathcal{T}(\mathcal{F}, \mathcal{V})$ into $\mathcal{T}(\mathcal{F}, \mathcal{V})$. Following common usage, we write $t\theta$ instead of $\theta(t)$.

Consider an extra constant $\Box$ called a hole. Then $C \in \mathcal{T}(\mathcal{F} \cup \{\Box\}, \mathcal{V})$ is called a context on $\mathcal{F}$. We use the notation $C[\,]$ for the context containing precisely one hole, and $C[t]$ denotes the result of placing a term t in the hole of $C[\,]$. A term s is called a subterm of $t = C[s]$. We denote $s \subseteq t$ if s is a subterm occurrence of t, and $s \subset t$ if $s \subseteq t$ and $s \neq t$.

A rewrite rule is a pair $\langle l, r\rangle$ of terms such that $l \notin \mathcal{V}$ and $\mathcal{V}(r) \subseteq \mathcal{V}(l)$. We write $l \to r$ for $\langle l, r\rangle$. A *term rewriting system* (TRS for short) $\mathcal{R}$ is a set of rewrite rules. The rewrite rules of a TRS $\mathcal{R}$ define a reduction relation $\to_\mathcal{R}$ on $\mathcal{T}(\mathcal{F}, \mathcal{V})$ as follow: $t \to_\mathcal{R} s$ iff there exist a rewrite rule $l \to r \in \mathcal{R}$, a context $C[\,]$ and a substitution θ such that $t = C[l\theta]$ and $s = C[r\theta]$. The transitive closure of $\to_\mathcal{R}$ is denoted by $\xrightarrow{+}_\mathcal{R}$. A TRS $\mathcal{R}$ is *terminating* if there exists no infinite reduction sequence $t_0 \to_\mathcal{R} t_1 \to_\mathcal{R} t_2 \to_\mathcal{R} \cdots$.

A (strict) partial order $\succ$ is a transitive and irreflexive relation. The reflexive closure of $\succ$ is denoted by $\succeq$. A partial order $\succ$ is well-founded if there is no infinite decreasing sequence $t_1 \succ t_2 \succ t_3 \succ \cdots$. A partial order $\succ$ is a *partial-well-order* iff for every infinite sequence $t_1, t_2, t_3, \cdots$ there exist indices $i < j$ such that $t_i \preceq t_j$.

A partial order $\succ$ on $\mathcal{T}(\mathcal{F}, \mathcal{V})$ is called a rewrite order if it is closed under context and substitution. A TRS $\mathcal{R}$ is compatible with a partial order $\succ$ if $l \succ r$ for every rule $l \to r \in \mathcal{R}$.

3 Simplification Order over Varyadic Signature

We now define the simplification ordering over varyadic signatures. Note that our definition lacks the deletion property, though it is necessary to relate the ordering with Kruskal's tree theorem in a varyadic setting [4, 5, 13].

Let $\mathcal{F}$ be a varyadic or fixed-arity signature. We say that a partial order $\succ$ on $\mathcal{T}(\mathcal{F}, \mathcal{V})$ has the subterm property if $s \succ t$ whenever $t \subset s$, and $\succ$ has the deletion property if $f(\cdots t \cdots) \succ f(\cdots\cdots)$ for all terms t whenever the arities of f allow it.

The TRS $\mathcal{R}_{emb}$ consists of all rewrite rules $f(x_1, \cdots, x_m) \to x_i$ with $f \in \mathcal{F}$ when the arities of f allow it. Here $x_1, \cdots, x_m$ are pairwise different variables. The embedding relation $\rhd_{emb}$ is defined by $\rhd_{emb} = \xrightarrow{+}_{\mathcal{R}_{emb}}$. The embedding relation $\rhd_{emb}$ is a partial order which has the subterm property.

Proposition 1 (Kruskal's Tree Theorem (Finite Version)). Let $\mathcal{F}$ be a fixed-arity signature and $\mathcal{F} \cup \mathcal{V}$ be finite. Then the embedding relation $\rhd_{emb}$ is a partial-well-order on $\mathcal{T}(\mathcal{F}, \mathcal{V})$ [2, 13, 15].

It should be noted that in the literature [4, 5] the embedding relation requires not only the subterm property but also the deletion property, i.e., $f(\cdots, t, \cdots) \rhd_{emb} f(\cdots\cdots)$, when function symbols are varyadic. This requirement is necessary for Kruskal's tree theorem. To see why, consider the infinite sequence $f(c), f(c,c), f(c,c,c), \cdots$ for a varyadic function symbol f and a constant c. This sequence shows that Kruskal's tree theorem fails over varyadic signatures if the embedding relation $\rhd_{emb}$ does not have the deletion property. However, the deletion property is incompatible with the lexicographic path ordering $\succ_{lpo}$ (which is just the subject studied through this paper); for instance, $f(c, b, b, b) \succ_{delete} f(b, b, b)$ but $f(c, b, b, b) \prec_{lpo} f(b, b, b)$ for $c < b$. Thus, we do not suppose the deletion property of $\rhd_{emb}$, but in order for $\rhd_{emb}$ to be a partial-well-order over a varyadic signature, we impose alternative restriction, namely that varyadic function symbols occurring in an infinite sequence of terms have their arities bounded by a natural number.

The maximum degree $D(t)$ of a term $t \in \mathcal{T}(\mathcal{F}, \mathcal{V})$ is defined by: (i) $D(x) = 0$ for a variable x, (ii) $D(c) = 0$ for a constant c, (iii) $D(f(t_1, \cdots, t_n)) = max\{n, D(t_1), \cdots, D(t_n)\}$ for $n > 0$. The set of terms having the maximum degree not more than m is defined by $\mathcal{T}_m(\mathcal{F}, \mathcal{V}) = \{t \mid D(t) \leq m\}$. Then, we have the following tree theorem over varyadic signatures as a corollary of Kruskal's tree theorem, even if the embedding relation $\rhd_{emb}$ does not have the deletion property.

Lemma 1 (Tree Theorem). Let $\mathcal{F}$ be a varyadic or fixed-arity signature and $\mathcal{F} \cup \mathcal{V}$ be finite. Then the embedding relation $\rhd_{emb}$ is a partial-well-order on $\mathcal{T}_m(\mathcal{F}, \mathcal{V})$ for any natural number m.

Proof. We code varyadic function symbols f into fixed-arity function symbols f^k by labeling f with its arity k as in Ferreira [5]. Since all function symbols occurring in terms over $\mathcal{T}_m(\mathcal{F}, \mathcal{V})$ have their arities bounded by m, we need to consider only finite coded function symbols f^k $(k \leq m)$ even if f has infinitely many arities. In this new setting, Kruskal's tree theorem remains valid for the embedding relation $\rhd_{emb}$ without the deletion property since the new signature is finite and fixed-arity. □

Definition 1. An order $\succ$ on $\mathcal{T}(\mathcal{F}, \mathcal{V})$ is called a *simplification order* if it is a rewrite order with the subterm property [2, 13, 15].

Lemma 2. A simplification order $\succ$ on $\mathcal{T}(\mathcal{F}, \mathcal{V})$ satisfies $\rhd_{emb} \subseteq \succ$ [2, 13, 15].

Proof. Trivial from the definitions of simplification order and of $\rhd_{emb}$. □

Definition 2. Let $\mathcal{F}$ be a varyadic or fixed-arity signature. A TRS $\mathcal{R}$ over $\mathcal{T}(\mathcal{F}, \mathcal{V})$ is *bounded* if for any infinite reduction sequence $t_0 \rightarrow_{\mathcal{R}} t_1 \rightarrow_{\mathcal{R}} t_2 \rightarrow_{\mathcal{R}} t_3 \rightarrow_{\mathcal{R}} \cdots$ there exist some natural number m and a finite set $\tilde{\mathcal{F}} \subseteq \mathcal{F}$ such that $D(t_i) \leq m$ and $\mathcal{F}(t_i) \subseteq \tilde{\mathcal{F}}$ for all i.

Lemma 3. Let $\mathcal{F}$ be a varyadic or fixed-arity signature and $\mathcal{R}$ be a finite TRS over $\mathcal{T}(\mathcal{F}, \mathcal{V})$. Then $\mathcal{R}$ is bounded.

Proof. Let $t_0 \to_{\mathcal{R}} t_1 \to_{\mathcal{R}} t_2 \to_{\mathcal{R}} t_3 \to_{\mathcal{R}} \cdots$ be an infinite reduction sequence. Taking $\mathcal{F}(t_0) \cup \bigcup_{l \to r \in \mathcal{R}} \mathcal{F}(r)$ as a finite set $\tilde{\mathcal{F}}$, we have $\mathcal{F}(t_i) \subseteq \tilde{\mathcal{F}}$ for all i. Let $p = max\{D(r) \mid l \to r \in \mathcal{R}\}$. Consider a reduction $s \to_{\mathcal{R}} t$ where $s = C[l\theta]$, $t = C[r\theta]$ and $l \to r \in \mathcal{R}$. Then we have

$$D(t) = D(C[r\theta]) =$$
$$max\{D(C[\ ]), D(r), D(x\theta) \mid x \in \mathcal{V}(r)\} \leq$$
$$max\{D(C[\ ]), D(l), D(r), D(x\theta) \mid x \in \mathcal{V}(r)\} \leq$$
$$max\{D(C[l\theta]), D(r)\} = max\{D(s), D(r)\} \leq max\{D(s), p\}.$$

Thus it holds that $D(t_{i+1}) \leq max\{D(t_i), p\}$ $(i = 0, 1, 2, \cdots)$. By taking $m = max\{D(t_0), p\}$ we conclude the claim. □

A TRS $\mathcal{R}$ is *right-bounded* if the sets $\bigcup_{l \to r \in \mathcal{R}} \mathcal{F}(r)$ and $\{D(r) \mid l \to r \in \mathcal{R}\}$ are finite. Note that $\mathcal{R}$ is right-bounded whenever $\mathcal{R}$ is finite. We can weaken a finite TRS to a right-bounded (infinite) TRS in Lemma 3 as follows.

Lemma 4. Let $\mathcal{F}$ be a varyadic or fixed-arity signature and $\mathcal{R}$ be a right-bounded TRS over $\mathcal{T}(\mathcal{F}, \mathcal{V})$. Then $\mathcal{R}$ is bounded.

Theorem 1. Let $\mathcal{F}$ be a varyadic or fixed-arity signature and $\mathcal{R}$ be a bounded TRS over $\mathcal{T}(\mathcal{F}, \mathcal{V})$ compatible with a simplification order. Then $\mathcal{R}$ is terminating.

Proof. Let $\mathcal{R}$ be compatible with a simplification order $\succ$. For a proof by contradiction, suppose an infinite reduction sequence $t_0 \to_{\mathcal{R}} t_1 \to_{\mathcal{R}} t_2 \to_{\mathcal{R}} t_3 \to_{\mathcal{R}} \cdots$. Then we have $\mathcal{V}(t_i) \subseteq \mathcal{V}(t_0)$ for all i. Since $\mathcal{R}$ is bounded, $t_0, t_1, t_2, \cdots \in \mathcal{T}_m(\tilde{\mathcal{F}}, \mathcal{V}(t_0))$ for some natural number m and a finite set $\tilde{\mathcal{F}}$. According to Lemma 1 (Tree Theorem), there exit $i < j$ such that $t_i \trianglelefteq_{emb} t_j$. From Lemma 2, we have $t_i \preceq t_j$; contradiction to $t_i \succ t_j$. Hence we conclude the claim. □

Corollary 1. Let $\mathcal{F}$ be a varyadic or fixed-arity signature and $\mathcal{R}$ be a finite (or right-bounded) TRS over $\mathcal{T}(\mathcal{F}, \mathcal{V})$ compatible with a simplification order. Then $\mathcal{R}$ is terminating.

4 Lexicographic Path Order over Varyadic Signature

The lexicographic path order over a fixed-arity signature was first described in Kamin and Lévy [7]. The following definition gives the lexicographic path order over varyadic signature.

Definition 3 (lpo for varyadic signature). Let $\mathcal{F}$ be a varyadic or fixed-arity signature and $>$ a precedence (i.e., a partial order) on $\mathcal{F}$. The lexicographic path order $\succ_{lpo}$ on $\mathcal{T}(\mathcal{F}, \mathcal{V})$ is recursively defined as follows: $s \succ_{lpo} t$ iff $s = f(s_1, \cdots, s_m)$, $t = g(t_1, \cdots, t_n)$ or $t \in \mathcal{V}$, and

(L1) $\exists i.\ s_i \succeq_{lpo} t$, or
(L2) $f > g$ and $\forall i.\ s \succ_{lpo} t_i$, or
(L3) $f = g$, $m > n$, and $s_1 = t_1, \cdots, s_n = t_n$, or
(L4) $f = g$ and $\exists i.\ s_1 = t_1, \cdots, s_{i-1} = t_{i-1}$, $s_i \succ_{lpo} t_i$, $s \succ_{lpo} t_{i+1}, \cdots, s \succ_{lpo} t_n$.

Lemma 5. The lexicographic path order $\succ_{lpo}$ is a simplification order.

Proof. It can be shown by structural induction. □

Note that the simplification order is not sufficient to guarantee well-foundedness of $\succ_{lpo}$; for example, we can have the infinite descending sequence $f(b) \succ_{lpo} f(c,b) \succ_{lpo} f(c,c,b) \succ_{lpo} f(c,c,c,b) \succ_{lpo} \cdots$ if f is varyadic and $b > c$. The problem arises from the fact that lexicographic sequences of unbounded size are not well-founded. Kamin and Lévy [7] proved that $\succ_{lpo}$ is well-founded when all function symbols are fixed-arity. Ferreira [5] weakened the fixed-arity restriction to the bounded varyadic restriction, i.e, the arity of every varyadic function is bounded by a natural number. In the following theorem we show that these restrictions can be weakened more.

Theorem 2. Let $\mathcal{F}$ be a varyadic or fixed-arity signature and $\mathcal{R}$ be a bounded TRS over $\mathcal{T}(\mathcal{F},\mathcal{V})$ compatible with the lexicographic path order $\succ_{lpo}$. Then $\mathcal{R}$ is terminating.

Proof. From Theorem 1 and Lemma 5 it follows. □

Corollary 2. Let $\mathcal{F}$ be a varyadic or fixed-arity signature and $\mathcal{R}$ be a finite (or right-bounded) TRS over $\mathcal{T}(\mathcal{F},\mathcal{V})$ compatible with the lexicographic path order $\succ_{lpo}$. Then $\mathcal{R}$ is terminating.

5 S-Expression Rewriting Systems

An S-expression, which is short for *symbolic expression*, is used for representing an expression or data in Lisp-like programming languages. As an S-expression uses the prefix notation, the term $f(a, g(b,c), d, e)$ is written in the S-expression $(f\ a\ (g\ b\ c)\ d\ e)$. A variable α at the prefix, like $(\alpha\ a\ (g\ b\ c)\ d\ e)$, works as a higher-order variable. This feature allows us to present higher-order rewrite rules as simple first-order rewrite rules. For example, the higher-order function *map* is presented by the following rewrite rules over S-expressions:

$$R \begin{cases} (map\ \alpha\ nil) \rightarrow nil \\ (map\ \alpha\ (cons\ x\ y)) \rightarrow (cons\ (\alpha\ x)\ (map\ \alpha\ y)) \end{cases}$$

where *map*, *cons*, *nil* are constants and α, x, y variables.

The most common method to treat higher-order rewriting without λ-*notation* is to use applicative terms, like $ap(ap(ap(f,a),b),c)$ for the term $f(a,b,c)$ with the application symbol ap [8]. The notion of applicative term can give a simple way for presenting higher-order rewrite rules just like S-expressions, but the left-associated sequence of ap is not convenient for proving termination of rewriting systems by simplification orderings. Consider two terms $g(b)$ and $f(g(a),b)$ where $g > f > b > a$. Then $g(b) \succ_{lpo} f(g(a),b)$ by the lexicographic path order $\succ_{lpo}$, but we have the reverse direction between the corresponding applicative terms, i.e., $ap(g,b) \prec_{lpo} ap(ap(f, ap(g,a)), b)$, because of $ap(g,b) \lhd_{emb} ap(ap(f, ap(g,a)), b)$.

The same problem arises from the usual syntax of S-expressions, by which $(f\ a\ b\ c)$ is expressed as $cons(f, cons(a, cons(b, cons(c, nil))))$ based on the constructor symbols $cons$ and nil. Thus apart from this syntax, we introduce a new syntax of S-expression based on a varyadic function symbol having infinitely many arities, on which the lexicographic path order works well like usual simplification orderings over first-order fixed-arity terms.

Definition 4 (S-Expression). Let $\mathcal{C}$ be a set of constants, $\circ \notin \mathcal{C}$ a varyadic function symbol with any natural numbers as its arities, and $\mathcal{V}$ a set of variables. Then the set of *S-expressions* built from $\mathcal{C}$ and $\mathcal{V}$, denoted by $\mathcal{S}(\mathcal{C}, \mathcal{V})$, is recursively defined as follows:

(i) $\mathcal{C} \cup \mathcal{V} \subseteq \mathcal{S}(\mathcal{C}, \mathcal{V})$.
(ii) $\circ(s_1, \cdots, s_n) \in \mathcal{S}(\mathcal{C}, \mathcal{V})$ if $s_1, \cdots, s_n \in \mathcal{S}(\mathcal{C}, \mathcal{V})$ $(n \geq 0)$.

From the definition it is trivial that $\mathcal{S}(\mathcal{C}, \mathcal{V}) = \mathcal{T}(\mathcal{C} \cup \{\circ\}, \mathcal{V})$. S-expressions $\circ(s_1, s_2, \cdots, s_n)$ are usually denoted by $(s_1\ s_2\ \cdots\ s_n)$, for short. For example, the S-expression $\circ(f, a, \circ(g, x), b, \circ())$ is written as $(f\ a\ (g\ x)\ b\ ())$.

The lexicographic path order on S-expressions is defined as follows.

Definition 5 (lps on S-expressions). Let $\mathcal{C}$ be a set of constants and $>$ a precedence (i.e., a partial order) on $\mathcal{C}$. The lexicographic path order $\succ_{lps}$ on $\mathcal{S}(\mathcal{C}, \mathcal{V})$ is recursively defined as follows: $s \succ_{lps} t$ iff $s = (s_1\ \cdots\ s_m)$ or $s \in \mathcal{C}$, $t = (t_1\ \cdots\ t_n)$ or $t \in \mathcal{C} \cup \mathcal{V}$, and

(S0) $s, t \in \mathcal{C}$ and $s > t$, or
(S1) $s \in \mathcal{C}$ and $\forall i.\ s \succ_{lps} t_i$, or
(S2) $\exists i.\ s_i \succeq_{lps} t$, or
(S3) $m > n$ and $s_1 = t_1, \cdots, s_n = t_n$, or
(S4) $\exists i.\ s_1 = t_1, \cdots, s_{i-1} = t_{i-1},\ s_i \succ_{lps} t_i,\ s \succ_{lps} t_{i+1}, \cdots, s \succ_{lps} t_n$.

Lemma 6. The lexicographic path order $\succ_{lps}$ on $\mathcal{S}(\mathcal{C}, \mathcal{V})$ is a simplification order.

Proof. We extend the precedence $>$ on $\mathcal{C}$ to on $\mathcal{C} \cup \{\circ\}$ by $a > \circ$ for all $a \in \mathcal{C}$. This extension gives the lexicographic path order $\succ_{lpo}$ on $\mathcal{T}(\mathcal{C} \cup \{\circ\}, \mathcal{V})$, i.e., $\succ_{lpo}$ on $\mathcal{S}(\mathcal{C}, \mathcal{V})$ since $\mathcal{S}(\mathcal{C}, \mathcal{V}) = \mathcal{T}(\mathcal{C} \cup \{\circ\}, \mathcal{V})$. It is proven that $\succ_{lps}$ and $\succ_{lpo}$ coincide. Thus from Lemma 5 the claim follows. □

An *S-expression rewriting system* (SRS for short) $\mathcal{R}$ is a term rewriting system on $\mathcal{S}(\mathcal{C}, \mathcal{V})$, i.e., the rewrite rules $l \to r \in \mathcal{R}$ consist of S-expressions $l, r \in \mathcal{S}(\mathcal{C}, \mathcal{V})$ and they define a reduction relation $\to_{\mathcal{R}}$ on $\mathcal{S}(\mathcal{C}, \mathcal{V})$.

Theorem 3. Let $\mathcal{R}$ be a bounded SRS over $\mathcal{S}(\mathcal{C}, \mathcal{V})$ compatible with the lexicographic path order $\succ_{lps}$. Then $\mathcal{R}$ is terminating.

Proof. From Theorem 1 and Lemma 6 it follows. □

Corollary 3. Let $\mathcal{R}$ be a finite (or right-bounded) SRS over $\mathcal{S}(\mathcal{C}, \mathcal{V})$ compatible with the lexicographic path order $\succ_{lps}$. Then $\mathcal{R}$ is terminating.

6 Termination of SRS

We verify termination of S-expression rewriting systems by the lexicographic path order $\succ_{lps}$. Various higher-order rewriting systems in the literature [1, 6, 10–12, 14] are naturally expressed in SRSs and termination of them is easily proven as that of SRSs without using the notion of type.

In the following examples, we denote variables playing like higher-order variables by $\alpha, \beta, \gamma, \cdots$ in distinction from $x, y, z, \cdots$ for readability.

Example 1 (map). Let $\mathcal{C} = \{map, cons, nil\}$ with $map > cons > nil$ and consider the following SRS $\mathcal{R}$.

$$\mathcal{R}\begin{cases} ((map\ \alpha)\ nil) \to nil \\ ((map\ \alpha)\ (cons\ x\ y)) \to (cons\ (\alpha\ x)\ ((map\ \alpha)\ y)) \end{cases}$$

For the first rule $((map\ \alpha)\ nil) \succ_{lps} nil$ is trivial by the subterm property. For the second rule we have (i) $(map\ \alpha) \succ_{lps} cons$ because $map > cons$, (ii) $((map\ \alpha)\ (cons\ x\ y)) \succ_{lps} (\alpha\ x)$ because $(map\ \alpha) \succ_{lps} \alpha$ and $((map\ \alpha)\ (cons\ x\ y)) \succ_{lps} x$, and (iii) $((map\ \alpha)\ (cons\ x\ y)) \succ_{lps} ((map\ \alpha)\ y)$ because $(cons\ x\ y) \succ_{lps} y$. From this we have $((map\ \alpha)\ (cons\ x\ y)) \succ_{lps} (cons\ (\alpha\ x)\ ((map\ \alpha)\ y))$. Thus $\mathcal{R}$ is terminating since it is compatible with $\succ_{lps}$. □

Note that in the above example, *map* is represented as a crimped notion $((map\ s)\ t)$ by the extra parentheses [11], instead of the usual flat notion $(map\ s\ t)$. This crimping is necessary in our ordering because $(map\ \alpha\ (cons\ x\ y)) \not\succ_{lps} (cons\ (\alpha\ x)\ (map\ \alpha\ y))$.

Example 2 (maplist). Let $\mathcal{C} = \{fmap, cons, nil\}$ with $fmap > cons > nil$ and consider the following SRS $\mathcal{R}$.

$$\mathcal{R}\begin{cases} ((fmap\ nil)\ x) \to nil \\ ((fmap\ (cons\ \alpha\ \beta))\ \ x) \to (cons\ (\alpha\ x)\ ((fmap\ \beta)\ x)) \end{cases}$$

Then $\mathcal{R}$ is terminating since it can be shown to be compatible with $\succ_{lps}$. □

Example 3 (twice). Let $\mathcal{C} = \{twice, \bullet\}$ with $twice > \bullet$ and consider the following SRS $\mathcal{R}$.

$$\mathcal{R}\begin{cases} ((\bullet\ \alpha\ \beta)\ x) \to (\alpha\ (\beta\ x)) \\ (twice\ \alpha) \to (\bullet\ \alpha\ \alpha) \end{cases}$$

Then $\mathcal{R}$ is terminating since it can be shown to be compatible with $\succ_{lps}$. □

Example 4 (filter). Let $\mathcal{C} = \{filter, if, true, false, cons, nil\}$ with $filter > if > cons > nil, true, false$ and consider the following SRS $\mathcal{R}$.

$$\mathcal{R}\begin{cases} ((filter\ \alpha)\ nil) \to nil \\ ((filter\ \alpha)\ (cons\ x\ y)) \to (if\ (\alpha\ x)\ (cons\ x\ ((filter\ \alpha)\ y))\ ((filter\ \alpha)\ y)) \\ \\ (if\ true\ x\ y) \to x \\ (if\ false\ x\ y) \to y \end{cases}$$

Then $\mathcal{R}$ is terminating since it can be shown to be compatible with $\succ_{lps}$. □

Example 5 (folding). Let $\mathcal{C} = \{sum, prod, fold, *, +, s, 0, cons, nil\}$ with sum, $prod > fold > * > + > s > 0 > cons > nil$ and consider the following SRS $\mathcal{R}$.

$$\mathcal{R}\begin{cases} ((fold\ \alpha\ x)\ nil) \to x \\ ((fold\ \alpha\ x)\ (cons\ y\ z)) \to (\alpha\ y\ ((fold\ \alpha\ x)\ z)) \\ \\ sum \to (fold\ +\ 0) \\ prod \to (fold\ *\ (s\ 0)) \\ \\ (+\ 0\ y) \to y \\ (+\ (s\ x)\ y) \to (s\ (+\ x\ y)) \\ (*\ 0\ y) \to 0 \\ (*\ (s\ x)\ y) \to (+\ (*\ x\ y)\ y) \end{cases}$$

Then $\mathcal{R}$ is terminating since it can be shown to be compatible with $\succ_{lps}$. □

Example 6 (recursor). Let $\mathcal{C} = \{rec, s, 0\}$ with $rec > s > 0$ and consider the following SRS $\mathcal{R}$.

$$\mathcal{R}\begin{cases} ((rec\ \alpha\ x)\ 0) \to x \\ ((rec\ \alpha\ x)\ (s\ y)) \to (\alpha\ (s\ y)\ ((rec\ \alpha\ x)\ y)) \end{cases}$$

Then $\mathcal{R}$ is terminating since it can be shown to be compatible with $\succ_{lps}$. □

Example 7 (sorting). Let $\mathcal{C} = \{asort, dsort, sort, ins, max, min, s, 0, cons, nil\}$ with $asort, dsort > sort > ins > max, min > s > 0 > cons > nil$ and consider the following SRS $\mathcal{R}$.

$$\mathcal{R}\begin{cases} ((sort\ \alpha\ \beta)\ nil) \to nil \\ ((sort\ \alpha\ \beta)\ (cons\ x\ y)) \to ((ins\ \alpha\ \beta)\ ((sort\ \alpha\ \beta)\ y)\ x) \\ ((ins\ \alpha\ \beta)\ nil\ y) \to (cons\ y\ nil) \\ ((ins\ \alpha\ \beta)\ (cons\ x\ z)\ y) \to (cons\ (\alpha\ x\ y)\ ((ins\ \alpha\ \beta)\ z\ (\beta\ x\ y))) \\ \\ (max\ 0\ y) \to y \\ (max\ x\ 0) \to x \\ (max\ (s\ x)\ (s\ y)) \to (s\ (max\ x\ y)) \\ (min\ 0\ y) \to 0 \\ (min\ x\ 0) \to 0 \\ (min\ (s\ x)\ (s\ y)) \to (s\ (min\ x\ y)) \\ \\ (asort\ z) \to ((sort\ min\ max)\ z) \\ (dsort\ z) \to ((sort\ max\ min)\ z) \end{cases}$$

Then $\mathcal{R}$ is terminating since it can be shown to be compatible with $\succ_{lps}$. □

7 Relation between TRS and SRS

A set of terms over a signature $\mathcal{F}$ can be embedded in a set of S-expressions by regarding a function symbol as a constant. In this section we discuss the relation between two lexicographic path orders $\succ_{lpo}$ on terms and $\succ_{lps}$ on S-expressions. We show that under isomorphic embedding, $\succ_{lpo}$ and $\succ_{lps}$ coincide if they are induced from the same precedence $>$ over a signature $\mathcal{F}$. Thus, the lexicographic path orders $\succ_{lps}$ on S-expressions is a conservative extension of $\succ_{lpo}$ on usual first-order terms.

Let $\mathcal{T}(\mathcal{F}, \mathcal{V})$ be a set of terms bult from a signature $\mathcal{F}$ and a set $\mathcal{V}$ of variables. The set $\mathcal{S}(\mathcal{F}, \mathcal{V})$ of S-expressions induced from a signature $\mathcal{F}$ is recursively defined by: (i) $x \in \mathcal{S}(\mathcal{F}, \mathcal{V})$ for $x \in \mathcal{V}$, (ii) $c \in \mathcal{S}(\mathcal{F}, \mathcal{V})$ for a constant c, and (iii) $(f\ s_1\ \cdots\ s_m)$ if $s_1, \cdots, s_m \in \mathcal{S}(\mathcal{F}, \mathcal{V})$ and $f \in \mathcal{F}$ has the arity m $(m \geq 1)$. Then $\mathcal{T}(\mathcal{F}, \mathcal{V})$ and $\mathcal{S}(\mathcal{F}, \mathcal{V})$ are isomorphic thanks to an isomorphism $\Psi : \mathcal{T}(\mathcal{F}, \mathcal{V}) \to \mathcal{S}(\mathcal{F}, \mathcal{V})$ such that $\Psi(f(t_1, \cdots, t_m)) = (f\ \Psi(t_1)\ \cdots\ \Psi(t_m))$ $(m \geq 1)$ and $\Psi(a) = a$ for a constant or variable a; for example, $\Psi(f(g(x, c), y)) = (f\ (g\ x\ c)\ y)$. Thus $\Psi(\mathcal{T}(\mathcal{F}, \mathcal{V})) = \mathcal{S}(\mathcal{F}, \mathcal{V})$.

Let $\mathcal{R}$ be a TRS over $\mathcal{T}(\mathcal{F}, \mathcal{V})$. Then an SRS over $\Psi(\mathcal{T}(\mathcal{F}, \mathcal{V}))$ induced from $\mathcal{R}$ is defined by $\Psi(\mathcal{R}) = \{\Psi(l) \to \Psi(r) \mid l \to r \in \mathcal{R}\}$. We remark that $s \to_{\mathcal{R}} t \Leftrightarrow \Psi(s) \to_{\Psi(\mathcal{R})} \Psi(t)$. Thus $\mathcal{R}$ is terminating if and only if $\Psi(\mathcal{R})$ is terminating.

A precedence $>$ on a signature $\mathcal{F}$ gives two lexicographic path orders $\succ_{lpo}$ on $\mathcal{T}(\mathcal{F}, \mathcal{V})$ and $\succ_{lps}$ on $\Psi(\mathcal{T}(\mathcal{F}, \mathcal{V}))$. The following result shows that $\succ_{lpo}$ and $\succ_{lps}$ are isomorphic.

Theorem 4. Two lexicographic path orders $\succ_{lpo}$ on $\mathcal{T}(\mathcal{F}, \mathcal{V})$ and $\succ_{lps}$ on $\Psi(\mathcal{T}(\mathcal{F}, \mathcal{V}))$ are isomorphic, i.e., $s \succ_{lpo} t \Leftrightarrow \Psi(s) \succ_{lps} \Psi(t)$.

Proof. By induction on the size of terms t and s, it can be proven. □

8 Typed S-Expressions and Currying

Transformation techniques [3] are widely used to prove termination of rewriting systems that cannot be proven by the classical orderings. In this section we propose a transformational method to strengthen the power of the lexicographic path order $\succ_{lps}$ based on the type information.

The set $\mathcal{T}$ of types built from a non-empty set of base types $\mathcal{B}$ is recursively defined by: (i) $\tau \in \mathcal{T}$ for $\tau \in \mathcal{B}$, and (ii) $\tau_1 \times \tau_2 \times \cdots \times \tau_n \to \tau_0$ $(n \geq 1)$ for $\tau_0, \ldots, \tau_n \in \mathcal{T}$. A type not in $\mathcal{B}$ is called function type. We use the symbols τ, σ, ... to denote types. For every type we assume a countably infinite set of variables of that type, written as $x, y, z, \ldots$ for base types and as $\alpha, \beta, \gamma, \ldots$ for functional types.

Let $\mathcal{C}^\tau$ be a set of constants of type τ, denoted by $f, g, h, \ldots$, and $\mathcal{V}^\tau$ a set of variables of type τ, and $\mathcal{C} = \bigcup_\tau \mathcal{C}^\tau$ and $\mathcal{V} = \bigcup_\tau \mathcal{V}^\tau$. Here, $\mathcal{C}^\tau \cap \mathcal{C}^\sigma = \phi$ and $\mathcal{V}^\tau \cap \mathcal{V}^\sigma = \phi$ if $\tau \neq \sigma$. The set $\mathcal{S}(\mathcal{C}, \mathcal{V})^\tau$ of terms (i.e., S-expressions) of type τ is recursively defined by: (i) $a \in \mathcal{S}(\mathcal{C}, \mathcal{V})^\tau$ for $a \in \mathcal{C}^\tau \cup \mathcal{V}^\tau$, and (ii) $(f\ t_1\ \cdots\ t_n) \in \mathcal{S}(\mathcal{C}, \mathcal{V})^\tau$ for $f \in \mathcal{C}^{\tau_1 \times \tau_2 \times \cdots \times \tau_n \to \tau}$ and $t_i \in \mathcal{S}(\mathcal{C}, \mathcal{V})^{\tau_i}$ $(i = 1, \cdots, n)$. We define $\mathcal{S}(\mathcal{C}, \mathcal{V}) = \bigcup_\tau \mathcal{S}(\mathcal{C}, \mathcal{V})^\tau$. The symbols $s, t, r, \ldots$ denote terms. The notation t^τ is sometimes used to indicate the type τ of a term t explicitly. In keeping well-typed structure we make use of the usual notions of substitution of a term t, denoted as $t\theta$, and replacement in a context, denoted as $C[t]$.

A rewrite rule, written as $l \to r$, is a pair of typed terms such that $l \notin \mathcal{V}$, $\mathcal{V}(r) \subseteq \mathcal{V}(l)$, and l and r have the same type. A *typed SRS* $\mathcal{R}$ is a set of rewrite rules. The rewrite rules of a typed SRS $\mathcal{R}$ define a reduction relation $\to_{\mathcal{R}}$ on $\mathcal{S}(\mathcal{C}, \mathcal{V})$ in the usual manner.

Definition 6 (currying). Let $\mathcal{T}$ be a set of types generated from a set $\mathcal{B}$ of base types. A *currying* Γ is a pair of types $\langle \tau_1 \times \cdots \times \tau_p \times \tau_{p+1} \times \cdots \times \tau_m \to \tau_0,\ \tau_1 \times \cdots \times \tau_p \to (\tau_{p+1} \times \cdots \times \tau_m \to \tau_0)\rangle$ where $1 \leq p < m$. A currying Γ can be extended to an endomorphism over $\mathcal{T}$ as follows.

$$\Gamma(\tau) = \begin{cases} \sigma & \text{if } \Gamma = \langle \tau, \sigma \rangle \\ \tau & \text{if } \tau \in \mathcal{B} \\ \Gamma(\rho_1) \times \cdots \times \Gamma(\rho_n) \to \Gamma(\rho) & \text{otherwise} \quad (\tau = \rho_1 \times \cdots \times \rho_n \to \rho) \end{cases}$$

Note that $\Gamma(\rho)$ is a type obtained from a type ρ by replacing all occurrences τ in ρ with a type σ when $\Gamma = \langle \tau, \sigma \rangle$. Thus $\Gamma(\rho) = \rho$ if no τ occurs in ρ.

Let $\mathcal{C}$ and $\mathcal{V}$ be sets of typed constants and variables respectively. By replacing the type τ of elements a^τ in $\mathcal{C}$ and $\mathcal{V}$ with $\Gamma(\tau)$, we obtain a set of constants $\Gamma(\mathcal{C}) = \{c^{\Gamma(\tau)} \mid c^\tau \in \mathcal{C}\}$ and a set of variables $\Gamma(\mathcal{V}) = \{x^{\Gamma(\tau)} \mid x^\tau \in \mathcal{V}\}$. Then each term in $\mathcal{S}(\mathcal{C}, \mathcal{V})$ can be converted into $\mathcal{S}(\Gamma(\mathcal{C}), \Gamma(\mathcal{V}))$ as follows.

Definition 7. Let $\mathcal{S}(\mathcal{C}, \mathcal{V})$ be a set of typed terms. Let Γ be a currying such that $\Gamma(\tau_1 \times \cdots \times \tau_p \times \tau_{p+1} \times \cdots \times \tau_m \to \tau_0) = \tau_1 \times \cdots \times \tau_p \to (\tau_{p+1} \times \cdots \times \tau_m \to \tau_0)$. Then a homomorphism from $\mathcal{S}(\mathcal{C}, \mathcal{V})$ to $\mathcal{S}(\Gamma(\mathcal{C}), \Gamma(\mathcal{V}))$ induced from Γ, denoted by the same letter Γ, is defined as follows.

$$\Gamma(t) = \begin{cases} ((\Gamma(t_0)\ \Gamma(t_1)\ \cdots\ \Gamma(t_p))\ \Gamma(t_{p+1})\ \cdots\ \Gamma(t_m)) \\ \qquad \text{if } t = (t_0{}^{\tau_1 \times \cdots \times \tau_p \times \tau_{p+1} \times \cdots \times \tau_m \to \tau_0}\ t_1{}^{\tau_1}\ \cdots\ t_p{}^{\tau_p}\ t_{p+1}{}^{\tau_{p+1}}\ \cdots\ t_m{}^{\tau_m}) \\ a^{\Gamma(\tau)} \qquad \text{if } t = a^\tau \in \mathcal{C} \cup \mathcal{V} \\ (\Gamma(t_0)\ \cdots\ \Gamma(t_n)) \qquad \text{otherwise} \quad (t = (t_0\ \cdots\ t_n)) \end{cases}$$

Example 8 (currying). Let $\mathcal{B} = \{N, L\}$ and $\mathcal{C} = \{map^{(N \to N) \times L \to L}, cons^{N \times L \to L}, nil^L\}$. Let $\Gamma((N \to N) \times L \to L) = (N \to N) \to (L \to L)$. Then we have the following curried term.

$$\Gamma((map^{(N \to N) \times L \to L}\ \alpha^{N \to N}\ (cons^{N \times L \to L}\ x^N\ y^L)^L)^L) =$$
$$((map^{(N \to N) \to (L \to L)}\ \alpha^{N \to N})^{L \to L}\ (cons^{N \times L \to L}\ x^N\ y^L)^L)^L \qquad \square$$

Lemma 7. Let t be a term of type σ in $\mathcal{S}(\mathcal{C}, \mathcal{V})$. Then $\Gamma(t)$ is a well-typed term of type $\Gamma(\sigma)$ in $\mathcal{S}(\Gamma(\mathcal{C}), \Gamma(\mathcal{V}))$.

Proof. By induction on the structure of t, we can prove the claim. $\square$

For a substitution θ, we define a substitution $\Gamma(\theta)$ by $\Gamma(\theta)(x^{\Gamma(\tau)}) = \Gamma(x^\tau\theta)$. The type τ of hole in a context $C[\ ]$ is denoted by $C[\ ]_\tau$. Then a context $\Gamma(C)[\ ]$ is defined by $\Gamma(C)[\ ]_{\Gamma(\tau)} = \Gamma(C[\ ]_\tau)$.

The following results are easily proven by induction on structure of a term.

Lemma 8. For a term t and a substitution θ we have $\Gamma(t\theta) = \Gamma(t)\Gamma(\theta)$.

Lemma 9. For a term t^τ and a context $C[\ ]_\tau$ we have $\Gamma(C[t]) = \Gamma(C)[\Gamma(t)]$.

For a typed SRS $\mathcal{R}$ on $\mathcal{S}(\mathcal{C}, \mathcal{V})$ we define a *curried* SRS $\Gamma(\mathcal{R})$ on $\mathcal{S}(\Gamma(\mathcal{C}), \Gamma(\mathcal{V}))$ by $\Gamma(\mathcal{R}) = \{\Gamma(l) \to \Gamma(r) \mid l \to r \in \mathcal{R}\}$. The next result combines termination of $\Gamma(\mathcal{R})$ with termination of $\mathcal{R}$.

Lemma 10. Let Γ be a currying and $\mathcal{R}$ a typed SRS. Then $\Gamma(s) \to_{\Gamma(\mathcal{R})} \Gamma(t)$ for $s \to_{\mathcal{R}} t$.

Proof. Let $s = C[l\theta]$ and $t = C[r\theta]$ for some $l \to r \in \mathcal{R}$, $C[\ ]$, θ. From Lemmas 8 and 9 we have $\Gamma(s) = \Gamma(C[l\theta]) = \Gamma(C)[\Gamma(l)\Gamma(\theta)]$ and $\Gamma(t) = \Gamma(C[r\theta]) = \Gamma(C)[\Gamma(r)\Gamma(\theta)]$. Thus it follows that $\Gamma(s) \to_{\Gamma(\mathcal{R})} \Gamma(t)$. □

Theorem 5. Let Γ be a currying and $\mathcal{R}$ a typed SRS. Then $\mathcal{R}$ is terminating if $\Gamma(\mathcal{R})$ is terminating.

Proof. For a proof by contradiction, suppose an infinite reduction sequence $t_0 \to_{\mathcal{R}} t_1 \to_{\mathcal{R}} t_2 \to_{\mathcal{R}} t_3 \to_{\mathcal{R}} \cdots$. Then from Lemma 10 there exists an infinite reduction sequence $\Gamma(t_0) \to_{\Gamma(\mathcal{R})} \Gamma(t_1) \to_{\Gamma(\mathcal{R})} \Gamma(t_2) \to_{\Gamma(\mathcal{R})} \Gamma(t_3) \to_{\Gamma(\mathcal{R})} \cdots$. This is a contradiction to termination of $\Gamma(\mathcal{R})$, and hence the claim holds. □

Note that termination of $\mathcal{R}$ does not guarantee that of $\Gamma(\mathcal{R})$. For example, consider $\mathcal{R} = \{(f\ ((\alpha\ x)\ y)) \to (f\ (a\ b\ c))\}$, where $f^{0\to 0}$, $a^{1\times 1\to 0}$, b^1, c^1 in $\mathcal{C}$ and $\alpha^{1\to 1\to 0}$, x^1, y^1 in $\mathcal{V}$. Take $\Gamma(1 \times 1 \to 0) = 1 \to (1 \to 1)$. We have $\Gamma(\mathcal{R}) = \{(f\ ((\alpha\ x)\ y)) \to (f\ ((a\ b)\ c))\}$. Then $(f\ ((a\ b)\ c)) \to_{\Gamma(\mathcal{R})} (f\ ((a\ b)\ c))$ but not $(f\ (a\ b\ c)) \to_{\mathcal{R}} (f\ (a\ b\ c))$.

In the following examples [1,6,10–12,14], we regard $\Gamma(\mathcal{R})$ as an untyped SRS and prove termination of it by applying the lexicographic path order $\succ_{lps}$. It should be noted that in our termination method the type information is used only for currying.

Example 9 ($map^{(N\to N)\times L\to L}$). Let $\mathcal{C} = \{map^{(N\to N)\times L\to L}, cons^{N\times L\to L}, nil^L\}$ with $map > cons > nil$ and consider the following typed SRS $\mathcal{R}$. (The type information is omitted in the subsequent examples, as it is easily derived.)

$$\mathcal{R}\begin{cases}(map\ \alpha\ nil) \to nil\\ (map\ \alpha\ (cons\ x\ y)) \to (cons\ (\alpha\ x)\ (map\ \alpha\ y))\end{cases}$$

Take $\Gamma((N \to N) \times L \to L) = (N \to N) \to (L \to L)$. Then we have the following curried SRS $\Gamma(\mathcal{R})$.

$$\Gamma(\mathcal{R})\begin{cases}((map\ \alpha)\ nil) \to nil\\ ((map\ \alpha)\ (cons\ x\ y)) \to (cons\ (\alpha\ x)\ ((map\ \alpha)\ y))\end{cases}$$

From Example 1, $\Gamma(\mathcal{R})$ is terminating. Thus, according to Theorem 5, $\mathcal{R}$ is terminating. □

Example 10 ($fmap^{L\times N\to L}$). Let $\mathcal{C} = \{fmap^{L\times N\to L}, cons^{(N\to N)\times L\to L}, nil^L\}$ with $fmap > cons > nil$ and consider the following typed SRS $\mathcal{R}$.

$$\mathcal{R}\begin{cases}(fmap\ nil\ x) \to nil\\ (fmap\ (cons\ \alpha\ \beta)\ \ x) \to (cons\ (\alpha\ x)\ (fmap\ \beta\ x))\end{cases}$$

Take $\Gamma(L \times N \to L) = L \to (N \to L)$. Then we have the following curried SRS $\Gamma(\mathcal{R})$.

$$\Gamma(\mathcal{R})\begin{cases}((fmap\ nil)\ x) \to nil\\ ((fmap\ (cons\ \alpha\ \beta))\ \ x) \to (cons\ (\alpha\ x)\ ((fmap\ \beta)\ x))\end{cases}$$

From Example 2, $\Gamma(\mathcal{R})$ is terminating. Thus, according to Theorem 5, $\mathcal{R}$ is terminating. □

Example 11 ($fold^{(N\times N\to N)\times N\times L\to N}$). Let $\mathcal{C} = \{fold^{(N\times N\to N)\times N\times L\to N}, cons^{N\times L\to L}, nil^{L}\}$ with $fold > cons > nil$ and consider the following typed SRS $\mathcal{R}$.

$$\mathcal{R}\begin{cases}(fold\ \alpha\ x\ nil) \to x\\ (fold\ \alpha\ x\ (cons\ y\ z)) \to (\alpha\ y\ (fold\ \alpha\ x\ z))\end{cases}$$

Take $\Gamma((N \times N \to N) \times N \times L \to N) = (N \times N \to N) \times N \to (L \to N)$. Then we have the following curried SRS $\Gamma(\mathcal{R})$.

$$\Gamma(\mathcal{R})\begin{cases}((fold\ \alpha\ x)\ nil) \to x\\ ((fold\ \alpha\ x)\ (cons\ y\ z)) \to (\alpha\ y\ ((fold\ \alpha\ x)\ z))\end{cases}$$

From Example 5, $\Gamma(\mathcal{R})$ is terminating. Thus, according to Theorem 5, $\mathcal{R}$ is terminating. □

Example 12 ($rec^{(N\times A\to A)\times A\times N\to A}$). Let $\mathcal{C} = \{rec^{(N\times A\to A)\times A\times N\to A}, s^{N\to N}, 0^{N}\}$ with $rec > s > 0$ and consider the following typed SRS $\mathcal{R}$.

$$\mathcal{R}\begin{cases}(rec\ \alpha\ x\ 0) \to x\\ (rec\ \alpha\ x\ (s\ y)) \to (\alpha\ (s\ y)\ (rec\ \alpha\ x\ y))\end{cases}$$

Take $\Gamma((N \times A \to A) \times A \times N \to A) = (N \times A \to A) \times A \to (N \to A)$. Then we have the following curried SRS $\Gamma(\mathcal{R})$.

$$\Gamma(\mathcal{R})\begin{cases}((rec\ \alpha\ x)\ 0) \to x\\ ((rec\ \alpha\ x)\ (s\ y)) \to (\alpha\ (s\ y)\ ((rec\ \alpha\ x)\ y))\end{cases}$$

From Example 6, $\Gamma(\mathcal{R})$ is terminating. Thus, according to Theorem 5, $\mathcal{R}$ is terminating. □

Example 13 ($sort^{(N\times N\to N)\times(N\times N\to N)\times L\to L}$). Let $\mathcal{C} = \{asort^{L\to L}, dsort^{L\to L}, sort^{(N\times N\to N)\times(N\times N\to N)\times L\to L}, ins^{(N\times N\to N)\times(N\times N\to N)\times L\times N\to L}, max^{N\times N\to N}, min^{N\times N\to N}, s^{N\to N}, 0^{N}, cons^{N\times L\to L}, nil^{L}\}$ with $asort, dsort > sort > ins > max, min > s > 0 > cons > nil$ and consider the following typed SRS $\mathcal{R}$.

$$\mathcal{R}\begin{cases}(sort\ \alpha\ \beta\ nil) \to nil\\ (sort\ \alpha\ \beta\ (cons\ x\ y)) \to (ins\ \alpha\ \beta\ (sort\ \alpha\ \beta\ y)\ x)\\ (ins\ \alpha\ \beta\ nil\ y) \to (cons\ y\ nil)\\ (ins\ \alpha\ \beta\ (cons\ x\ z)\ y) \to (cons\ (\alpha\ x\ y)\ (ins\ \alpha\ \beta\ z\ (\beta\ x\ y)))\\ \\ (max\ 0\ y) \to y\\ (max\ x\ 0) \to x\\ (max\ (s\ x)\ (s\ y)) \to (s\ (max\ x\ y))\\ (min\ 0\ y) \to 0\\ (min\ x\ 0) \to 0\\ (min\ (s\ x)\ (s\ y)) \to (s\ (min\ x\ y))\\ \\ (asort\ z) \to (sort\ min\ max\ z)\\ (dsort\ z) \to (sort\ max\ min\ z)\end{cases}$$

Take two curryings $\Gamma_1((N \times N \to N) \times (N \times N \to N) \times L \to L)) = (N \times N \to N) \times (N \times N \to N) \to (L \to L)$ and $\Gamma_2((N \times N \to N) \times (N \times N \to N) \times L \times N \to L)) = (N \times N \to N) \times (N \times N \to N) \to (L \times N \to L)$. Currying $\mathcal{R}$ twice, we have the following curried SRS $\Gamma_2(\Gamma_1(\mathcal{R}))$.

$$\Gamma_2(\Gamma_1(\mathcal{R}))\begin{cases} ((sort\ \alpha\ \beta)\ nil) \to nil \\ ((sort\ \alpha\ \beta)\ (cons\ x\ y)) \to ((ins\ \alpha\ \beta)\ ((sort\ \alpha\ \beta)\ y)\ x) \\ ((ins\ \alpha\ \beta)\ nil\ y) \to (cons\ y\ nil) \\ ((ins\ \alpha\ \beta)\ (cons\ x\ z)\ y) \to (cons\ (\alpha\ x\ y)\ ((ins\ \alpha\ \beta)\ z\ (\beta\ x\ y))) \\ \\ (max\ 0\ y) \to y \\ (max\ x\ 0) \to x \\ (max\ (s\ x)\ (s\ y)) \to (s\ (max\ x\ y)) \\ (min\ 0\ y) \to 0 \\ (min\ x\ 0) \to 0 \\ (min\ (s\ x)\ (s\ y)) \to (s\ (min\ x\ y)) \\ \\ (asort\ z) \to ((sort\ min\ max)\ z) \\ (dsort\ z) \to ((sort\ max\ min)\ z) \end{cases}$$

From Example 7, $\Gamma_2(\Gamma_1(\mathcal{R}))$ is terminating. Thus, according to Theorem 5, $\mathcal{R}$ is terminating. □

Acknowledgments

The author would like to thank Keiichiro Kusakari and Takahito Aoto for valuable discussion. Thanks also due to the referees for useful comments and suggestions.

References

1. T. Aoto and T. Yamada, Proving termination of simply typed term rewriting systems automatically, *IPSJ Trans. on Prog. 44, SIG-4* (2003) 67-77, *in Japanese.*
2. F. Baader and T. Nipkow, Term Rewriting and All That, Cambridge University Press (1998).
3. F. Bellegarde and P. Lescanne, Transformation orderings, *Lecture Notes in Comput. Sci. 249* (1987) 69-80.
4. N. Dershowitz, Ordering for term-rewriting systems, *Theoretical Comput. Sci. 17* (1982) 279-301.
5. M. Ferreira, Termination of term rewriting: Well-foundedness, totality and transformation, *PhD thesis*, Dep. of Comput. Sci., Utrecht University (1995).
6. J.-P. Jouannaud and A. Rubio, Higher-Order Recursive Path Orderings, *Procs. 14th IEEE Symp. on Logic in Comput. Sci.* (1999) 402–411.
7. S. Kamin and J.-J. Levy. Two generalizations of the recursive path ordering, *Unpublished Manuscript*, University of Illinois (1980).
8. R. Kennaway, J. W. Klop, M. R. Sleep, F.-J. de Vries, Comparing Curried and Uncurried Rewriting, *J. Symb. Comput. 21(1)* (1996) 15-39.
9. B. Kruskal, Well quasi ordering, the tree theorem and Vazsonyi's conjecture, *Trans. Am. Math. Soc. 95* (1960) 210-225.

10. K. Kusakari, On Proving Termination of Term Rewriting Systems with Higher-Order Variables, *IPSJ Trans. on Prog. 42, SIG-7* (2001) 35-45.
11. M. Linfantsev and L. Bachmair, An LPO-based termination ordering for higher-order terms without λ-abstraction, *Lecture Notes in Comput. Sci. 1479* (1998) 277-293.
12. O. Lysne and J. Piri,. A termination ordering for higher order rewrite systems, *Lecture Notes in Comput. Sci. 914* (1995) 26-24.
13. A. Middeldorp and H. Zantema, Simple termination of rewrite systems, *Theoretical Comput. Sci. 175* (1997) 127-158.
14. F. van Raamsdonk, On Termination of Higher-Order Rewriting, *Lecture Notes in Comput. Sci. 2051* (2001) 261–275.
15. Terese, Term Rewriting Systems, Cambridge University Press (2003).

Monadic Second-Order Unification Is NP-Complete*

Jordi Levy[1], Manfred Schmidt-Schauß[2], and Mateu Villaret[3]

[1] IIIA, CSIC, Campus de la UAB, Barcelona, Spain
http://www.iiia.csic.es/~levy
[2] Institut für Informatik, Johann Wolfgang Goethe-Universität,
Postfach 11 19 32, D-60054 Frankfurt, Germany
http://www.ki.informatik.uni-frankfurt.de/persons/schauss/schauss.html
[3] IMA, UdG, Campus de Montilivi, Girona, Spain
http://ima.udg.es/~villaret

Abstract. Monadic Second-Order Unification (MSOU) is Second-Order Unification where all function constants occurring in the equations are unary. Here we prove that the problem of deciding whether a set of monadic equations has a unifier is NP-complete. We also prove that Monadic Second-Order Matching is also NP-complete.

1 Introduction

Monadic Second-Order Unification (MSOU) is Second-Order Unification (SOU) where all function constants occurring in the problem are at most unary. It is well-known that the problem of deciding if a SOU problem has a solution is undecidable [Gol81,Far91,Lev98,LV00], whereas, in the case of MSOU, the problem is decidable [Hue75,Zhe79,Far88]. It is not a restriction to assume for this discussion that second order variables are also unary. In [SSS98], it is proved that the problem is NP-hard. In this paper, we prove that it is in NP.

MSOU can be decided by first making a guess for every variable, whether it uses its argument or not, and then calling string unification. This shows again that MSOU is decidable since string unification is decidable [Mak77], and also that MSOU is in PSPACE by using the result that string unification is in PSPACE [Pla99]. Since this is the currently known upper bound for string unification, our result that MSOU is NP-complete gives a sharp bound that (currently) cannot be obtained from results on string unification.

MSOU is a specialization of bounded second order unification (BSOU)[1]. BSOU is decidable [SS04], which provides another proof of decidability of MSOU, but no tight upper complexity bound. On the other hand, our proof and results

* This research has been partially supported by the CICYT Research Projects CAD-VIAL (TIC2001-2392-C03-01) and LOGFAC (TIC2001-1577-C03-01).

[1] Accordingly to Property 2 we can restrict variables to be unary, as constants. Then in instantiations $\lambda x.t$ for variables X of the problem, the variable x can occur at most once in t, as in BSOU.

V. van Oostrom (Ed.): RTA 2004, LNCS 3091, pp. 55–69, 2004.

suggest an application to BSOU, which may result in proving a precise upper complexity bound for BSOU.

To prove that MSOU is in NP, first, we show how, for any solvable set of equations, we can represent (at least) one of the solution (unifiers) in polynomial space. Then, we prove that we can check if a substitution (written in such representation) is a solution in polynomial time.

There are two key results to obtain this sharp bound: One is the result on the exponential upper bound on the exponent of periodicity of size-minimal unifiers[Mak77,KP96,SSS98] (see Lemma 3). This upper bound allows us to represent exponents in linear space. The other key is a result of Plandowski [Pla94,Pla95] (see Theorem 1) where he proves that, given two context-free grammars with just one rule for every non-terminal symbol, we can check if they define the same (singleton) language in polynomial time (on the size of the grammars).

This paper proceeds as follows. After some preliminary definitions, in Section 3, we define lazy unifiers and prove that size-minimal unifiers are lazy unifiers. We prove some properties of singleton CFG in Section 4. We use a graph in order to describe the instance of some variable (Section 5). Sometimes, we need to rewrite such graph (Section 6). Based in this graph, we prove that, for any size-minimal lazy unifier, we can represent the value of some variable instance using a polynomial singleton grammar (Theorem 2). In Section 7, we extend this result to the whole unifier, and conclude the NP-ness of the MSOU problem.

2 Preliminary Definitions

Like in general second-order unification, we deal with a signature $\Sigma = \bigcup_{i\geq 0} \Sigma_i$, where constants of Σ_i are i-ary, and a set of variables $\mathcal{X} = \bigcup_{i\geq 0} \mathcal{X}_i$, where variables of $\mathcal{X}_i$ are also i-ary. Variables of $\mathcal{X}_0$ are therefore first-order typed and those of $\mathcal{X}_i$, with $i \geq 1$ are second-order typed. Well-typed terms are built as usual. We notate free variables with capital letters X, Y,... and bound variables and constants with lower-case letters x, y,... and a, b,... Terms are written in $\beta\eta$-normal form, thus arities can be inferred from the context. As far as we do not consider third or higher-order constants, first-order typed terms (in normal form) do not contain λ-abstractions, and second-order typed terms only contain λ-abstractions in topmost positions. The *size* of a term t is noted $|t|$ and defined as its number of symbols. A term is said to be *monadic* if it is built without using constants of arity greater than one, i.e. on a signature with $\Sigma_i = \emptyset$, for any $i \geq 2$. Notice that there is no restriction on the arity of variables.

Second-order substitutions are functions from terms to terms, defined as usual. For any substitution σ, the set of variables X, such that $\sigma(X) \neq X$, is finite and is called the *domain* of the substitution, and noted $\mathtt{Dom}(\sigma)$. A substitution σ can be presented as $[X_1 \to t_1, \ldots, X_n \to t_n]$, where $X_i \in \mathtt{Dom}(\sigma)$, t_i has the same type as X_i, and satisfies $t_i = \sigma(X_i)$. Given two substitutions σ and ρ, their composition is defined by $(\sigma \circ \rho)(t) = \sigma(\rho(t))$, for any term t, and is also a substitution, i.e. $\mathtt{Dom}(\sigma \circ \rho)$ is finite. We say that a substitution σ is *more general* than another substitution ρ, noted $\sigma \preceq \rho$, if there exists a substitution τ such that $\rho(X) = \tau(\sigma(X))$, for any variable $X \in \texttt{Dom}(\sigma)$. This defines a preorder relation on substitutions. An equivalence relation can also be defined as $\sigma \approx \rho$ if $\sigma \preceq \rho$ and $\rho \preceq \sigma$. A substitution σ is said to be *ground* if $\sigma(X)$ is a closed term, i.e. it does not contains free occurrences of variables, for any $X \in \texttt{Dom}(\sigma)$.

An instance of the MSOU problem is a *set of equations* $\{t_1 \stackrel{?}{=} u_1, \ldots, t_n \stackrel{?}{=} u_n\}$ where t_i's and u_i's are monadic terms with the same first-order type, i.e. not containing λ-abstractions. A *solution* or *unifier* is a second-order substitution σ solving all equations: $\sigma(t_i) = \sigma(u_i)$ modulo $\beta\eta$-equality. A unifier σ of E is said to be *ground* if $\sigma(t_i)$ does not contain free occurrences of variables, for any $t_i \stackrel{?}{=} u_i \in E$. (Notice that not all ground substitutions that are unifiers are ground unifiers, because t_i may contain variables not instantiated by the unifier). A ground unifier is said to be *size-minimal* if it minimizes $\sum_{i=1}^{n} |\sigma(t_i)|$ among all ground unifiers. (Notice that w.l.o.g. size-minimal unifiers are required to be ground, because if a problem has a non-ground unifier, then it has also a ground unifier of equal or smaller size). *Most general unifiers* are defined as usual.

Notice that instances of the problem are required to be build from monadic terms, but there are not restrictions on the solutions. However, the following property ensures that most general unifiers instantiate variables by monadic terms.

Property 1. For any set of second-order unification equations E, and most general unifier σ, all constants occurring in σ also occur in E.

This property does not hold for variables. Even if the set of equations is built from unary second-order variables, most general unifiers can introduce fresh n-ary variables with $n \geq 2$. For instance, the set of equations $\{X(a) \stackrel{?}{=} Y(b)\}$ has only a most general unifier $[X \mapsto \lambda x.Z(x,b), Y \mapsto \lambda x.Z(a,x)]$, that introduces a binary second-order variable Z. Fortunately, non-unary variables do not give rise to undecidability, we need non-unary constants. In fact, one single binary constant is enough to generate a class of undecidable second-order unification problems [Far88,LV02].

3 Lazy Unifiers

We can restrict instances of second-order variables to ones that do not use their arguments, *whenever this is possible.* In this way we obtain what we call *lazy unifiers.*

Definition 1. *A substitution σ is said to be a* lazy unifier *if*

1. *it is a ground unifier, and,*
2. *it can be decomposed as $\sigma = \rho \circ \tau$, where τ is a most general unifier and ρ has the form $[X_1 \mapsto \lambda x_1.\cdots.\lambda x_{n_1}.a_1, \ldots, X_m \mapsto \lambda x_1.\cdots.\lambda x_{n_m}.a_m]$, where $a_1, \ldots, a_m \in \Sigma_0$ are first-order constants.*

Lemma 1. *For any solvable set of MSOU equations E containing at least a 0-ary constant, there exists a lazy unifier σ that does not introduce constants not occurring in E.*

Lemma 2. *Any size-minimal unifier is a lazy unifier.*

From now on, when we say a *solution* of a set of equations, we mean a *lazy unifier.* We also assume that any set of equations contains, at least, a 0-ary constant. The following property allows us to go a step further and assume that all variables are unary.

Property 2. Monadic SOU is NP-reducible to monadic SOU where all variable are second-order typed and unary.

From now on, since all symbols are unary or zero-ary, we can avoid parenthesis, and represent the term $a(X(b))$ as the word $a\,X\,b$, and $\lambda x.a(x)$ as $\lambda x.a\,x$.

We know that the size-minimal ground unifiers of a monadic second-order unification problem satisfies the exponent of periodicity lemma [Mak77, KP96, SSS98, SS04]. Therefore, from Lemma 2, we can conclude that it also holds for lazy unifiers:

Lemma 3 ([SS04]). *There exists a constant $\alpha \in \mathbb{R}$ such that, for any solvable monadic second-order unification problem E, there exists a lazy unifier σ such that, for any variable X, and words w_1, w_2 and w_3,*

$$\sigma(X) = \lambda x.w_1\,w_2^n\,w_3\,x \text{ and } w_2 \text{ not empty implies } n \leq 2^{\alpha|E|}.$$

4 Singleton Context Free Grammars

A *context-free grammar (CFG)* is a 4-tuple (Σ, N, P, s), where Σ is an alphabet of *terminal* symbols, N is an alphabet of *non-terminal* symbols (contrarily to the standard conventions, and in order to avoid confusion between free variables and non-terminal symbols, all terminal and non-terminal symbols are denoted by lower-case letters), P is a finite set of rules, and $s \in N$ is the *start symbol.* We will not distinguish a particular start symbol, and we will represent a context free grammars as a 3-tuple (Σ, N, P). Moreover, we will use Chomsky grammars with at most two symbols on the right hand side of the rules.

Definition 2. *We say that a context free grammar $G = (\Sigma, N, P)$* generates *a word $w \in \Sigma^*$ if there exists a non-terminal symbol $a \in N$ such that w belongs to the language defined by (Σ, N, P, a). In such case, we also say that a generates w.*

We say that a context free grammar is a singleton CFG *if it is not recursive and every non-terminal symbol occurs in the left-hand side of exactly one rule. Then, every non-terminal symbol $a \in N$ generates just one word, denoted w_a, and we say that a* defines *w_a. In general, for any sequence $\alpha \in (\Sigma \cup N)^*$, $w_\alpha \in \Sigma^*$ denotes the word generated by α.*

Plandowski [Pla94,Pla95] defines singleton grammars, but he calls them *grammars defining set of words*. He proves the following result.

Theorem 1 ([Pla95], Theorem 33). *The word equivalence problem for singleton context-free grammars is defined as follows: Given a grammar and two non-terminal symbols a and b, to decide whether $w_a = w_b$. This problem can be solved in polynomial worst-case time on the size of the grammar.*

Definition 3. *Let $G = (\Sigma, N, P)$ be a singleton CFG.*
For any terminal symbol $a \in \Sigma$, we define $\mathtt{depth}(a) = 0$*, and for any non-terminal symbol $a \in N$ we define*

$$\mathtt{depth}(a) = \max\{\mathtt{depth}(b_i) + 1 \mid a \to b_1\, b_2 \in P \ \wedge\ i = 1, 2\}$$

We define the depth of G as $\mathtt{depth}(G) = \max\{\mathtt{depth}(a) \mid a \in N\}$.
We define the size of G as its number of rules. (Notice that this definition is for Chomsky grammars).

We can enlarge a grammar in order to define concatenation, exponentiation and prefixes of words already defined by the grammar. We use these operation in the next sections to build the grammar defining some lazy unifier of the unification problem. The following three lemmas state how the size and the depth of the grammar are increased with these transformations.

Lemma 4. *Let G be a singleton grammar defining the words $w_1, \dots, w_n$. There exists a singleton grammar $G' \supseteq G$ that defines the word $w = w_1 \dots w_n$ and satisfies*

$$\begin{aligned} |G'| &\leq |G| + n - 1 \\ \mathtt{depth}(G') &\leq \mathtt{depth}(G) + \lceil \log n \rceil \end{aligned}$$

Lemma 5. *Let G be a singleton grammar defining the word w. For any n, there exists a singleton grammar $G' \supseteq G$ that defines the word w^n and satisfies*

$$\begin{aligned} |G'| &\leq |G| + 2 \lfloor \log n \rfloor \\ \mathtt{depth}(G') &\leq \mathtt{depth}(G) + \lceil \log n \rceil \end{aligned}$$

Lemma 6. *Let G be a singleton grammar defining the word w. For any prefix or suffix w' of w, there exists a singleton grammar $G' \supseteq G$ that defines w' and satisfies*

$$\begin{aligned} |G'| &\leq |G| + \mathtt{depth}(G) \\ \mathtt{depth}(G') &= \mathtt{depth}(G) \end{aligned}$$

Definition 4. *Consider a signature composed by a nonempty set Σ_0 of first-oder constants, a set Σ_1 of second-order and unary constants, a set N of non-terminal symbols, and a set $\mathcal{X}_1$ of second-order and unary variables.*

A generalized set of equations *is a pair $\langle E, G \rangle$, where E is a set of equations of the form $\{t_1 \stackrel{?}{=} u_1, \dots, t_n \stackrel{?}{=} u_n\}$ where terms t_i and u_i, for $i = 1, \dots, n$, are sequences of $(\Sigma_1 \cup \mathcal{X}_1 \cup N)^* \, \Sigma_0$, and $G = \langle \Sigma_1, N, P \rangle$ is a singleton context free* *grammar with just one production for every non-terminal symbol of N occurring in E.*

A generalized unifier *of $\langle E, G\rangle$ is a pair $\langle \sigma, G'\rangle$, where σ is a mapping that assigns either $[X \mapsto \lambda x.a\,x]$ or $[X \mapsto \lambda x.a\,b]$ or $[X \mapsto \lambda x.x]$ or $[X \mapsto \lambda x.b]$ to each variable X, for some $a \in N \cup \Sigma_1$ and $b \in \Sigma_0$, and $G' \supseteq G$ is a singleton grammar that contains a production for every non-terminal symbol of E or σ, such that, replacing every variable X by its instance in E, $\beta\eta$-normalizing both sides of each equation, and then replacing every nonterminal symbol a by the word w_a that it defines, all the equations of E are satisfied as equalities.*

Notice that non-terminal symbols derive into sequences of second-order constants, and that we do not consider first-order variables.

Example 1. Consider the generalized set of equations $\langle E, G\rangle$, defined by

$$E = \{b\,X\,X\,f\,a \stackrel{?}{=} Y\,Y\,Y\,a\}$$
$$G = \{b \to c\,c,\ c \to f\,f\}$$

Then, the pairs $\langle \sigma, G'\rangle$, defined by

$$\sigma = [X \mapsto \lambda x.c\,x,\ Y \mapsto \lambda x.d\,x] \quad G' = \{b \to c\,c,\ c \to f\,f,\ d \to c\,f\}$$

and

$$\sigma = [X \mapsto \lambda x.a,\ Y \mapsto \lambda x.b\,a] \quad G' = \{b \to c\,c,\ c \to f\,f\}$$

are generalized unifiers. In fact, these would be the only two lazy unifiers found by our algorithm. Notice that the second one is a lazy unifier corresponding to the most general unifier $[X \mapsto \lambda x.Z,\ Y \mapsto \lambda x.f\,f\,f\,f\,Z]$.

From now on, an instance of a MSOU problem will be a generalized set of equations. Let σ assign $[X \mapsto \lambda x.a\,x]$. We will use $\sigma(X)$ to denote indistinctly the functions $\lambda x.a\,x$ or $\lambda x.w_a\,x$, or the word w_a, being its meaning clear from the context.

Notice that any monadic set of equations E is equivalent to the generalized set of equations $\langle E, \emptyset\rangle$, and vice versa, any generalized set of equations is equivalent to the monadic set of equations that we obtain by replacing every non-terminal symbol by the word that it defines. Therefore, solvability of monadic set of equations and of generalized set of equations are, with respect to decidability, equivalent problems. With respect to their complexity, we will prove that solvability of generalized sets of equations can be decided in NP-time. This implies that solvability MSOU is also in NP.

5 The Graph of Surface Dependencies

In this Section we define the graph of surface dependencies. The purpose of this graph is to describe, for a given lazy unifier σ, the instance $\sigma(X)$ of some variable X of the problem. In some cases, the ones not covered by Lemmas 8, 9 and 10, the graph is not able to describe such instances, and it becomes necessary to rewrite it to obtain a new graph with this capability. This graph rewriting process will be described in Section 6.

The graph of surface dependencies is defined only for *simplified* equations (not containing rigid-rigid pairs, i.e. pairs with constants in the head of both sides of the equation). In case we have such kind of equations we can simplify them. This can increase the size of the associated grammar as the following Lemma states. After this Lemma, if nothing is said, we will assume that all sets of equations are simplified.

Lemma 7. *Given a generalized set of equations* $\langle E, G\rangle$*, where* $E = \{t_1 \stackrel{?}{=} u_1, \dots, t_n \stackrel{?}{=} u_n\}$*, we can get an equivalent* simplified *problem* $\langle E', G'\rangle$*, where* $E' = \{t'_1 \stackrel{?}{=} u'_1, \dots, t'_m \stackrel{?}{=} u'_m\}$*, i.e. a problem with exactly the same set of solutions, where, for every equation* $t'_i \stackrel{?}{=} u'_i$*, either* t'_i *or* u'_i *has a variable in the head (all rigid-rigid pairs have been removed), and*

$$|E'| \leq |E| \qquad |G'| \leq |G| + n\,\mathtt{depth}(G)$$
$$m \leq n \qquad \mathtt{depth}(G') = \mathtt{depth}(G)$$

Definition 5. *Let* $\langle E, G\rangle$ *be a simplified generalized set of equations, the graph of surface dependencies of* $\langle E, G\rangle$ *is defined as follows.*

Let $\approx$ *be the minimal equivalence relation defined by: if* E *contains an equation of the form* $X\,w_1 \stackrel{?}{=} Y\,w_2$*, then* $X \approx Y$*. This defines a partition on the variables of* E*.*

Every node of the graph is labeled by an $\approx$*-equivalence class of variables, the empty set, or a first-order constant, and every edge is labeled by either a terminal or a second-order constant. Then:*

- *We add just one node for every* $\approx$*-equivalence class of variables.*
- *For every equation of the form* $X\,w_1 \stackrel{?}{=} a_1 \cdots a_n\,Y\,w_2$*, where* $a_1 \cdots a_n \in (N \cup \Sigma_1)^*$*, we add a sequence of nodes with the empty set as labels, and a sequence of labeled edges of the form*

where $X \in L_1$ *and* $Y \in L_2$*.*
- *For every equation of the form* $X\,w_1 \stackrel{?}{=} a_1 \cdots a_n\,b$*, where* $a_1 \cdots a_n \in (N \cup \Sigma_1)^*$ *and* $b \in \Sigma_0$*, we add a sequence of nodes with the empty set as label, and a sequence of labeled edges of the form*

where $X \in L$*.*

Notice that, for every variable X, there is just one node with label L satisfying $X \in L$. This is called its *corresponding* node.

The cycles of this graph describe the base of *some* exponentiation occurring in the instance of some variables. For instance, the solutions of the equation $X\,f\,a \stackrel{?}{=} f\,X\,a$ have the form $[X \mapsto \lambda x.f^n\,x]$, for some $n \geq 0$. The base of this power is described by a cycle in its graph of dependencies:

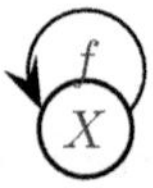

We prove that, if one of the following conditions holds:

1. the graph contains a cycle (Lemma 8),
2. there is a node with two exiting edges with distinct and *divergent* labels (Lemma 9), or
3. there is at most one exiting edge, for every node (Lemma 10),

then the graph of surface dependencies describes the instance of some variable. In the rest of cases it may be necessary to rewrite the graph in order to obtain the desired description.

Lemma 8. *Given a generalized set of equations $\langle E, G\rangle$, let D be its graph of surface dependencies. If D contains a cycle, then, for every lazy unifier σ, whose exponent of periodicity does not exceed k, there exists a variable X such that its corresponding node is inside the cycle, and, if $\alpha \in (N \cup \Sigma_1)^*$ is the sequence of transitions completing the cycle from this node, for some $0 \le n \le k$, and some prefix w' of w_α, we have $\sigma(X) = \lambda x.(w_\alpha)^n\, w'\, x$.*

Moreover, there exists a singleton context-free grammar $G' \supseteq G$ that generates $(w_\alpha)^n\, w'$ and satisfies

$$|G'| \le |G| + \texttt{depth}(G) + |D| + \lceil \log |D| \rceil + 2\lfloor \log k \rfloor$$
$$\texttt{depth}(G') \le \texttt{depth}(G) + \lceil \log |D| \rceil + \lceil \log k \rceil + 1$$

Definition 6. *A dependence graph D is said to contain a* divergence $L_2 \stackrel{a}{\leftarrow} L_1 \stackrel{b}{\rightarrow} L_3$, *if it contains a subgraph of the form:*

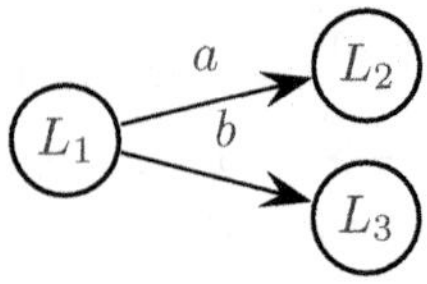

where neither w_a is a prefix of w_b, nor w_b a prefix of w_a.

Lemma 9. *Given a generalized set of equations $\langle E, G\rangle$, let D be its graph of surface dependencies. If D contains a divergence $L_2 \stackrel{a}{\leftarrow} L_1 \stackrel{b}{\rightarrow} L_3$, then, for any lazy unifier σ of E, there exists a variable $X \in L_1$ and some common prefix w' of w_a and w_b, such that $\sigma(X) = \lambda x.w'\,x$.*

Moreover, there exists a singleton context-free grammar $G' \supseteq G$ that generates w', has the same depth as G, and satisfies $|G'| \le |G| + \texttt{depth}(G)$.

Lemma 10. *Given a generalized set of equations $\langle E, G\rangle$, let D be its graph of surface dependencies. If D contains at most one exiting edge, for every node, and D does not contain cycles, then, for any size-minimal lazy unifier σ of E, one of the following properties holds:*

1. *for some node L without exiting edges, some variable $X \in L$, and some first-order constant $b \in \Sigma_0$, we have $\sigma(X) = \lambda x.b$, or*
2. *for some node L, let α be the unique path starting at L and finishing in a node without exiting edges, then, either*
   (a) the sequence α ends in a node labeled with a first-order constant $b \in \Sigma_0$, and $\sigma(X) = \lambda x.w_\alpha\, b$, or
   (b) for some proper prefix w' of w_α, we have $\sigma(X) = \lambda x.w'\, x$.

Moreover, there exists a singleton context-free grammar $G' \supseteq G$ that, in each case, defines w_α or w', and satisfies

$$|G'| \leq |G| + \mathtt{depth}(G) + |D| + \lceil \log |D| \rceil - 1$$
$$\mathtt{depth}(G') \leq \mathtt{depth}(G) + \lceil \log |D| \rceil$$

Remark 1. Notice that this Lemma, contrarily to Lemmas 8 and 9, only applies to *size-minimal* lazy unifiers. Notice also that in case 1, it forces some variable to *forget* its argument, and therefore, applies to lazy unifiers, but not to most general unifiers. In this point is where the search of a size-minimal lazy unifier differs from the search of a most-general unifier, and in fact, where our algorithm looses its completeness and soundness when applied to word unification. (Otherwise this paper would prove NP-completeness of word unification!!).

For instance, in Example 1, the graph of surface dependencies is

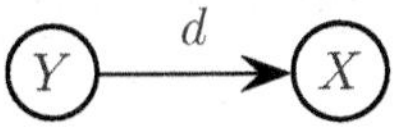

Therefore, we are in the conditions of Lemma 10. Applying sub-case 1, we find $\sigma(X) = \lambda x.a$ (and later, $\sigma(Y) = \lambda x.d\, a$). Applying sub-case 2b, we find, among others, $\sigma(Y) = \lambda x.f\, f\, f\, x$ (and later, $\sigma(X) = \lambda x.f\, f\, x$). The first one is a lazy, but not a most general unifier, whereas the second one is a ground and most general (and therefore lazy) unifier. Notice that there are other most general unifiers of the form $[X \mapsto \lambda x.f^{2+3n}\, x,\ Y \mapsto \lambda x.f^{3+2n}\, x]$ that are not found by our algorithm for $n \geq 1$.

6 Rewriting the Graph of Dependencies

There are graphs not satisfying any of the conditions of Lemmas 8, 9 and 10. These graphs contain a node with two *compatible* exiting edges. In other words, these graphs contain a subgraph, used as redex in our transformation rules, of the form $L_2 \stackrel{a}{\leftarrow} L_1 \stackrel{b}{\rightarrow} L_3$, where w_a is a prefix of w_b, or w_b is a prefix of w_a. An example of such kind of graphs is shown in Example 2. In these cases, in order to obtain a description of some variable instantiation, it can be necessary to transform the graph of dependencies using the following graph rewriting system. These rules transform the redexes described above.

Definition 7. *We consider a transformation system described by rules that work on pairs of the form $\langle D, G \rangle$, where D is a dependence graph and G is a singleton grammar. The transformation on the dependence graph is interpreted as a graph rewriting system.*

Rule 1:

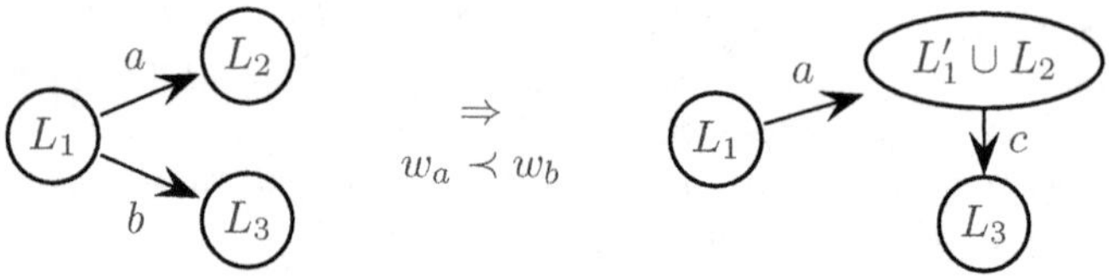

where c is a fresh non-terminal symbol, and $L'_1 = \{X' \mid X \in L_1\}$ is the set of labels of L_1 where we have added a quote to every variable name.

Rule 2:

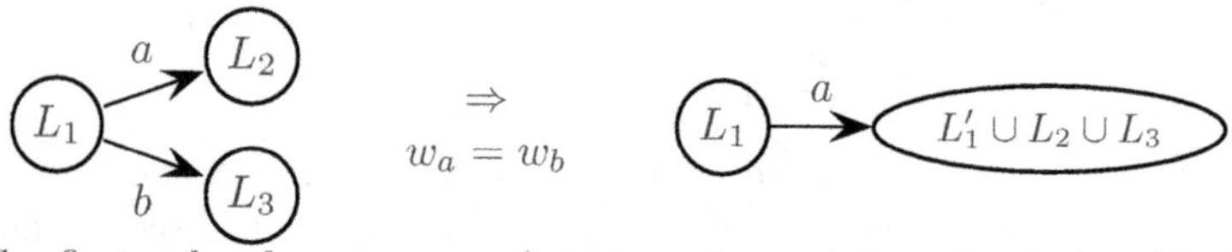

In the first rule, the grammar G is transformed in order to be able to define the word w_c satisfying $w_b = w_a\, w_c$. According to Lemma 6, we can obtain such grammar G' satisfying

$$|G'| \le |G| + \mathtt{depth}(G)$$
$$\mathtt{depth}(G') = \mathtt{depth}(G)$$

In the second rule, the grammar is not modified.

These rules can only be applied if the graph has no cycles and there are no divergences.

Example 2. Consider the following set of equations, and their set of unifiers, for $n \ge 0$.

$$X_1\, c_1 \stackrel{?}{=} a\, b\, X_2\, c_1 \qquad X_1 \mapsto \lambda x.(a\, b)^{n+2}\, x$$
$$X_1\, c_2 \stackrel{?}{=} a\, X_3\, a\, b\, c_2 \qquad X_2 \mapsto \lambda x.(a\, b)^{n+1}\, x$$
$$X_2\, c_3 \stackrel{?}{=} a\, X_3\, c_3 \qquad X_3 \mapsto \lambda x.b\, (a\, b)^n\, x$$

The graph of surface dependencies is

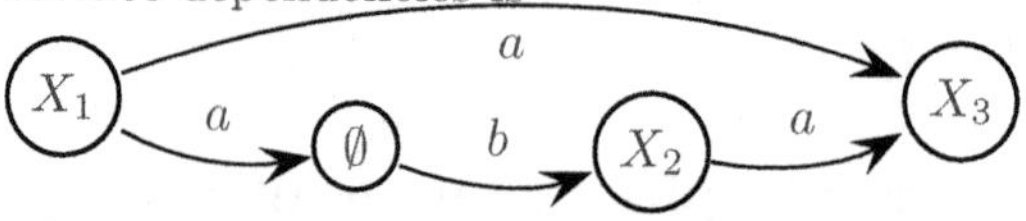

Using the second rule of the graph rewriting system we get

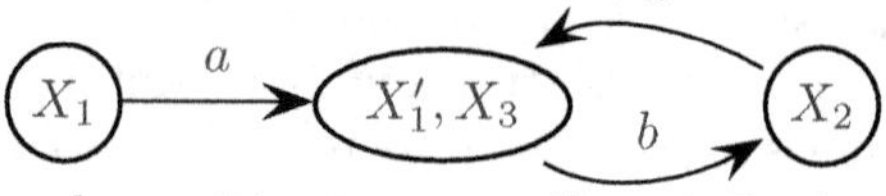

Lemma 11. *Any graph rewriting sequence $D_1 \Rightarrow^* D_n$ has length at most $n \le |D_1|^2$, where $|D_1|$ is the number of edges of D_1.*

As we have said, if the graph of surface dependencies does not contain redexes, then it describes a variable instance. Moreover, depending on the lazy unifier, even if the graph contains redexes, it can also describe some variable instance. We distinguish, according to the lazy unifier, between *incompatible* and *compatible* redexes. If the graph contains an incompatible redex, it already describes a variable instance, and must not be rewritten. Thus, we only rewrite a graph if all redexes are compatible.

Definition 8. *Given a generalized set of equations $\langle E, G\rangle$, its graph of dependencies D, and a lazy unifier σ, we say that a graph rewriting step with redex $L_2 \overset{a}{\leftarrow} L_1 \overset{b}{\rightarrow} L_3$ is* incompatible with σ *if, for some variable $X \in L_1$, we have $\sigma(X) = \lambda x.w'\,x$, where w' is a proper prefix of w_a and w_b. Otherwise, it is said to be* compatible.

Lemma 12. *Given a generalized set of equations $\langle E, G\rangle$, its graph of dependencies D, and a lazy unifier σ, if there exists an incompatible with σ redex $L_2 \overset{a}{\leftarrow} L_1 \overset{b}{\rightarrow} L_3$ in D, i.e. there exists a variable $X \in L_1$ with $\sigma(X) = \lambda x.w'\,x$, where w' is a proper prefix of w_a and w_b, then there exist a singleton context-free grammar $G' \supseteq G$ that defines w' and satisfies $|G'| \leq |G| + \mathtt{depth}(G)$ and $\mathtt{depth}(G') = \mathtt{depth}(G)$.*

When we rewrite a graph, the new graph does not describe *exactly* a variable instance of the lazy unifier, but an instance of a *modification* of this unifier. This new unifier is also a solution of a *modification* of the set of equations. Therefore, when we rewrite the graph of surface dependencies, apart from the associated grammar, we have to transform the set of equations and its lazy unifier. The next Lemma describes how we have to make such modifications.

Lemma 13. *Given a generalized set of equations $\langle E, G\rangle$, its graph of dependencies D, a lazy unifier σ, and a compatible rewriting step $\langle D, G\rangle \Rightarrow \langle D', G'\rangle$, there exist a generalized set of equations $\langle E', G'\rangle$ and a substitution σ' such that*

1. *σ' is a lazy unifier of $\langle E', G'\rangle$,*
2. *D' is the graph of dependencies of E', and*
3. *σ' extends σ as $\sigma'(X) = \sigma(X)$, for any variable occurring in E, and satisfies $\sigma(X) = \sigma'(X) = \lambda x.w_a\,\sigma'(X')\,x$, for any variable $X \in L_1$ occurring in the redex of the rewriting step.*

Graphically we can represent this Lemma as a category-like commutative diagram:

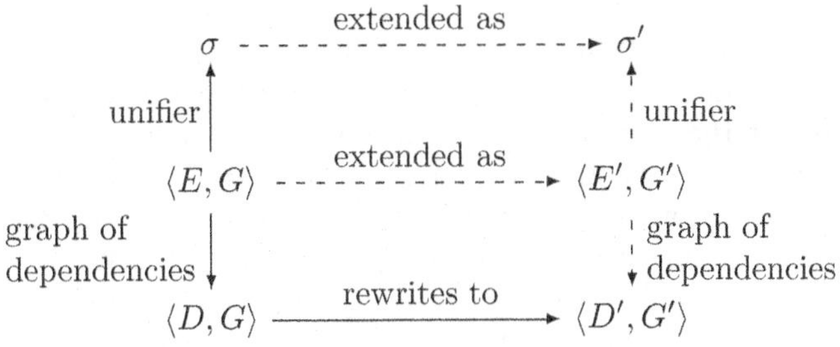

Example 3. Consider the following set of equations

$$X_1\,Y_1\,d \stackrel{?}{=} a\,c\,X_3\,d \qquad X_1\,Y_2\,d \stackrel{?}{=} a\,X_2\,Y_3\,d$$
$$X_2\,Y_6\,d \stackrel{?}{=} b\,c\,X_3\,d \qquad X_2\,Y_4\,d \stackrel{?}{=} b\,X_1\,Y_5\,d$$

The graph of surface dependencies is

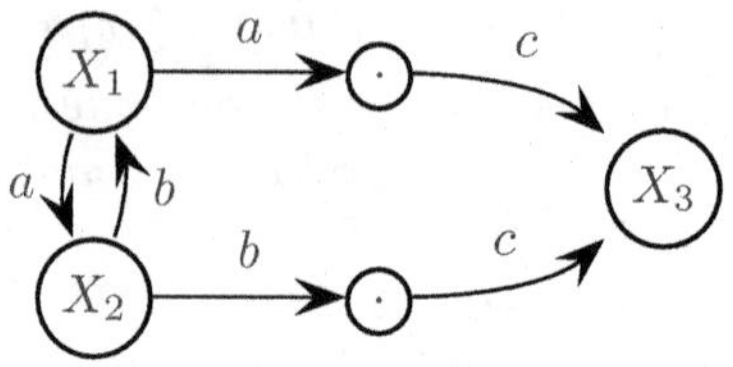

Now either $\sigma(X_1) = \lambda x.x$, and we already have the description of a variable instance, or we can apply a compatible rewriting step, to the redex $\emptyset \overset{a}{\leftarrow} \{X_1\} \overset{a}{\rightarrow} \{X_2\}$, to obtain the following graph:

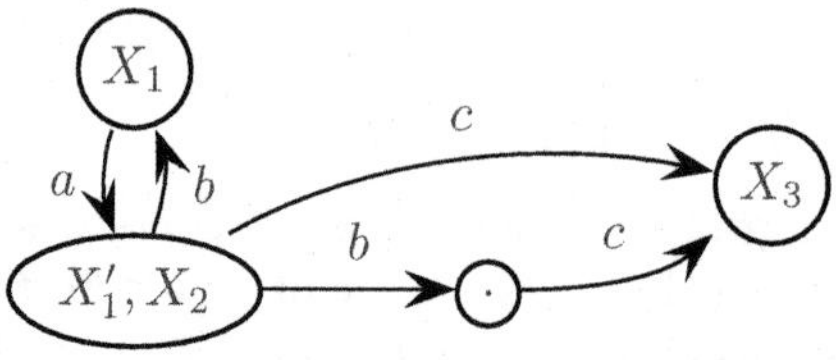

Now, we have an inconsistency, and we can apply Lemma 9 to ensure that either $\sigma'(X_1') = \lambda x.x$ or $\sigma'(X_2) = \lambda x.x$. Applying Lemma 13, in the first case, we obtain $\sigma(X_1) = \lambda x.a\,x$, and in the second case $\sigma(X_2) = \lambda x.x$. Now, we have the instance of some variable that can be instantiated in the equations, and we can repeat the process to obtain instances of other variables.

Theorem 2. *Let σ be a size-minimal lazy unifier of $\langle E, G\rangle$ with exponent of periodicity not exceeding k. Then there exist a variable X in E, and a singleton grammar G', deriving $\sigma(X)$ and such that:*

$$|G'| \leq |G| + \mathcal{O}(|E|^2\,\texttt{depth}(G) + \log k)$$
$$\texttt{depth}(G') \leq \texttt{depth}(G) + \mathcal{O}(\log k + \log |E|)$$

Proof. If the set of equations E is not simplified, we can apply Lemma 7 in order to obtain an equivalent set of simplified equations. This transformation implies a worst-case increase of order $\mathcal{O}(|E|\,\texttt{depth}(G))$ on the size of G, which is compensated by the increase of order $\mathcal{O}(|E|^2\,\texttt{depth}(G) + \log k)$ stated on the Theorem.

Let $\langle E_1, G_1\rangle = \langle E, G\rangle$, $\sigma_1 = \sigma$, and D_1 be the graph of dependencies of E_1. If Lemmas 8, 9 and 10 are not applicable, then there exists a redex in the graph D_1. Then either there exists a redex incompatible with σ_1, or all redexes are compatible. In the second case, we can rewrite $D_1 \Rightarrow D_2$, and use Lemma 13 to find a new substitution σ_2, and set of generalized equations $\langle E_2, G_2\rangle$. Repeating this argument, we can obtain a diagram of the form:

$$\begin{array}{ccccccc}
\sigma_1 & \longrightarrow & \sigma_2 & \cdots & \sigma_{n-1} & \longrightarrow & \sigma_n \\
\uparrow & & \uparrow & & \uparrow & & \uparrow \\
\langle E_1, G_1\rangle & \longrightarrow & \langle E_2, G_2\rangle & \cdots & \langle E_{n-1}, G_{n-1}\rangle & \longrightarrow & \langle E_n, G_n\rangle \\
\downarrow & & \downarrow & & \downarrow & & \downarrow \\
\langle D_1, G_1\rangle & \Longrightarrow & \langle D_2, G_2\rangle & \cdots & \langle D_{n-1}, G_{n-1}\rangle & \Longrightarrow & \langle D_n, G_n\rangle
\end{array}$$

where D_n either satisfies Lemmas 8, 9 or 10, or contains a redex incompatible with σ_n. Now, either using Lemmas 8, 9 or 10, if D_n does not contain redexes, or using Lemma 12, if D_n contains an incompatible redex, we can find the instance $\sigma_n(X)$ of some variable X of E_n. Using the bounds of Lemmas 8, 9, 10 and 12, there exists a singleton context-free grammar G'_n that generates $\sigma_n(X)$ and satisfies:

$$|G'_n| \leq |G_n| + \texttt{depth}(G_n) + |D_n| + \lceil \log |D_n| \rceil + 2 \lfloor \log k \rfloor$$
$$\texttt{depth}(G'_n) \leq \texttt{depth}(G_n) + \lceil \log k \rceil + \lceil \log |D_n| \rceil + 1$$

Notice that the worst bounds are given by Lemma 8.

By Lemma 11 we have $n \leq |D_1|^2$, and using Lemma 6 (see Definition 7), we have

$$|G_n| \leq |G_1| + |D_1|^2 \, \texttt{depth}(G_1)$$
$$\texttt{depth}(G_n) = \texttt{depth}(G_1)$$

Moreover, the size of the dependence graph does not increase during the rewriting steps, therefore $|D_n| \leq |D_1|$. Notice that it is possible that X would not be a variable of E_1. In this case, X will be a variable with primes, say $X^{(m)}$ with $m \leq n$. Then it is possible to construct the instance $\sigma_1(X)$ from $\sigma_n(X^{(m)})$ as $\sigma_1(X) = \lambda x.w_{a_1} \cdots w_{a_m} \sigma_n(X^{(m)})\, x$, where w_{a_i} is the word generated by a_i, and this is a non-terminal symbol of $G_i \subseteq G_n$. Therefore, if we already have a grammar generating $\sigma_n(X^{(m)})$, we can construct a grammar generating $\sigma_1(X)$ by simply adding new rules. In the worst case, this increases the depth of the grammar by $\lceil \log(m+1) \rceil$, and its size by m.

Summarizing, we can find a grammar $G' \supseteq G$ generating $\sigma_1(X)$, for some variable X of E, and satisfying:

$$|G'| \leq |G| + |D|^2 \, \texttt{depth}(G) + |D| + \texttt{depth}(G) + \lceil \log |D| \rceil + 2 \lfloor \log k \rfloor + |D|^2$$
$$\texttt{depth}(G') \leq \texttt{depth}(G) + \lceil \log k \rceil + \lceil \log |D| \rceil + \lceil \log(|D|^2 + 1) \rceil$$

Using orders:

$$|G'| \leq |G| + \mathcal{O}(|D|^2 \, \texttt{depth}(G) + \log k)$$
$$\texttt{depth}(G') \leq \texttt{depth}(G) + \mathcal{O}(\log k + \log |D|)$$

Since $|D|$ is the number of edges in the graph of dependencies, and $|E|$ the number of symbols in the equations, by construction of the graph of dependencies from the equations, we have $|D| \leq |E|$. □

7 Main Results and Some Remarks

Theorem 2 states that, given a generalized set of equations $\langle E, G \rangle$, we can build a new grammar defining the instance of some variable of E. Then we can instantiate this variable in the equations in order to obtain a new set of equations with one variable less. This process does not increase the size of the equations, since we use just one non-terminal symbol on the grammar to describe the instance of the variable. We can repeat N times this process, being N the number of variables, bounded by the original size of the problem $|E|$. The increase on the depth of the grammar is $N\,\mathcal{O}(\log k)$ (being $k = 2^{\mathcal{O}(|E|)}$ the bound on the exponent of periodicity), thus $\mathcal{O}(|E|^2)$. The increase on the size of the grammar is

$N\,\mathcal{O}(|E|^2\,\mathtt{depth}(G) + \log k)$. Although it depends on the depth of the grammar (see Remark 2), it has order $\mathcal{O}(|E|^5)$. This allows us to conclude:

Theorem 3. *For any solvable generalized set of equations $\langle E, \emptyset\rangle$, there exists a lazy unifier $\langle \sigma, G\rangle$ such that the size of $\langle \sigma, G\rangle$ is polynomially bounded on the size of E, in fact $|\sigma| = \mathcal{O}(|E|)$, $|G| = \mathcal{O}(|E|^5)$, and $\mathtt{depth}(G) = \mathcal{O}(|E|^2)$.*

Theorem 3 proves the existence of a polynomially bounded solution for every solvable MSOU problem. Now we have to prove that checking if a substitution is a solution can be performed in polynomial time. Given a substitution, we instantiate the equations. This will remove all variable occurrences, and it will not increase their sizes, because every variable is replaced by just one symbol of the grammar (in some cases, their argument are removed, but this decreases the size). With a small increase of $|E|$ (according to Lemma 4) on the size of the grammar, we can obtain a new grammar defining both sides of every equation, and use Plandowski's Theorem 1 to check their syntactic equality.

Corollary 1. *Monadic Second-Order Unification is NP-complete.*

Theorem 4. *Monadic Second-Order Matching is NP-complete.*

Note that the proof of NP-hardness of MSOU in [SS04] is not a MSO-matching problem, so the proof of Theorem 4 requires a different encoding. We also use the ONE-IN-THREE-SAT problem, which is known to be NP-complete. We associate a pair of second-order variables X_p, Y_p to every propositional variable p, and use a pair of equations $X_p\,Y_p\,b \stackrel{?}{=} a\,b$ and $X_p\,Y_p\,c \stackrel{?}{=} a\,c$, to ensure that their values are $\lambda x.a\,x$, interpreted as true, or $\lambda x.x$, interpreted as false. Then, we encode every clause $p \vee q \vee r$ as $X_p\,X_q\,X_r\,b \stackrel{?}{=} a\,b$.

Remark 2. Theorem 3 clarifies the increase of the size of the grammar representing the solution of a set of equations, after instantiating N variables, according to Theorem 2. This Theorem fixes this increase with respect to the size of the equations, the logarithm of the upper bound on the exponent of periodicity, and the *depth* of the grammar. The question is then: Could we avoid the use of the depth of the grammar? The answer is no. For instance, Lemma 6 says that, if we want to define a prefix of some word defined by a grammar G, in the worst case, we can keep the depth, but we may need to increase the size of G' as $|G'| \leq |G| + \mathtt{depth}(G)$. If we only use the size of the grammar to characterize it, then in the worst case we may be forced to duplicate the size of the grammar $|G'| \leq 2\,|G|$. Each time that we instantiate a variable, it can be necessary to define a new prefix, therefore, in the worst case, the size of the resulting grammar would be 2^N, being $N \leq |E|$ the number of variables. This would result in an exponential upper bound on the size of the grammar.

8 Conclusions

In this paper we prove that Monadic Second-Order Unification (MSOU) is in NP using a result of Plandowski about context-free grammars [Pla94,Pla95].

This result, together the NP-hardness of the problem [SS04] proves its NP-completeness. As we mention in the introduction, MSOU is a specialization of Bounded Second-Order Unification (BSOU). This suggests us that some of the ideas contained in this paper could be used to try to prove that BSOU is in NP.

References

[Far88] W. M. Farmer. A unification algorithm for second-order monadic terms. *Annals of Pure and Applied Logic*, 39:131–174, 1988.

[Far91] W. M. Farmer. Simple second-order languages for wich unification is undecidable. *Theoretical Computer Science*, 87:173–214, 1991.

[GJ79] M. Garey and D. Johnson. *"Computers and Intractability": A guide to the theory of NP-completeness.* W.H. Freeman and Co., San Francisco, 1979.

[Gol81] W. D. Goldfarb. The undecidability of the second-order unification problem. *Theoretical Computer Science*, 13:225–230, 1981.

[Hue75] G. Huet. A unification algorithm for typed λ-calculus. *Theoretical Computer Science*, 1:27–57, 1975.

[KP96] A. Kościelski and L. Pacholski. Complexity of Makanin's algorithm. *Journal of the ACM*, 43(4):670–684, 1996.

[Lev98] J. Levy. Decidable and undecidable second-order unification problems. In *Proceedings of the 9th Int. Conf. on Rewriting Techniques and Applications (RTA'98)*, volume 1379 of *LNCS*, pages 47–60, Tsukuba, Japan, 1998.

[LV00] J. Levy and M. Veanes. On the undecidability of second-order unification. *Information and Computation*, 159:125–150, 2000.

[LV02] J. Levy and M. Villaret. Currying second-order unification problems. In *Proc. of the 13th Int. Conf. on Rewriting Techniques and Applications (RTA'02)*, volume 2378 of *LNCS*, pages 326–339, Copenhagen, Denmark, 2002.

[Mak77] G. S. Makanin. The problem of solvability of equations in a free semigroup. *Math. USSR Sbornik*, 32(2):129–198, 1977.

[Pla94] W. Plandowski. Testing equivalence of morphisms in context-free languages. In J. van Leeuwen, editor, *Proc. of the 2nd Annual European Symposium on Algorithms (ESA'94)*, volume 855 of *LNCS*, pages 460–470, 1994.

[Pla95] W. Plandowski. *The Complexity of the Morphism Equivalence Problem for Context-Free Languages.* PhD thesis, Department of Mathematics, Informatics and Mechanics, Warsaw University, 1995.

[Pla99] W. Plandowski. Satisfiability of word equations with constants is in PSPACE. In *Proc. of the 40th IEEE Annual Symposium on Foundations of Computer Science (FOCS'99)*, pages 495–500, 1999.

[SS01] M. Schmidt-Schauß. Stratified context unification is in PSPACE. In L. Fribourg, editor, *Proc. of the 15th Int. Workshop in Computer Science Logic (CSL'01)*, volume 2142 of *LNCS*, pages 498–512, 2001.

[SS04] M. Schmidt-Schauß. Decidability of bounded second order unification. *Information and Computation*, 188(2):143–178, 2004.

[SSS98] M. Schmidt-Schauß and K. U. Schulz. On the exponent of periodicity of minimal solutions of context equations. In *Proc. of the 9th Int. Conf. on Rewriting Techniques and Applications (RTA'98)*, volume 1379 of *LNCS*, pages 61–75, Tsukuba, Japan, 1998.

[Zhe79] A. P. Zhezherun. Decidability of the unification problem for second order languages with unary function symbols. *Kibernetika (Kiev)*, 5:120–125, 1979. Translated as *Cybernetics* 15(5):735–741, 1980.

A Certified AC Matching Algorithm

Evelyne Contejean

LRI, CNRS UMR 8623 & LogiCal Project, INRIA Futurs
Bât. 490, Université Paris-Sud, Centre d'Orsay
91405 Orsay Cedex, France
contejea@lri.fr

Abstract. In this paper, we propose a matching algorithm for terms containing some function symbols which can be either free, commutative or associative-commutative. This algorithm is presented by inference rules and these rules have been formally proven sound and complete, and decreasing in the COQ proof assistant while the corresponding algorithm is implemented in the C*i*ME system. Moreover some preparatory work has been done in COQ, such as proving that checking the equality of two terms modulo some commutative and associative-commutative theories is decidable.

1 Introduction

Nowadays, the communities of automated deduction and of proof theory are quite close, and it is possible to consider the integration of rewriting modulo AC into proof assistants. In this setting, it is crucial to use a *reliable* AC matching algorithm for the rewriting embedded in the assistant in order to retain the confidence in the validated proofs. In this paper, we present such an algorithm, which is based on inference rules. These rules are sound and complete, and a formal proof of this has been done in the COQ proof assistant [9]. Moreover, each rule has been proven strictly decreasing with respect to the multiset extension of a well-founded ordering[1]. As a preliminary work, terms, substitutions, and equality modulo AC have been defined and equality modulo AC has been shown to be decidable in COQ. The whole COQ development is more that 10000 lines long, and it could probably be shortened by using some more well-chosen lemmas.

The corresponding algorithm is used in C*i*ME for one year, and seems reasonably efficient. A key point is that no copy of the subject, nor of the pattern terms has to be done before running the algorithm itself, in order to avoid mixing pattern variables which can be instantiated and subject variables which should be considered as constants. Compared with some previous version of the algorithm which needed copying the subject, the benchmarks times has been divided by 2, even if a lazy copy was used.

The paper is organized as follows: first some basic notions are recalled in Section 2, and then the inference rules are presented in section 3. In Section 4, we discuss the main difficulties encountered during the formal proof, and finally we conclude in Section 5.

[1] However, the fact that the multiset extension itself is well-founded (Dickson's Lemma) has not been formally proven (yet).

V. van Oostrom (Ed.): RTA 2004, LNCS 3091, pp. 70–84, 2004.

2 Preliminaries

We assume that the reader is familiar with the basic notions of term algebra, substitution and equational theory as surveyed in [4]. In the following we are given $\mathcal{F}$, a signature, $\mathcal{X}$ an infinite set of variables, and $\mathcal{E}$ an equational theory defined over the term algebra $\mathcal{T}(\mathcal{F}, \mathcal{X})$ by a disjoint union of set of axioms of the form

$$C(+) = \{x + y = y + x\} \text{ or } AC(+) = \left\{ \begin{array}{l} x + y = y + x \\ (x + y) + z = x + (y + z) \end{array} \right\},$$

where $+$ is a binary symbol which belongs to $\mathcal{F}$.

An equational step is a pair of terms (s, t) such that there exists an axiom $l = r$ in the above set, a position p in s and a substitution σ such that the subterm of s at position p is equal to $l\sigma$, and $t = s[r\sigma]_p$. This is denoted by $s \longleftrightarrow t$. $\mathcal{E}$ is the reflexive, symmetric and transitive closure of $\longleftrightarrow$. In the following $(s, t) \in \mathcal{E}$ will be denoted as $s =_{\mathcal{E}} t$, and even $s = t$ when there is no ambiguity.

It is well-known folklore that $=_{\mathcal{E}}$ is decidable when the underlying set of axioms contains only C and AC. The most obvious proof is that all the terms in an equivalence class have the same size and the same multiset of function symbols and variables, hence every equivalence class is finite, and can be computed by saturation. However, in practice, this is not how one checks that two terms are equal (modulo): terms are recursively flattened under the AC function symbols and the arguments under C and AC symbols are sorted with respect to a total ordering. Two terms are equal if and only their canonical forms are identical. This has been formally proven by the author within the COQ proof assistant, see the section 4 for some more details.

The problem we are tackling with is, given two terms p and s, to find the substitutions σ such that $p\sigma = s$. Since the class of s is finite, so is the set of solutions. Several algorithms have been proposed to compute it [6, 7, 3, 1, 2, 5, 8]. The one which is presented here is based on inference rules and its core has been proven to be sound and complete in COQ.

We use here a slightly more general definition, where a matching problem M is considered as a conjunction of n pairs of terms $p_1 = s_1 \wedge p_2 = s_2 \wedge \ldots \wedge p_n = s_n$, and a substitution σ is a solution of P whenever for all i such that $1 \leq i \leq n$, $p_i\sigma = s_i$, where $=$ denotes the equality modulo (C or AC). The rules presented below are using a more detailed structure, as defined below:

Definition 1. *A* simple matching problem *M consists in 3 parts:*

- *the unsolved part U, a conjunction of pairs of terms,*
- *the partly solved part P, a conjunction of quadruples $(x, +, y, s)$, where x and y are variables, $+$ an AC function symbol and s is a term. Sometimes, this quadruple will also be denoted by $x = y + s$,*
- *the solved part S, a conjunction of pairs of terms, where the first term is actually a variable.*

Init $\dfrac{p_1 = s_1 \wedge p_2 = s_2 \wedge \ldots \wedge p_n = s_n}{p_1 = s_1 \wedge p_2 = s_2 \wedge \ldots \wedge p_n = s_n, \emptyset, \emptyset}$

Extract$_i$ $\dfrac{\exists \boldsymbol{y}\ (\emptyset, P, S)}{\exists \boldsymbol{y}\ (\mathbf{id}, (\emptyset, P, S))}$ where **id** denotes the identity substitution.

Extract$_S$ $\dfrac{\exists \boldsymbol{y}\ (\sigma, (\emptyset, P, x = s \wedge S))}{\exists \boldsymbol{y}\ (\{x \mapsto s\} \oplus \sigma, (\emptyset, P, S)}$

where $\{x \mapsto s\} \oplus \sigma$ denotes the substitution whose domain is the domain of σ union $\{x\}$, and which is equal to σ, excepts that it maps x onto s.

Extract$_P$ $\dfrac{\exists y \exists \boldsymbol{y}\ (\sigma, (\emptyset, x = y + s \wedge P, \emptyset))}{\exists \boldsymbol{y}\ (\{x \mapsto y\sigma + s\} \oplus \bigoplus_{x' \in \mathcal{D}om(\sigma), x' \neq y} \{x' \mapsto x'\sigma\}, (\emptyset, P, \emptyset))}$

Extract $\dfrac{(\sigma, (\emptyset, \emptyset, \emptyset))}{\sigma}$

Fail $\dfrac{\exists \boldsymbol{y}(p = s \wedge U, P, S)}{\perp}$ if no other rule can apply.

Fig. 1. The rules for the initialization, the extraction of solutions, and failure cases.

A substitution σ is a solution *of M whenever*

- $\forall (p, s) \in U(M)\ \ p\sigma = s,$
- $\forall (x, +, y, s) \in P(M)\ \ x\sigma = y\sigma + s,$
- $\forall (x, s) \in S(M)\ \ x\sigma = s.$

A matching problem *is a pair consisting of a list* $\boldsymbol{y}$ *of existentially quantified variables and a simple matching problem M, and is denoted as $\exists \boldsymbol{y} M$.*

A substitution σ is a solution *of $\exists \boldsymbol{y} M$, whenever there exists a substitution σ', such that σ' is a solution of M and $\forall x\ \ x \notin \boldsymbol{y} \Longrightarrow x\sigma = x\sigma'$.*

3 Rules

Since AC equality is checked via the identity of canonical forms, the terms handled by the rules are in canonical form. Moreover, for sake of readability, when $+$ is an AC function symbol and $t_1 + \ldots + t_n$ is the canonical form of a well-formed term, we allow to write the term $t_1 + \ldots + t_{k-1} + t_{k+1} + \ldots + t_n$, with the convention that when the list of subterms $t_1, \ldots, t_{k-1}, t_{k+1}, \ldots, t_n$ is reduced to a singleton t, $t_1 + \ldots + t_{k-1} + t_{k+1} + \ldots + t_n$ actually denotes t. Notice that this list cannot be empty. The rules are split into several subsets:

- in Figure 1, the rule **Init** builds a matching problem from a set of equations to be solved, the rules **Extract** build a solution from a problem with an empty solved part, and the rule **Fail** apply whenever no other rule can be applied.

Merge$_S$ $\dfrac{\exists \boldsymbol{y}(x = s \wedge U, P, x = s' \wedge S)}{\exists \boldsymbol{y}(U, P, x = s' \wedge S)}$ if $s = s'$

Clash$_S$ $\dfrac{\exists \boldsymbol{y}(x = s \wedge U, P, x = s' \wedge S)}{\bot}$ if $s \neq s'$

Merge$_P$ $\dfrac{\exists \boldsymbol{y}(x = s_1 + \ldots + s_n \wedge U, x = y + s' \wedge P, S)}{\exists \boldsymbol{y}(y = s_1 + \ldots + s_{k-1} + s_{k+1} + \ldots + s_n \wedge U, x = y + s' \wedge P, S)}$
if $+$ is AC, $s_k = s'$

Clash$_P$ $\dfrac{\exists \boldsymbol{y}(x = s \wedge U, x = y + s' \wedge P, S)}{\bot}$ if $s(\epsilon) \neq +$

Solve $\dfrac{\exists \boldsymbol{y}(x = s \wedge U, P, S)}{\exists \boldsymbol{y}(U, P, x = s \wedge S)}$
if x does not occur as a left member of a pair in S, nor in P

Fig. 2. The rules for the case where the pattern term of the first unsolved equation is a variable.

- in Figure 2, the rules apply to problems where the pattern term of the first unsolved equation $p = s$ is a variable x. **Merge**$_S$ and **Clash**$_S$ (resp. **Merge**$_P$ and **Clash**$_P$) check the compatibility of s with the assigned plain (resp. partial) value of x. **Solve** simply moves the equation in the solved part whenever x has no value yet.
- in Figure 3, the rules apply on problems where the pattern term of the first unsolved equation is not a variable. **Clash** and **Dec** are the usual rules, and **Dec**$_C$ is the syntactic decomposition modulo commutativity. All the other rules apply to an equation $p_1 + \ldots + p_m = s_1 + \ldots + s_n$ where $+$ is AC: when p_1 is a variable x, **AC**$_{S=}$ and **AC**$_{S\neq}$ (resp. **AC**$_{P=}$ and **AC**$_{P\neq}$) ensure the compatibility of $p_1 + \ldots + p_m = s_1 + \ldots + s_n$ with the assigned plain (resp. partial) value of x. **AC**$_{\neq}$ and **AC**$_{=}$ respectively assign a single s_i and several s_is to x when x has no value yet. When p_1 is not a variable, **AC** non-deterministically assigns a single s_i to it.

Most of the rules are deterministic, but a few of the last subset are not: **Dec**$_C$, **AC**, **AC**$_{P\neq}$, **AC**$_{\neq}$ and **AC**$_{=}$. In these latter cases, all possibilities have to be developed in order to get a complete algorithm (don't-know non-determinism).

3.1 The Algorithm

The corresponding algorithm is quite simple:

- from a set of equations to be solved, build a matching problem using the rule **Init**.

Clash $\dfrac{\exists \boldsymbol{y}(f(p_1,\ldots,p_n)=g(s_1,\ldots,s_m)\wedge U,P,S)}{\bot}$ if $f \neq g$

Dec $\dfrac{\exists \boldsymbol{y}(f(p_1,\ldots,p_n)=f(s_1,\ldots,s_n)\wedge U,P,S)}{\exists \boldsymbol{y}(p_1=s_1\wedge\ldots\wedge p_n=s_n\wedge U,P,S)}$ if f is free or a constant (that is $n=0$)

Dec$_C$ $\dfrac{\exists \boldsymbol{y}(f(p_1,p_2)=f(s_1,s_2)\wedge U,P,S)}{\exists \boldsymbol{y}(p_1=s_i\wedge p_2=s_{(i \bmod 2)+1}\wedge U,P,S)}$ if f is C, $i=1$ or $i=2$

AC$_{S\neq}$ $\dfrac{\exists \boldsymbol{y}(x+p_2+\ldots+p_m=s_1+\ldots+s_n\wedge U,P,x=s\wedge S)}{\exists \boldsymbol{y}(p_2+\ldots+p_m=s_1+\ldots+s_{k-1}+s_{k+1}+\ldots+s_n\wedge U,P,x=s\wedge S)}$

if $+$ is AC, $m\leq n$, $s_k=s$

AC$_{S=}$ $\dfrac{\exists \boldsymbol{y}(x+p_2+\ldots+p_m=s_1+\ldots+s_n\wedge U,P,x=s'_1+\ldots+s'_{n'}\wedge S)}{\exists \boldsymbol{y}(p_2+\ldots+p_m=s''_1+\ldots+s''_{n''}\wedge U,P,x=s'_1+\ldots+s'_{n'}\wedge S)}$

if $+$ is AC, $1<n'$m $m-1\leq n-n'$, $\{s''_1\ldots s''_{n''}\}\cup\{s'_1\ldots s'_{n'}\}=\{s_1\ldots s_n\}$, where $\cup$ is the union of multisets.

AC$_{P\neq}$ $\dfrac{\exists \boldsymbol{y}(x+p_2+\ldots+p_m=s_1+\ldots+s_n\wedge U,x=y*s\wedge P,S)}{\exists \boldsymbol{y}(y=s'_2*\ldots*s'_{n'}\wedge p_2+\ldots+p_m=s_1+\ldots+s_{k-1}+s_{k+1}+\ldots+s_n\wedge U,x=y*s\wedge P,S)}$

$+$ is AC, $m\leq n$, $+\neq *$, $s_k=s*s'_2*\ldots*s'_{n'}$

AC$_{P=}$ $\dfrac{\exists \boldsymbol{y}(x+p_2+\ldots+p_m=s_1+\ldots+s_n\wedge U,x=y+s\wedge P,S)}{\exists \boldsymbol{y}(y+p_2+\ldots+p_m=s_1+\ldots+s_{k-1}+s_{k+1}+\ldots+s_n\wedge U,x=y+s\wedge P,S)}$

$+$ is AC, $m<n$, $s_k=s$

AC$_{\neq}$ $\dfrac{\exists \boldsymbol{y}(x+p_2+\ldots+p_m=s_1+\ldots+s_n\wedge U,P,S)}{\exists \boldsymbol{y}(p_2+\ldots+p_m=s_1+\ldots+s_{k-1}+s_{k+1}+\ldots+s_n\wedge U,P,x=s_k\wedge S)}$

$+$ is AC, $m\leq n$, x does not occur as a left member of a pair in S, nor in P.

AC$_=$ $\dfrac{\exists \boldsymbol{y}(x+p_2+\ldots+p_m=s_1+\ldots+s_n\wedge U,P,S)}{\exists y\exists \boldsymbol{y}(y+p_2+\ldots+p_m=s_1+\ldots+s_{k-1}+s_{k+1}+\ldots+s_n\wedge U,x=y+s_k\wedge P,S)}$

$+$ is AC, $m<n$, x does not occur as a left member of a pair in S, nor in P, and y is a new (left) variable.

AC $\dfrac{\exists \boldsymbol{y}(g(\boldsymbol{p'})+p_2+\ldots+p_m=s_1+\ldots+s_n\wedge U,P,S)}{\exists \boldsymbol{y}(g(\boldsymbol{p'})=s_k\wedge p_2+\ldots+p_m=s_1+\ldots+s_{k-1}+s_{k+1}+\ldots+s_n\wedge U,P,S)}$

$+$ is AC, $1<m\leq n$, $s_k(\epsilon)=g$

Fig. 3. The rules for the case when the pattern term of the first unsolved equation is not a variable.

- apply the rules **Merge**$_S$, **Clash**$_S$, **Merge**$_P$, **Clash**$_P$, **Solve, Clash, Dec, Dec**$_C$, **AC**$_{S\neq}$, **AC**$_{S=}$, **AC**$_{P\neq}$, **AC**$_{P=}$, **AC**$_{\neq}$, **AC**$_{=}$, **AC** as far as possible. If no rule is applicable, and the unsolved part is not empty, then apply **Fail**.
- if the unsolved part is empty, apply **Extract**$_i$ in order to build the solution, then **Extract**$_S$, **Extract**$_P$ as far as possible, and eventually **Extract**.

Since some of the rules have to be applied non-deterministically, the algorithm actually handles disjunction of problems, and a rule transforms a disjunction of problems by replacing one of the problems by all the problems it can produce by the non-deterministic choices (or even none in the case of failure rules). This process can be seen as developing a (search) tree; hence there are many possible strategies, such as breadth-first, depth-first search, and so on.

Moreover, notice that the unsolved equations can actually be considered as *a set*, that is, one may pick any equation, and consider it as the first one, which enlarges the possible strategies. The solved and partly solved equations are not plain sets, since any variable can occur only at most as a left-hand side in one of them. In order to properly extract the solutions from a matching problem where the unsolved part is empty, it is important to keep record of the ordering of the introduction of the existentially quantified variables, hence the partly solved part should be seen as a stack.

By definition of the solutions of a matching problem, it is clear that they are an invariant of the rule **Init**.

3.2 Correctness of the Second Step of the Algorithm

The rules **Merge**$_S$, **Clash**$_S$, **Merge**$_P$, **Clash**$_P$, **Solve, Clash, Dec, Dec**$_C$, **AC**$_{S\neq}$, **AC**$_{S=}$, **AC**$_{P\neq}$, **AC**$_{P=}$, **AC**$_{\neq}$, **AC**$_{=}$, **AC** and **Fail** do not build some ill-formed problems, they are decreasing, sound and complete. The first, the second and the third properties seem quite trivial. We shall give some hints for the proof of the last property.

Lemma 1. *If a rule among* **Merge**$_S$, **Clash**$_S$, **Merge**$_P$, **Clash**$_P$, **Solve, Clash, Dec, Dec**$_C$, **AC**$_{S\neq}$, **AC**$_{S=}$, **AC**$_{P\neq}$, **AC**$_{P=}$, **AC**$_{\neq}$, **AC**$_{=}$, **AC** *can apply to a matching problem* $\exists \boldsymbol{y}(U, P, S)$ *such that*

- *every term occurring in* U, P *and* S *is the canonical form of a well-formed term,*
- *any variable occurs at most once as a left value in* P *and* S,
- *all the quadruples* $(x, +, y, s)$ *in* P *are such that* $+$ *is AC and* $s(\epsilon) \neq +$,

so is (are) the resulting problem(s).

Lemma 2. *The rules* **Merge**$_S$, **Clash**$_S$, **Merge**$_P$, **Clash**$_P$, **Solve, Clash, Dec, Dec**$_C$, **AC**$_{S\neq}$, **AC**$_{S=}$, **AC**$_{P\neq}$, **AC**$_{P=}$, **AC**$_{\neq}$, **AC**$_{=}$, **AC** *are decreasing with respect to (the multiset extension of) the measure* $\mathcal{M}$*:*

$$\mathcal{M}(p_1 = s_1 \wedge \ldots \wedge p_n = s_n, P, S) = \sum_{i=1}^{n} size(p_i) + size(s_i)$$

Lemma 3. *Let M be a well-formed matching problem in the sense of Lemma 1. If σ is a solution of M', and M' is obtained from M by applying one of the rules* **Merge$_S$, Clash$_S$, Merge$_P$, Clash$_P$, Solve, Clash, Dec, Dec$_C$, AC$_{S\neq}$, AC$_{S=}$, AC$_{P\neq}$, AC$_{P=}$, AC$_{\neq}$, AC$_=$, AC**, *then σ is a solution of M.*

Sketch of the proof.
The proof is done by cases on the rule which has been applied. We shall examine in details only one case, the others being similar. Let us assume that the rule is **AC$_{P\neq}$**, that is

$$M \equiv \exists \boldsymbol{y}(x + p_2 + \ldots + p_m = s_1 + \ldots + s_n \wedge U, x = y * s \wedge P, S)$$

$$M' \equiv \exists \boldsymbol{y}(y{=}s'_2{*}\ldots{*}s'_{n'} \wedge p_2{+}\ldots{+}p_m{=}s_1{+}\ldots{+}s_{k-1}{+}s_{k+1}{+}\ldots{+}s_n \wedge U, \quad x{=}y * s \wedge P, S),$$

and $+$ is AC, $m \leq n$, $+ \neq *$, $s_k = s * s'_2 * \ldots * s'_{n'}$. By definition σ is a solution of M' means that

1. $y\sigma = s'_2{*}\ldots{*}s'_{n'}$,
2. $(p_2{+}\ldots{+}p_m)\sigma = s_1{+}\ldots{+}s_{k-1}{+}s_{k+1}{+}\ldots{+}s_n$,
3. $\forall p = s \in U, \quad p\sigma = s$,
4. $x\sigma = y\sigma * s$,
5. $\forall (x', +', y', s') \in P, \quad x'\sigma = y'\sigma +' s'$,
6. $\forall x' = s' \in S, \quad x'\sigma = s'$.

We shall prove that σ is a solution of M, that is

- $(x + p_2{+}\ldots{+}p_m)\sigma = s_1 + \ldots + s_n$,
- $\forall p = s \in U, \quad p\sigma = s$,
- $x\sigma = y\sigma * s$,
- $\forall (x', +', y', s') \in P, \quad x'\sigma = y'\sigma +' s'$,
- $\forall x' = s' \in S, \quad x'\sigma = s'$.

Let us consider the first item, the others being obvious:

$$\begin{aligned}
(x + p_2{+}\ldots{+}p_m)\sigma &= x\sigma + (p_2{+}\ldots{+}p_m)\sigma \\
&= (y\sigma * s) + (p_2{+}\ldots{+}p_m)\sigma && \text{by (4)} \\
&= ((s'_2{*}\ldots{*}s'_{n'}) * s) + (p_2{+}\ldots{+}p_m)\sigma && \text{by (1)} \\
&= (s * s'_2{*}\ldots{*}s'_{n'}) + (p_2{+}\ldots{+}p_m)\sigma && * \text{ is AC, by Lemma 1} \\
&= s_k + (p_2{+}\ldots{+}p_m)\sigma && \text{by hypothesis} \\
&= s_k + s_1{+}\ldots{+}s_{k-1}{+}s_{k+1}{+}\ldots{+}s_n && \text{by (2)} \\
&= s_1 + \ldots + s_n && + \text{ is AC, by hypothesis}
\end{aligned}$$

□

Lemma 4. *Let M be a well-formed matching problem in the sense of Lemma 1. If the unsolved part of M is not empty and if σ is a solution of M, then one of the rules* **Merge$_S$, Clash$_S$, Merge$_P$, Clash$_P$, Solve, Clash, Dec, Dec$_C$, AC$_{S\neq}$, AC$_{S=}$, AC$_{P\neq}$, AC$_{P=}$, AC$_{\neq}$, AC$_=$, AC** *can be applied, and there exists M' among the resulting problems such that σ is a solution of M'.*

Proof. By hypothesis, the unsolved part of M is not empty, hence let $p = s$ be the first unsolved equation of M. We shall reason by cases over the form of $p = s$. It should be noticed that once the first unsolved equation of M is fixed (or chosen), only one rule can apply. In the following, we assume that σ is actually a solution of the underlying simple matching problem of M. We shall discuss the problem of existentially quantified variables only when the rule $\mathbf{AC}_=$ can be applied, since all the other rules do not modify the list of existentially quantified variables.

1. If p is a variable x, there are three disjoint sub-cases:
   (a) If there is an equation of the form $x = s'$ in S, then either s and s' are equal (modulo AC), or they are not.
   In the first case, the rule $\mathbf{Merge}_S$ applies and it is obvious that σ is a solution of the resulting problem.
   In the second case, the rule $\mathbf{Clash}_S$ applies, and σ cannot be a solution of M, since this would imply $s = x\sigma = s'$.
   (b) If there is an equation of the form $x = y + s'$ in P, since M is a well-formed problem, $+$ is an AC function symbol. Either the top symbol of s is equal to $+$ or not. In the first case, if the requested conditions are fulfilled, the rule $\mathbf{Merge}_P$ applies, else **Fail**:
      – In the $\mathbf{Merge}_P$ case, there exists a subterm s_k of $s \equiv s_1 + \ldots + s_n$ which is equal to s', $s \equiv s_1 + \ldots + s_n = x\sigma = y\sigma + s_k$. Hence $s_1 + \ldots + s_{k-1} + s_{k+1} + \ldots + s_n = y\sigma$, and σ is a solution of the resulting problem.
      – In the **Fail** case, there does not exist any subterm s_k of $s \equiv s_1 + \ldots + s_n$ which is equal to s', which contradicts the fact that σ is a solution, in particular that $s = y\sigma + s_k$.
   In the second case the rule $\mathbf{Clash}_P$ applies, and σ cannot be a solution of M, since this would imply $s = x\sigma = y\sigma + s'$, and two terms equal modulo AC have the same top symbol.
   (c) otherwise, **Solve** applies, and σ is obviously a solution of the resulting problem.
2. If p is not a variable term, then either the top symbol of p and s are equal or not. In this latter case, the rule **Clash** applies, and σ cannot be a solution of M, since two terms equal modulo AC have the same top symbol, and p and $p\sigma$ have the same top symbol.
   Otherwise,
   (a) if the top is a free function symbol, then the rule **Dec** applies, and obviously σ is a solution of the resulting problem, since no C or AC axioms can be applied at the top of the terms in an equational proof of $p\sigma = s$.
   (b) if the top is a C function symbol, $\mathbf{Dec}_C$ applies and since commutativity is a syntactic, then σ is a solution of one of the two resulting problems.
   (c) if the top is an AC function symbol $+$, then $p \equiv p_1 + \ldots + p_m$ and $s \equiv s_1 + \ldots + s_n$. There are two sub-cases, according to p_1 the first subterm of p:

– if p_1 is a variable x, then we split into tree disjoint sub-cases, as for the case when p is variable term.
  - if there is an equation of the form $x = s'$ in S, since σ is a solution of M,

$$\begin{aligned} s \equiv s_1 + \ldots + s_n = p\sigma &\equiv p_1\sigma + p_2\sigma + \ldots + p_m\sigma \\ &= x\sigma + p_2\sigma + \ldots + p_m\sigma \\ &= s' + p_2\sigma + \ldots + p_m\sigma \end{aligned}$$

    There two disjoint cases: either the top symbol of s' is equal to $+$, or it is different.
    In the first case, s' is of the form $s'_1 + \ldots + s'_{n'}$, and $s_1 + \ldots + s_n = s'_1 + \ldots + s'_{n'} + p_2\sigma + \ldots + p_m\sigma$. Since we consider only flattened terms, the s_i's and the s'_i's do not have $+$ as top symbol, hence the multiset $\{s_1, \ldots, s_n\}$ contains $\{s'_1, \ldots, s'_{n'}\}$: there exists a multiset $\{s''_1, \ldots, s''_{n''}\}$ such that

$$\{s_1, \ldots, s_n\} = \{s'_1, \ldots, s'_{n'}\} \cup \{s''_1, \ldots, s''_{n''}\}$$

    This means that the rule $\mathbf{AC}_{S=}$ applies. Moreover $s''_1 + \ldots + s''_{n''} = p_2\sigma + \ldots + p_m\sigma$, hence σ is a solution of the resulting problem.
    In the second case, $s'(\epsilon) \neq +$, since we consider only flatten terms, the s_i's do not have $+$ as top symbol, hence the multiset $\{s_1, \ldots, s_n\}$ contains s', that is, there exists k such that $s' = s_k$. Moreover, after flattening the $p_i\sigma$'s, the number of subterms of $s' + p_2\sigma + \ldots + p_m\sigma$ may only increase, which means that $m \leq n$. Hence the rule $\mathbf{AC}_{S\neq}$ applies to M. Since $s_1 + \ldots + s_{k-1} + s_{k+1} + \ldots + s_n = p_2\sigma + \ldots + p_m\sigma$, σ is also a solution of the resulting problem.
  - if there is a quadruple $(x, *, y, s')$ in P, since σ is a solution of M,

$$\begin{aligned} s \equiv s_1 + \ldots + s_n = p\sigma &\equiv p_1\sigma + p_2\sigma + \ldots + p_m\sigma \\ &= x\sigma + p_2\sigma + \ldots + p_m\sigma \\ &= (y\sigma * s') + p_2\sigma + \ldots + p_m\sigma \end{aligned}$$

    If $* = +$, $s_1 + \ldots + s_n = y\sigma + s' + p_2\sigma + \ldots + p_m\sigma$, and the top symbol of s' is not $+$ (M is well-formed). This implies that s' is contained in $\{s_1, \ldots, s_n\}$, that is there exists k such that $s_k = s'$. Moreover, after flattening $y\sigma$ and the $p_i\sigma$'s, the number of subterms of $y\sigma + s' + p_2\sigma + \ldots + p_m\sigma$ may only increase, which means that $m < n$. Hence the rule $\mathbf{AC}_{P=}$ applies to M. Since $s_1 + \ldots + s_{k-1} + s_{k+1} + \ldots + s_n = y\sigma + p_2\sigma + \ldots + p_m\sigma$, σ is also a solution of the resulting problem.
    If $* \neq +$, $y\sigma * s'$ is contained in $\{s_1, \ldots, s_n\}$, that is there exists k such that $s_k = y\sigma * s'$. Moreover, after flattening the $p_i\sigma$'s, the number of subterms of $(y\sigma * s') + p_2\sigma + \ldots + p_m\sigma$ may only increase, which means that $m \leq n$. Hence the rule $\mathbf{AC}_{P\neq}$ applies

to M. The top symbol of s' is not equal to $*$, hence s' is contained in the multiset of the direct subterms of $s_k \equiv s'_1 * \ldots * s'_{n'}$. We assume that $s' = s'_1$ (without loss of generality, after reordering the subterms). Hence $y\sigma = s'_2 * \ldots * s'_{n'}$. We have also $s_1 + \ldots + s_{k-1} + s_{k+1} + \ldots + s_n = p_2\sigma + \ldots + p_m\sigma$, hence σ is a solution of the resulting problem.

- otherwise x does not occur as a left-hand side of an equation neither in S nor in P. σ is a solution of M:

$$\begin{aligned} s \equiv s_1 + \ldots + s_n = p\sigma &\equiv p_1\sigma + p_2\sigma + \ldots + p_m\sigma \\ &= x\sigma + p_2\sigma + \ldots + p_m\sigma \end{aligned}$$

After flattening $x\sigma$ and the $p_i\sigma$'s, the number of subterms in $x\sigma + p_2\sigma + \ldots + p_m\sigma$ may only increase, hence $m \leq n$.
If the top symbol of $x\sigma$ is equal to $+$, then flattening $x\sigma$ actually strictly increases this number of subterms, hence $m < n$. Both rules $\mathbf{AC}_=$ and $\mathbf{AC}_{\neq}$ non-deterministically apply. We shall prove that σ is a solution of one of the resulting problems of $\mathbf{AC}_=$. The multiset $\{s'_1, \ldots, s'_{n'}\}$ of direct subterms of $x\sigma$ is included in $\{s_1, \ldots, s_n\}$: there exists k such that $s'_1 = s_k$, and σ is a solution of M'_k, the k-th problem yielded by $\mathbf{AC}_=$: y is a new variable which does not occur in M, and is existentially quantified in M'_k. This means that we shall prove that there exists σ', a solution of M'_k, which is equal to σ, except maybe on y. The chosen value for $y\sigma'$ is $s'_2 + \ldots + s'_{n'}$. It is clear that with this value,

$$\begin{aligned} &y\sigma' + p_1\sigma' + p_2\sigma' + \ldots + p_m\sigma' = s_1 + \ldots + s_{k-1} + s_{k+1} + \ldots + s_n \\ &x\sigma' = y\sigma' + s_k, \end{aligned}$$

that is, that σ' is a solution of M'_k.
If the top symbol of $x\sigma$ is not equal to $+$, the rule $\mathbf{AC}_{\neq}$ non-deterministically applies (and maybe also $\mathbf{AC}_=$). We shall prove that σ is a solution of one of the resulting problems of $\mathbf{AC}_{\neq}$. $x\sigma$ is included in the multiset $\{s_1, \ldots, s_n\}$, there exists k such that $x\sigma = s_k$. Hence $p_2\sigma + \ldots + p_m\sigma = s_1 + \ldots + s_{k-1} + s_{k+1} + \ldots + s_n$. σ is a solution of the k-th problem yielded by $\mathbf{AC}_{\neq}$.

– if the first subterm of p is not a variable, it is a term $g(\mathbf{p'})$, where g is distinct from $+$, since p is a flatten term. σ is a solution of M:

$$\begin{aligned} s \equiv s_1 + \ldots + s_n = p\sigma &\equiv p_1\sigma + p_2\sigma \ldots + p_m\sigma \\ &= g(\mathbf{p'\sigma}) + p_2\sigma \ldots + p_m\sigma \end{aligned}$$

After flattening the $p_i\sigma$'s, the number of subterms in $g(\mathbf{p'\sigma}) + p_2\sigma + \ldots + p_m\sigma$ may only increase, hence $m \leq n$, hence the rule **AC** non-deterministically applies for the terms s_k such that $s_k(\epsilon) = g$. We shall prove that σ is a solution of one of the resulting problems. $g(\mathbf{p'\sigma})$ is included in the multiset $\{s_1, \ldots, s_n\}$: there exists k such that $g(\mathbf{p'\sigma}) = s_k$ (this implies that $s_k(\epsilon) = g$) and σ is a solution of M'_k, the k-th problem yielded by **AC**. □

3.3 Correctness of the Third Step of the Algorithm

So far, we have proven that the first two steps of the algorithm are terminating, sound and complete. From a conjunction of equations to be solved, they build a disjunction of well-formed matching problems where the unsolved part is empty. It remains to prove that the last step, which extracts a substitution from such problems is also terminating, sound and complete. For that purpose, we need to define what is a solution of a pair made of a substitution and a matching problem:

Definition 2. *Let (π, M) be a pair of substitution and simple matching problem. A substitution σ is a* solution *of (π, M) if*

- *σ extends π, that is the domain of σ contains the domain of π, and for all x in the domain of π, $x\pi = x\sigma$.*
- *σ is a solution of the matching problem M.*

A substitution σ is a solution *of $\exists \boldsymbol{y}(\pi, M)$ whenever there exists a substitution σ', such that*

- *σ' is a solution of (π, M),*
- *$\forall x \quad x \notin \boldsymbol{y} \Longrightarrow x\sigma = x\sigma'$.*

The fact that the rules for extracting the solutions are sound relies on the following

Lemma 5. *Let M' be a matching problem obtained from $(U_0, \emptyset, \emptyset)$ by applying the rules* **Merge$_S$**, **Merge$_P$**, **Solve**, **Dec**, **Dec$_C$**, **AC$_{S\neq}$**, **AC$_{S=}$**, **AC$_{P\neq}$**, **AC$_{P=}$**, **AC$_{\neq}$**, **AC$_{=}$** *and* **AC**: *M' is of the form*

$$\exists y_n \ldots \exists y_k \ldots \exists y_1 (U, x_n = y_n +_n s_n \wedge \ldots \wedge x_k = y_k +_k s_k \wedge \ldots \wedge x_1 = y_1 +_1 s_1, S)$$

and $\forall k, k' \quad k' \leq k \Longrightarrow x_{k'} \neq y_k$.

Sketch of the proof.
The proof is by induction over the number of applications of the rules. The only rule which modifies the existentially quantified variables and the partly solved part of a problem is **AC$_{=}$**, which transforms

$$\exists y_{n-1} \ldots \exists y_1 (U, x_{n-1} = y_{n-1} +_{n-1} s_{n-1} \wedge \ldots \wedge x_1 = y_1 +_1 s_1, S)$$

into

$$\exists y_n \exists y_{n-1} \ldots \exists y_1 (U', x_n = y_n +_n s_n \wedge x_{n-1} = y_{n-1} +_{n-1} s_{n-1} \wedge \ldots \wedge x_1 = y_1 +_1 s_1, S')$$

and y_n is a new variable, which means in particular that $\forall k' \quad k' \leq n \Longrightarrow x_{k'} \neq y_n$. By induction hypothesis, we have

$$\forall k, k' \quad k' \leq k \leq n-1 \Longrightarrow x_{k'} \neq y_k,$$

which completes the proof. □

The fact that the rules **Extract$_S$** and **Extract$_P$** are terminating is obvious, since they are decreasing with respect to the measure defined in Lemma 2. The rules **Extract$_i$** and **Extract$_S$** are sound and complete since they obviously do not alter the set of solutions of a matching problem.

Lemma 6. *Let $\exists \boldsymbol{y}'(\pi', M')$ be the result of the application of the rule* **Extract**$_P$ *to $\exists y \exists \boldsymbol{y}(\pi, M)$, obtained at the third step of the matching algorithm. A substitution is a solution of $\exists y \exists \boldsymbol{y}(\pi, M)$ if and only if it is a solution of $\exists \boldsymbol{y}'(\pi', M')$.*

Proof. In order to have the rule **Extract**$_P$ applicable to $\exists y \exists \boldsymbol{y}(\pi, M)$, M is of the form $(\emptyset, x = y+s \wedge P, \emptyset)$. Hence $\pi' = \{x \mapsto y\pi + s\} \oplus \bigoplus_{x' \in \mathcal{D}om(\pi), x' \neq y} \{x' \mapsto x'\pi\}$, $\boldsymbol{y}' = \boldsymbol{y}$ and $M' = (\emptyset, P, \emptyset)$. It should be noticed that $x \notin \mathcal{D}om(\pi)$, since a variable occurs at most once either in the solved part, or in the partly solved part, or in the domain of the substitution.

1. Let us assume that σ is a solution of $\exists y \exists \boldsymbol{y}(\pi, M)$, we shall prove that σ is a solution of $\exists \boldsymbol{y}(\pi', M')$. Since σ is a solution of $\exists y \exists \boldsymbol{y}(\pi, M)$, there exists σ' a solution of (π, M), such that σ and σ' are equal except maybe on y and the elements of $\boldsymbol{y}$. We define σ'' as σ' except for y: $y\sigma'' = y\sigma$.
   (a) The domain of π' is equal to the domain of π, minus y, plus x. By hypothesis, σ' extends π; let x' be a variable such that $x' \in \mathcal{D}om(\pi), x' \neq y$:

$$\begin{aligned} x'\pi' &= x'\pi && \text{by definition of } \pi' \\ &= x'\sigma' && \sigma' \text{ extends } \pi \\ &= x'\sigma'' && x' \neq y \text{ by hypothesis} \end{aligned}$$

   Concerning x:

$$\begin{aligned} x\pi' &= y\pi + s && \text{by definition of } \pi' \\ &= y\sigma' + s && \sigma' \text{ extends } \pi \\ &= x\sigma' && \sigma' \text{ is a solution of } M \\ &= x\sigma'' && x \neq y \text{ by Lemma 5} \end{aligned}$$

   Hence, σ'' extends π'.
   (b) σ'' is obviously a solution of $(\emptyset, P, \emptyset)$, since σ' is a solution of $(\emptyset, x = y + s \wedge P, \emptyset)$, and y does not occur in P by Lemma 5.

   As a conclusion, σ'' is a solution of (π', M'), σ and σ'' are equal except maybe on the elements of $\boldsymbol{y}$, hence σ is a solution of $\exists \boldsymbol{y}(\pi', M')$.
2. Let us assume that σ is a solution of $\exists \boldsymbol{y}(\pi', M')$, we shall prove that σ is a solution of $\exists y \exists \boldsymbol{y}(\pi, M)$. Since σ is a solution of $\exists \boldsymbol{y}(\pi', M')$, there exists σ' a solution of (π', M'), such that σ and σ' are equal except maybe on the elements of $\boldsymbol{y}$. We define σ'' as σ' except for y: $y\sigma'' = y\pi$.
   (a) It is clear that σ'' extends π since
      - $y\pi = y\sigma''$,
      - let x' be a variable in $\mathcal{D}om(\pi)$, which is different from y; since $x' \in \mathcal{D}om(\pi)$, $x' \neq x$.

$$\begin{aligned} x'\pi &= x'\pi' && x' \neq x,\ x' \neq y \text{ and by definition of } \pi' \\ &= x'\sigma' && \text{since } \sigma' \text{ extends } \pi' \\ &= x'\sigma'' && x' \neq y \text{ and by definition of } \sigma'' \end{aligned}$$

(b) σ'' is a solution of P since σ' is a solution of P, and y does not occur in P. σ'' is also a solution of $x = y + s$:

$$\begin{aligned} x\sigma'' &= x\sigma' && x \neq y \text{ and by definition of } \sigma'' \\ &= x\pi' && \sigma' \text{ extends } \pi' \text{ and } x \in \mathcal{D}om(\pi') \\ &= y\pi + s && \text{definition of } \pi' \\ &= y\sigma'' + s && \text{definition of } \sigma'' \end{aligned}$$

Hence σ'' is a solution of $x = y + s \wedge P$.

As a conclusion, σ'' is a solution of (π, M), and σ and σ'' are equal, except maybe on y and $\boldsymbol{y}$, hence σ is a solution of $\exists y \exists \boldsymbol{y}(\pi, M)$.

Theorem 1. *The above matching algorithm is terminating, sound and complete.*

4 Formal Proof

We now discuss the main difficulties encountered during the formal proof of the algorithm.

4.1 Terms and Equality Modulo

The terms are defined in COQ in a very natural way, from a set of symbols and a set of variables:

```
Parameter symbol : Set.
Parameter variable : Set.
Inductive term : Set := | Var : variable -> term
                        | Term : symbol -> list term -> term.
```

The symbols are equipped with an arity, and the notion of well-formed term is easily defined. The equality modulo is defined in several stages,

- first a single equational step at the top of the terms, but where the associativity axiom can be used in both ways,
- then, a single equational step at any position of the terms,
- and finally, the reflexive and transitive closure of a single step. The closure does not need to be symmetric, since the basic relation is itself symmetric.

Then, the notion of canonical form of a term is defined, with respect to the arity: the direct subterms of a term are recursively put in a canonical form, and if the top symbol is commutative, they are sorted, if the top symbol is AC, they are flattened and sorted:

```
Fixpoint canonical_form (t : term) : term :=
  match t with
  | Var _ => t
  | Term f l =>
    Term f (match arity f with
            | Free _ => map canonical_form l
            | C => quicksort (map canonical_form l)
            | AC => quicksort (flatten f (map canonical_form l) nil)
      end)
end.
```

It was quite easy to prove that when two terms are AC equal, then their canonical forms are identical. In order to prove the converse, however, we needed an extra assumption, that is that the terms are well-formed. Moreover, the proof was very long (2500 lines of COQ script vs 250 lines for the other direction of the equivalence), and uses 7 lemmas, each one of them corresponding with an axiom of the syntactic presentation of AC. It was then quite easy to prove that checking AC equality of two well-formed terms is decidable, by using the fact that checking the syntactic equality of two terms is decidable. The proof uses two extra assumptions, which are quite reasonable: the equality of variables and the equality of symbols are decidable.

4.2 Matching Modulo

The modeling part was quite short (300 lines of COQ script) for defining a single application of the rules $\mathbf{Merge}_S$, $\mathbf{Clash}_S$, $\mathbf{Merge}_P$, $\mathbf{Clash}_P$, **Solve**, **Clash**, **Dec**, $\mathbf{Dec}_C$, $\mathbf{AC}_{S\neq}$, $\mathbf{AC}_{S=}$, $\mathbf{AC}_{P\neq}$, $\mathbf{AC}_{P=}$, $\mathbf{AC}_{\neq}$, $\mathbf{AC}_{=}$, **AC** and **Fail**. The formal proof follows exactly the pattern given in Section 3, that is, a sequence of lemmas in order to show that the rules preserve the fact that a problem is well-formed (600 lines), they are decreasing (500 lines), sound (1000 lines) and complete (1700 lines).

Most of the proofs actually concern the permutations of the lists of arguments of the terms, using lemmas such as "if two non empty lists are equal up to permutation, after removing the same element from them, the results are still equal up to permutation". The proofs are quite long since there are many subcases, even more than expected from the large number of inference rules. For example, for an informal proof, it is enough to distinguish whether or not a term has a top symbol equal to the AC function symbol of interest. In a formal proof, we have first to handle the case of a variable term, and then the non-variable case which split itself into two cases according to the top symbol. The convention that an AC function symbol applied to a list of arguments actually denotes the element of the list, if the list has length 1 also double the number of cases in a formal proof.

There are also some technicalities due to the fact that we have model the non-deterministic rules such as **AC** in order to avoid selecting several the same subterm of the subject term multiple times should it occur more than once, since this would produce multiple instances of the same new problem.

The extraction part has not been formally proven yet.

5 Conclusion

We have proposed here an AC matching algorithm presented by inference rules. An intrinsic advantage of this algorithm is that it can handle directly a subject term which contains some variables, even when these variables are shared in the pattern: no preliminary renaming has to be done. This is important in the framework of completion, for example, since the rewriting for normalization is performed over non-ground terms which may share some variables with the rewrite rules. An other advantage of this algorithm is that it is simple enough to be modeled in COQ and its core is certified: what remains to be proved is the extraction step, and the Dickson's lemma in order to embed the application of a rule into a loop. Once this will be done, one will be able to extract a certified code of the formal proof. The algorithm has been implemented in C*i*ME, and behaves reasonably well in practice. It has to be mentioned that even if the code was quite short and well-understood, the formal proof of the algorithm has revealed that the implemented version was not complete, since the part corresponding to the rule $\mathbf{AC}_{P=}$ was missing. As a conclusion, this paper presents the first certified AC matching algorithm, at least to the author's knowledge. This should be a first step to a certified AC unification algorithm, since a lot of preliminary work has been done.

References

1. Leo Bachmair, Ta Chen, and I.V. Ramakrishnan. Associative-commutative discrimination nets. In M. C. Gaudel and J.-P. Jouannaud, editors, *4th International Joint Conference on Theory and Practice of Software Development*, volume 668 of *Lecture Notes in Computer Science*, pages 61–74, Orsay, France, April 1993. Springer-Verlag.
2. Ta Chen and Siva Anantharaman. Storm: A many-to-one associative-commutative matcher. In Jieh Hsiang, editor, *6th International Conference on Rewriting Techniques and Applications*, volume 914 of *Lecture Notes in Computer Science*, Kaiserslautern, Germany, April 1995. Springer-Verlag.
3. Jim Christian. Flatterms, Discrimination Nets, and Fast Term Rewriting. *Journal of Automated Reasoning*, 10(1):95–113, February 1993.
4. Nachum Dershowitz and Jean-Pierre Jouannaud. Rewrite systems. In J. van Leeuwen, editor, *Handbook of Theoretical Computer Science*, volume B, pages 243–320. North-Holland, 1990.
5. Steven Eker. Associative-Commutative Matching Via Bipartite Graph Matching. *Computer Journal*, 38(5):381–399, 1995.
6. J.-M. Hullot. Associative commutative pattern matching. In *Proc. 6th IJCAI (Vol. I), Tokyo*, pages 406–412, August 1979.
7. Emmanuel Kounalis and Denis Lugiez. Compiling pattern matching with associative-commutative functions. In S. Abramsky and T. S. E. Maibaum, editors, *Colloquium on Trees in Algebra and Programming*, volume 493 of *Lecture Notes in Computer Science*, pages 57–73. Springer-Verlag, April 1991.
8. Pierre-Etienne Moreau and Hélène Kirchner. A Compiler for Rewrite Programs in Associative-Commutative Theories. In C. Palamidessi, H. Glaser, and K. Meinke, editors, *ALP/PLILP: Principles of Declarative Programming*, volume 1490 of *Lecture Notes in Computer Science*, pages 230–249, Pisa (Italy), September 1998. Springer-Verlag.
9. The Coq Development Team. *The Coq Proof Assistant Reference Manual – Version V7.4*, January 2003. `http://coq.inria.fr`.

Matchbox: A Tool for Match-Bounded String Rewriting

Johannes Waldmann

Hochschule für Technik, Wirtschaft und Kultur (FH) Leipzig
Fachbereich IMN, Postfach 30 11 66, D-04251 Leipzig, Germany
waldmann@imn.htwk-leipzig.de

Abstract. The program Matchbox implements the exact computation of the set of descendants of a regular language, and of the set of non-terminating strings, with respect to an (inverse) match-bounded string rewriting system. Matchbox can search for proof or disproof of a Boolean combination of match-height properties of a given rewrite system, and some of its transformed variants. This is applied in various ways to search for proofs of termination and non-termination. Matchbox is the first program that delivers automated proofs of termination for some difficult string rewriting systems.

1 Introduction

The theory of match bounded string rewriting, recently developed in [4, 6–8], allows to apply methods from formal language theory to problems of string rewriting. The basic result is a decomposition theorem. It implies effective preservation of regular languages, and effective regularity of the set of non-terminating strings w.r.t. the inverse system. The program Matchbox contains implementations of all of the fundamental algorithms, and allows to apply them in various settings. Given a rewrite system, Matchbox searches for proofs and disproofs of a Boolean combination of match-height properties of the input, and possibly transformed versions of it.

Since match-bounded systems are terminating, a verification of match-boundedness is a proof of termination. This method can be extended by verifying match-boundedness for right hand sides of forward closures only. Another way to use match-bounds for deciding termination is to verify match-boundedness for the inverse system (with left and right hand sides swapped). For inverse match-bounded systems, the set of non-terminating strings is effectively regular, so termination can be decided. Again, this idea can be extended by using the inverse of the ancestor system instead.

For the exact computation of the set of descendants, represented as a finite automaton, the decomposition theorem gives a procedure that has high complexity (both in the run time and in the source text). Recently, Zantema developed a very efficient algorithm [14] that approximates the rewrite closure of an automaton. If this computation halts, termination can be inferred as well – even if the approximated match-bound is too large. Matchbox implements a variant of this algorithm.

V. van Oostrom (Ed.): RTA 2004, LNCS 3091, pp. 85–94, 2004.

2 Theoretical Background

We cite here the basic definitions and results on match-bounded string rewriting. Standard notation from rewriting and formal language theory is assumed. If R is a rewriting system over Σ, we often denote its rewrite relation $\to_R$ on Σ^* by R as well. If the inverse R^- of a rewriting system R has property p, we also say that R has property *inverse-p*.

We annotate positions in strings by natural numbers that indicate their *matching height*. Positions in a reduct will get height $h+1$ if the minimal height of all positions in the corresponding redex is h.

Define the morphisms $\text{lift}_c : \Sigma^* \to (\Sigma \times \mathbb{N})^*$ for $c \in \mathbb{N}$ by $\text{lift}_c : a \mapsto (a, c)$, $\text{base} : (\Sigma \times \mathbb{N})^* \to \Sigma^*$ by $\text{base} : (a, c) \mapsto a$, and $\text{height} : (\Sigma \times \mathbb{N})^* \to \mathbb{N}^*$ by $\text{height} : (a, c) \mapsto c$.

For a string rewriting system R over an alphabet Σ with $\epsilon \notin \text{lhs}(R)$ we define the system

$$\text{match}(R) = \{\ell' \to \text{lift}_c(r) \mid (\ell \to r) \in R, \text{base}(\ell') = \ell, c = 1 + \min(\text{height}(\ell'))\}$$

over alphabet $\Sigma \times \mathbb{N}$. For example, the system $\text{match}(R)$ for $R = \{aa \to aba\}$ contains the rules $\{a_0a_0 \to a_1b_1a_1, a_0a_1 \to a_1b_1a_1, a_1a_0 \to a_1b_1a_1, a_1a_1 \to a_2b_2a_2, a_0a_2 \to a_1b_1a_1, \ldots\}$, where x_c is shorthand for $(x, c) \in \Sigma \times \mathbb{N}$. Note that $\text{match}(R)$ is an infinite system.

A string rewriting system R over Σ with $\epsilon \notin \text{lhs}(R)$ is called *match-bounded* for $L \subseteq \Sigma^*$ by $c \in \mathbb{N}$ if $\max(\text{height}(x)) \leq c$ for every $x \in \text{match}(R)^*(\text{lift}_0(L))$. If we omit L, then it is understood that $L = \Sigma^*$.

Proposition 1 (Decomposition Theorem, [10]). *For each match-bounded string rewriting system R over Σ there effectively are an alphabet $\Gamma \supseteq \Sigma$, a finite substitution $s : \Sigma^* \to \Gamma^*$, and a context-free string rewriting system C such that $R^* = (s \circ C^{-*})|_\Sigma$.*

Here, we call C context-free if for each rule $(l, r) \in C$ we have $|l| \leq 1$. The decomposition implies that R^* effectively preserves regularity of languages, and that R^{-*} effectively preserves context-freeness. As an application, we get:

Proposition 2 ([4]). *Given a string rewriting system R, a regular language L, and $c \in \mathbb{N}$, it is decidable whether R is match-bounded by c for L.*

For a relation ρ over U, let $\text{Inf}(\rho) = \{x \in U \mid \exists^\infty y : \rho(x, y)\}$ denote the set of elements with infinitely many descendants.

Proposition 3 ([4, 7]). *If R is inverse match-bounded, then* $\text{Inf}(R^*)$ *is effectively regular, and termination of R is decidable.*

For a string rewriting system R over Σ, the set of *forward closures* [11, 2] $\text{FC}(R) \subseteq \Sigma^* \times \Sigma^*$ is defined as the least set containing R such that

- if $(u, v) \in \text{FC}(R)$ and $v \to_R w$, then $(u, w) \in \text{FC}(R)$ (*inside reduction*), and
- if $(u, v\ell_1) \in \text{FC}(R)$ and $(\ell_1\ell_2 \to r) \in R$ for strings $\ell_1, \ell_2 \in \Sigma^+$, then $(u\ell_2, vr) \in \text{FC}(R)$ (*right extension*).

Let $\mathrm{RFC}(R) = \{y \mid \exists x : (x, y) \in \mathrm{FC}(R)\}$ denote the set of right hand sides of forward closures. It is known [1] that a string rewriting system R is terminating on Σ^* if and only if R is terminating on $\mathrm{RFC}(R)$.

We can obtain $\mathrm{RFC}(R)$ as a set of descendants modulo the rewriting system

$$\mathrm{rfc}(R) = R \cup \{\ell_1\# \to r \mid (\ell_1\ell_2 \to r) \in R, \ell_1, \ell_2 \in \Sigma^+\}$$

over $\Sigma \cup \{\#\}$, where right extension is simulated via the end-marker $\#$. This leads to the following termination criterion.

Proposition 4 ([5, 6]). *If* $\mathrm{rfc}(R)$ *is match-bounded for* $\mathrm{rhs}(R) \cdot \#^*$, *then* R *is terminating.*

Likewise, we can restrict the language when computing heights for the inverse system. The *ancestor* system of R is defined as

$$\begin{array}{rcl} & & \cup \{ \ell \to \langle r_2 \mid (\ell \to r_1 r_2) \in R, \quad r_1, r_2 \in \Sigma^+ \} \\ \mathrm{ancestor}(R) & = & R \cup \{ \ell \to r_1\rangle \mid (\ell \to r_1 r_2) \in R, \quad r_1, r_2 \in \Sigma^+ \} \\ & & \cup \{ \ell \to \langle r_2\rangle \mid (\ell \to r_1 r_2 r_3) \in R, r_1, r_2, r_3 \in \Sigma^+ \}, \end{array}$$

where $\rangle$ and $\langle$ are two fresh letters.

Proposition 5 ([8]). *If* $\mathrm{ancestor}(R)$ *is inverse match-bounded for* $\langle^* \cdot \mathrm{lhs}(R) \cdot \rangle^*$, *then termination of* R *is decidable.*

3 How to Use Matchbox

As core functionalities, `Matchbox` inputs a string rewriting system R over Σ and a regular expression for a language L over Σ, and it computes (an approximation to) an automaton for $\mathrm{match}(R)^*(\mathrm{lift}_0(L))$, as well as an automaton for $\mathrm{Inf}(R^{-*})$.

These basic computations actually consist of a sequence of steps. Basic computations can be modified and combined, using a very expressive control language.

The output can be rendered as an HTML document that also includes visualizations of automata graphs, obtained by tools from the GraphViz suite.

3.1 Stepwise Computation

Given R and L, `Matchbox` computes a sequence of finite automata $A_0, A_1, \ldots$ over $\Sigma \times \mathbb{N}$ such that $L_0 = L(A_0) \subseteq L(A_1) \subseteq \ldots \subseteq \mathrm{match}(R)^*(L_0)$, writing L_0 for $\mathrm{lift}_0(L)$. Here, $L(A_{k+1}) = R_k^*(L(A_k))$ where R_k consists of exactly those rules of $\mathrm{match}(R)$ whose left hand side occurs in $L(A_k)$, viz.

$$R_k = \mathrm{match}(R) \cap (\mathrm{factors}(L(A_k)) \times (\Sigma \times \mathbb{N})^*).$$

Note that each R_k is finite and terminating.

If $L(A_n) = \mathrm{match}(R)^*(\mathrm{lift}_0(L))$ for some n, then R is match-bounded for L, and A_n is a certificate.

`Matchbox` also computes $I_k = \mathrm{Inf}(R_k^{-*})$ by Proposition 3. A non-empty I_k certifies non-termination of R^-. If $I_k = \emptyset$ and A_k is closed under rewriting, then this proves termination of R^-.

3.2 Example: Computing the Set of Descendants

In the following examples, we describe how `Matchbox` is called from the command line. Equivalent input options are available in the web interface.

Here is how to compute $\text{match}(R)^*(L_0)$ for Zantema's System $R = \{a^2b^2 \to b^3a^3\}$ and $L = (a^2b^4)^*$:

```
Matchbox  aabb bbbaaa --control="bound match" --for="(a^2 b^4)^*"
```

(Note that arguments have to be quoted in order to protect them from being evaluated by the operating system shell.) The output is

```
[("aabb","bbbaaa")]  : True
    value of 'bound match'
    is True, because
        R is match bounded by 2
            for R = [ ( "aabb", "bbbaaa")]
            and L = (a^2 b^4)^*
            with size of closure automaton (non-det, det): ( 21, 21)
```

The following automaton for $\text{match}(R)^*(L_0)$ is obtained, which is rendered in a compressed notation where arrows are labelled by strings (instead of letters). The start and end state is double-circled. By inspection, it can be verified that the automaton is closed under rewriting.

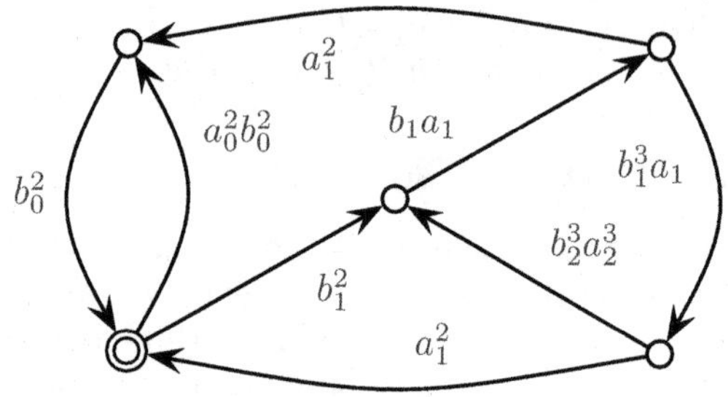

In many cases, match-boundedness can be constructively disproved by regular sets of self-embedding witnesses [4]. Option `--embed=`n (for $n > 0$) instructs `Matchbox` to try and construct such a set. If one is found, `bound match` returns `False`. Otherwise, the proof attempt will eventually fail (after breaking the limits that have been set as configuration parameters).

3.3 Regular Expressions

The following notation is available for regular expressions:

- atomic expressions:
  - lowercase letters and digits stand for themselves
  - there are some identifiers, starting with an upper case letter:
    - `Eps` denotes ϵ,
    - `Sigma` stands for the set Σ of letters occuring in the rewriting system,
    - and `All` denotes Σ^*
  - parenthesized compound expressions

- compound expressions, using binary operators (listed in order of precedence):
  - powers: L `^` e for e a natural number, or `*`, or `+`,
  - concatenation: L_1 `*` L_2, but the `*` operator can also be omitted,
  - left and right quotient, shuffle: `\`, `/`, `$`
  - intersection, union, difference and symmetric difference: `&`, `+`, `-`, `<>`

`Matchbox` computes the minimal deterministic automaton for the corresponding language.

3.4 Example: Computing Inf(R^*)

`Matchbox` can check whether $\emptyset = \text{Inf}(R^*)$ as well.

```
Matchbox aba baaab --control="empty-inf match"
```

The output is

```
[("aba","baaab")] : False
    value of ‘empty-inf match’
    is False, because
        Inf (R^-*) contains Sigma^* (aaba + abaa) Sigma^*
            for R = [ ( "baaab", "aba")]
            and L = Sigma^*
```

The language Inf(R^*) has been determined via the decomposition theorem, but it can be verified independently, since any non-empty regular language I with $I \subseteq R^-(I)$ is an effective certificate of non-termination, because it contains only non-terminating words.

To obtain a representation as in the example, `Matchbox` computes the set K of factor-minimal words of I (no proper factor of a K word is in I) by $K = I \setminus (\Sigma \cdot I \cup I \cdot \Sigma)$. (Often K is found to be finite.) Then $\Sigma^* K \Sigma^*$ is a nice representation for I (or for a superset of I that still is a certificate).

3.5 Atomic Computations: Predicates and Transformers

With `bound match` and `empty-inf match`, we have seen two examples of atomic computations. In general, an *atomic computation* is specified by

- a *predicate* that is to be applied to a pair (R, L) of a string rewriting system R and regular language L,
- and a (possibly empty) sequence of transformations that are applied to (R, L) beforehand.

Available predicates are

- `bound match`: prove or disprove that R is match-bounded for L (for length-preserving systems, also `bound change` is available [13])
- `empty-inf match`: prove or disprove that Inf(R^{-*}) is empty
- `loop`: prove that R has a loop (by enumeration of forward closures)

The following transformers can be applied:

- `reverse`: reverse the rewriting system, replace R by reverse(R)
- `inverse`: invert the rewriting system, replace R by R^-
- `rfc`: replace R by the system rfc(R), and L by rhs(R) $\cdot \#^*$
- `ancestor`: replace R by the system ancestor(R), and L by $\langle^* \cdot \mathrm{lhs}(R) \cdot \rangle^*$

For instance, match heights for right hand sides of forward closures of the reversed system are computed by

```
Matchbox babaaa aaababab --control="bound match . rfc . reverse"
```

Here, reverse(R) is RFC-match-bounded by 1, but R does not seem to be RFC-match-bounded (altough `Matchbox` cannot prove this).

3.6 The Control Language

The argument to the `--control` option is a Boolean expression that describes a combination of properties of a rewrite system. `Matchbox` tries to decide the truth value of the expression.

For instance, the command

```
Matchbox -E Nogrid --control="not bound match . rfc  \
     and approx match . rfc . reverse"
```

enumerates one-rule non-grid systems [3] and produces output like this:

```
2 : [("aba","aaabb")] : False
    value of 'not bound match . rfc and approx match . rfc . reverse'
    is False, because
        value of 'not bound match . rfc'
        is False, because
            value of 'bound match . rfc'
            is True, because
                R is match bounded by 0
                   for R = [ ( "aba", "aaabb")
                           , ( "a#", "aaabb") , ( "ab#", "aaabb")]
                   and L = aaabb #^*
                   with size of closure automaton (non-det, det): ( 6, 6)
```

Any number of atomic `Matchbox` proof searches can be combined, using the standard boolean connectives (not, and, or). Combinations are evaluated concurrently. The scheduling is done in such a way that each (interesting) subcomputation executes one step in turn.

The evaluation uses shortcuts: as soon as one `False` occurs in a conjunction, the result is `False` and all its subcomputations are discarded; likewise for a `True` in an alternative.

A special *parallel* connective `par` is available. In p `par` q, both proof attempts are started, and as soon as one of them returns a value, this is taken as the value of the connective, and the other attempt is discarded.

This will give a well-defined result only in case computations are compatible, i. e. if it cannot happen that one returns `True`, and the other would later return `False`. Typically, one would use `par` to combine attempts to prove the same thing (e. g., termination) via different methods, as in

```
Matchbox aabb bbbaaa --control="(not loop) par \
   (bound match . rfc) par (bound match . rfc . reverse)"
```

Note that `not loop` will return `False` as soon as it finds a loop (but it will never say `True` because a simple enumeration cannot prove absence of loops), and the `bound match . rfc` proof attempts will return `True` on success.

The following transformation is also provided: on input R, `Matchbox` can try to prove termination of R by looking at all systems obtained by dropping common prefixes and suffixes. For instance, the command

```
Matchbox abab aaa aaaababab --factors
   --control="bound match . rfc  par  bound match . rfc . reverse"
```

will find the reverse-RFC-match-bound 1 for $babaaa \to aaababab$. Note that in this case, neither R nor reverse(R) are RFC-match-bounded.

3.7 Approximate Computations of Termination Certificates

For a string rewriting system R over Σ, and a language L over Σ, a finite automaton A over $\Sigma \times \mathbb{N}$ can be employed as a certificate of termination if

- (start condition): $\text{lift}_0(L) \subseteq L(A)$, and
- (closure under match(R)): for each rule $(l \to r) \in \text{match}(R)$, and for each redex path $p \xrightarrow{l} q$ in A, there is a corresponding reduct path $p \xrightarrow{r} q$ between the same pair of states.

As proven by Zantema's recent implementation in TORPA, such automata can be constructed directly (i. e., without using the decomposition theorem for some subset of match(R)) from an automaton for $\text{lift}_0(L)$. Following the above definition, if a reduct path is not there, it has to be added. In general, this will require to add states as well, so the process is not guaranteed to stop (even if R is match-bounded).

When approximating, `Matchbox` tries to re-use existing states as much as possible. It is looking for reduct decompositions $r = xyz$ such that there are paths $p \xrightarrow{x} p'$ and $q' \xrightarrow{z} q$ in A, so that only the path $p' \xrightarrow{y} q'$ has to be added.

`Matchbox` adds a shortest non-empty such path $p' \xrightarrow{y} q'$ for which the following restriction holds: if at least one of x or z is non-empty, then it is not allowed that there already is a path $p' \xrightarrow{y'} q'$ in A with $\text{base}(y) = \text{base}(y')$. It is felt that the presence of such a path (that usually has lower height annotations) leads to non-termination of the algorithm.

While Zantema's approximation seems very well suited to RFC-match-bounds, the `Matchbox` approximation described here is also applicable to standard match bounds, as examples show. For instance,

```
Matchbox aabb bbbaaa --control="approx match" --chunk=2
```

computes a termination certificate. The option `chunk=`n instructs `Matchbox` to add at most n reduct paths in one approximation step. (If `--chunk` is missing, then all missing reduct paths are added in one step.)

It is interesting to note that in the above example, a certificate automaton with 51 states is obtained, while the exact automaton for match$(R)^*(\Sigma_0^*)$ has 85 states.

3.8 Configuration Parameters

`Matchbox` understands many more options. The command `Matchbox -h` gives a complete list.

To control the output, `-T` requests a computation trace (as plain text to stdout), and `-D` will write HTML output, including automata graphs that are rendered with a graph layout program, which can be set by `--layout=`p, where $p \in \{$`dot`,`neato`$\}$.

To control the computation, the maximum number of iterations (for each subcomputation) can be set with `-c` n, and the maximum number of states of intermediate automata by `-p` n. Proof attempts will fail if they reach these bounds.

4 Software Engineering Aspects

`Matchbox` has been designed for flexibility and extendibility. For the user, it provides a rather expressive control language, and for the programmer, its source text (approx. 10 kLOC) is factorized into approx. 200 hierarchical modules.

Polymorphic data types and higher order functions are used to support orthogonal design and re-usability. In fact, half of the modules (those for input (parsing), (web) output, and finite automata) are being developed and used in another large project [12].

`Matchbox` relies on standard libraries and tools: the `Parsec` parser combinator library by Daan Leijen; the pretty printing library by John Hughes and Simon Peyton Jones; the `Network.CGI` library by Erik Meijer, Sven Panne and Andy Gill; the `DrIFT` generic programming tool by Noel Winstanley. Malcolm Wallace and John Meacham; and the GraphViz toolkit by Emden Gansner et. al.

`Matchbox` is implemented in Haskell, the standard polymorphic higher order lazy functional programming language, using some extensions (existential types, multi parameter type classes, and functional dependencies). It can be compiled with `ghc`, the Glasgow Haskell Compiler (version 6.0 or later), so it runs on any `ghc`-supported platform. (It was tested on GNU Linux and Solaris.)

The `Matchbox` web page is at `http://theo1.informatik.uni-leipzig.de/matchbox/`. It contains instructions for downloading, compiling, and using the program, and there is also a web interface to an online evaluation copy.

5 Conclusion: Related Work and Further Research

Matchbox has been written to explore match-bounded string rewriting, and it is not intended to be a general-purpose termination prover. So a direct comparison between such provers on the one hand, and Matchbox's ability to generate termination proofs on the other, would miss the point.

Match-bounded rewriting can be seen as a method that analyzes local overlap patterns of rule applications with utmost accuracy. Therefore it "wins" w.r.t. other methods that cannot use such information. On the other hand, Matchbox "loses" in many termination problems that can be handled by standard methods like path orders, interpretations, labellings, or dependency pairs.

As match bounds seem orthogonal to standard methods, a termination prover that combines both should be rather more powerful – as TORPA [14] shows.

We plan to extend Matchbox to term rewriting; mainly to compute sets of descendants of regular tree languages, but hopefully giving new ways of proving termination on the way.

Acknowledgements

Matchbox accompanies joint research work with Alfons Geser and Dieter Hofbauer. We thank Hans Zantema for implementing RFC-match-bounds in TORPA, and for telling us of his approximation algorithm. On the implementation side, I gratefully acknowledge the support by the Haskell community, providing compilers and libraries. I especially thank Mirko Rahn for testing Matchbox's finite automata routines.

References

1. N. Dershowitz. Termination of linear rewriting systems. In S. Even and O. Kariv (Eds.), *Proc. 8th Int. Coll. Automata, Languages and Programming ICALP-81, Lecture Notes in Comput. Sci.* Vol. 115, pp. 448–458. Springer-Verlag, 1981.
2. N. Dershowitz. Termination of rewriting. *J. Symbolic Comput.*, 3(1–2):69–115, 1987.
3. A. Geser. Decidability of Termination of Grid String Rewriting Rules. *SIAM J. Comput.* 31(4): 1156-1168, 2002.
4. A. Geser, D. Hofbauer and J. Waldmann. Match-bounded string rewriting systems. In B. Rovan and P. Vojtas (Eds.), *Proc. 28th Int. Symp. Mathematical Foundations of Computer Science MFCS-03, Lecture Notes in Comput. Sci.* Vol. 2747, pp. 449-459. Springer-Verlag, 2003.
5. A. Geser, D. Hofbauer, and J. Waldmann. Match-bounded string rewriting systems and automated termination proofs. In A. Rubio (Ed.), *Proc. 6th Int. Workshop on Termination WST-03*, Technical Report DSIC-II/15/03, Universidad Politécnica de Valencia, Spain, pp. 19–22, 2003.
6. A. Geser, D. Hofbauer and J. Waldmann. Match-bounded string rewriting systems. NIA Report 2003-09, National Institute of Aerospace, Hampton, VA, USA. Available at http://research.nianet.org/~geser/papers/nia-matchbounded.html.

7. A. Geser, D. Hofbauer and J. Waldmann. Termination proofs for string rewriting systems via inverse match-bounds. NIA Report 2003-XX, National Institute of Aerospace, Hampton, VA, USA. Available at `http://research.nianet.org/~geser/papers/nia-inverse.html`.
8. A. Geser, D. Hofbauer and J. Waldmann. Deciding Termination for Ancestor Match-Bounded String Rewriting Systems. submitted to RTA-04.
9. T. N. Hibbard. Context-limited grammars. *J. ACM*, 21(3):446–453, 1974.
10. D. Hofbauer and J. Waldmann. Deleting string rewriting systems preserve regularity. In Z. Ésik and Z. Fülöp (Eds.), *Proc. 7th Int. Conf. Developments in Language Theory DLT-03*, *Lecture Notes in Comput. Sci.* Vol. 2710, pp. 337-348. Springer-Verlag, 2003.
11. D. S. Lankford and D. R. Musser. A finite termination criterion. Technical Report, Information Sciences Institute, Univ. of Southern California, Marina-del-Rey, CA, 1978.
12. M. Rahn and J. Waldmann. The Leipzig autotool System for Grading Student Homework In M. Hanus, S. Krishnamurthi, and S. Thompson (Eds.), *Proc. Functional and Declarative Programming in Education FDPE-02*. Technical Report No. 0210, Universitt Kiel, 2002.
13. B. Ravikumar. Peg-solitaire, string rewriting systems and finite automata. In H.-W. Leong, H. Imai, and S. Jain (Eds.), *Proc. 8th Int. Symp. Algorithms and Computation ISAAC-97*, *Lecture Notes in Comput. Sci.* Vol. 1350, pp. 233–242. Springer-Verlag, 1997.
14. H. Zantema, TORPA: Termination Of Rewriting Proved Automatically (version 1.2). Available at `http://www.win.tue.nl/~hzantema/torpa.html`, 2004.

TORPA:
Termination of Rewriting Proved Automatically

Hans Zantema

Department of Computer Science, TU Eindhoven
P.O. Box 513, 5600 MB, Eindhoven, The Netherlands
h.zantema@tue.nl

Abstract. The tool TORPA (Termination of Rewriting Proved Automatically) can be used to prove termination of string rewriting systems (SRSs) fully automatically. The underlying techniques include semantic labelling, polynomial interpretations, recursive path order, the dependency pair method and match bounds of right hand sides of forward closures.

1 Introduction

Lots of techniques have been developed for proving termination of rewriting. In the last few years work in this area concentrates on proving termination automatically: the development of tools by which a rewrite system can be entered and by which a termination proof is generated fully automatically.

The tool TORPA has been developed by the author. The present version only works on string rewriting. On the one hand this is a strong restriction compared to general term rewriting. On the other hand string rewriting is a natural and widely accepted paradigm with full computational power.

There are many small SRSs from a wide variety of origins for which termination is proved fully automatically by TORPA within a fraction of a second. For some of them all other techniques for (automatically) proving termination seem to fail.

The main feature of TORPA is that an SRS is given as input and that TORPA generates a proof that this SRS is terminating. This proof is given in text. It is given in such a way that any human familiar with the basic techniques used in TORPA as they are described here, can read and check the proof. The five basic techniques used are

- polynomial interpretations ([10]),
- recursive path order ([5]),
- semantic labelling ([12]),
- dependency pairs ([1]), and
- RFC-match-bounds ([7]).

Polynomial interpretations, recursive path order and RFC-match-bounds are direct techniques to prove termination, while semantic labelling and dependency

V. van Oostrom (Ed.): RTA 2004, LNCS 3091, pp. 95–104, 2004.

pairs are techniques for transforming an SRS to another one in such a way that termination of the original SRS can be concluded from (relative) termination of the transformed SRS. Semantic labelling is the most significant transformation used in TORPA. For very small SRSs, in particular single rules, RFC-match-bounds provide the most powerful technique. In this system description we describe how the given five techniques are applied, and give some examples of proofs as they are generated by TORPA. For more examples and a full exposition of the theory including proofs of the theorems, but excluding the most recent part on RFC-match-bounds, we refer to [14].

Other tools like TTT ([9]), CiME ([4]) and AProVE ([6]) combine dependency pairs and path orders, too, and apply them much more involved than we do. The tool Termptation ([3]) is based on the semantic path order. They turn out to be the best tools for proving termination of term rewriting of the moment. However, applied to SRSs all of these tools are much weaker than TORPA.

A completely different approach is RFC-match-boundedness as introduced in [7], and implemented in the tool Matchbox by Johannes Waldmann. However, Matchbox only involves techniques related to match-boundedness, and for typically hard examples like $aabb \to bbbaaa$ TORPA is much more efficient than the Matchbox version from before January 2004 when the authors of [7] were informed about the heuristics implemented in TORPA.

TORPA is freely available in two versions:

- A full executable version written in Delphi with a graphical user interface, including facilities for editing SRSs. This runs directly in any Windows environment.
- A plain version written in standard Pascal to be used on other platforms, or for running batches.

Downloading is done from

`http://www.win.tue.nl/~hzantema/torpa.html`

where also some more detailed information is given. The present version is version 1.2. In the earlier version 1.1 RFC-match-boundedness was not implemented.

The structure of this paper is as follows. First we give preliminaries of string rewriting and relative termination. Then in five consecutive sections we discuss each of the basic techniques. In the final section we give conclusions and discuss further research. To distinguish text generated by TORPA from the text of the paper the text generated by TORPA is always given in `typewriter font`.

2 Preliminaries

A string rewrite system (SRS) over an alphabet Σ is a set $R \subseteq \Sigma^+ \times \Sigma^*$. Elements $(l, r) \in R$ are called *rules* and are written as $l \to r$; l is called the left hand side (lhs) and r is called the right hand side (rhs) of the rule. In TORPA format the arrow $\to$ is written by the two symbols `->`. A string $s \in \Sigma^*$ rewrites to a string $t \in \Sigma^*$ with respect to an SRS R, written as $s \to_R t$ if strings $u, v \in \Sigma^*$ and a rule $l \to r \in R$ exist such that $s = ulv$ and $t = urv$.

An SRS R is called *terminating* (strongly normalizing, $\mathsf{SN}(R)$) if no infinite sequence $t_1, t_2, t_3, \ldots$ exists such that $t_i \to_R t_{i+1}$ for all $i = 1, 2, 3, \ldots$. An SRS R is called *terminating relative to* an SRS S, written as $\mathsf{SN}(R/S)$, if no infinite sequence $t_1, t_2, t_3, \ldots$ exists such that

- $t_i \to_{R \cup S} t_{i+1}$ for all $i = 1, 2, 3, \ldots$, and
- $t_i \to_R t_{i+1}$ for infinitely many values of i.

The notation R/S is also used for the rewrite relation $\to_S^* \cdot \to_R \cdot \to_S^*$; clearly $\mathsf{SN}(R/S)$ coincides with termination of this rewrite relation. By definition $\mathsf{SN}(R/S)$ and $\mathsf{SN}(R/(S \setminus R))$ are equivalent. Therefore we will use the notation $\mathsf{SN}(R/S)$ only for R and S being disjoint. In writing an SRS $R \cup S$ for which we want to prove $\mathsf{SN}(R/S)$ we write the rules of R by $l \to r$ and the rules of S by $l \to^= r$. In TORPA format the arrow $\to^=$ is written by the three symbols `->=`. The rules from R are called *strict* rules; the rules from S are called *non-strict* rules. Clearly $\mathsf{SN}(R/\emptyset)$ and $\mathsf{SN}(R)$ coincide.

Our first theorem is very fruitful for stepwise proving (relative) termination.

Theorem 1. *Let R, S, R' and S' be SRSs for which*

- $R \cup S = R' \cup S'$ *and* $R \cap S = R' \cap S' = \emptyset$,
- $\mathsf{SN}(R'/S')$ *and* $\mathsf{SN}((R \cap S')/(S \cap S'))$.

Then $\mathsf{SN}(R/S)$.

Theorem 1 is applied in TORPA as follows. In trying to prove $\mathsf{SN}(R/S)$ it is tried to split up $R \cup S$ into two disjoint parts R' and S' for which $R' \neq \emptyset$ and $\mathsf{SN}(R'/S')$. If this succeeds then the proof obligation $\mathsf{SN}(R/S)$ is weakened to $\mathsf{SN}((R \cap S')/(S \cap S'))$, i.e., all rules from R' are removed. This process is repeated as long as it is applicable. If after a number of steps $R \cap S' = \emptyset$ then $\mathsf{SN}((R \cap S')/(S \cap S'))$ trivially holds and the desired proof has been given.

Next we consider reversing strings. For a string s write s^{rev} for its reverse. For an SRS R write $R^{\mathsf{rev}} = \{\, l^{\mathsf{rev}} \to r^{\mathsf{rev}} \mid l \to r \in R \,\}$.

Lemma 1. *Let R and S be disjoint SRSs. Then* $\mathsf{SN}(R/S)$ *if and only if* $\mathsf{SN}(R^{\mathsf{rev}}/S^{\mathsf{rev}})$.

Lemma 1 is strongly used in TORPA: if $\mathsf{SN}(R/S)$ has to be proved then all techniques are not only applied on R/S but also on $R^{\mathsf{rev}}/S^{\mathsf{rev}}$.

3 Polynomial Interpretations

The ideas of (polynomial) interpretations go back to [10, 2]. First we give the underlying theory for doing this for string rewriting.

Let A be a non-empty set and Σ be an alphabet. Let ϵ denote the empty string in Σ^*. If $f_a : A \to A$ has been defined for every $a \in \Sigma$ then $f_s : A \to A$ is defined for every $s \in \Sigma^*$ inductively as follows:

$$f_\epsilon(x) = x, \quad f_{as}(x) = f_a(f_s(x)), \quad \text{for every } x \in A, a \in \Sigma, s \in \Sigma^*.$$

Theorem 2. *Let A be a non-empty set and let $>$ be a well-founded order on A. Let $f_a : A \to A$ be strictly monotone for every $a \in \Sigma$, i.e., $f_a(x) > f_a(y)$ for every $x, y \in A$ satisfying $x > y$. Let R and S be disjoint SRSs over Σ such that $f_l(x) > f_r(x)$ for all $x \in A$ and $l \to r \in R$, and $f_l(x) \geq f_r(x)$ for all $x \in A$ and $l \to r \in S$. Then* $\mathsf{SN}(R/S)$.

In the general case this approach is called *monotone algebras* ([11, 13]). In case A consists of all integers $> N$ with the usual order for some number N, and the functions f_a are polynomials this approach is called *polynomial interpretations.*

In TORPA only three distinct polynomials are used: the identity, the successor $\lambda x \cdot x + 1$, and $\lambda x \cdot 10x$. For every symbol a one of these three polynomials is chosen, and then it is checked whether using Theorem 2 gives rise to $\mathsf{SN}(R/S)$ for some non-empty R, where R consists of the rules for which '$>$' is obtained. If so, then by using Theorem 1 the proof obligation is weakened and the process is repeated until no rules remain, or only non-strict rules. As a first example consider the two rules $ab \to ba, a \to^= ca$, i.e., $\mathsf{SN}(R/S)$ has to be proved where R consists of the rule $ab \to ba$ and S consists of the rule $a \to ca$. Now TORPA yields:

```
Choose polynomial interpretation:
a: lambda x.10x,   b: lambda x.x+1,   rest identity
remove: ab -> ba
Relatively terminating since no strict rules remain.
```

Here for f_a and f_b respectively $\lambda x \cdot 10x$ and the successor are chosen. Since $f_{ab}(x) = f_a(f_b(x)) = 10(x+1) > 10x+1 = f_b(f_a(x)) = f_{ba}(x)$ for every x indeed the first rule may be removed due to Theorem 2, and relative termination may be concluded due to Theorem 1.

Checking whether $f_l(x) > f_r(x)$ or $f_l(x) \geq f_r(x)$ for all x for some rule $l \to r$ is easily done due to our restriction to linear polynomials. However, for n distinct symbols there are 3^n candidate interpretations, which can be too big. Therefore a selection is made: only choices are made for which at least $n - 2$ symbols have the same interpretations. In this way the number of candidates is quadratic in n. Attempts to prove (relative) termination are done both for the given SRS and its reverse. For instance, for the single rule $ab \to baa$ no polynomial interpretation is possible (not even an interpretation in $\mathbf{N}$ as is shown in [11]), but for its reverse TORPA easily finds one. TORPA applied on the three rules $a \to fb$, $bd \to cdf$, $dc \to adfd$ yields

```
Reverse every lhs and rhs of the system and choose polynomial
interpretation:    f: identity, d: lambda x.x+1, rest lambda x.10x
remove: dc -> adfd
Choose polynomial interpretation   a: lambda x.x+1,   rest identity
remove: a -> fb
Choose polynomial interpretation   b: lambda x.x+1,   rest identity
remove: bd -> cdf
Terminating since no rules remain.
```

4 Recursive Path Order

Recursive path order is an old technique too; it was introduced by Dershowitz [5]. Restricted to string rewriting it means that for a fixed order $>$ on the finite alphabet Σ, called the *precedence*, there is an order $>_{rpo}$ on Σ^* called *recursive path order*. The main property of this order is that if $l >_{rpo} r$ for all rules $l \to r$ of an SRS R, then R is terminating. This order $>_{rpo}$ has the following defining property: $s >_{rpo} t$ if and only if s can be written as $s = as'$ for $a \in \Sigma$, and either

- $s' = t$ or $s' >_{rpo} t$, or
- t can be written as $t = bt'$ for $b \in \Sigma$, and either
  - $a > b$ and $s >_{rpo} t'$, or
  - $a = b$ and $s' >_{rpo} t'$.

For further details we refer to [13]. To avoid branching in the search for a valid precedence in TORPA a slightly weaker version is used. On the other hand, the basic order is also used in combination with removing symbols and reversing. For details see [14]. As an example we give TORPA's result on the single rule $abc \to bacb$:

```
Terminating by recursive path order with precedence:   a>b  b>c
```

5 Semantic Labelling

The technique of semantic labelling was introduced in [12]. Here we restrict to the version for string rewriting in which every symbol is labelled by the value of its argument. For this version we present the theory for relative termination; TORPA only applies this for (quasi-)models containing only two elements. This approach grew out from [8].

Fix a non-empty set A and maps $f_a : A \to A$ for all $a \in \Sigma$ for some alphabet Σ. Let f_s for $s \in \Sigma^*$ be defined as before. Let $\overline{\Sigma}$ be the alphabet consisting of the symbols a_x for $a \in \Sigma$ and $x \in A$. The *labelling function* $\mathsf{lab} : \Sigma^* \times A \to \overline{\Sigma}^*$ is defined inductively as follows:

$$\mathsf{lab}(\epsilon, x) = \epsilon, \quad \mathsf{lab}(sa, x) = \mathsf{lab}(s, f_a(x))a_x, \qquad \text{for } s \in \Sigma^*, a \in \Sigma, x \in A.$$

For an SRS R define $\mathsf{lab}(R) = \{ \mathsf{lab}(l, x) \to \mathsf{lab}(r, x) \mid l \to r \in R, x \in A \}$.

Theorem 3. *Let R and S be two disjoint SRSs over an alphabet Σ. Let $>$ be a well-founded order on a non-empty set A. Let $f_a : A \to A$ be defined for all $a \in \Sigma$ such that*

- *$f_a(x) \geq f_a(y)$ for all $a \in \Sigma, x, y \in A$ satisfying $x > y$, and*
- *$f_l(x) \geq f_r(x)$ for all $l \to r \in R \cup S, x \in A$.*

Let Dec *be the SRS over $\overline{\Sigma}$ consisting of the rules $a_x \to a_y$ for all $a \in \Sigma, x, y \in A$ satisfying $x > y$. Then* $\mathsf{SN}(R/S)$ *if and only if* $\mathsf{SN}(\mathsf{lab}(R)/(\mathsf{lab}(S) \cup \mathsf{Dec}))$.

In case the relation $>$ is empty the set A together with the functions f_a for $a \in \Sigma$ is called a *model* for the SRS, otherwise it is called a *quasi-model*. It is called a model since then for every rule $l \to r$ the interpretation f_l of l is equal to the interpretation f_r of r. Note that $\mathsf{Dec} = \emptyset$ in case of a model. On the single rule $aa \to aba$ TORPA may yield:

```
Apply labelling with the following interpretation in {0,1}:
  a: constant 0
  b: constant 1
and label every symbol by the value of its argument.
This interpretation is a model.
Labelled system:
    a0 a0  -> a1 b0 a0
    a0 a1  -> a1 b0 a1
Choose polynomial interpretation  a0 : lambda x.x+1,  rest identity
```

by which both rules are removed. In the notation of Theorem 3 this means that $A = \{0,1\}$, $f_a(x) = 0$ and $f_b(x) = 1$ for $x \in A$, $R = \{aa \to aba\}$, $S = \mathsf{lab}(S) = \mathsf{Dec} = \emptyset$. Since $\mathsf{lab}(aa, x) = a_0 a_x$ and $\mathsf{lab}(aba, x) = a_1 b_0 a_x$ for $x = 0, 1$, the labelled system $\mathsf{lab}(R)$ is as indicated, for which indeed a termination proof is found.

For $A = \{0,1\}$ for every symbol a there are four possibilities for $f_a : A \to A$: $f_a = \lambda x \cdot x$, $f_a = \lambda x \cdot 0$, $f_a = \lambda x \cdot 1$, $f_a = \lambda x \cdot 1 - x$. Up to renaming $A = \{0,1\}$ admits only two strict orders $>$: $> = \emptyset$ and $> = (1,0)$. For the first one (the model case) for all symbols a all four interpretations for f_a are allowed, and the only restriction is that $f_l = f_r$ for all rules $l \to r \in R \cup S$. For the second order (the quasi-model case) for all symbols a only the first three interpretations for f_a are allowed, since $f_a = \lambda x \cdot 1 - x$ does not satisfy the requirement that $f_a(x) \geq f_a(y)$ for $x > y$. On the other hand, now the restriction on the rules is weaker: rather than $f_l(x) = f_r(x)$ it is only required that $f_l(x) \geq f_r(x)$ for all rules $l \to r \in R \cup S$ and $x \in A$.

In TORPA first the model approach is tried for random choices of the functions f_a until the model requirements hold. Then polynomial interpretations and recursive path order are applied on the labelled systems. If this succeeds the desired proof is generated, otherwise the whole procedure is repeated. There is a basic maximal number of attempts to be done. The default of this number is 100. Subsequent attempts to prove termination by TORPA may yield different solutions, due to the use of the random generator.

In case this first series of attempts was not yet successful a similar procedure is applied for quasi-models. For both the model case and the quasi-model case everything is done twice while no solution is found: once for R/S and once for $R^{\mathsf{rev}}/S^{\mathsf{rev}}$. On the four rules $a \to bc$, $ab \to ba$, $dc \to da$, $ac \to ca$ TORPA yields:

```
Reverse every lhs and rhs of the system.
Apply labelling with the following interpretation in {0,1}:
  a: identity              c: identity
  b: constant 0            d: constant 1
and label every symbol by the value of its argument.
This is a quasi-model for  1 > 0.
```

and for the resulting labelled system a termination proof is given by polynomial interpretations. We do not know any other way to prove termination of this SRS.

In case some attempt to prove termination of a labelled system fails, but applying polynomial interpretations succeeds in removing some rules, then in the default setting of TORPA it is checked whether after removing all labels a strict subset of the original SRS is obtained. If so, then the whole procedure starts again on this smaller SRS. In this way TORPA finds a termination proof for $aba \to abba$, $bab \to baab$ by first finding a labelling by which one rule can be removed, and then find another labelling for the remaining rule.

6 Dependency Pairs

The technique of dependency pairs was introduced in [1] and is extremely useful for automatically proving termination of term rewriting. Here we only use a mild version of it, without explicitly doing argument filtering or dependency graph approximation. It turns out that often the same reduction of the problem caused by these more involved parts of the dependency pair technique is done by applying our versions of labelling and polynomial interpretations.

For an SRS R over an alphabet Σ let Σ_D be the set of *defined symbols* of R, i.e., the set of symbols occurring as the leftmost symbol of the left hand side of a rule in R. For every defined symbol $a \in \Sigma_D$ we introduce a fresh symbol $\overline{a}$. TORPA follows the convention that if a is a lowercase symbol then its capital version is used as the notation for $\overline{a}$. Write $\overline{\Sigma} = \Sigma \cup \{\overline{a} \mid a \in \Sigma_D\}$. The SRS $DP(R)$ over $\overline{\Sigma}$ is defined to consist of all rules of the shape $\overline{a}l' \to \overline{b}r''$ for which $al' = l$ and $r = r'br''$ for some rule $l \to r$ in R and $a, b \in \Sigma_D$. Rules of $DP(R)$ are called *dependency pairs*. Now the main theorem of dependency pairs reads as follows.

Theorem 4. *Let R be any SRS. Then* $\mathsf{SN}(R)$ *if and only if* $\mathsf{SN}(DP(R)/R)$.

It is used in TORPA as follows: if proving $\mathsf{SN}(R)$ does not succeed by the earlier techniques, then the same techniques are applied for trying to prove $\mathsf{SN}(DP(R)/R)$ or $\mathsf{SN}(DP(R^{\mathsf{rev}})/R^{\mathsf{rev}})$. In fact the desire for being able to do so was one of the main reasons to generalize the basic methods to relative termination and design TORPA to cover relative termination.

Applying TORPA on the two rules $ab \to c, c \to ba$ yields:

```
Dependency pair transformation:
  ab ->= c
  c ->= ba
  Ab -> C
  C -> A
```

followed by a simple proof by polynomial interpretations.

On the four rules $bca \to ababc$, $b \to cc$, $cd \to abca$, $aa \to acba$ TORPA yields the following remarkable termination proof. First the third rule is removed by a polynomial interpretation. Then the remaining SRS is reversed and the dependency pair transformation is applied, yielding 10 rules. By polynomial interpretations three rules are removed, and on the remaining 7 rules a quasi-model labelling is found, giving a labelled SRS of 20 rules. By polynomial interpretations 10 of them are removed. By removing all labels it suffices to prove relative termination of the five rules $acb \to^= cbaba$, $b \to^= cc$, $aa \to^= abca$ $Aa \to Abca$, $Aa \to A$. By again finding a quasi-model labelling, applying polynomial interpretations and removing labels one strict rule is removed. Next lhs's and rhs's are reversed, and for the resulting SRS a third quasi-model labelling is found by which the remaining strict rule is removed, proving relative termination and concluding the termination proof of the original SRS. The full proof was found completely automatically by TORPA in one or two seconds.

7 RFC-Match-Bounds

A recent very elegant and powerful approach for proving termination of string rewriting is given in [7]. The strongest version is proving match bounds of right hand sides of forward closures, shortly RFC-match-bounds. Here we present the main theorem as it is used in TORPA; for the proof and further details we refer to [7]. For an SRS R over an alphabet Σ we define the SRS $R_\#$ over $\Sigma \cup \{\#\}$ by $R_\# = R \cup \{ l_1\# \to r \mid l \to r \in R \wedge l = l_1 l_2 \wedge l_1 \neq \epsilon \neq l_2 \}$. For an SRS R over an alphabet Σ we define the infinite SRS $\mathsf{match}(R)$ over $\Sigma \times \mathbf{N}$ to consist of all rules $(a_1, n_1) \cdots (a_p, n_p) \to (b_1, m_1) \cdots (b_q, m_q)$ for which $a_1 \cdots a_p \to b_1 \cdots b_q \in R$ and $m_i = 1 + \min_{j=1,\ldots,p} n_j$ for all $i = 1, \ldots, q$.

Theorem 5. *Let R be an SRS and let $N \in \mathbf{N}$ such that for all rhs's $b_1 \cdots b_q$ of R and all $k \in \mathbf{N}$ and all reductions*

$$(b_1, 0) \cdots (b_q, 0)(\#, 0)^k \to^*_{\mathsf{match}(R_\#)} (c_1, n_1) \cdots (c_r, n_r)$$

it holds that $n_i \leq N$ for all $i = 1, \ldots, r$. Then R is terminating.

The minimal number N satisfying the condition in Theorem 5 is called the corresponding *match bound.* The way to verify the condition of Theorem 5 is to construct a *certificate*, being a finite automaton M over the alphabet $(\Sigma \cup \{\#\}) \times \mathbf{N}$, where Σ is the alphabet of R, satisfying:

- for every rhs $b_1 \cdots b_q$ of R and every $k \in \mathbf{N}$ the automaton M accepts $(b_1, 0) \cdots (b_q, 0)(\#, 0)^k$, and
- M is closed under $\mathsf{match}(R_\#)$, i.e., if M accepts v and $v \to_{\mathsf{match}(R_\#)} u$ then M accepts u too.

The pair $(a, k) \in (\Sigma \cup \{\#\}) \times \mathbf{N}$ will shortly be written as a_k, and the number k is called the *label* of this pair. It is easy to see that if a (finite) certificate M has been found then for N being the biggest label occurring in M the condition of Theorem 5 holds. Hence the only thing to be done for proving termination by this approach is finding a certificate. All of these observations are found in [7]; the only new contribution of TORPA is the much more powerful heuristic of searching for such a certificate.

As an example we consider the single rule $aba \to abbba$. None of the earlier techniques works, but TORPA yields the following certificate:

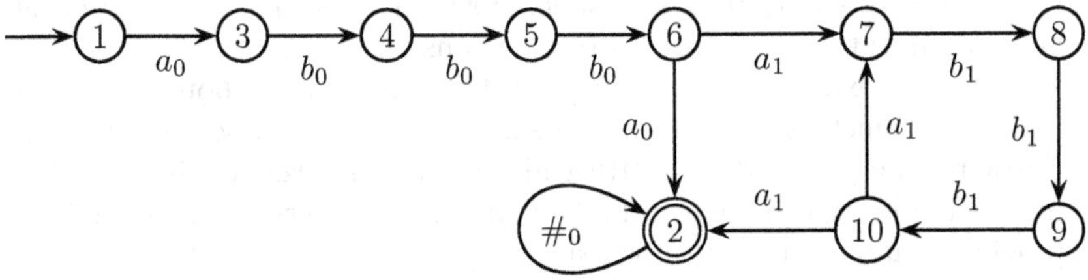

Rather than giving such a picture TORPA yields a list of all transitions, containing all information about the automaton. Indeed this automaton is a certificate: it accepts $a_0 b_0^3 a_0 \#_0^k$ for every k, and it is closed under $\mathsf{match}(R_\#)$, hence proving termination. For instance, it accepts $a_0 b_0^3 a_0 \#_0$ which rewrites to $a_0 b_0^3 a_1 b_1^3 a_1$ by the rule $a_0 \#_0 \to a_1 b_1^3 a_1$ of $\mathsf{match}(R_\#)$, also accepted by the automaton.

TORPA tries to construct a certificate if the earlier techniques fail, both for the SRS and its reverse. This construction starts by an automaton exactly accepting $(b_1, 0) \cdots (b_q, 0)(\#, 0)^k$ for every rhs $b_1 \cdots b_q$ of R and every $k \in \mathbf{N}$. Then for every path in the automaton labelled by a lhs of $\mathsf{match}(R_\#)$ it is checked whether there exists a path between the same two nodes labelled by the corresponding rhs. If not, then such a path has to be constructed. Here the heuristic comes in. If a path from n_1 to n_2 has to be constructed labelled by the string au for $a \in (\Sigma \cup \{\#\}) \times \mathbf{N}$ and $u \in ((\Sigma \cup \{\#\}) \times \mathbf{N})^*$, then it is checked whether a node n exists for which a path from n to n_2 exists labelled by u. If so, then an edge from n_1 to n is added labelled by a, if not, then a completely fresh path from n_1 to n_2 is constructed labelled by au. This process is repeated until either no edges need to be added or overflow occurs. In the first case the resulting automaton is a certificate by construction, proving termination. Overflow occurs if the automaton contains 800 edges.

In the above example nodes 7, 8, 9, 10 are added for making a path from 6 to 2 labelled by $a_1 b_1^3 a_1$. Then the edge from 10 to 7 is added for making a path from 10 to 2 labelled by $a_1 b_1^3 a_1$, and then nothing is to be added any more. This very simple heuristic was found after trying many other and more involved heuristics that turned out to be much less powerful.

Termination of the single rule $aabb \to bbbaaa$ is easily proved by this approach, yielding exactly the same automaton as given in [7] having 42 nodes and match bound 4. However, there it was found after an extensive process on intermediate automata of thousands of nodes, while in our approach no intermediate automata exceeding the final result occurred. The main difference is that in [7] an exact automaton is computed while TORPA computes an approximation. However, for nearly all examples both automata coincide.

8 Conclusions and Further Research

For many small SRSs TORPA automatically finds a termination proof, but finding a human proof allowing any presently known technique seems to be a really hard job. Usually, the generated proofs are not more than a few pages of text, including many details. For people familiar with the underlying theory verifying the generated proofs is always feasible. However, this may be very boring, and redundant since the proofs are correct by construction.

Due to the extension in January 2004 by RFC-match-bounds, the present version 1.2 is much stronger than the earlier version 1.1 described in [14].

Most techniques used in TORPA also apply for term rewriting rather than string rewriting. Hence a natural follow up will be a version of TORPA capable of proving termination of term rewriting.

References

1. T. Arts and J. Giesl. Termination of term rewriting using dependency pairs. *Theoretical Computer Science*, 236:133–178, 2000.
2. A. Ben-Cherifa and P. Lescanne. Termination of rewriting systems by polynomial interpretations and its implementation. *Science of Computer Programming*, 9:137–159, 1987.
3. C. Boralleras and A. Rubio. Termptation: termination proof techniques automation. Available at `www.lsi.upc.es/~albert/`.
4. E. Contejean, C. Marché, B. Monate, and X. Urbain. The CiME rewrite tool. Available at `http://cime.lri.fr/`.
5. N. Dershowitz. Orderings for term-rewriting systems. *Theoretical Computer Science*, 17:279–301, 1982.
6. J. Giesl et al. Automated program verification environment (AProVE). Available at `http://www-i2.informatik.rwth-aachen.de/AProVE/`.
7. A. Geser, D. Hofbauer, and J. Waldmann. Match-bounded string rewriting. Technical Report 2003-09, National Institute of Aerospace, Hampton, VA, 2003. Submitted for publication in a journal.
8. J. Giesl and H. Zantema. Liveness in rewriting. In R. Nieuwenhuis, editor, *Proceedings of the 14th Conference on Rewriting Techniques and Applications (RTA)*, volume 2706 of *Lecture Notes in Computer Science*, pages 321–336. Springer, 2003.
9. N. Hirokawa and A. Middeldorp. Tsukuba termination tool. In R. Nieuwenhuis, editor, *Proceedings of the 14th Conference on Rewriting Techniques and Applications (RTA)*, volume 2706 of *Lecture Notes in Computer Science*, pages 311–320, 2003.
10. D.S. Lankford. On proving term rewriting systems are noetherian. Technical report MTP 3, Louisiana Technical University, 1979.
11. H. Zantema. Termination of term rewriting: Interpretation and type elimination. *Journal of Symbolic Computation*, 17:23–50, 1994.
12. H. Zantema. Termination of term rewriting by semantic labelling. *Fundamenta Informaticae*, 24:89–105, 1995.
13. H. Zantema. Termination. In *Term Rewriting Systems, by Terese*, pages 181–259. Cambridge University Press, 2003.
14. H. Zantema. Termination of string rewriting proved automatically. Technical Report CS-report 03-14, Eindhoven University of Technology, 2003. Submitted, available via `http://www.win.tue.nl/inf/onderzoek/en_index.html` .

Querying Unranked Trees with Stepwise Tree Automata

Julien Carme, Joachim Niehren, and Marc Tommasi

Mostrare project, INRIA Futurs, Lille, France

Abstract. The problem of selecting nodes in unranked trees is the most basic querying problem for XML. We propose *stepwise tree automata* for querying unranked trees. Stepwise tree automata can express the same monadic queries as monadic Datalog and monadic second-order logic. We prove this result by reduction to the ranked case, via a new systematic correspondence that relates unranked and ranked queries.

1 Introduction

Querying semi-structured documents is a base operation for information extraction from the Web or semi-structured databases. It requires expressive query languages whose queries can be answered efficiently [8]. The most widely known querying language these days is the W3C standard XPath (see e.g. [10, 9]).

Semi-structured documents in XML or HTML form *unranked trees* whose nodes may have an unbounded list of children. The most basic querying problem is to select sets of nodes in unranked trees. *Monadic queries* approach this problem declaratively. They specify sets of nodes in a tree that can then be computed by a generic algorithm.

We are interested in query languages that can describe all regular sets of nodes in trees. This property is satisfied by three classes of queries, those represented by tree automata [16, 12, 3, 13, 6], *monadic second-order logic* (MSO) [16, 8] and *monadic Datalog* [1, 7] over trees. Automata and Datalog queries can be answered in linear time. They are satisfactory in efficiency and expressiveness, in theory and practice [11].

Unranked trees are problematic in that they may be recursive in depth and breadth, in contrast to ranked trees. This additional level of recursion needs to be accounted for by recursive queries. In MSO and monadic Datalog, breadth recursion can be programmed from the next_sibling relation. Unfortunately, this relation cannot be expressed in WSωS, so that traditional results on ranked trees don't carry over for free. Selection automata [6] reduce breadth recursion to depth recursion, by operating on binary encodings of unranked trees. Encodings are problematic in that they alter locality and path properties; furthermore the close relationship to unranked Datalog queries gets lost. Hedge automata [16, 12, 3] express horizontal recursion by an extra recursion level in transition rules. This syntactic extension leads to numerous technical problems [13, 7] that one might prefer to avoid.

V. van Oostrom (Ed.): RTA 2004, LNCS 3091, pp. 105–118, 2004.

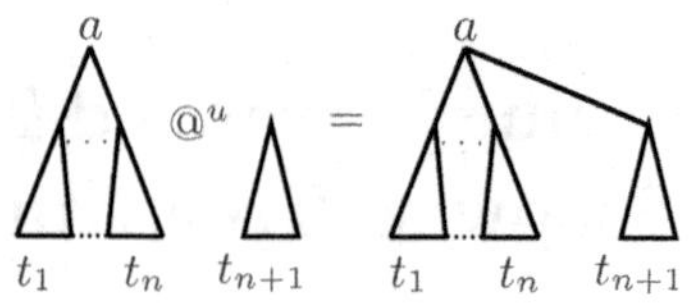

Fig. 1. Tree extension

In this paper, we propose *stepwise tree automata* for querying unranked trees. Stepwise tree automata are traditional tree automata that can either operate on unranked or ranked trees. They combine the advantages of selection and hedge automata. They model horizontal recursion by traversing siblings stepwise from the left to the right.

The algebraic approach behind stepwise tree automata yields a new systematic correspondence between queries for unranked and ranked trees. We elaborate this correspondence for monadic queries. We show that stepwise tree automata, monadic Datalog programs, and MSO can express the same monadic queries over unranked trees. We reduce this result to the case of ranked trees due to our new systematic correspondence. Specific proofs for unranked queries are not needed in contrast to [13, 7].

2 The Algebras of Unranked and Ranked Trees

An *unranked signature* Σ is a set of *symbol* ranged over by a, b. An ordered *unranked tree* or *u-tree* t over Σ satisfies the following abstract syntax:

$$t ::= a(t_1, \ldots, t_n) \qquad \text{where } n \geq 0.$$

Unordered unranked trees were investigated in [14, 15, 18]. We identify an unranked tree $a()$ with the symbol a. We write tree^u for the set of unranked trees. The extension operator $@^u : \mathsf{tree}^u \times \mathsf{tree}^u \to \mathsf{tree}^u$ for unranked trees is depicted in Fig. 1. The extended tree $t@^u t'$ is obtained from t by adjoining t' as next sibling of the last child of t:

$$a(t_1, \ldots, t_n)@^u t' = a(t_1, \ldots, t_n, t')$$

Note that $a(t_1, \ldots, t_n) = a@^u t_1 @^u \ldots @^u t_n$ with parenthesis set from left to right. Tree extension $@^u$ is neither associative nor commutative.

Let $\Sigma_@ = \Sigma \cup \{@\}$ be the *ranked signature* of function symbols with constants in Σ and a single binary function symbol $@$. Ranked trees over $\Sigma_@$ are ground terms over $\Sigma_@$, i.e., binary trees that satisfy the grammar:

$$t ::= a \mid t_1 @ t_2$$

We omit parenthesis as in the λ-calculus; $a@b@(c@b@a)$ for instance is $(a@b)@((c@b)@a)$. We write tree^r for the set of ranked trees over $\Sigma_@$. Ranked trees can

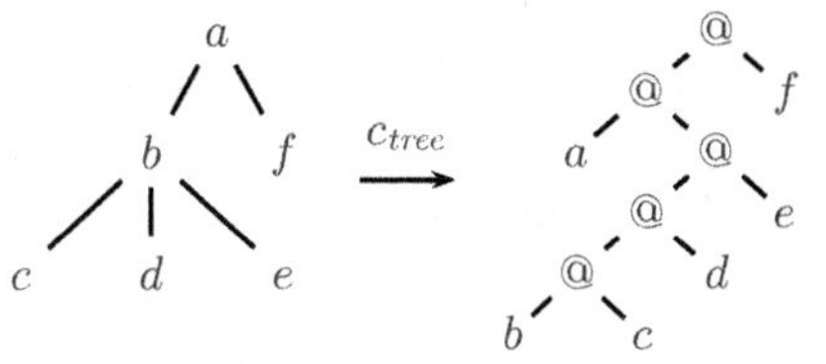

Fig. 2a. An unranked tree $a(b(c,d,e),f)$, its ranked construction $a@(b@c@d@e)@f$

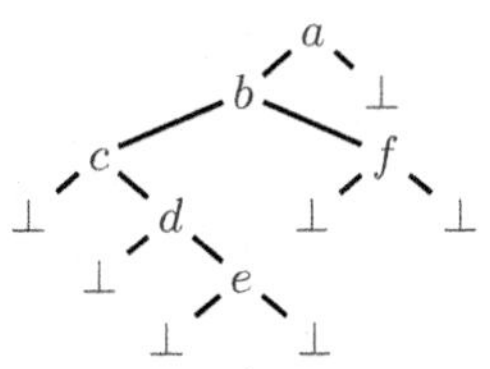

Fig. 2b. First-child next-sibling encoding

be constructed by the binary operator $@^r : \mathsf{tree}^r \times \mathsf{tree}^r \to \mathsf{tree}^r$ which satisfies for all ranked trees t_1, t_2:

$$t_1 @^r t_2 = t_1 @ t_2$$

Unranked trees correspond precisely to ranked constructions with respect to the function $c_{tree} : \mathsf{tree}^u \to \mathsf{tree}^r$ which satisfies for all unranked trees t_1, t_2 and symbols $a \in \Sigma$:

$$c_{tree}(t_1 @^u t_2) = c_{tree}(t_1) @^r c_{tree}(t_2) \quad \text{and} \quad c_{tree}(a) = a$$

The idea of this binary construction is known from Currying. An unranked tree describes an application of a function to a list of arguments. Its binary construction represents the Curried version of this function, receiving arguments one by one. We therefore write tree extension as function application @.

The set tree^u of unranked trees over Σ with the extension operation $@^u$ is a $\Sigma_@$ algebra, as well as the set tree^r of ranked trees over $\Sigma_@$ with the operation $@^r$.

Proposition 1. *The construction function* $c_{tree} : \mathsf{tree}^u \to \mathsf{tree}^r$ *is an isomorphism between* $\Sigma_@$*-algebras.*

Ranked and unranked trees thus have the same algebraic properties and the same finite automata [17, 5, 14].

Ranked constructions are binary representations of unranked trees. Previous approaches towards querying unranked trees rely on a different binary representation [8, 6], which encodes first-child and next-sibling relations. An example is given in Fig. 2b. The new binary construction, however, permits to carry over traditional results from ranked to unranked trees more systematically.

3 Stepwise Tree Automata

Stepwise tree automata A over signature Σ are traditional tree automata ([4]) over the signature $\Sigma_@$. They consist of a finite set $\mathsf{states}(A)$ of *states*, a set $\mathsf{final}(A) \subseteq \mathsf{states}(A)$ of *final states*, and a finite set of $\mathsf{rules}(A)$ of *transition rules* of two forms, where $a \in \Sigma$ and $q, q_1, q_2 \in \mathsf{states}(A)$:

$$a \to q \qquad \text{or} \qquad q_1 @ q_2 \to q$$

$$\mathsf{eval}^{\alpha}_{A}(a) = \{q \mid a \to q \in \mathsf{rules}(A)\}$$
$$\mathsf{eval}^{\alpha}_{A}(t_1 @^{\alpha} t_2) = \{q \mid q_1 \in \mathsf{eval}^{\alpha}_{A}(t_1),\ q_2 \in \mathsf{eval}^{\alpha}_{A}(t_2), q_1 @ q_2 \to q \in \mathsf{rules}(A)\}$$

Fig. 3. Ranked and unranked evaluation

Stepwise tree automata can evaluate unranked and ranked trees, i.e. α-trees where $\alpha \in \{u, r\}$. The *α-evaluator* of A is the function $\mathsf{eval}^{\alpha}_{A} : \mathsf{tree}^{\alpha} \to 2^{\mathsf{states}(A)}$ defined in Fig. 3. Both evaluators only differ in the interpretation of the symbol @.

Lemma 1. $\mathsf{eval}^{u}_{A}(t) = \mathsf{eval}^{r}_{A}(c_{tree}(t))$ *for all unranked trees* t.

Stepwise tree automata A recognize all α-trees t that can be evaluated into a final state, i.e., $\mathsf{eval}^{\alpha}_{A}(t) \cap \mathsf{final}(A) \neq \emptyset$. The α-language $L^{\alpha}(A)$ consists of all α-trees recognized by A.

Proposition 2. *A stepwise tree automaton accepts an unranked tree if and only if it accepts its ranked construction, i.e., for all A:*

$$L^{u}(A) = c_{tree}^{-1}(L^{r}(A))$$

Proof. By Lemma 1 all unranked tree t satisfy: $t \in L^{u}(A)$ iff $\mathsf{final}_A \cap \mathsf{eval}^{u}_{A}(t) \neq \emptyset$ iff $\mathsf{final}_A \cap \mathsf{eval}^{r}_{A}(c_{tree}(t)) \neq \emptyset$ iff $c_{tree}(t) \in L^{r}(A)$.

As a consequence, stepwise tree automata inherit numerous properties from traditional tree automata, even if interpreted over unranked trees. Recognizable unranked tree languages are closed under intersection, union, and complementation. Emptiness can be checked in linear time, and membership $t \in L^{u}(A)$ in linear time $O(|t| * |A|)$.

Example. We define a stepwise tree automaton A with signature $\Sigma = \{a, b\}$ that recognizes all unranked trees with at least one a-labeled leaf.

Automaton A is illustrated to the right. It has three states 0, 1, 2 two of which are final: 1, 2. A successful run of A on an unranked tree assigns state 1 in a non deterministic way to a unique a-leaf and labels by 2 all edges visited afterwards. All other nodes and edges are labeled by 0.

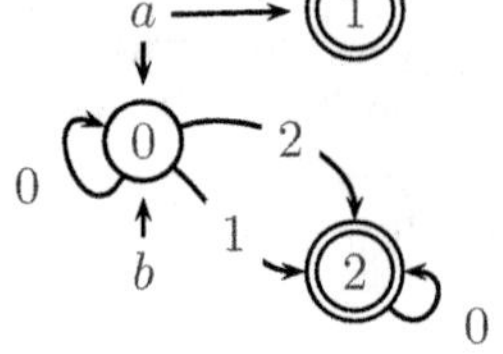

a. An a-node can be the selected a-leaf: $a \to 1$.
b. Any node can be assigned to state 0: $a \to 0$, $b \to 0$, $0@0 \to 0$.
c. The edge pointing to the selected a-leaf may go into state 2: $0@1 \to 2$.
d. All edges visited later on may go into state 2, too: $2@0 \to 2$, $0@2 \to 2$.

In Fig. 4, we show the unique successful run of A on the unranked tree $a(b, a(a, b))$ and on its construction $a@b@(a@a@b)$. Both runs bisimulate each other. They evaluate the respective tree into the final state 2.

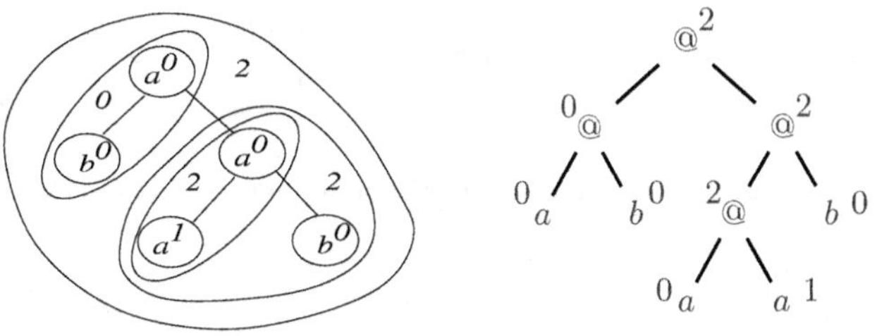

Fig. 4. An example run on an unranked tree and its construction

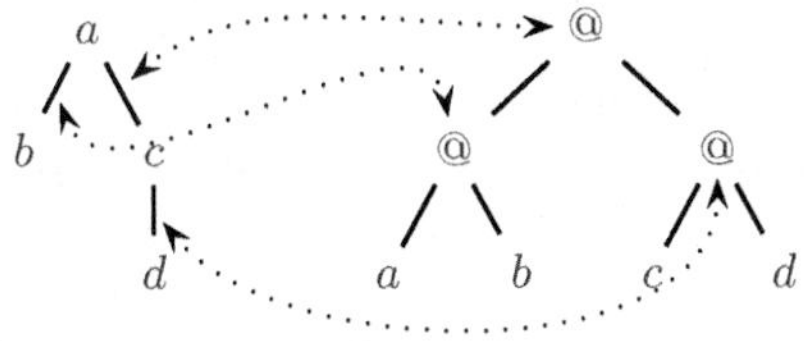

Fig. 5. Correspondence between edges and application nodes

4 Monadic Queries

Queries on unranked trees should correspond precisely to their counterparts on ranked constructions. This requires a precise correspondence between the domains of unranked trees and their ranked constructions. The domain of a ranked tree t is the set of its nodes:

$$\begin{aligned} \mathsf{dom}^r(t_1 @^r t_2) &= \mathsf{dom}^r(t_1) \uplus \mathsf{dom}^r(t_2) \uplus \{\mathsf{root}\} \\ \mathsf{dom}^r(a) &= \{\mathsf{root}\} \end{aligned}$$

Disjoint union $A \uplus B$ can be implemented by $\{1\}\times A \cup \{2\}\times B$. The ranked domain of $a@(a@a)$ is then implemented by:

$$\{\mathsf{root}, (1, \mathsf{root}), (2, \mathsf{root}), (2, (1, \mathsf{root})), (2, (2, \mathsf{root}))\}$$

As usual in mathematics, we will abstract from this implementation and talk about disjoint unions as if they were simple unions.

The nodes of unranked trees correspond to leaves in ranked constructions. But what do application nodes in construction trees correspond to? The example in Fig. 5 illustrates that they correspond precisely to edges of unranked trees. Every edge was added by some application step, and vice versa, every application step adds some edge. We therefore define the domain of an unranked tree as the union of its nodes and edges.

$$\begin{aligned} \mathsf{dom}^u(t_1 @^u t_2) &= \mathsf{dom}^u(t_1) \uplus \mathsf{dom}^u(t_2) \uplus \{\mathsf{last\_edge}\} \\ \mathsf{dom}^u(a) &= \{\mathsf{root}\} \end{aligned}$$

The $\mathsf{last\_edge}$ of $t_1 @^u t_2$ links the root of t_1 to the root of t_2. Note that all nodes of $t_1 @^u t_2$ either belong to t_1 or t_2; the root of $t_1 @^u t_2$ is that of t_1.

$$c_{dom}(a)(\mathsf{root}) = \mathsf{root}, \quad c_{dom}(t_1 @^u t_2)(\pi) = \begin{cases} \mathsf{root} & \text{if } \pi = \mathsf{last\_edge} \\ c_{dom}(t_1)(\pi) & \text{if } \pi \in \mathsf{dom}^u(t_1) \\ c_{dom}(t_2)(\pi) & \text{if } \pi \in \mathsf{dom}^u(t_2) \end{cases}$$

Fig. 6. Definition of the correspondence on domains c_{dom}

The *correspondence* $c_{dom}(t)$ for an unranked tree t is a function between the domains of t and its ranked construction defined in Fig. 6 by recursion over the construction of t:

$$c_{dom}(t) : \mathsf{dom}^u(t) \to \mathsf{dom}^r(c_{tree}(t))$$

Definition 1. *A* monadic query *is a function q that maps trees to subsets of their domain. This definition applies to ranked and unranked trees, i.e., for both $\alpha \in \{u, r\}$. A monadic α-query q satisfies for all $t \in \mathsf{tree}^\alpha$:*

$$q(t) \subseteq \mathsf{dom}^\alpha(t)$$

More general n-ary queries map to n-tuples of elements of the domain [2]. Monadic queries q on unranked trees correspond to ranked monadic queries $c_{query}(q)$ such that for all $t \in \mathsf{tree}^r$:

$$c_{query}(q)(t) = c_{dom}(t)(q(c_{tree}^{-1}(t)))$$

All previous query notions for unranked trees [8, 6] only talk about nodes. Our extension with edges, however, is necessary to keep the symmetry to ranked queries, the reason for the simplicity or our approach.

5 Automata Queries

In the remainder of the paper, we will discuss regular monadic queries. These can be defined by tree automata, monadic second-order logic, and monadic Datalog. Here, we start with tree automata.

We call a ranked monadic query q is *regular* if all $q(t)$ are regular unary relations where $t \in \mathsf{tree}^r$. Let us recall the definition of regular node relations in ranked trees [17]. Given a ranked tree t over the signature $\Sigma \times \mathsf{Bool}$, and $i \in \{1, 2\}$, let $\mathsf{proj}^i(t)$ be the ranked tree obtained by projecting all labels in t to their i'th component. For a monadic query q and ranked tree t let $\mathsf{zip}(t, q)$ be the ranked tree over the extended signature $\Sigma \times \mathsf{Bool}$ with $\mathsf{proj}^1(\mathsf{zip}(t, q)) = t$ and $\mathsf{proj}^2(\mathsf{zip}(t, q)) = q$.

Definition 2. *A monadic query q for ranked trees is* regular *if the set $\{\mathsf{zip}(t, q) \mid t \in \mathsf{tree}^r\}$ can be recognized by a tree automaton. A monadic query q for unranked trees is regular if $c_{query}(q)$ is.*

This traditional definition of regular monadic queries is simple for ranked trees but has drawbacks otherwise. First, it is not obvious how to compute such

$$\frac{a \to r(\mathsf{root}) \in \mathsf{rules}(A)}{r \in \mathsf{runs}^\alpha_A(a)} \qquad \frac{r_{|\mathsf{dom}^\alpha(t_1)} \in \mathsf{runs}^\alpha_A(t_1) \qquad r_{|\mathsf{dom}^\alpha(t_1)} \in \mathsf{runs}^\alpha_A(t_2) \qquad r(\mathsf{head}^\alpha(t_1))@r(\mathsf{head}^\alpha(t_2)) \to r(\mathsf{head}(t_1@^\alpha t_2)) \in \mathsf{rules}(A)}{r \in \mathsf{runs}^\alpha_A(t_1@^\alpha t_2)}$$

Fig. 7. Runs $\mathsf{runs}^\alpha_A(t)$ of stepwise tree automata A on α-trees t

queries efficiently, second, it is not obvious how to express them in monadic Datalog, and third, the definition of regular unranked queries depends on the correspondence to ranked queries.

Run-based queries [15, 6] with stepwise tree automata resolve these problems. We define them parametically for ranked and unranked trees. Runs of stepwise tree automaton on α-trees t associate states to all elements of the domain of t. Sequences of children in unranked trees are visited *stepwise* from the left to the right, while annotating edges to the children by states. See Fig. 4 for an example. More formally, let the *head* of a ranked tree be its root; the head of a non-constant unranked tree is last_edge, and the head of a constant unranked tree the root:

$$\mathsf{head}^r(t) = \mathsf{root}, \quad \mathsf{head}^u(t_1@^u t_2) = \mathsf{last\_edge}, \quad \mathsf{head}^u(a) = \mathsf{root}.$$

A *run* of a tree automaton A on an α-tree t is a function labeling elements of the domain of t by states of A:

$$r : \mathsf{dom}^\alpha(t) \to \mathsf{states}(A)$$

such that all transitions are licensed by rules of A. If $t = t_1@^u t_2$ then the restrictions of r to the domains of t_1 and t_2 must be runs and the annotation of the head of t must be justified. Furthermore, annotations of constants must be licensed. These conditions are captured by the inference rules in Fig. 7.

Lemma 2. *Let A be a stepwise tree automaton and t an α-tree, then:*

$$\mathsf{eval}^\alpha_A(t) = \{r(\mathsf{head}^\alpha(t)) \mid r \in \mathsf{runs}^\alpha_A(t)\}$$

Proof. By induction on the construction of unranked trees. If $t = a$ then $\mathsf{eval}^\alpha(a) = \{q \mid a \to q \in \mathsf{rules}(A)\} = \{r(\mathsf{root}) \mid r \in \mathsf{runs}^\alpha_A(a)\}$. For $t = t_1@^\alpha t_2$ we have $\mathsf{eval}^\alpha(t) = \{q \mid q_i \in \mathsf{eval}^\alpha_A(t_i),\ q_1@q_2 \to q \in \mathsf{rules}(A)\}$ which is equal to $\{q \mid r_i \in \mathsf{runs}^\alpha_A(t_i), r_1(\mathsf{head}^\alpha(t_1))@r_2(\mathsf{head}^\alpha(t_1)) \to q \in \mathsf{rules}(A)\}$ by induction hypothesis. The definition of runs in Fig. 7 yields $\{r(\mathsf{head}^\alpha_i(t)) \mid r \in \mathsf{runs}^\alpha_A(t)\}$.

A run r of an automaton A on an α-tree t is *successful* if $r(\mathsf{head}^\alpha(t)) \in \mathsf{final}(A)$. Let $\mathsf{succ\_runs}^\alpha_A(t)$ be the set of successful runs of A on $t \in \mathsf{tree}^\alpha$.

Definition 3. *A pair of a tree automaton A and a set $Q \subseteq \mathsf{states}(A)$ defines a monadic query which selects all elements from the domain of a tree t that are labeled by a state in Q in some successful run of A on t:*

$$\mathit{query}^\alpha_{A,Q}(t) = \{\pi \in \mathsf{dom}^\alpha(t) \mid r \in \mathsf{succ\_runs}^\alpha_A(t), r(\pi) \in Q\}$$

Selection automata [6] similarly express queries for unranked trees, but rely on universal quantification over successful runs, and use a binary encoding of unranked trees in contrast to the definition above.

Example. Reconsider the automaton A from Section 3: $\mathsf{query}^u_{A,\{1\}}$ defines the set of all a-leaves in unranked trees. Note that no automaton query with a bottom-up deterministic automaton can compute the same query, since it couldn't distinguish different a-nodes.

Proposition 3. *A monadic query on ranked trees is regular if and only if it is equal to some* $\mathsf{query}^r_{A,Q}$.

The proof relies on two standard automata transformations. The idea of the transformation from regular to run based queries $\mathsf{query}^r_{A,Q}$ is memorize the Boolean values in $\mathsf{zip}(t,q)$ in automata states. In order to generalize Proposition 3 to unranked trees, we establish the correspondence between unranked and ranked run-based queries.

Theorem 1. *Queries with stepwise tree automata on ranked and unranked trees correspond:*

$$\mathsf{query}^r_{A,Q} = c_{query}(\mathsf{query}^u_{A,Q})$$

Proof. We first note that runs of stepwise automata on unranked trees and ranked constructions correspond. For all $t \in \mathsf{tree}^r$, we can prove:

$$\mathsf{runs}^u_A(c^{-1}_{tree}(t)) = \{r \circ c_{dom}(t) \mid r \in \mathsf{runs}^r_A(t)\}$$

The theorem follows from straightforward calculations. For all $t \in \mathsf{tree}^r$:

$$\begin{aligned} c_{query}(\mathsf{query}^u_{A,Q})(t) &= c_{dom}(t)(\mathsf{query}^u_{A,Q}(c^{-1}_{tree}(t))) \\ &= \{\pi \mid r \in \mathsf{runs}^u_A(c^{-1}_{tree}(t)),\ r(c_{dom}(t)^{-1}(\pi)) \in Q\} \\ &= \{\pi \mid r' \in \mathsf{runs}^r_A(t).\ r'(c_{dom}(t)(c_{dom}(t)^{-1}(\pi))) \in Q\} \\ &= \mathsf{query}^r_{A,Q}(t) \end{aligned}$$

□

6 Monadic Second Order Logic

We next represent regular monadic queries in ranked and unranked trees in monadic second-order logic (MSO).

The domain of the logical structure induced by an α-tree t is $\mathsf{dom}^\alpha(t)$. The signature R^r for structures of ranked trees contains the follwing relation symbols:

$$R^r = \{\mathsf{child}_1, \mathsf{child}_2, \mathsf{root}, \mathsf{leaf}\} \cup \{\mathsf{label}_a \mid a \in \Sigma_@\}$$

The binary relations child_1 and child_2 relate nodes to their first resp. second child. Unary relations label_a hold for all nodes labeled by $a \in \Sigma$. Furthermore, we permit the unary relations root and leaf.

Logical structures for unranked trees have the following signature:

$$R^u = \{\mathsf{first\_edge}, \mathsf{next\_edge}, \mathsf{target}, \mathsf{root}, \mathsf{leaf}\} \cup \{\mathsf{label}_a \mid a \in \Sigma_@\}$$

The binary relation $\mathsf{first\_edge}$ holds between a node and the edge to its first child. The $\mathsf{next\_edge}$ relation links an edge with target π to the edge whose target is the next sibling of π. The target relation holds between an edge and its target node.

Let x, y, z range over an infinite set Vars of node variables and p, q over an infinite set Preds of monadic predicates. The logics MSO^α have the following formulas:

$$\phi ::= B_n(x_1, \ldots, x_n) \mid p(x) \mid \phi \wedge \phi' \mid \neg\phi \mid \exists x\phi \mid \exists p\phi$$

where $B_n \in R^\alpha$ is a predicate with fixed tree interpretation of arity n. Note that the relations root and leaf could be expressed by the remaining relations in MSO^α. We add them anyway, as they will be needed in monadic Datalog later on.

Let ϕ be an MSO^α-formula, t an α-tree, and σ an assignment of variables into the domain of t and of predicates into the powerset of this domain. We write $t, \sigma \models_{\mathsf{MSO}^\alpha} \phi$ if ϕ becomes true in t under σ. Every formula $\phi(x)$ with a single variable x defines monadic query:

$$\mathsf{query}^\alpha_{\phi(x)}(t) \;=\; \{\sigma(x) \mid t, \sigma \models_{\mathsf{MSO}^\alpha} \phi\}$$

Theorem 2. *[Thatcher & Wright [17]] Monadic queries expressed in monadic second order logic over ranked trees* MSO^r *are regular.*

Theorem 3. *Ranked and unranked monadic queries expressed in monadic second-order logic correspond, and are thus regular; corresponding queries can be computed in linear time:*

$$\{query^r_{\phi(x)} \mid \phi \in \mathsf{MSO}^r\} = \{c_{query}(query^u_{\phi'(x)}) \mid \phi' \in \mathsf{MSO}^u\}$$

Fig. 8 presents forth and back translations between MSO^u and MSO^r. We have to show for every $\phi \in \mathsf{MSO}^u$ that $c_{query}(\mathsf{query}^u_{\phi(x)}) = \mathsf{query}^r_{[\![\phi(x)]\!]_r}$, and the analoguous property for the back translation. We proceed by structural induction over formulas. The base cases contains the difficulty, the induction step being straightforward.

We sketch the proof for formula $\mathsf{first\_edge}(x, y)$ to illustrate the principles. Consider an u-tree t and a variable assignment σ under which $\mathsf{first\_edge}(x, y)$ becomes true. There exists a u-tree $t_0 = t_1@t_2$ involved in the construction of t such that $\sigma(x)$ is the root of t_1 and $\sigma(y)$ is the edge from t_1 to t_2. Since this is the first edge, t_1 is a constant. Therefore, in the corresponding ranked tree, the node $c_{dom}(\sigma(x))$ is a leaf and we have $\mathsf{child}_1(c_{dom}(\sigma(y)), c_{dom}(\sigma(x)))$. The converse is proved in a similar way.

The case of $\mathsf{target}(x, y)$ is more tedious as it relies on the recursive $\mathsf{lar}(x, y)$ formula, stating that x is a leaf, whose last ancestor to the right is y. This means that y denotes the up most node with $\mathsf{child}_1^*(y, x)$. A model of $\mathsf{target}(x, y)$ and its translation $[\![\phi(x)]\!]_r$ in Fig.10.

Auxiliary predicates:

$$\begin{aligned}
\mathsf{ar}'(x,p) &=_{\mathsf{def}} p(x) \wedge \forall y \forall z((\mathsf{child}_1(y,z) \wedge p(z)) \rightarrow p(y)) \\
\mathsf{ar}(x,p) &=_{\mathsf{def}} \mathsf{ar}'(x,p) \wedge \forall p'(\mathsf{ar}'(x,p') \rightarrow \mathsf{subset}(p,p')) \\
\mathsf{lar}(x,y) &=_{\mathsf{def}} \mathsf{leaf}(x) \wedge \exists p.p(y) \wedge \mathsf{ar}(x,p) \wedge (\mathsf{root}(y) \vee \exists y'\ \mathsf{child}_2(y',y))
\end{aligned}$$

Logical connectives

$$\begin{aligned}
[\![\exists x\ \psi]\!]_r &=_{\mathsf{def}} \exists x\ [\![\psi]\!]_r \\
[\![\exists p\ \psi]\!]_r &=_{\mathsf{def}} \exists p\ [\![\psi]\!]_r \\
[\![\psi \wedge \psi']\!]_r &=_{\mathsf{def}} [\![\psi]\!]_r \wedge [\![\psi']\!]_r \\
[\![\neg\psi]\!]_r &=_{\mathsf{def}} \neg[\![\psi]\!]_r \\
[\![p(x)]\!]_r &=_{\mathsf{def}} p(x)
\end{aligned}$$

Node relations

$$\begin{aligned}
[\![\mathsf{first_edge}(x,y)]\!]_r &=_{\mathsf{def}} \mathsf{child}_1(y,x) \wedge \mathsf{leaf}(x) \\
[\![\mathsf{next_edge}(x,y)]\!]_r &=_{\mathsf{def}} \exists z\ (\mathsf{child}_1(y,x) \wedge \mathsf{child}_1(x,z)) \\
[\![\mathsf{target}(x,y)]\!]_r &=_{\mathsf{def}} \exists z\ (\mathsf{child}_2(x,z) \wedge \mathsf{lar}(y,z)) \\
[\![\mathsf{label}_a(x)]\!]_r &=_{\mathsf{def}} \mathsf{label}_a(x) \quad (a \in \Sigma) \\
[\![\mathsf{last_edge}(x,y)]\!]_r &=_{\mathsf{def}} \exists z\ (\mathsf{lar}(x,y) \wedge \mathsf{child}_1(y,z)) \\
[\![\mathsf{root}(x)]\!]_r &=_{\mathsf{def}} \exists y\ (\mathsf{lar}(x,y) \wedge \mathsf{root}(y)) \\
[\![\mathsf{leaf}(x)]\!]_r &=_{\mathsf{def}} \exists y\ \mathsf{child}_2(y,x) \wedge \mathsf{leaf}(x)
\end{aligned}$$

Fig. 8. Unranked into ranked MSO

$$\begin{aligned}
[\![\exists x\ \psi]\!]_u &=_{\mathsf{def}} \exists x\ [\![\psi]\!]_u \\
[\![\exists p\ \psi]\!]_u &=_{\mathsf{def}} \exists p\ [\![\psi]\!]_u \\
[\![\psi \wedge \psi']\!]_u &=_{\mathsf{def}} [\![\psi]\!]_u \wedge [\![\psi']\!]_u \\
[\![\neg\psi]\!]_u &=_{\mathsf{def}} \neg[\![\psi]\!]_u \\
[\![p(x)]\!]_u &=_{\mathsf{def}} p(x)
\end{aligned}$$

$$\begin{aligned}
[\![\mathsf{child}_1(x,y)]\!]_u &=_{\mathsf{def}} \mathsf{first_edge}(y,x) \vee \mathsf{next_edge}(y,x) \\
[\![\mathsf{child}_2(x,y)]\!]_u &=_{\mathsf{def}} (\mathsf{target}(x,y) \wedge \mathsf{leaf}(y)) \\
&\qquad \vee \exists z\ (\mathsf{target}(x,z) \wedge \mathsf{last_edge}(z,y)) \\
[\![\mathsf{label}_a(x)]\!]_u &=_{\mathsf{def}} \mathsf{label}_a(x) \quad (a \in \Sigma) \\
[\![\mathsf{root}(x)]\!]_u &=_{\mathsf{def}} (\mathsf{leaf}(x) \wedge \mathsf{root}(x)) \\
&\qquad \vee \exists z\ (\mathsf{root}(z) \wedge \mathsf{last_edge}(x,z)) \\
[\![\mathsf{leaf}(x)]\!]_u &=_{\mathsf{def}} \mathsf{root}(x) \vee \exists z\ \mathsf{target}(z,x)
\end{aligned}$$

Fig. 9. Ranked into unranked MSO

7 Monadic Datalog

We next express regular monadic queries in Monadic Datalog, a logic programming language, and discuss the expressive power compared to automata and MSO queries, for ranked and unranked trees.

We consider monadic Datalog in trees without negation. The languages $\mathsf{Datalog}^\alpha$ have the same signatures as MSO^α. The programs of $\mathsf{Datalog}^\alpha$ are logic program without function symbols, predefined n-ary predicates in R^α, and free monadic predicates $p, q \in \mathsf{Preds}$. More precisely, a program $P \in \mathsf{Datalog}^\alpha$ is a finite set of rules of the form:

$$p(x) :-\ \mathit{Body}$$

where *Body* is a sequence of goals with n-ary predicates $B_n \in R^\alpha$:

$$\mathit{Body} ::= B_n(x_1, \ldots, x_n) \ \mid\ p(x) \ \mid\ \mathit{Body}, \mathit{Body}'$$

Every program of $\mathsf{Datalog}^\alpha$ can be seen as a formula of MSO^α; sets of clauses are conjunctions, clauses $p(x) :-\ \mathit{Body}$ are universally quantified implications $\forall\mathsf{Vars}.\ p(x) \leftarrow \mathit{Body}$, and bodies *Body* conjunctions of goals.

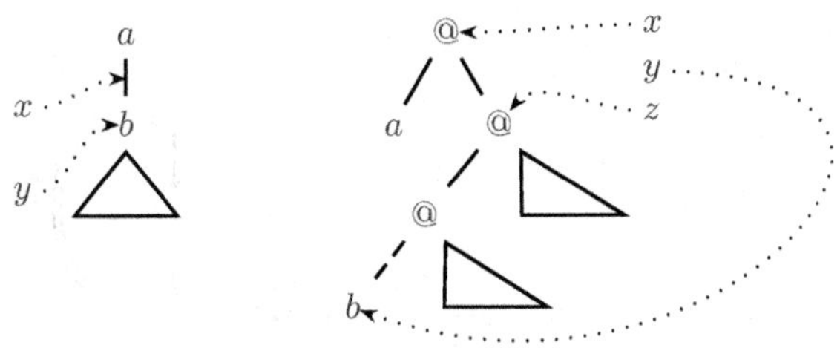

Fig. 10. A solution of $\mathsf{target}(x,y)$ on the left; a solution of its ranked translation $[\![\mathsf{target}(x,y)]\!]_r = \exists z\ (\mathsf{child}_2(x,z) \wedge \mathsf{lar}(z,y))$ on the right

We interpret programs in $\mathsf{Datalog}^\alpha$ in the least fixed point semantics over the structures of MSO^α. For every program $P \in \mathsf{Datalog}^\alpha$, predicate $p \in \mathsf{Preds}$, and $t \in \mathsf{tree}^\alpha$ let $T^\omega_{P,t}(p)$ be the least solution of P over the tree structure of the α-tree t for predicate p. This yields a notion of monadic queries:

$$\mathsf{query}^\alpha_{P(p)}(t) = T^\omega_{P,t}(p)$$

Least fixed points can be expressed in MSO. As a consequence, every query in $\mathsf{Datalog}^\alpha$ can be expressed in linear time in MSO^α. This shows that ranked queries in $\mathsf{Datalog}^r$ are regular (Theorem 2).

Ranked monadic run-based automata queries can always be expressed in ranked monadic Datalog; the reduction is in linear time. The resulting Datalog program models the two phases of the linear time algorithm for answering automata queries: the first bottom up phase computes all states of all nodes seen in all runs of the automaton, and a top down phase selects all nodes labeled by selection states in successful runs.

Theorem 4. *Ranked and unranked monadic queries expressed in monadic Datalog correspond, and are thus regular:*

$$\{\mathit{query}^r_{P(p)} \mid P \in \mathsf{Datalog}^r\} = \{c_{query}(\mathit{query}^u_{P'(p)}) \mid P' \in \mathsf{Datalog}^u\}$$

Corresponding unranked queries can be computed from ranked queries in linear time; the converse is not true.

It suffices to encode ranked Datalog queries into corresponding unranked queries in linear time. The translation basically refines the encoding from MSO^r into MSO^u: roughly, a rule $p(x)$:– *Body* is translated into $p(x)$:– $[\![Body]\!]_u$. Conjunctions in $[\![Body]\!]_u$ can be replaced by commas, existential quantifications can be ommited, i.e., replaced by implicit universal quantification in rules. Disjunctions as in the definitions of $[\![\mathsf{child}_1(x,y)]\!]_u$, $[\![\mathsf{child}_2(x,y)]\!]_u$, and $[\![\mathsf{root}(x)]\!]_u$ can expressed by multiple rules. Such a rewriting, however, spoils linear time. We can circumvent this problem following Gottlob and Koch [7]: we normalise programs of $\mathsf{Datalog}^r$ into *tree marking normal form* (TMNF) in linear time before translation. TMNF programs have of forms:

$$p(x) :- B_1(x). \qquad p(x) :- q(y), B_2(y,x).$$
$$p(x) :- p_0(x), p_1(x). \qquad p(x) :- q(y), B_2(x,y).$$

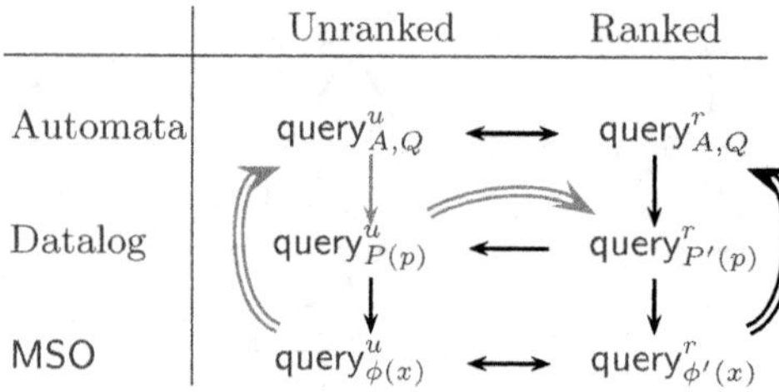

Fig. 11. Summary of reductions. Solid lines are in linear time, double lines are non elementary. Black lines are proved, red lines induced

$$\frac{t = a(t_1, \ldots, t_n) \qquad \forall 1 \leq i \leq n : r_{|\mathsf{nodes}(t_i)} \in \mathsf{run}_H(t_i) \qquad a(L) \to r(\mathsf{root}(t)) \in \mathsf{rules}(H) \qquad r(\mathsf{root}(t_1)) \ldots r(\mathsf{root}(t_n)) \in L}{r \in \mathsf{run}_H(t)}$$

Fig. 12. Runs of Hedge Automata

where B_n is a n-ary predicate of R^r. On ranked TMNF programs the reduction $[\![\text{-}]\!]_u$ can clearly be done in linear time.

The inverse translation can be composed from our translations so far, which are summarized in Fig. 11. We first reduce unranked Datalog queries to MSO^u, then to MSO^r, move to ranked automata queries and then to ranked Datalog queries. The overall reduction has nonelementary complexity.

Note that we cannot specialize the translation $[\![\text{-}]\!]_r$ from MSO^u to MSO^r into a translation from $\mathsf{Datalog}^u$ to $\mathsf{Datalog}^r$. The problem is that we cannot express the auxilary binary predicate $\mathsf{lar}(x, y)$. Its definition is recursive, but only monadic recursive predicates can be defined in monadic Datalog.

Corollary 1. *Stepwise tree automata, monadic Datalog, and* MSO *capture the class of regular monadic queries over unranked trees.*

This is a corollary of our correspondences between ranked and unranked queries and traditional result on ranked trees. All unranked automata queries $\mathsf{query}^u_{A,Q}$ can be expressed in linear time in $\mathsf{Datalog}^u$, by indirection over ranked automata queries and $\mathsf{Datalog}^r$. Unranked queries in MSO^u can be expressed by unranked automata queries by reduction to the ranked case.

8 Hedge Automata

We finally show how to express monadic queries with hedge automata [16, 3] in linear time with stepwise tree automata. A hedge automaton H over Σ consists of a set $\mathsf{states}(H)$ of states, a set $\mathsf{final}(H)$ of final states, and set set of transition rules of the form $a(L) \to q$ where L is a regular set of words over states.

Runs of hedge automata H on unranked trees t are functions $r : \mathsf{nodes}(t) \to \mathsf{states}(H)$ that satisfy the inference rule in Fig 12. A hedge automaton H and a set of states $Q \subseteq \mathsf{states}(H)$ defines a monadic query for unranked trees:

$$\mathsf{query}_{H,Q}(t) = \{\pi \in \mathsf{nodes}(t) \mid r \in \mathsf{succ\_runs}_H(t), r(\pi) \in Q\}$$

$$\begin{aligned}
\mathsf{states}(\mathsf{step}(H)) &= \mathsf{states}(H) \uplus \biguplus_{a\in\Sigma,q\in\mathsf{states}(H)} \mathsf{states}(H_{a,q}) \\
\mathsf{rules}(\mathsf{step}(H)) &= \bigcup_{a\in\Sigma,q\in\mathsf{states}(H)} \mathsf{rules}(H_{a,q}) \\
&\cup \{p \xrightarrow{\epsilon} q \mid p \in \mathsf{final}(H_{a,q}), q \in \mathsf{states}(H), a \in \Sigma\} \\
&\cup \{a \to p \mid p \in \mathsf{init}(H_{a,q}), q \in \mathsf{states}(H)\} \\
\mathsf{final}(\mathsf{step}(H)) &= \mathsf{final}(H)
\end{aligned}$$

Fig. 13. Hedge automata into stepwise tree automata

For translating hedge automata into stepwise tree automata, we need to represent all regular language L in transition rules explicitly. We use a sequence of finite word automata $(H_{a,q})_{a\in\Sigma,q\in\mathsf{states}(H)}$ over the alphabet $\mathsf{states}(H)$ to do so.

Proposition 4. *Queries by hedge automata can be translated in linear time to queries by stepwise tree automata.*

Proof. Given an hedge automaton H we define a stepwise tree automaton $\mathsf{step}(H)$ by unifying all subautomata $H_{a,q}$ into a single finite automaton. We then add all states of $\mathsf{states}(A)$ to this automaton, and link them to all final states p of $H_{a,q}$ through ϵ-transitions $p \xrightarrow{\epsilon} q$. We add rules $a \to q'$ for all initial states q' of some $H_{a,q}$. The final states of $\mathsf{step}(H)$ are those in $\mathsf{final}(H)$, not those in $\mathsf{final}(H_{a,q})$. The complete construction is detailed in Fig. 13. It remains to show that every run of H can be simulated by a run of $\mathsf{step}(H)$. □

Acknowledgments

We would like to thank our collegues from the Mostrare project for their continuous support, particularly R. Gilleron and S. Tison, and the anonymous referees for hinting us towards Currification.

References

1. R. Baumgartner, S. Flesca, and G. Gottlob. Visual web information extraction with Lixto. In *The Very Large Data Bases Journal*, pages 119–128, 2001.
2. A. Berlea and H. Seidl. Binary queries. In *Proceedings of Extreme Markup Languages*, Montreal, 2002.
3. A. Brüggemann-Klein, M. Murata, and D. Wood. Regular tree and regular hedge languages over unranked alphabets. Technical report, 2001.
4. H. Comon, M. Dauchet, R. Gilleron, F. Jacquemard, D. Lugiez, S. Tison, and M. Tommasi. Tree automata techniques and applications. Online book, 450 pages. Available at: http://www.grappa.univ-lille3.fr/tata, 1997.
5. B. Courcelle. On recognizable sets and tree automata. In H. Ait-Kaci and M. Nivat, editors, *Resolution of Equations in Algebraic Structures, Algebraic Techniques*, volume 1, chapter 3, pages 93–126. Academic Press, 1989.
6. M. Frick, M. Grohe, and C. Koch. Query evaluation on compressed trees. In *Proceedings of the IEEE Symposium on Logic In Computer Sciences*, Ottawa, 2003.

7. G. Gottlob and C. Koch. Monadic datalog and the expressive power of languages for web information extraction. In *Proceedings of the ACM Symposium on Principle of Databases Systems*, pages 17–28, 2002.
8. G. Gottlob and C. Koch. Monadic queries over tree-structured data. In *Proceedings of the IEEE Symposium on Logic In Computer Sciences*, Copenhagen, 2002.
9. G. Gottlob, C. Koch, and R. Pichler. The complexity of XPATH query evaluation. In *Proceedings of the ACM Symposium on Principle of Databases Systems*, pages 179–190. 2003.
10. G. Gottlob, C. Koch, and R. Pichler. XPATH processing in a nutshell. *ACM SIGMOD Record*, 32(2):21–27, 2003.
11. C. Koch. Efficient processing of expressive node-selecting queries on XML data in secondary storage: A tree automata-based approach. In *Proceedings of the International Conference on Very Large Data Bases*, 2003.
12. A. Neumann, H. Seidl. Locating matches of tree patterns in forests. In *Foundations of Software Technology and Theoretical Computer Science*, pages 134–145, 1998.
13. F. Neven and T. Schwentick. Query automata over finite trees. *Theoretical Computer Science*, 275(1-2):633–674, 2002.
14. J. Niehren and A. Podelski. Feature automata and recognizable sets of feature trees. In *Proceedings of TAPSOFT'93*, volume 668 of *LNCS*, pages 356–375, 1993.
15. H. Seidl, T. Schwentick, and A. Muscholl. Numerical document queries. In *Proc. of the IEEE Symposium on Principles of Database Systems*, pages 155–166, 2003.
16. J. W. Thatcher. Characterizing derivation trees of context-free grammars through a generalization of automata theory. *J. of Comp. and Syst. Sci.*, 1:317–322, 1967.
17. J. W. Thatcher, J. B. Wright. Generalized finite automata with an application to a decision problem of second-order logic. *Math. System Theory*, 2:57–82, 1968.
18. S. D. Zilio and D. Lugiez. XML schema, tree logic and sheaves automata. In R. Nieuwenhuis, editor, *Proceedings of RTA*, volume 2706 of *LNCS*, pages 246–263. 2003.

A Verification Technique Using Term Rewriting Systems and Abstract Interpretation*

Toshinori Takai

CREST, Japan Science and Technology Agency (JST)
t-takai@aist.go.jp

Abstract. Verifying the safety property of a transition system given by a term rewriting system is an undecidable problem. In this paper, we give an abstraction for the problem which is automatically generated from a given TRS by using abstract interpretation. Then we show that there are some cases in which the problem can be decided. Also we show a new decidable subclass of term rewriting systems which effectively preserves recognizability.

1 Introduction

Term rewriting systems (TRSs) can be used for modelling infinite state transition systems, e.g. cryptographic protocols, and some verification techniques for TRSs are proposed. Because of the universal computational power of TRSs, even the safety property for a TRS is undecidable. There are two approaches:

1. to find a decidable sufficient condition [12, 13] and
2. to give an abstraction [5, 8].

If a given TRS satisfies the condition of the first approach, then we can automatically verify, for example, safety property. However, some simple TRSs do not satisfy any of the conditions. On the other hand, in order to give an abstraction, one have to analyse an instance of the problem and give a special abstraction. In this paper, we give an abstraction for the problem which is automatically generated by using *abstract interpretation*[4, 9].

Abstract interpretation is a typical method to give an abstraction for data domains of programs. We give an abstraction for the set of all terms by giving an equational theory. The state space (the set of all terms) is abstracted to equivalence classes induced by a given equational theory. A similar idea can be found in [7]. The purpose of this study is to give a construction of an appropriate equational theory which defines an abstract domain for the state space.

We also propose a new decidable sufficient condition for a verification problem to be decidable. The decidable class is given by the condition that no abstraction is needed in our method.

* This work is done as a part of CREST project of Japan Science and Technology Agency.

V. van Oostrom (Ed.): RTA 2004, LNCS 3091, pp. 119–133, 2004.

We use the usual notions of *a (ground, linear) term, a position, a context, a TRS, a rewrite relation, a (recognizable) tree language, a tree automaton, etc.* See [1, 3] for details. In the following, we only deal with linear TRSs. For a signature Σ and a set of variables $\mathcal{V}$, $\mathcal{T}(\Sigma, \mathcal{V})$ denotes the set of all terms constructed from Σ and $\mathcal{V}$ and $\mathcal{T}(\Sigma)$ denotes the set of all ground terms constructed from Σ. In this paper, we use $f, g, h, \ldots$ as function symbols, $x, y, z, \ldots$ as variables and $a, b, c, \ldots$ as constant symbols. For a term t, let $\mathcal{P}os(t)$ be the set of all positions of t and $\mathcal{P}os_{\mathcal{V}}(t) = \{p \in \mathcal{P}os(t) \mid t/p \in \mathcal{V}\}$ where t/p is the subterm of t at p. A TRS constructed from terms in $\mathcal{T}(\Sigma, \mathcal{V})$ will be said to be *a TRS on* $\mathcal{T}(\Sigma, \mathcal{V})$. For a TRS $\mathcal{R}$, $\to_{\mathcal{R}}$ denotes the rewrite relation induced by $\mathcal{R}$ and its reflexive and transitive closure is denoted by $\to_{\mathcal{R}}^*$. For a tree language L_1 and a TRS $\mathcal{R}$, the image of L_1 by the relation $\to_{\mathcal{R}}^*$ is denoted by $(\to_{\mathcal{R}}^*)(L_1)$, i.e. $(\to_{\mathcal{R}}^*)(L_1) = \{t \mid s \to_{\mathcal{R}}^* t, \ s \in L_1\}$. A tree automaton (TA) is a 4-tuple $(\Sigma, \mathcal{Q}, \mathcal{Q}_f, \Delta)$ where Σ is a signature, $\mathcal{Q}$ is a set of states, $\mathcal{Q}_f$ is a set of final states and Δ is a set of transition rules. Transition rules are given as rewrite rules and the transition relation is defined as the rewrite relation of transition rules as in [3]. For a TA $\mathcal{A}$, let $\mathcal{L}(\mathcal{A})$ denote the accepting language of $\mathcal{A}$.

2 Verification Problem

In this section, we define the verification problem which we deal with throughout this paper. The problem is defined as follows.

Problem 1. For given a signature Σ, a set $\mathcal{V}$ of variables, a TRS $\mathcal{R}$ on $\mathcal{T}(\Sigma, \mathcal{V})$ and two recognizable tree languages $L_1, L_2 \subseteq \mathcal{T}(\Sigma)$, decide whether or not $(\to_{\mathcal{R}}^*)(L_1) \subseteq L_2$. □

Usually each recognizable tree language is given by a TA. Problem 1 can be regarded as a verification problem as follows. For a TRS $\mathcal{R}$ on $\mathcal{T}(\Sigma, \mathcal{V})$, we can consider a transition system whose state space is $\mathcal{T}(\Sigma)$ and the transition relation is defined by the rewrite relation by $\mathcal{R}$. When we regard terms (states) in L_1 as the initial states, $(\to_{\mathcal{R}}^*)(L_1)$ means the set of all reachable states of the transition system. Once a safety property or an invariant is given by the set L_2, we can see that the transition system is safe if and only if $(\to_{\mathcal{R}}^*)(L_1) \subseteq L_2$. The situation mentioned above is shown in Fig. 1.

Problem 1 is undecidable; this fact can easily be seen from the fact that the reachability problem of a TRS is undecidable in general.

3 Abstract Interpretation

Abstract interpretation[4, 9] can deal with the following problem.

Problem 2. For given an ordered set $(C, \leq)$ and elements $c_1, c_2 \in C$, decide $c_1 \leq c_2$ or not. □

When deciding $c_1 \leq c_2$ in the domain C is difficult or undecidable, abstract interpretation can be used as follows.

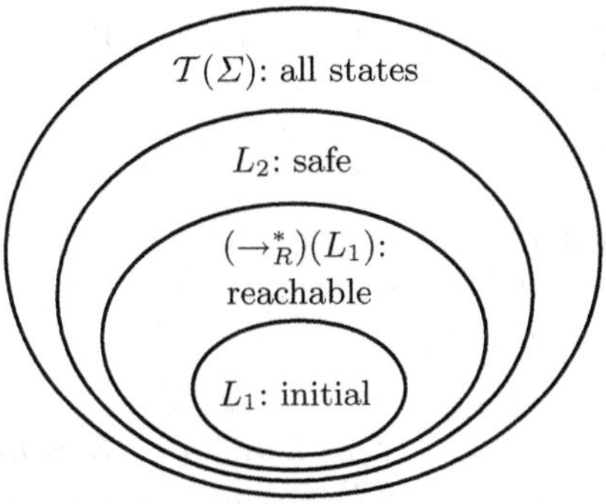

Fig. 1. A verification problem

1. Find another ordered set $(A, \leq_A)$ and mappings $\alpha\colon C \to A$ and $\gamma\colon A \to C$ satisfying the following condition.
$$\forall c \in C \forall a \in A.\ c \leq \gamma(a) \text{ iff } \alpha(c) \leq_A a. \tag{1}$$
The mappings α and γ satisfying the condition above is called *Galois connection*. In this situation, $(C, \leq)$ is often called the *concrete domain* and $(A, \leq_A)$ the *abstract domain*.
2. Find $a_1 \in A$ satisfying the following condition.
$$\alpha(c_1) \leq a_1 \text{ and } \gamma(a_1) \leq c_2. \tag{2}$$
From the condition (1), if there is such a_1, we can conclude that $c_1 \leq c_2$.

Since the original problem Problem 2 has been divided into two problems as (2), one may think that the problem becomes more difficult. But in some special cases as mentioned below, we can easily find a_1 satisfying (2). For functions f and g, let $lfp(\mathsf{f})$ denote the least fixed point of f and $\mathsf{f} \circ \mathsf{g}$ denote the composition of f and g. If $c_1 = lfp(\mathsf{f})$ for a monotone function $\mathsf{f}\colon C \to C$, let $\mathsf{g}\colon A \to A$ be a function defined as $\mathsf{g} = \alpha \circ \mathsf{f} \circ \gamma$. If $lfp(\mathsf{g}) \in A$ exists, then we can see $\alpha(lfp(\mathsf{f})) \leq_A lfp(\mathsf{g})$ from the condition (2).

4 Abstraction for Tree Languages

In this section we define an abstraction for verification problems by adapting Problem 1 to Problem 2. First, let $(C, \leq)$ be $(2^{\mathcal{T}(\Sigma)}, \subseteq)$, the set of all tree languages and the inclusion order, c_1 and c_2 be $(\to_{\mathcal{R}}^*)(L_1)$ and L_2, respectively. Then Problem 1 can be seen as an instance of Problem 2. In the following, we assume that Σ, $\mathcal{V}$, $\mathcal{R}$ and L_1, L_2 as in Problem 1 are given. Let $\mathsf{f}_{\mathcal{R}}\colon 2^{\mathcal{T}(\Sigma)} \to 2^{\mathcal{T}(\Sigma)}$ be a function defined as $\mathsf{f}_{\mathcal{R}}(L) = \{t \mid s \to_{\mathcal{R}} t, s \in L\} \cup L_1$, then we obtain that $lfp(\mathsf{f})$ is the greatest lower bound of the sequence $(\mathsf{f}_{\mathcal{R}}^n(\emptyset))_n$ with $n \geq 0$ and thus $lfp(\mathsf{f}_{\mathcal{R}}) = (\to_{\mathcal{R}}^*)(L_1)$. We define an abstract domain and mappings α and γ to satisfy that
$$\gamma(lfp(\mathsf{g}_{\mathcal{R}})) \text{ is recognizable} \tag{3}$$

if $lfp(\mathsf{g}_\mathcal{R})$ is obtained where $\mathsf{g}_\mathcal{R} = \alpha \circ \mathsf{f}_\mathcal{R} \circ \gamma$. If an abstraction satisfies the condition (3), then we can check

$$\gamma(lfp(\mathsf{g}_\mathcal{R})) \subseteq L_2 \tag{4}$$

or not and if (4) holds, then we can see that $(\to^*_\mathcal{R})(L_1) \subseteq L_2$.

The abstraction used in this paper is defined by an equational theory. A similar idea can be found in [7].

Definition 1. *For a set $\mathcal{E}$ of equations, let $A_\mathcal{E}$ be $A_\mathcal{E} = 2^{\mathcal{T}(\Sigma)/\mathcal{E}}$ and $\eta\colon \mathcal{T}(\Sigma) \to \mathcal{T}(\Sigma)/\mathcal{E}$ be $t \mapsto [t]_{\approx_\mathcal{E}}$ for $t \in \mathcal{T}(\Sigma)$ where $\mathcal{T}(\Sigma)/\mathcal{E}$ is the equivalence classes of the equational theory induced by $\mathcal{E}$ and $\approx_\mathcal{E}$ denotes the equational theory induced by $\mathcal{E}$. We call $A_\mathcal{E}$ the* equation-based abstraction *by $\mathcal{E}$ with the mappings $\alpha_\mathcal{E}\colon 2^{\mathcal{T}(\Sigma)} \to 2^{\mathcal{T}(\Sigma)/\mathcal{E}}$ and $\gamma_\mathcal{E}\colon 2^{\mathcal{T}(\Sigma)/\mathcal{E}} \to 2^{\mathcal{T}(\Sigma)}$ which are induced from η, i.e. $\alpha_\mathcal{E}(L) = \{\eta(t) \mid t \in L\}$ and $\gamma_\mathcal{E}(L) = \{t \mid \eta(t) \in L\}$.*

Example 1. Let $\mathcal{R}_1$ be $\{f(x,y) \to f(g(x),h(y))\}$, $\mathcal{E}$ be a set of equations consisting of $\{g(g(x)) = g(x), h(h(x)) = h(x)\}$ and L_1 be a recognizable tree language $\{f(a,b)\}$. Then, we have

$$(\to^*_{\mathcal{R}_1})(L_1) = \{f(g^n(a),h^n(b)) \mid n \geq 0\} \text{ and} \tag{5}$$

$$\gamma_\mathcal{E} \circ \alpha_\mathcal{E}((\to^*_{\mathcal{R}_1})(L_1)) = \{f(g^n(a),h^m(b)) \mid n,m \geq 1\} \cup \{f(a,b)\}. \tag{6}$$

In fact, (6) is recognizable whereas (5) is not and (6) includes (5). □

In general, for a set $\mathcal{E}$ of equations, the condition (3) does not holds by the equation-based abstraction by $\mathcal{E}$. From the next section, we present a technique to obtain a set of equations which can be used for solving verification problems in the sense that (3) holds as shown in Example 1.

5 Verification Procedure

For the function $\mathsf{f}_\mathcal{R}$ in the previous section ($\mathsf{f}_\mathcal{R}(L) = \{t \mid s \to_\mathcal{R} t, s \in L\} \cup L_1$), we adopt the following procedures. In the following, a state of TAs may have a tree structure. A state having a tree structure t may be written as $\langle t \rangle$ to emphasize that t is a state. We call a tree structured state $\langle t \rangle$ of the form $t = f(t_1,\ldots,t_n)$ a *term state*, otherwise a *constant state*. See the following example.

Example 2. Let $\mathcal{A}$ be a TA consisting of $\{c \to q,\ g(q) \to q\}$ with the final state q and $\mathcal{R} = \{g(x) \to h(x)\}$ be a TRS. If we add new transition rules $h(q) \to \langle h(q) \rangle$ and $\langle h(q) \rangle \to q$ with a new state $\langle h(q) \rangle$ to $\mathcal{A}$, we can obtain a TA accepting $(\to^*_\mathcal{R})(\mathcal{L}(\mathcal{A}))$. Here, the state q is a constant state and $\langle h(q) \rangle$ is a term state. □

Procedure 1. (addtrans) This procedure takes a term t on $\mathcal{T}(\mathcal{F} \cup \mathcal{Q})$. If t has already been defined as a state, then the procedure defines no transitions. Otherwise the procedure defines new states and transition rules of $\mathcal{A}$ as follows: (i) If $t = c$ with c a constant, then define $c \to \langle c \rangle$ as a transition rule. (ii) If $t = f(t_1,\ldots,t_n)$ with f a function symbol of arity n, then define $f(\langle t_1 \rangle,\ldots,\langle t_n \rangle) \to \langle t \rangle$ and execute $\mathsf{addtrans}(t_i)$ for $1 \leq i \leq n$. □

Procedure 2. (modify) This procedure takes a TA $\mathcal{A} = (\mathcal{F}, \mathcal{Q}, \mathcal{Q}_f, \Delta)$ and a linear TRS $\mathcal{R}$ on $\mathcal{T}(\Sigma, \mathcal{V})$ as inputs and outputs TA $\mathcal{A}'$ as follows: For any rule $l \to r \in \mathcal{R}$, any substitution $\sigma\colon \mathcal{V} \to \mathcal{Q}$ and any state $q \in \mathcal{Q}$, if $l\sigma \to^*_{\mathcal{A}} q$ and $r\sigma \not\to^*_{\mathcal{A}} q$, then construct $\mathcal{A}'$ by adding $\langle r\sigma \rangle \to q$ to $\mathcal{A}$ and execute $\mathsf{addtrans}(r\sigma)$ where $\langle r\sigma \rangle$ is a term state. That is, construct a new TA by adding new transition rules and states to satisfy that $r\sigma \to^* q$ holds. A transition move caused by $\langle r\sigma \rangle \to q$ is called a *rewriting move.* □

We assume that any state not defined in addtrans is a constant state. Let *Rec* be the class of recognizable tree languages. In the following, we regard a TA $\mathcal{A}$ as a tree language $\mathcal{L}(\mathcal{A})$ and for a TRS $\mathcal{R}$ the procedure modify as a function $\mathsf{modify}_{\mathcal{R}}\colon Rec \to Rec$ defined by $\mathcal{A} \mapsto \mathsf{modify}(\mathcal{A}, \mathcal{R})$. Although the function $\mathsf{modify}_{\mathcal{R}}$ is not a strict implementation of $\mathsf{f}_{\mathcal{R}}$, it is proved in [11, 12] that $lfp(\mathsf{f}_{\mathcal{R}}) = lfp(\mathsf{modify}_{\mathcal{R}})$. From this fact and the properties of recognizable tree languages, we can obtain the following theorem. For an integer $k \geq 0$, let $\mathsf{modify}^k_{\mathcal{R}}(\mathcal{A})$ be a TA obtained from a TA $\mathcal{A}$ by applying $\mathsf{modify}_{\mathcal{R}}$ for k times.

Theorem 1. *[11, 12] For a verification problem $(\mathcal{R}, L_1, L_2)$ (Problem 1), if there is an integer $k \geq 0$ such that $\mathsf{modify}^{k+1}_{\mathcal{R}}(\mathcal{A}) = \mathsf{modify}^k_{\mathcal{R}}(\mathcal{A})$ where $\mathcal{L}(\mathcal{A}) = L_1$, then $(\to^*_{\mathcal{R}})(\mathcal{L}(\mathcal{A})) = \mathcal{L}(\mathsf{modify}^k_{\mathcal{R}}(\mathcal{A}))$ and it is recognizable.* □

This theorem says that if the procedure terminates for a given input, then the verification problem given by the input can be solved effectively.

6 Verification Procedure with Abstraction

In this section, first we give a verification procedure with abstraction. Then we consider a class of TRSs which do not need any abstraction.

The basic idea of the construction of equations is to find contexts which may appear in terms in $lfp(\mathsf{f}_{\mathcal{R}})$ repeatedly. For example, in Example 1, a context $g(\Box)$ can be such a context. For such a context C, define an equation $C[C[x]] = C[x]$, which intuitively means that once a term t containing C, i.e. $t = C'[C[t']]$ where C' is a context and t' a term, appears in $lfp(\mathsf{f}_{\mathcal{R}})$, we approximate $lfp(\mathsf{f}_{\mathcal{R}})$ by a tree language including terms of the forms $C'[C^n[t']]$ for $n \geq 1$. We try to find such contexts by analyzing a kind of overlapping relation between rewrite rules in TRS. We define a graph (Definition 3) for the analysis constructed from $\mathcal{R}$. The overlapping relation is defined as follows.

Definition 2. *Let λ denote the root position.* A term s sticks-out of a term t at *(p_1, p_2) with $p_1 \in \mathcal{P}os_{\mathcal{V}}(t)$ and $p_2 \in \mathcal{P}os_{\mathcal{V}}(s)$ if for any position p with $\lambda \preceq p \prec p_1$ we have $p \in \mathcal{P}os(s)$ and the function symbol of s at p and the function symbol of t at p are the same and $p_1 \preceq p_2$. If s sticks-out of t at (p_1, p_2) and $p_1 \prec p_2$, then we say that s* properly sticks-out of t.

Example 3. Let s and t be terms $f_3(h(x), g(h(y)), b)$ and $f_3(a, g(x), g(y))$, respectively. Then s properly sticks-out of t at $(2.1, 2.1.1)$. □

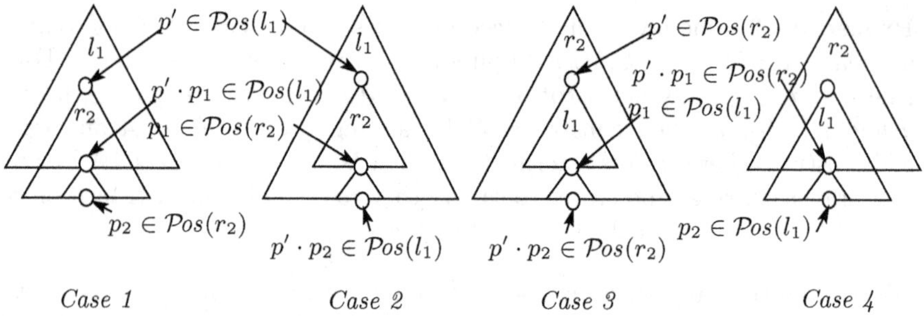

Fig. 2. Conditions for four cases

Definition 3. *The* generalized sticking-out graph *of a TRS $\mathcal{R}$ is a directed graph $G = (V, E)$ where $V = \{(l \to r, v) \mid l \to r \in \mathcal{R}, v \in \mathcal{V}ar(l) \cup \mathcal{V}ar(r)\}$ and each edge in E, defined as follows, has a weight. Let $l_1 \to r_1$ and $l_2 \to r_2$ be (possibly identical) rewrite rules in $\mathcal{R}$. The following situations from 1 to 4 are illustrated in Fig. 2.*

1. *If r_2 properly sticks-out of l_1/p' at (p_1, p_2) with $p' \in \mathcal{P}os(l_1)$, then E contains an edge from $(l_2 \to r_2, r_2/p_2)$ to $(l_1 \to r_1, l_1/(p' \cdot p_1))$ with weight one.*
2. *If l_1/p' sticks-out of r_2 at (p_1, p_2) with $p' \in \mathcal{P}os(l_1)$, then E contains an edge from $(l_2 \to r_2, r_2/p_1)$ to $(l_1 \to r_1, l_1/(p' \cdot p_2))$ with weight zero.*
3. *If r_2/p' at properly sticks-out of l_1 at (p_1, p_2) with $p' \in \mathcal{P}os(r_2)$, E contains an edge from $(l_2 \to r_2, r_2/(p' \cdot p_2))$ to $(l_1 \to r_1, l_1/p_1)$ with weight one.*
4. *If l_1 sticks-out of r_2/p' at (p_1, p_2) with $p' \in \mathcal{P}os(r_2)$, then E contains an edge from $(l_2 \to r_2, r_2/(p' \cdot p_1))$ to $(l_1 \to r_1, l_1/p_2)$ with weight zero.*

Example 4. Let $\mathcal{R}_2, \mathcal{R}_3$ and $\mathcal{R}_4$ be TRSs defined as follows.

$$\mathcal{R}_2 \left\{ v\colon h(f_2(x, h(g(y)))) \to f_2(g(k(y)), h(x)) \right.$$

$$\mathcal{R}_3 \begin{cases} v_1\colon g(x) \to k(l(x)) \\ v_2\colon k(x) \to f(g(h(x))) \\ v_3\colon f(x) \to k(l'(x)) \end{cases} \mathcal{R}_4 \begin{cases} u_1\colon g(f(k(k(x)))) \to g(f(k(h(x)))) \\ u_2\colon h(f(x)) \to f(k(x)) \end{cases}$$

Then the generalized sticking out graphs $G_{\mathcal{R}_2}, G_{\mathcal{R}_3}$ and $G_{\mathcal{R}_4}$ of $\mathcal{R}_2, \mathcal{R}_3$ and $\mathcal{R}_2$ respectively are shown in Fig. 3. Let us see the construction of the generalized sticking-out graph $G_{\mathcal{R}_2}$. Since the right-hand side of v properly sticks-out of $h(f_2(x, h(g(y))))/1$, a subterm of the left-hand side, at $(1, 1.1.1)$, which is the case 1 of Definition 3, there is an edge from (v, y) to (v, x) with weight 1. Moreover, since $h(f_2(x, h(g(y))))/1$, a subterm of the left-hand side, sticks-out of the right-hand side at $(2.1, 2.1.1)$, there is an edge from (v, x) to (v, y) with weight 0 due to case 2 of Definition 3. □

For defining $\mathsf{g}_{\mathcal{R}}$ in Sect. 3, we consider a function $\triangledown\colon Rec \to Rec$, called an *abstraction operator* defined as $\triangledown = \gamma_E \circ \alpha_E$ where γ_E and α_E are defined by a set E of equations as in Definition 1. Once we obtain an abstraction operator

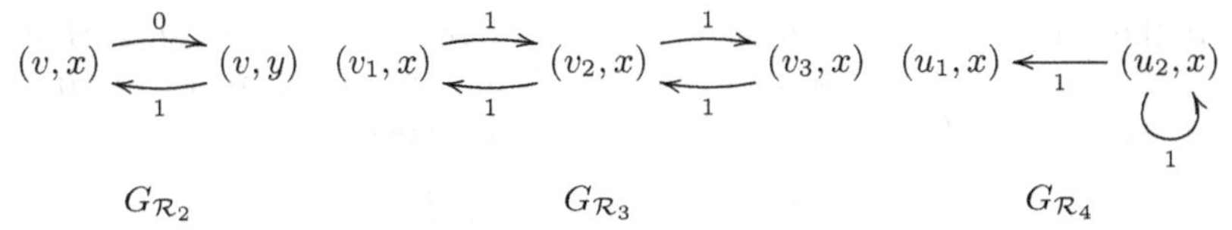

Fig. 3. Generalized sticking-out graphs

$\triangledown$, by using $\mathsf{modify}_{\mathcal{R}}$ and $\triangledown$, we can define the sequence of TAs $(\mathcal{A}^n_{\mathcal{R},\triangledown})_{n\geq 0}$ for a TA $\mathcal{A}$ and a TRS $\mathcal{R}$ as $\mathcal{A}^n_{\mathcal{R},\triangledown} = (\triangledown \circ \mathsf{modify}_{\mathcal{R}})^n(\mathcal{A})$. If there is m such that $\mathcal{A}^{m+1}_{\mathcal{R},\triangledown} = \mathcal{A}^m_{\mathcal{R},\triangledown}$, we can take $\mathcal{A}^m_{\mathcal{R},\triangledown}$ as an abstraction of $lfp(\mathsf{f}_{\mathcal{R}})$.

In our technique described below, the equations for abstraction are dynamically obtained during the iteration of procedure modify. In other words, instead of finding a set of equations statically from a given TRS and a TA, we use a sequence $(E_n)_{n\geq 0}$ of sets of equations with $E_n \subseteq E_{n+1}$ for $n \geq 0$. Although it may not be necessary to construct a set of equations dynamically, in order to prove that an obtained set of equations satisfies the condition (3), we give a verification procedure by modifying the procedure modify. Let $\triangledown_n = \gamma_{E_n} \circ \alpha_{E_n}$ be an abstraction operator defined from E_n for $n \geq 0$. Each E_n for $n \geq 0$ is obtained during the computation of $(\mathcal{A}^n_{\mathcal{R},\triangledown_n})_{n\geq 0}$ where $\mathcal{A}^n_{\mathcal{R},\triangledown_n}$ is defined as $\mathcal{A}^n_{\mathcal{R},\triangledown_n} = (\triangledown_n \circ \mathsf{modify}_{\mathcal{R}})^n(\mathcal{A})$. In this case, we still have that if there is m such that $\mathcal{A}^{m+1}_{\mathcal{R},\triangledown_{m+1}} = \mathcal{A}^m_{\mathcal{R},\triangledown_m}$, then we can take $\mathcal{A}^m_{\mathcal{R},\triangledown_m}$ as an abstraction of $lfp(\mathsf{f}_{\mathcal{R}})$.

For constructing a set of equations for abstraction, we modify the procedure modify by adding supplementary information S as an argument. The new argument S is a set and an element in S is intuitively a heap of contexts. The heap increases or decreases according to the edges of the generalized sticking-out graph. If the generalized sticking-out graph has a cycle with weight one or more, the procedure introduces an equation by using the information in S. The procedure $\triangledown_E$, defined below, corresponds to introducing an equation for abstraction.

Procedure 3. (e-modify) This procedure takes a TA $\mathcal{A} = (\mathcal{F}, \mathcal{Q}, \mathcal{Q}_f, \Delta)$, a linear TRS $\mathcal{R}$ on $\mathcal{T}(\Sigma, \mathcal{V})$ and S and S_e of tuples $(\langle t\rangle, o, H)$ of a state $\langle t\rangle \in \mathcal{Q}$, a position o of t and a sequence H of vertices of the generalized sticking-out graph $G_{\mathcal{R}}$ of $\mathcal{R}$ as inputs and outputs TA $\mathcal{A}'$ and sets S' and S'_e of the same things of S and S_e which are defined by adding something to $\mathcal{A}$ and sets S and S_e respectively as follows. For any rewrite rule $l \to r \in \mathcal{R}$, any substitution $\sigma\colon \mathcal{V} \to \mathcal{Q}$ and any state $q \in \mathcal{Q}$, if

$$l\sigma \to^*_{\mathcal{A}} q \text{ and } r\sigma \not\to^*_{\mathcal{A}} q, \tag{7}$$

then (1) construct $\mathcal{A}'$ by adding $\langle r\sigma\rangle \to q$ to $\mathcal{A}$, (2) execute $\mathsf{addtrans}(r\sigma)$ and (3) do the following. For any variable $x \in \mathcal{V}ar(l)$, let o_x, o'_x be the positions of x in l and r respectively and p_x be the state associated with x, i.e. $l/o_x = x$, $r/o'_x = x$ and $x\sigma = p_x$ and do the following.

1. For any position o' with $\lambda \prec o' \prec o'_x$ add $((\langle r\sigma/o'\rangle, o'', (l \to r, x))$ to S where o'' is a position such that $o' \cdot o'' = o'_x$.

2. If there is an element $(p_x, o_h, H) \in S$ for some o_h and H such that $H = H' \cdot (l' \to r', x')$ for some $H', l' \to r'$ and x' and there is an edge from $(l' \to r', x')$ to $(l \to r, x)$ in $G_{\mathcal{R}}$ with weight one or

$$\text{with weight zero and } |H'| \geq 1, \tag{8}$$

then (a) if H has a form $H = (l \to r, x) \cdot H''$ for some sequence H'', then add $(p_x, o_h, H \cdot (l \to r, x))$ to S_e and (b) for any position o' with $\lambda \prec o' \prec o'_x \cdot o_h$ add $(\langle r\sigma/o' \rangle, o'', H \cdot (l \to r, x))$ to S where o'' is a position such that $o' \cdot o'' = o'_x \cdot o_h$. □

Procedure 4. ($\triangledown_E$) The abstraction operator $\triangledown_E$ takes a TA $\mathcal{A} = (\mathcal{F}, \mathcal{Q}, \mathcal{Q}_f, \Delta)$, a linear TRS $\mathcal{R}$ on $\mathcal{T}(\Sigma, \mathcal{V})$ and sets S and S_e of the same things as in e-modify as inputs and outputs TA $\mathcal{A}'$ which is defined from $\mathcal{A}$ as follows.

1. For any element (p_x, o_h, H) in S_e, (a) add an ε-transition rule $\langle p_x\{o_h \leftarrow p_x\}\rangle \to p_x$ to $\mathcal{A}$ and (b) execute addtrans($\langle p_x\{o_h \leftarrow p_x\}\rangle$) for defining $\mathcal{A}'$. The ε-rule $\langle p_x\{o_h \leftarrow p_x\}\rangle \to p_x$ defined above is called an *abstraction rule* and moves caused by an abstraction move is called an *abstraction move.*
2. Finally, for any term state $\langle t \rangle$, if $t \to^*_{\mathcal{A}'} \langle t' \rangle$ for some term state $\langle t' \rangle$ with $\langle t' \rangle \neq \langle t \rangle$ and $t \not\to^*_{\mathcal{A}'} \langle t' \rangle$, add an ε-transition rule $\langle t \rangle \to \langle t' \rangle$ to $\mathcal{A}'$. □

We mention about the equation $C[C[x]] = C[x]$ more precisely. We remark that if a state p_x is a term state, p_x can be written as $p_x = C_{p_x}[q_1, \ldots, q_n]$ where q_i is a constant state ($1 \leq i \leq n$). Also remark that a TA $\mathcal{A}$ can be seen as an order-sorted signature [2] by regarding each state as a sort and the set of terms of sort q (q is a state of $\mathcal{A}$) is defined by $\{t \in \mathcal{T}(\mathcal{F}) \mid t \to^*_{\mathcal{A}} q\}$. We assume $C_{p_x}[q_1, \ldots, q_n]/o_h = q_m$ for some m and write x_p for a variable of sort p. The corresponding equation for the transition $\langle p_x\{o_h \leftarrow p_x\}\rangle \to p_x$ is $C_{p_x}[x'_{q_1}, \ldots, x'_{q_{m-1}}, C_{p_x}[x_{q_1}, \ldots, x_{q_n}], x'_{q_{m+1}}, \ldots, x_{q_n}] = C_{p_x}[x_{q_1}, \ldots, x_{q_n}]$.

Let e-modify$_{\mathcal{R}}$ be a map $(\mathcal{A}, S, S_e) \mapsto (\mathcal{A}', S', S'_e)$ where $(\mathcal{A}', S', S'_e)$ is defined as e-modify$(\mathcal{A}, \mathcal{R}, S, S_e)$ and $\triangledown^{\mathcal{R}}_E$ be $(\mathcal{A}, S, S_e) \mapsto (\mathcal{A}', S, S_e)$ where $\mathcal{A}' = \triangledown_E(\mathcal{A}, \mathcal{R}, S, S_e)$. We will write $(\triangledown^{\mathcal{R}}_E \circ \text{e-modify}_{\mathcal{R}})^n(\mathcal{A})$ for the first component of $(\triangledown^{\mathcal{R}}_E \circ \text{e-modify}_{\mathcal{R}})^n(\mathcal{A}, \emptyset, \emptyset)$. In this case, we still have that if there is k such that $(\triangledown^{\mathcal{R}}_E \circ \text{e-modify}_{\mathcal{R}})^{k+1}(\mathcal{A}) = (\triangledown^{\mathcal{R}}_E \circ \text{e-modify}_{\mathcal{R}})^k(\mathcal{A})$, we can take the TA as an abstraction of $lfp(\mathsf{f}_{\mathcal{R}})$. Summarizing the discussion above, we have a sound solution to Problem 1.

Theorem 2. *For a verification problem $(\mathcal{R}, L_1, L_2)$ (Problem 1) if there is k such that $(\triangledown^{\mathcal{R}}_E \circ \text{e-modify}_{\mathcal{R}})^{k+1}(\mathcal{A}) = (\triangledown^{\mathcal{R}}_E \circ \text{e-modify}_{\mathcal{R}})^k(\mathcal{A})$ where $\mathcal{A}$ is a TA such that $\mathcal{L}(\mathcal{A}) = L_1$, then $(\to^*_{\mathcal{R}})(\mathcal{L}(\mathcal{A})) \subseteq \mathcal{L}((\triangledown^{\mathcal{R}}_E \circ \text{e-modify}_{\mathcal{R}})^k(\mathcal{A}))$ and $\mathcal{L}((\triangledown^{\mathcal{R}}_E \circ \text{e-modify}_{\mathcal{R}})^k(\mathcal{A}))$ satisfies (3) in Sect. 4 (i.e. it is recognizable).* □

If $(\triangledown^{\mathcal{R}}_E \circ \text{e-modify}_{\mathcal{R}})^k(\mathcal{A})$ in the theorem above satisfies $\mathcal{L}((\triangledown^{\mathcal{R}}_E \circ \text{e-modify}_{\mathcal{R}})^k(\mathcal{A})) \subseteq L_2$ (the inclusion problem can be solved effectively due to the properties of recognizable tree languages), then we can conclude that $(\to^*_{\mathcal{R}})(L_1) \subseteq L_2$.

Example 5. Let $\mathcal{A}$ be a TA consisting of $\{a \to q_a, b \to q_b, f(q_a, q_b) \to q_f\}$ with a final state q_f, i.e. $\mathcal{L}(\mathcal{A}) = \{f(a, b)\}$ and $\mathcal{R}_1$ be the TRS $\{v\colon f(x, y) \to f(g(x),$

$h(y))\}$ in Example 1. The generalized sticking-out graph $G_{\mathcal{R}_1}$ of $\mathcal{R}_1$ has cycles of weight one.

$$_1\circlearrowleft (v,x) \qquad (v,y) \circlearrowright_1 \tag{9}$$

As mentioned in Sect. 4, $lfp(\mathsf{f}_{\mathcal{R}_1})$ is not recognizable (see (5)). Let us apply e-modify. Since $f(q_a, q_b) \rightarrow^*_{\mathcal{A}} q_f$ and $f(g(q_a), h(q_b)) \not\rightarrow^*_{\mathcal{A}} q_f$, we have the following new transition rules.

$$\begin{array}{ll} g(q_a) \rightarrow \langle g(q_a)\rangle & h(q_b) \rightarrow \langle h(q_b)\rangle \\ f(\langle g(q_a)\rangle, \langle h(q_b)\rangle) \rightarrow \langle f(g(q_a), h(q_b))\rangle & \langle f(g(q_a), h(q_b))\rangle \rightarrow q_f \end{array}$$

The procedure also adds elements $(\langle g(q_a)\rangle, 1, (v,x))$ and $(\langle h(q_b)\rangle, 1, (v,y))$ to the set S at 1 in e-modify. Since there is no element (p, o, H) in S such that H has the form $v \cdot H' \cdot v$ for some v and H', the abstraction operator ∇_E does nothing. For the second step, since $f(\langle g(q_a)\rangle, \langle h(q_b)\rangle) \rightarrow^*_{\mathcal{A}} \langle f(g(q_a), h(q_b))\rangle$ and $f(g(\langle g(q_a)\rangle), h(\langle h(q_b)\rangle)) \not\rightarrow^*_{\mathcal{A}} \langle f(g(q_a), h(q_b))\rangle$, the procedure e-modify generates new transition rules and new elements for the set S as follows.

$$\begin{array}{ll} g(\langle g(q_a)\rangle) \rightarrow \langle g(g(q_a))\rangle & h(\langle h(q_b)\rangle) \rightarrow \langle h(h(q_b))\rangle \\ \multicolumn{2}{l}{f(\langle g(g(q_a))\rangle, \langle h(h(q_b))\rangle) \rightarrow \langle f(g(g(q_a)), h(h(q_b)))\rangle} \\ \multicolumn{2}{l}{\langle f(g(g(q_a)), h(h(q_b)))\rangle \rightarrow \langle f(g(q_a), h(q_b))\rangle} \\ (\langle g(g(q_a))\rangle, 1, (v,x)) & (\langle h(h(q_b))\rangle, 1, (v,y)) \\ (\langle g(g(q_a))\rangle, 1\cdot 1, (v,x)\cdot(v,x)) & (\langle h(h(q_b))\rangle, 1\cdot 1, (v,y)\cdot(v,y)) \\ (\langle g(q_a)\rangle, 1, (v,x)\cdot(v,x)) & (\langle h(q_b)\rangle, 1, (v,y)\cdot(v,y)) \end{array}$$

Here we also obtain S_e as follows.

$$(\langle g(q_a)\rangle, 1, (v,x)\cdot(v,x)) \quad (\langle h(q_b)\rangle, 1, (v,y)\cdot(v,y))$$

By ∇_E of Procedure 4 and the set S_e above, we obtain the following transitions.

$$\langle g(g(q_a))\rangle \rightarrow \langle g(q_a)\rangle \quad \langle h(h(q_b))\rangle \rightarrow \langle h(q_b)\rangle \tag{10}$$

Due to the abstraction moves by (10), there are no transition rules and no states to add. Finally, we have a TA whose accepting tree language is

$$\{f(g^n(a), h^m(b)) \mid n, m \geq 1\} \cup \{f(a,b)\}. \tag{11}$$

In fact, (11) is recognizable and (11) includes (5). □

Example 6. Consider the TRS $\mathcal{R}_2 = \{v\colon l_v \rightarrow r_v\}$ where $l_v = f_2(x, h(g(y)))$ and $r_v = f_2(g(k(y)), h(x))$ in Example 4 and TA $\mathcal{A}$ consisting of the transition rules $\{a \rightarrow q_a, b \rightarrow q_b, g(q_a) \rightarrow q_1, g(q_b) \rightarrow q_2, h(q_2) \rightarrow q_3, f_2(q_1, q_3) \rightarrow q_f\}$. The accepting language is $\mathcal{L}(\mathcal{A}) = \{f_2(g(a), h(g(b)))\}$. By procedure e-modify, since $l_v\sigma_1 \rightarrow^* q_f$ and $r_v\sigma_1 \not\rightarrow^* q_f$ with $\sigma_1 = \{x \mapsto q_1, y \mapsto q_b\}$, we have the following new transition rules

$$\begin{array}{lll} h(q_1) \rightarrow \langle h(q_1)\rangle & k(q_b) \rightarrow \langle k(q_b)\rangle & g(\langle k(q_b)\rangle) \rightarrow \langle g(k(q_b))\rangle \\ \multicolumn{2}{l}{f_2(\langle g(k(q_b))\rangle, \langle h(q_1)\rangle) \rightarrow \langle r_v\sigma_1\rangle} & \langle r_v\sigma_1\rangle \rightarrow q_f \end{array}$$

and elements $(\langle h(q_1)\rangle, 1, x)$, $(\langle k(q_b)\rangle, 1, y)$ and $(\langle g(k(q_b))\rangle, 1.1, y)$ in S where the name of the rewrite rule is omitted. For the second step, consider the sequence $l_v\sigma_2 \to^* \langle r_v\sigma_1\rangle$ with $\sigma_2 = \{x \mapsto \langle g(k(q_b))\rangle, y \mapsto q_a\}$. Since $r_v\sigma_2 \not\to^* \langle r_v\sigma_1\rangle$, we have the following new transition rules

$$\begin{array}{lll} h(\langle g(k(q_b))\rangle) \to \langle h(g(k(q_b)))\rangle & k(q_a) \to \langle k(q_a)\rangle & g(\langle k(q_a)\rangle) \to \langle g(k(q_a))\rangle \\ f_2(\langle g(k(q_a))\rangle, \langle h(g(k(q_b)))\rangle) \to \langle r_v\sigma_2\rangle & \langle r_v\sigma_2\rangle \to \langle r_v\sigma_1\rangle & \end{array}$$

and elements $(\langle h(g(k(q_b)))\rangle, 1, x)$, $(\langle k(q_a)\rangle, 1, y)$ and $(\langle g(k(q_a))\rangle, 1.1, y)$ of S.

For the third step, consider the sequence $l_v\sigma_3 \to^* \langle r_v\sigma_2\rangle$ with $\sigma_3 = \{x \mapsto \langle g(k(q_a))\rangle, y \mapsto \langle k(q_b)\rangle\}$. Since $r_v\sigma_3 \not\to^* \langle r_v\sigma_2\rangle$, we have the following new transition rules and S

$$\begin{array}{lll} h(\langle g(k(q_a))\rangle) \to \langle h(g(k(q_a)))\rangle & k(\langle k(q_b)\rangle) \to \langle k(k(q_b))\rangle & \langle r_v\sigma_3\rangle \to \langle r_v\sigma_2\rangle \\ f_2(\langle g(k(k(q_b)))\rangle, \langle h(g(k(q_a)))\rangle) \to \langle r_v\sigma_3\rangle & g(\langle k(k(q_b))\rangle) \to \langle g(k(k(q_b)))\rangle & \\ (\langle h(g(k(q_a)))\rangle, 1, x) & (\langle k(k(q_b))\rangle, 1, y) & (\langle g(k(k(q_b)))\rangle, 1.1, y) \\ (\langle k(q_b)\rangle, 1, y \cdot y) & (\langle k(k(q_b))\rangle, 1.1, y \cdot y) & (\langle g(k(k(q_b)))\rangle, 1.1.1, y \cdot y) \end{array}$$

with an element $(\langle k(q_b)\rangle, 1, y \cdot y)$ of S_e. By the procedure ∇_E, the following transition rules corresponding to an abstraction are added.

$$\langle k(k(q_b))\rangle \to \langle k(q_b)\rangle \quad \langle g(k(k(q_b)))\rangle \to \langle g(k(q_b))\rangle$$

For the forth step, consider the sequence $l_v\sigma_4 \to^* \langle r_v\sigma_3\rangle$ with $\sigma_4 = \{x \mapsto \langle g(k(k(q_b)))\rangle, y \mapsto \langle k(q_a)\rangle\}$. Since $r_v\sigma_4 \not\to^* \langle r_v\sigma_3\rangle$, we have the following new transition rules and S

$$\begin{array}{lll} h(\langle g(k(k(q_b)))\rangle) \to \langle h(g(k(k(q_b))))\rangle & k(\langle k(q_a)\rangle) \to \langle k(k(q_a))\rangle & \langle r_v\sigma_4\rangle \to \langle r_v\sigma_3\rangle \\ f_2(\langle g(k(k(q_a)))\rangle, \langle h(g(k(k(q_b))))\rangle) \to \langle r_v\sigma_4\rangle & g(\langle k(k(q_a))\rangle) \to \langle g(k(k(q_a)))\rangle & \\ (\langle h(g(k(k(q_b))))\rangle, 1, x) & (\langle k(k(q_a))\rangle, 1, y) & (\langle g(k(k(q_a)))\rangle, 1.1, y) \\ (\langle k(q_b)\rangle, 1, y \cdot y \cdot x) & (\langle k(k(q_b))\rangle, 1.1, y \cdot y \cdot x) & (\langle g(k(k(q_b)))\rangle, 1.1.1, y \cdot y \cdot x) \\ (\langle k(q_a)\rangle, 1, y \cdot y) & (\langle k(k(q_a))\rangle, 1.1, y \cdot y) & (\langle g(k(k(q_a)))\rangle, 1.1.1, y \cdot y) \end{array}$$

with an element $(\langle k(q_a)\rangle, 1, y \cdot y)$ of S_e. By the procedure ∇_E, the following transition rules corresponding to an abstraction are added.

$$\langle k(k(q_a))\rangle \to \langle k(q_a)\rangle \quad \langle g(k(k(q_a)))\rangle \to \langle g(k(q_a))\rangle$$

Due to those abstraction rules, the procedure terminates and we obtain the following language.

$$\begin{aligned} (\nabla_E^{\mathcal{R}} \circ \textsf{e-modify}_{\mathcal{R}})^4(\mathcal{A}) = {} & \{f_2(g(k^n(a)), h(g(k^m(b)))) \mid n, m \geq 1\} \cup \\ & \{f_2(g(k^{n+1}(b)), k(g(k^m(a)))) \mid n, m \geq 1\} \cup \\ & \{f_2(g(a), h(g(b))), f_2(g(k(b)), h(g(a)))\} \end{aligned} \tag{12}$$

On the other hand, we can observe the following fact.

$$\begin{aligned} \mathit{lfp}(\mathsf{f}_{\mathcal{R}_2}) = {} & \{f_2(g(k^n(a)), h(g(k^n(b)))) \mid n \geq 0\} \cup \\ & \{f_2(g(k^{n+1}(b)), k(g(k^n(a)))) \mid n \geq 0\} \end{aligned} \tag{13}$$

We can see (12) is recognizable whereas (13) is not and (12) includes (13). □

In the rest of this section, we will characterize a class of TRSs in which any abstraction is not needed to compute $lfp(\mathsf{f})$. As a consequence, we will give a new decidable subclass of TRSs which effectively preserve recognizability. The outline is as follows. First, we show that if no abstraction is added during the iteration of e-modify, then the process terminates; this means that we can obtain the strict implementation for $lfp(\mathsf{f})$ by Theorem 1. Then we show that if an input TRS satisfies a certain condition (L-GFPO-TRS, Definition 4), no abstraction rule is added.

For a state $q \in \mathcal{Q}$ which may have a tree structure (i.e. the state q may be defined in addtrans), define the *number of layers of* q, denoted $\mathsf{layer}(q)$ as follows: (i) if q is a constant state, then $\mathsf{layer}(q) = 0$ and (ii) if $q = \langle r\sigma/o \rangle$ with $l \to r \in \mathcal{R}$, $o \in \mathcal{P}os(r)$, r/o is not a variable, $\mathcal{V}ar(r/o) = \{x_1, \ldots, x_n\}$ and $\sigma = \{x_i \mapsto q_i \mid 1 \le i \le n\}$, then $\mathsf{layer}(q) = 1 + \max\{\mathsf{layer}(q_i) \mid 1 \le i \le n\}$. For a TRS $\mathcal{R}$ and a TA $\mathcal{A}$, we will consider a sequence of TAs $(\mathcal{A}'^{n}_{\mathcal{R},\nabla})_{n \ge 0}$ with $\mathcal{A}'^{n}_{\mathcal{R},\nabla} = (\mathcal{F}, \mathcal{Q}_n, \mathcal{Q}_f, \Delta_n)$ where $\mathcal{A}'^{n}_{\mathcal{R},\nabla} = (\nabla^{\mathcal{R}}_{E} \circ \text{e-modify}_{\mathcal{R}})^n(\mathcal{A})$.

Lemma 1. *Assume $(\mathcal{F}, \mathcal{Q}_k, \mathcal{Q}_f, \Delta_k)$ is defined as above for $k \ge 0$ with an input TRS $\mathcal{R}$. Also assume that Δ_k does not contain any abstraction rule. Let $l \to r$ be a rewrite rule in $\mathcal{R}$, σ be a substitution and q be a state which are used at (7) of* e-modify. *For a variable $x \in \mathcal{V}ar(l)$ such that $x\sigma$ is a term state, then there is an element $(x\sigma, o_h, H' \cdot (l' \to r', x'))$ in S such that there is an edge from $(l' \to r', x')$ to $(l \to r, x)$ in the generalized sticking-out graph $G_{\mathcal{R}}$ with weight one or the edge and H' are satisfying condition (8) in* e-modify. *This situation implies that new elements are added to S at 2 of (3) in* e-modify.

Proof. The proof is shown by induction on the number of k. The base case holds since any state in $\mathcal{Q}_0$ is a constant state. Assume the lemma holds for $k = n - 1$ and consider the case when $k = n$. Let o_x be the position of x in l and p_x be a term state such that $x\sigma = p_x$. Let us consider the sequence $l\sigma \to^* q$ at (7) of e-modify and how the number of layers changes from o_x to the root. There are four cases:

1. A rewriting move is caused at a certain position from o_x to the root in the sequence $l\sigma \to^* q$. Let o be the outer most position among them. There are two different subcases:
   (a) The number of layers does not increase at any o' with $o \prec o' \prec o_x$.
   (b) There is a position o' with $o \prec o' \prec o_x$ such that the number of layers increases at o'.
2. There are no rewriting moves from o_x to the root in the sequence $l\sigma \to^* q$. There are two subcases:
   (a) The number of layers does not increase at any o' with $o \prec o' \prec o_x$.
   (b) There is a position o' with $\lambda \prec o' \prec o_x$ such that the number of layers increases at o'.

Assume case 1(a). The transition rule which causes the rewriting move at o can be written as $\langle r'\sigma' \rangle \to q'$ for some $l' \to r' \in \mathcal{R}$, σ' and q'. When the rewrite $\langle r'\sigma' \rangle \to q'$ is introduced, $\langle l'\sigma' \rangle \to q'$ holds. Let o'' be a position such that $o \cdot o'' =$

o_x and x' be a variable such that $r'/o'' = x'$. By inductive hypothesis, there is an (p_x, o_h, H) in S for some o_h and H and an element $(p_x, o_h, H \cdot (l' \to r', x'))$ is added to S. Since there is no rewriting move from o_x to o and since p_x is a term state, we can see that r' properly sticks-out of l/o, case 1 in Definition 3.

Assume case 1(b), i.e. there is a position o' with $o \prec o' \prec o_x$ such that the number of layers increases at o'. Let o'_i be the inner-most position among them. Let $l' \to r'$ be the rewrite rule which is used for defining the rewrite move at o. Then the state just before the rewriting move at o can be written as $\langle r'\sigma' \rangle$ since the transition rule which causes the rewrite move has the form $\langle r'\sigma' \rangle \to q'$ for some state q'. Let o_1 and o'_1 be the positions such that $o \cdot o_1 = o_x$ and $o \cdot o'_1 = o'_i$ respectively. By the definition of o'_i, we have $r'/o'_1 \in \mathcal{V}$. From the fact that there is no rewriting move from o'_i to o, the rule used there is defined in addtrans, which implies that each function symbol from o' to o is coincide in l/o and r'. This means that l/o sticks-out r at (o'_1, o_1), case 1 in Definition 3. On the other hand, the state p_x can be written as the form $\langle r''\sigma''/o'' \rangle$ for some $l'' \to r'' \in \mathcal{R}$, a substitution σ'' and a position o''. Since there is no rewriting move from o_x to o'_i, the state at o'_i can be written as $\langle r''\sigma''/o''_1 \rangle$ with $o''_1 \prec o''$. Let $p' = \langle r''\sigma''/o''_1 \rangle$, $x' = r'/o'_1$ and o''' be a position with $o''_1 \cdot o''' = o''$. Remark that $x'\sigma' = p'$ and $p'/o''' = \langle r''\sigma''/o'' \rangle (= p_x)$. When the transition rule $\langle r'\sigma' \rangle \to q'$ is introduced, $l'\sigma' \to^* q'$ holds. By inductive hypothesis, there is an element of the form $(p', o'_h, H'' \cdot (l'' \to r'', x''))$ for some o'_h, H'' and x'' in S and an element

$$(p_x, o''_h, H'' \cdot (l'' \to r'', x'') \cdot (l' \to r', x')) \tag{14}$$

is added to S with $o''' \cdot o''_h = o'_h$ since $o''_1 \prec o''$ and $p_x = p'/o'''$. According to the fact that there is an edge from $(l' \to r', x')$ to $(l \to r, x)$ with weight zero and (14), which satisfies condition (8) in e-modify, the lemma holds.

Assume case 2(a). The proof of this case is similar to case 1(a). Since there is no rewriting move from o_x to the root, the state attached at the root position can be rewritten as $\langle r'\sigma'/o' \rangle$ for some $l' \to r' \in \mathcal{R}$, σ' and $o' \in \mathcal{P}os(r')$ and p_x can be written as $\langle r'\sigma'/o'' \rangle$ with $o' \cdot o_x = o''$. Moreover, we can see that the function symbols in l and r'/o' on the path from o_x to the root are coincide with each other, which is in case 3 of Definition 3. By using inductive hypothesis, in the same way of case 1(a), we can see that the lemma holds.

Assume case 2(b), i.e. there is a position o' with $o \prec o' \prec o_x$ such that the number of layers increases at o'. The proof of this case is similar to case 1(b). Let o'_i be the inner-most position among them. Since the number of layers is increased at o'_i, the state attached at the root position can be written as $\langle r'\sigma'/o' \rangle$ for some $l' \to r' \in \mathcal{R}$, σ' and $o' \in \mathcal{P}os(r')$. Since there is no rewriting move from o'_i to the root, the function symbols in l and r'/o' on the path from o'_i to the root are coincide with each other, which is in case 4 of Definition 3. By using inductive hypothesis, in the same way of case 1(b), we can see that the lemma holds. □

The following proposition says that if no abstraction rule is added during the iteration of e-modify, then the process must terminate.

Proposition 1. *If there is no $k \geq 0$ such that ${\mathcal{A}'}^k_{\mathcal{R},\nabla}$ contains an abstraction rule, then there is $m \geq 0$ such that ${\mathcal{A}'}^{m+1}_{\mathcal{R},\nabla} = {\mathcal{A}'}^m_{\mathcal{R},\nabla}$.*

Proof. The proof is by contradiction. Assume there is no m such that ${\mathcal{A}'}^{m+1}_{\mathcal{R},\nabla} = {\mathcal{A}'}^m_{\mathcal{R},\nabla}$, which implies that the number of the states of the TAs during the construction is unbounded. In other words, there is no bound on the size of term states in the constructed TAs. A state introduced during the construction has the form $\langle C_n[\cdots[C_1[q]]]\rangle$ where q is a state of the original automaton and C_i has the form $r_i\sigma_i/o_i$ for some rewrite rule $l_i \to r_i \in \mathcal{R}$, a substitution σ_i and a position o_i $(1 \leq i \leq n)$. So there are states of the form $\langle C_n[\cdots[C_1[q]]]\rangle$ for any $n \geq 1$. On the other hand, when the state $\langle C_1[q]\rangle$ is introduced, an element $(\langle C_1[q]\rangle, o_1, (l_1 \to r_1, x_1))$ is added to S at 1 of (3) in the procedure e-modify for some position o_1 and variable x_1. Moreover, for any $n \geq 2$, due to Lemma 1, an element $(\langle C_n[\cdots[C_1[q]]]\rangle, o_n, (l_1 \to r_1, x)\cdots(l_n \to r_n, x_n))$ is added to S for some $(l_i \to r_i, x_i)$ and position o_i with $2 \leq i \leq n$ and $C_n[\cdots[C_1[q]]]/o'_n = C_1[q]$ at 2(b) of (3) in e-modify. Since the numbers of the rewrite rules and the variables of in $\mathcal{R}$ are finite, there are integers j_1 and j_2 such that $j_1 < j_2$ and $(\langle C_{j_2}[\cdots[C_1[q]]]\rangle, o_j, (l_1 \to r_1, x_1)\cdots(l_{j_2} \to r_{j_2}, x_{j_2}))$ with $(l_{j_1} \to r_{j_1}, x_{j_1}) = (l_{j_2} \to r_{j_2}, x_{j_2})$. In this situation, from the construction 1 of e-modify, we can see that there is an element $(\langle C_{j_2}[\cdots[C_{j_1}[q']]]\rangle, o', (l_{j_1} \to r_{j_1}, x_{j_1})\cdots(l_{j_2} \to r_{j_2}, x_{j_2}))$ for some q' and o'. In this case, since $(l_{j_1} \to r_{j_1}, x_{j_1}) = (l_{j_2} \to r_{j_2}, x_{j_2})$, at 2(a) of (3) in e-modify, an element is added to S_e and an abstraction rule is added in the procedure ∇_E, a contradiction. □

The following proposition says about the relation between abstraction rules and the generalized sticking-out graph.

Proposition 2. *If an abstraction rule is added in the procedure ∇_E, then in the generalized sticking-out graph $G_\mathcal{R}$, there is a cycle with weight one or more.*

Proof. An abstraction rules is added if and only if there is an element in S of the form $(p_x, o_h, v \cdot H' \cdot v)$ for some p_x, o_h and v, which is also an element of S_e. Since an elements of S is added according to $G_\mathcal{R}$, we can see that if there is no cycle in $G_\mathcal{R}$, no element is added to S_e. Let (p, o, H) be an arbitrary element in S with $H = (l_1 \to r_1, x_1) \cdot \cdots \cdot (l_n \to r_n, x_n)$ and $n \geq 2$. Due to the condition (8) at 2 in e-modify, there is an integer $i \geq 1$ such that there is an edge from $(l_i \to r_i, x_i)$ to $(l_{i+1} \to r_{i+1}, x_{i+1})$ with weight one. This means that if an element is added to S_e, then there is a cycle with weight one or more. □

Definition 4. *A TRS $\mathcal{R}$ is a* generalized finitely path-overlapping term rewriting system (GFPO-TRS) *if the generalized sticking-out graph of $\mathcal{R}$ has no cycle of weight one or more.*

We write L-GFPO-TRS for the class of linear GFPO-TRSs. As a corollary of Propositions 1 and 2 and Theorem 1, we obtain the following theorem.

Theorem 3. *A TRS in L-GFPO-TRS effectively preserves recognizability.* □

Example 7. Let $\mathcal{R} = \{v\colon f_2(x,y) \to f_2(a, g(x))\}$ be a TRS. The generalized sticking-out graph $G_{\mathcal{R}}$ of $\mathcal{R}$ has two vertices $(v,x), (v,y)$ and one edge $(v,x) \to (v,y)$ with weight 1. Since $G_{\mathcal{R}}$ has no cycle, $\mathcal{R}$ belongs to GFPO-TRS. □

Here, we compare GFPO-TRS with the class right-linear FPO-TRS (RL-FPO-TRS) [12], that is known as the widest decidable subclass of TRSs which effectively preserves recognizability. The class RL-FPO-TRS includes both *right-linear monadic TRSs* [10] and *linear generalized semi-monadic TRSs* [6]. The class FPO-TRS is defined by *sticking-out graphs* [12], which can be re-defined by using the notion of a generalized sticking-out graph. Let ϕ be a graph homomorphism defined for generalized sticking-out graphs as follows. For a TRS $\mathcal{R}$, a vertex (v,x) of the generalized sticking-out graph $G_{\mathcal{R}}$ of $\mathcal{R}$ with $v \in \mathcal{R}$ and $x \in \mathcal{V}ar$, define $\phi(v,x)$ be v and for an edge $e\colon (v,x) \to (v',x')$ with weight w, define $\phi(e)$ be an edge $v \to v'$ with weight w. For a TRS $\mathcal{R}$, the sticking-out graph of $\mathcal{R}$ is $\phi(G_{\mathcal{R}})$ where $G_{\mathcal{R}}$ is the generalized sticking-out graph of $\mathcal{R}$.

Definition 5. *[12] A TRS $\mathcal{R}$ is* finitely path overlapping (FPO-TRS) *if the sticking-out graph of $\mathcal{R}$ has no cycle with weight one or more.*

According to the definition of the (generalized) sticking-out graph, we can see that FPO-TRS $\subseteq$ GFPO-TRS. Moreover, by Example 7, we obtain the following inclusion relation.

Proposition 3. *FPO-TRS $\subset$ GFPO-TRS.* □

By the previous proposition and the fact that RL-FPO-TRS includes non-left-linear TRSs, we obtain a new result on a decidable subclass of TRSs which effectively preserves recognizability.

Proposition 4. *L-GFPO-TRS and RL-FPO-TRS are incomparable.* □

7 Discussion

There are (even if linear) TRSs for which an abstraction cannot be obtained by our verification technique proposed in this paper. In other words, the verification procedure with abstractions (e-modify) does not always terminate. More precisely, the abstraction rules proposed in this paper correspond to equations of the form $C[C[x]] = C[x]$ for a context C. For the TRS $\mathcal{R}_3$ in order to obtain abstraction by equation-based abstraction, we may need equations of the form $C_1[C_2[x]] = C_2[C_1[x]]$ where C_1 and C_2 are contexts. In general a TRS whose generalized sticking-out graph has two cycles of weight one or more which share the same vertex cannot be abstracted by our technique. To overcome such cases, a new abstraction operation ∇'_E instead of ∇_E may be able to used where ∇'_E is defined by replacing (1) and (2) of ∇_E by (1') adding an ε-transition rule $\langle p_x \rangle \to p_x/o_h$ to $\mathcal{A}$. This move corresponds to an equation of the form $C[x] = x$ where we can regard x as a variable of sort p_x/o_h as mentioned in Sect. 5.

Acknowledgements

The author thanks to the members of the research center for verification and semantics of AIST for discussions. The author also thanks to the anonymous reviewers for giving a lot of helpful suggestions.

References

1. F. Baader and T. Nipkow: *Term Rewriting and All That*, Cambridge University Press, 1998.
2. H. Comon: "Equational Formulas in Order-Sorted Algebras," *Proc. of 4th RTA*, LNCS 443, pp. 674–688, 1990.
3. H. Comon, M. Dauchet, R. Gilleron, F. Jacquemard, D. Lugiez, S. Tison and M. Tommasi: *Tree Automata Techniques and Applications*, 1997. `http://www.grappa.univ-lille3.fr/tata/`
4. P. Cousot and R. Cousot: "Abstract Interpretation: a Unified Lattice Model for Static Analysis of Programs by Construction or Approximation of Fixpoints," *Proc. of 4th POPL*, pp. 238–252, ACM Press, 1977.
5. T. Genet and F. Klay: "Rewriting for Cryptographic Protocol Verification," *Proc. of 17th CADE*, LNAI 1831, pp. 271–290, 2000.
6. P. Gyenizse and S. Vágvölgyi: "Linear Generalized Semi-Monadic Rewrite Systems Effectively Preserve Recognizability," Theoretical Computer Science 194(1–2), pages 87–122, 1998.
7. J. Meseguer, M. Palomino and N. Martí-Oliet: "Equational Abstractions," *Proc. of 19th CADE*, LNCS 2741, pp. 2–16, 2003.
8. D. Monniaux: "Abstracting Cryptographic Protocols with Tree Automata," *Proc. of 6th SAS*, LNCS 1694, pp. 149–163, 1999.
9. F. Nielson, H. R. Nielson and C. Hankin: "Abstract Interpretation," In *Principles of Program Analysis*, Chapter 4, Springer-Verlang, 1999.
10. K. Salomaa: "Decidability of Confluence and Termination of Monadic Term Rewriting Systems," *Proc. of 4th RTA*, LNCS 488, pages 275–286, 1991.
11. T. Takai, Y. Kaji and H. Seki: "Right-linear Finite Path Overlapping Term Rewriting Systems Effectively Preserve Recognizability," *Proc. of 11th RTA*, LNCS 1833, pp.246–260, 2000.
12. T. Takai, Y. Kaji and H. Seki: "Right-linear Finite-Path Overlapping Term Rewriting Systems Effectively Preserve Recognizability," Scienticae Mathematicae Japonicae. To appear.
13. T. Takai, H. Seki, Y. Fujinaka and Y. Kaji: "Layered Transducing Term Rewriting System and Its Recognizability Preserving Property," IEICE Transactions on Information and Systems E86-D(2), pp. 285–295, 2003.

Rewriting for Fitch Style Natural Deductions

Herman Geuvers[1] and Rob Nederpelt[2]

[1] Nijmegen University, The Netherlands
[2] Eindhoven University of Technology, The Netherlands

Abstract. Logical systems in natural deduction style are usually presented in the Gentzen style. A different definition of natural deduction, that corresponds more closely to proofs in ordinary mathematical practice, is given in [Fitch 1952]. We define precisely a Curry-Howard interpretation that maps Fitch style deductions to simply typed terms, and we analyze why it is not an isomorphism. We then describe three reduction relations on Fitch style natural deductions: one that removes *garbage* (subproofs that are not needed for the conclusion), one that removes *repeats* and one that *unshares* shared subproofs. We also define an equivalence relation that allows to *interchange* independent steps. We prove that two Fitch deductions are mapped to the same λ-term if and only if they are equal via the congruence closure of the aforementioned relations (the reduction relations plus the equivalence relation). This gives a Curry-Howard isomorphism between equivalence classes of Fitch deductions and simply typed λ-terms. Then we define the notion of cut-elimination on Fitch deductions, which is only possible for deductions that are completely unshared (normal forms of the unsharing reduction). For conciseness, we restrict in this paper to the implicational fragment of propositional logic, but we believe that our results extend to full first order predicate logic.

1 Introduction

For Gentzen style natural deduction, ([Gentzen 1969]) there is a well-known notion of rewriting: cut-elimination. This is a procedure for eliminating 'detours' in a logical derivation that arise from first applying an introduction rule and then an elimination rule for a connective. This notion of reduction can be defined more concisely by associating typed λ-terms to natural deductions: there is the Curry-Howard isomorphism between natural deductions and simply typed terms and cut-elimination in the first corresponds to β-reduction in the latter (see [Howard 1980]). A different definition of natural deduction is *flag style natural deduction* defined by [Fitch 1952]. See [Bornat and Sufrin 1999] for a nice implementation of a proof assistant based on flag deductions. Here, deductions are linear, written vertically where every line consists of a formula and a motivation for the derivability of that formula (referring to previous lines). Furthermore, there is a notion of 'scope' (of an assumption) which is indicated by a flag. Apart from the closer correspondence to proofs in mathematical practice, a positive aspect is that subproofs can be shared. A negative aspect is that, due to the sharing, the notion of cut-elimination is blurred. Also, the order of the

V. van Oostrom (Ed.): RTA 2004, LNCS 3091, pp. 134–154, 2004.

steps in a Fitch style deduction is somewhat arbitrary: a flag deduction can be seen as a linearization of a Gentzen style natural deduction tree and this linearization involves some arbitrary (bureaucratic) choices. This implies that one Gentzen style *tree deduction* corresponds to many flag deductions. We make this precise by using simply typed λ-calculus (see [Barendregt 1992]) and to define a Curry-Howard *formulas-as-types* (and *proofs-as-terms*) interpretation from flag deductions to typed terms. The Curry-Howard interpretation that we define here is not the only possible one. We will come back to this point briefly in Section 5. The interest of our interpretation lies in the fact that it ignores those aspects of flag deductions that can be seen as 'bureaucracy'. To keep things simple and for space restrictions, we restrict here to simply typed λ-calculus with just arrow types and for the logic to propositional logic with just implication.

2 Flag Style Natural Deduction

We consider the implicational fragment, so the set of formulas, Form, is built up from a set of parameters, Par, using the implication $\rightarrow$. The rules have the same style as for Gentzen style tree deduction, fixing the meaning of a connective by saying how to eliminate (an E-rule) it and how to introduce it (an I-rule).

Flag deduction, a first definition

Definition 2.1. *The rules for natural deductions in flag style are the following.*

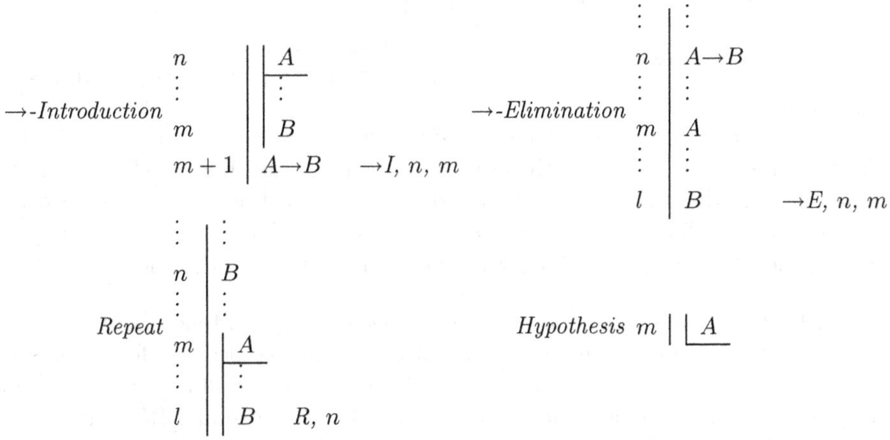

Remark 2.2. The order in which the lines appear in a flag deduction should be exactly as suggested by the above diagrams, except for the rule $\rightarrow$-E, where $A \rightarrow B$ and A may be interchanged. All rules can be applied 'under a flag'. Furthermore, the repeat rule and the $\rightarrow$-E rule can take their premise (the B, resp. the A and the $A \rightarrow B$) from arbitrarily high, with the proviso that B must be 'in scope', i.e. it must not be 'under a closed flag'.

The repeat rule allows to use one sub-deduction several times. Its use will be discussed later. Some aspects of the definition are vague: we have not defined what it means for a formula to be 'in scope'. The idea is that the →-I rule closes a sub-deduction and that the formulas in that sub-deduction are not in scope anymore. The definition of flag deductions above lacks precision, especially if we want to study their structure in detail. We will therefore make the definition more precise in Definition 2.6, but first we give some examples.

Example 2.3. The following four examples give an impression of flag deductions.

D_1

1	$A{\to}B$ (flag)	
2	$\quad A$ (flag)	
3	$\quad\quad A$ (flag)	
4	$\quad\quad B$	→E, 1, 2
5	$\quad A{\to}B$	→I, 3, 4
6	$A{\to}A{\to}B$	→I, 2, 5
7	$(A{\to}B){\to}A{\to}A{\to}B$	→I, 1, 6

D_2

1	A (flag)	
2	$\quad B$ (flag)	
3	$\quad A$	R, 1
4	$B{\to}A$	→I, 2, 3
5	$A{\to}B{\to}A$	→I, 1, 4

D_3

n	A	
⋮	⋮	
m	$A{\to}A{\to}B$ (flag)	
$m+1$	A	R, n
$m+2$	$A{\to}B$	→E, m, $m+1$
$m+3$	B	→E, $m+2$, $m+1$

D_4

1	$A{\to}B{\to}C$ (flag)	
2	$\quad A{\to}B$ (flag)	
3	$\quad\quad A$ (flag)	
4	$\quad\quad B$	→E, 2, 3
5	$\quad\quad B{\to}C$	→E, 1, 3
6	$\quad\quad C$	→E, 5, 4
7	$\quad A{\to}C$	→I, 3, 6
8	$(A{\to}B){\to}A{\to}C$	→I, 2, 7
9	$(A{\to}B{\to}C){\to}(A{\to}B){\to}A{\to}C$	→I, 1, 8

In D_1, there are two possible ways of deriving B on line 4 →-E, $1, 2$ or $E, 1, 3$. As it is recorded in the motivation at the end of the line (→-E, $1, 2$ in this case), we can distinguish these two deductions: we will consider these two flag deductions as different. In D_2 and D_3 we see two possible uses of the repeat rule. D_4 is a derivation of a well-known axiom of Hilbert style deduction.

In D_3 we can avoid the use of (repeat), because we can take premises for the →-E rule from arbitrary high; the use of an explicit (repeat) is for readability only. In D_2, we can only avoid the use of (repeat) if we allow →-introduction without explicitly writing the conclusion as the last formula. It's a matter of taste (and choice) whether one wants to allow this. If one does, we can omit the use of (repeat) completely from our deductions. We choose for this option, so we can change D_2 into the following correct deduction.

D_2

1	A (flag)	
2	$\quad B$ (flag)	
3	$B{\to}A$	→I, 2, 1
4	$A{\to}B{\to}A$	→I, 1, 3

Flag deduction, an improved definition. We have already pointed out that Definition 2.1 is a bit informal, especially as to which formulas are 'in scope' (and may henceforth be repeated or used in the $\rightarrow$-E rule). To define mappings between flag deductions and typed λ-terms, we give a more precise definition of flag deductions. In this definition, a flag deduction consists of a sequence of tuples $\langle l_0, A_0, m_0\rangle, \langle l_1, A_1, m_1\rangle, \ldots, \langle l_n, A_n, m_n\rangle$. Here each tuple is of the form $\langle l, A, m\rangle$, where $l \in \mathcal{I}$ is the *label*, taken from a countable set $\mathcal{I}$ (usually $\mathbb{N}$) A is a formula (the formula derived at that line) and m is the *motivation* for the derivation. This motivation contains the following information: which rule has been applied to derive the formula and on which lines the derivation is based. The motivations m will be of the following forms. (We write – to mean 'nothing', to be used for hypotheses.)

m	meaning
-	*hypothesis* ("raising" a flag)
$\rightarrow$-E, l_1, l_2	$\rightarrow$-*elimination* on the formulas on the lines with labels l_1, l_2
$\rightarrow$-I, l_1, l_2	$\rightarrow$-*introduction* on the formulas on the lines with labels l_1, l_2
R, l	*repeat* the formula on the line with label l

Definition 2.4. *1. If Σ is a sequence of tuples $\langle l, A, m\rangle$ as above where each label occurs at most once as a line number, we call it a* pre-deduction.

2. Given a pre-deduction Σ and an $l' \in \mathcal{I}$, we say that l' is Σ-fresh *if l' does not yet occur in any of the tuples $\langle l, A, m\rangle$ of Σ.*

3. We say that the formula A is on line l in Σ *(or just* A is on line l, *if Σ is clear from the context), if the tuple $\langle l, A, \ldots\rangle$ occurs in Σ.*

We will write a pre-deduction in the shape of a flag deduction, like the ones in Example 2.3. So a line $\langle m, A, -\rangle$ is depicted as

$$m \;\Big|\; \underline{\big| A\;}$$

This raises a flag, whose "flagpole" we extend to the *last* line with and $\rightarrow$-I, m, l motivation.

Crucial notions in the definition of flag deductions are the *scope* of a deduction and the *flag* of a deduction. Given a deduction Σ, Scope(Σ) will be the set of lines in Σ from which we can 'use' formulas (at the end of Σ). Flag(Σ) is the line l of the 'last open flag'.

Definition 2.5. *For a pre-deduction Σ, we define the* scope *of Σ,* Scope(Σ), *as follows*[1]

$$\text{Scope}\left(\begin{array}{c|lc} \vdots & \Sigma_1 & \\ m & \;\big|\;\underline{A} & \\ \vdots & \;\big|\;\Sigma_2 & \\ l & B & \rightarrow\text{-}I,\ m{,}n \end{array}\right) = \text{Scope}(\Sigma_1) \cup \{(l, B)\}$$

[1] For clarity, we give this definition by writing the deductions graphically. We could equivalently give it on the basis of the pre-deductions of Definition 2.4, but that just blurs the presentation.

$$\mathrm{Scope}\left(\begin{array}{c|ll} \vdots & \Sigma & \\ l & A & m \end{array}\right) = \mathrm{Scope}(\Sigma) \cup \{(l, A)\} \textit{ if } m \neq [\to\text{-}I, \ldots, \ldots]$$

In the first clause, we have $[\to\text{-}I, m, n]$ *on the last line and we "search" upward for the line* m *containing a hypothesis (flag); if there is no line* m *or if line* m *contains no hypothesis, the scope is* $\emptyset$*.*

For Σ *a pre-deduction, we define* Flag(Σ) *as follows.*

$$\mathrm{Flag}\left(\begin{array}{c|l} \vdots & \Sigma \\ l & \underline{|\,A\quad} \end{array}\right) = (l, A)$$

$$\mathrm{Flag}\left(\begin{array}{c|ll} \vdots & \Sigma_1 & \\ m & \underline{|\,A\quad} & \\ \vdots & |\,\Sigma_2 & \\ l & B & \to\text{-}I,\ m,n \end{array}\right) = \mathrm{Flag}(\Sigma_1)$$

$$\mathrm{Flag}\left(\begin{array}{c|l} \vdots & \Sigma \\ l & A \end{array}\right) = \mathrm{Flag}(\Sigma) \textit{ otherwise}$$

Note that Scope *may yield an empty set and that* Flag *may be undefined.*

Definition 2.6. *We inductively define the notion of* flag deduction.

1. *(Flag raising) If* Σ *is a flag deduction or* $\Sigma = \emptyset$ *and* l *is a fresh label, then*

$$\begin{array}{c|l} \vdots & \Sigma \\ l & \underline{|\,A\quad} \end{array}$$

   is a flag deduction.
2. *(*$\to$*-E) If* Σ *is a flag deduction with* $(l_1, A\to B), (l_2, A) \in \mathrm{Scope}(\Sigma)$ *and* l *is fresh, then*

$$\begin{array}{c|ll} \vdots & \Sigma & \\ l & B & \to\text{-}E, l_1, l_2 \end{array}$$

   is a flag deduction.
3. *(*$\to$*-I) If* Σ *is a flag deduction with* $\mathrm{Flag}(\Sigma) = (l_1, A)$*,* $(l_2, B) \in \mathrm{Scope}(\Sigma)$ *and* l *is a fresh label, then*

$$\begin{array}{c|ll} \vdots & \Sigma & \\ l & A\to B & \to\text{-}I, l_1, l_2 \end{array}$$

   is a flag deduction.
4. *(Repeat) If* Σ *is a flag deduction with* $(l', A) \in \mathrm{Scope}(\Sigma)$ *and* l *is a fresh label, then*

$$\begin{array}{c|ll} \vdots & \Sigma & \\ l & A & R, l' \end{array}$$

   is a flag deduction.

Definition 2.7. *Let Σ be a flag deduction.*

1. *The* conclusion *of Σ is the formula on the last line of Σ.*
2. *The* assumptions *of Σ are the formulas in the flags in Σ that are open (i.e. that have not been closed by an $\to$-I rule). More formally, the assumptions of Σ are the formulas A for which*

$$\exists l \in \mathcal{I}(\langle l, A, - \rangle \in \Sigma \wedge (l, A) \in \mathrm{Scope}(\Sigma)).$$

3. *For Δ a set of formulas and A a formula, $\Delta \vdash_f A$ if there is a flag-deduction Σ with conclusion A and assumptions in Δ.*

2.1 The Problem with Cut-Elimination

In natural deduction, we speak of a *cut* if we first introduce a connective (via an intro rule) and then immediately eliminate it (via an elim rule for that same connective). In our case, that means: doing an $\to$-I to introduce, say $A \to B$ and then doing an $\to$-E to derive B. (In full proposition logic there are more notions of cut: "commuting cuts", see [Prawitz 1965]) that allow deduction rules to be interchanged, but that's not of interest here.) In Fitch style, this would amount to the following, where we denote the process of cut-elimination by $\Longrightarrow_c$.

$$\begin{array}{l|l|ll}
1 & & \underline{A} & \\
\vdots & & \Sigma & \\
n & & B & \\
n+1 & \multicolumn{2}{l}{A \to B} & \to\mathrm{I},\ 1,\ n \\
\vdots & \multicolumn{2}{l}{\Theta} & \\
m & \multicolumn{2}{l}{A} & \\
\vdots & \multicolumn{2}{l}{\Pi} & \\
l & \multicolumn{2}{l}{B} & \to\mathrm{E},\ n+1,\ m
\end{array}
\quad \Longrightarrow_c \quad
\begin{array}{l|l}
\vdots & \Theta \\
m & A \\
\vdots & \Sigma[m/1] \\
n & B \\
\vdots & \Pi
\end{array}$$

The problem here is that the right derivation may not be well-formed: if in the sub-derivations Θ or Π the formula $A \to B$ on line $n+1$ is used, the right hand side derivation has invalid line references (referring to non-existent lines). The problem with cut-elimination is due to *sharing* of sub-derivations. To give a precise definition of cut-elimination we first define the formulas-as-types interpretation from flag deductions to simply typed λ-calculus.

3 Formulas-as-Types for Fitch Style Natural Deduction

For flag deductions, we define an interpretation into the set of simply typed λ-terms. For the mapping back, we will need *labeled* simply typed λ-terms (to be able to restore all the labels in the flag deduction). Labeled terms are terms where all sub-terms (except the variables) are labeled with a label from $\mathcal{I}$:

$$\mathsf{LTerm} ::= \mathsf{Var}_{\mathcal{I}}^{\mathsf{Typ}} \mid (\mathsf{LTerm}\ \mathsf{LTerm})^{\mathcal{I}} \mid (\lambda \mathsf{Var}_{\mathcal{I}}^{\mathsf{Typ}}.\mathsf{LTerm})^{\mathcal{I}}$$

The typing rules are the usual: we allow any $l \in \mathcal{I}$ to be used as a label. So we have

- $x_i^A : A$,
- $(MN)^l : B$ if $M : A \rightarrow B$, $N : A$ and $l \in \mathcal{I}$,
- $(\lambda x_i^A.M)^l : A \rightarrow B$ if $M : B$ and $l \in \mathcal{I}$.

If we want to denote the label of a sub-term explicitly (in an application or an abstraction) we write $(M^{l_1} N^{l_2})^l$ or $(\lambda x_i^A.M^{l_1})^l$.

Definition 3.1. *The set of labels and variable indices of a labeled λ-term M will be denoted by $lab(M)$. So,*

- $lab(x_i^A) = \{i\}$,
- $lab((MN)^l) = \{l\} \cup lab(M) \cup lab(N)$,
- $lab((\lambda x_i^A.M)^l) = \{l, i\} \cup lab(M)$.

Similarly, we will denote the set of lines of a flag deduction Σ by $lab(\Sigma)$. (These are the 'line numbers' that occur in Σ, which are taken from the same index set $\mathcal{I}$ of the labels and indices of λ-terms.)

There is a straightforward mapping from the labeled to the unlabeled terms, by just erasing labels. We denote it by $|\_|$, so if $M \in \mathsf{LTerm}$, then $|M| \in \mathsf{Term}$. We view a labeled term as a specific representation of an unlabeled term. Stated otherwise, we consider the terms *modulo relabeling.* However, we do not just allow any kind of relabeling, but only *injective* ones, because we want to be able to distinguish, e.g. $\lambda f^{A \rightarrow A \rightarrow B}.\lambda g^{B \rightarrow A}.\lambda x^B.((f(gx)^4)^6(gx)^5)^7$ and $\lambda f^{A \rightarrow A \rightarrow B}.\lambda g^{B \rightarrow A}.\lambda x^B.((f(gx)^4)^6(gx)^4)^7$.

Definition 3.2. *A* relabeling *is an injective map $r : \mathcal{I} \rightarrow \mathcal{I}$. By* adding a labeling *to a simply typed λ-term N, we mean to construct a labeled simply typed term M such that $|M| = N$ and all labels and variable-indices in M are unique (i.e. distinct sub-terms of M have different labels and similar for variables).*

A relabeling $r : \mathcal{I} \rightarrow \mathcal{I}$ extends immediately to a map on labeled simply typed terms and to a map on flag deductions. If the labeled term M arises from P as a result of some relabeling r, we say that M is a *relabeling of P*.

For mapping simply typed λ-terms to flag deductions, we first add a labeling and then we define the associated flag deduction. The mapping of flag deductions to simply typed λ-terms can be defined directly (without using labeled terms). As we will be using the labeled terms later, we define the map from flag deductions to labeled terms.

Definition 3.3. *The interpretation of a flag deduction Σ as a labeled simply typed λ-term, $[\![\Sigma]\!]^L$, is defined as follows. (We use the cases of Definition 2.6.)*

1. *(Flag raising) If the last rule in Σ is a flag, then $[\![\Sigma]\!]^L = x_l^A$.*
2. *($\rightarrow$-E) If the last rule in Σ is $\rightarrow$-E, then then $[\![\Sigma]\!]^L = ([\![\Sigma{\leq}l_1]\!]^L [\![\Sigma{\leq}l_2]\!]^L)^l$.*
3. *($\rightarrow$-I) If the last rule in Σ is $\rightarrow$-I then $[\![\Sigma]\!]^L = (\lambda x_{l_1}^A.[\![\Sigma{\leq}l_2]\!]^L)^l$.*
4. *(Repeat) If the last rule in Σ is R, then $[\![\Sigma]\!]^L = [\![\Sigma{\leq}l']\!]^L$.*

Here $\Sigma{\leq}l$ denotes the pre-deduction Σ up to and including l. *The interpretation of a flag deduction as a simply typed term is defined by just erasing the labels, so*

$$[\![\Sigma]\!] := |[\![\Sigma]\!]^L|.$$

When treating examples, we will be using the usual notation for flag deductions, with real flags and flag poles to indicate the scope of a flag. It should be clear how we can transform the formal notation of flag deductions (as used above) into a deduction with 'real' flags and flag poles.

Example 3.4. We show the interpretation of two flag deductions as typed λ-terms. For readability, we write the type labels of the variables only in the λ-abstraction.

1	$A \to A \to B$ (flag)	
2	$\quad A$ (flag)	
3	$\quad A \to B$	$\to$E, 1, 2
4	$\quad B$	$\to$E, 3, 2
5	$A \to B$	$\to$I, 2, 4
6	$(A \to A \to B) \to (A \to B)$	$\to$I, 1, 5

1	A (flag)	
2	$\quad B$ (flag)	
3	$\quad A$	R, 1
4	$B \to A$	$\to$I, 2, 3
5	$A \to B \to A$	$\to$I, 1, 4

The typed λ-term associated to these terms are $\lambda x_1^{A \to A \to B}.\lambda x_2^A.(x_1 x_2)x_2$ and $\lambda x_1^A.\lambda x_2^B.x_1$.

Theorem 3.5 (Soundness of $[\![-]\!]$). *If Σ is a flag deduction with conclusion A, then $[\![\Sigma]\!] : A$. Moreover, if the assumptions of Σ are $B_1, \ldots, B_n$, with labels $i_1, \ldots, i_n$, respectively, then $FV([\![\Sigma]\!]) = \{x_{i_1}^{B_1}, \ldots, x_{i_n}^{B_n}\}$.*

Proof. By induction on the flag deduction Σ. □

In the definition of the opposite embedding, we assume that the terms are *uniquely labeled*, that is, distinct sub-terms of M have different labels and all the indices of the bound variables are distinct and distinct from the labels. To define the embedding, we first ignore the free variables (i.e. we don't raise flags), defining $(\!|-|\!)^p$. Then we raise flags for all free variables, defining $(\!|-|\!)$. There are basically two versions of the embedding: a simple minded one that just creates a sub-derivation for every sub-term and uses a lot of instances of the repeat rule, and a more sophisticated one that does not create any repeat rule. We only treat the second one.

Definition 3.6 (Simply typed terms $\to$ Flag deductions). *The interpretation of a labeled simply typed λ-term M as a pre-deduction, $(\!|M|\!)^p$, is defined as follows.*

1. *For* $\llparenthesis (M^{l_1} N^{l_2})^l \rrparenthesis^p$ *distinguish cases according to the shape of* M *and* N*. Let* B *be the type of* $(M^{l_1} N^{l_2})^l$*.*
   (a) M *and* N *are variables:*

$$\llparenthesis (x_i^{A\to B}\, x_j^{A})^l \rrparenthesis^p = l \mid B \qquad \to\text{-}E,i,j$$

   (b) M *is a variable,* N *is not:*

$$\llparenthesis (x_i^{A\to B} N^{l_2})^l \rrparenthesis^p = \begin{array}{r|l} \vdots & \llparenthesis N^{l_2} \rrparenthesis^p \\ l & B \end{array} \qquad \to\text{-}E,i,l_2$$

   (c) N *is a variable,* M *is not:*

$$\llparenthesis (M^{l_1} x_j^{A})^l \rrparenthesis^p = \begin{array}{r|l} \vdots & \llparenthesis M^{l_1} \rrparenthesis^p \\ l & B \end{array} \qquad \to\text{-}E,l_1,j$$

   (d) *Neither* M *nor* N *is a variable:*

$$\llparenthesis (M^{l_1} N^{l_2})^l \rrparenthesis^p = \begin{array}{r|l} \vdots & \llparenthesis M^{l_1} \rrparenthesis^p \\ \vdots & \llparenthesis N^{l_2} \rrparenthesis^p \\ l & B \end{array} \qquad \to\text{-}E,l_1,l_2$$

2. *For* $\llparenthesis (\lambda x_i^A.M)^l \rrparenthesis^p$ *distinguish cases according to whether* M *is a variable or not. Let* B *be the type of* M*.*
   (a)

$$\llparenthesis (\lambda x_i^A.x_j^B)^l \rrparenthesis^p = \begin{array}{r|l} i & \underline{\lfloor A} \\ l & A\to B \end{array} \qquad \to\text{-}I,i,j$$

   (b)

$$\llparenthesis (\lambda x_i^A.M^{l_1})^l \rrparenthesis^p = \begin{array}{r|l} i & \begin{array}{|l} \underline{A} \\ \llparenthesis M^{l_1} \rrparenthesis^p \end{array} \\ \vdots & \\ l & A\to B \end{array} \qquad \to\text{-}I,i,l_1$$

The interpretation of a labeled simply typed λ*-term* M *as a full flag deduction,* $\llparenthesis M \rrparenthesis$*, is defined by adding flags for the free variables:*
If $FV(M) = \{x_{i_1}^{B_1}, \ldots, x_{i_n}^{B_n}\}$*, then* $\llparenthesis M \rrparenthesis$ *is*

$$\begin{array}{r|l} i_1 & \begin{array}{|l} \underline{B_1} \\ \vdots \\ \begin{array}{|l} \underline{B_n} \\ \llparenthesis M \rrparenthesis^p \end{array} \end{array} \\ \vdots & \\ i_n & \\ \vdots & \end{array}$$

Example 3.7. Compare with 3.4. We define the interpretation of two labeled λ-terms as flag deductions. For readability, we write the type labels only in the λ-abstractions. Consider the labeled typed λ-terms

$$(\lambda x_1^{A \to A \to B}.(\lambda x_2^A.((x_1\, x_2)^3\, x_2)^4)^5)^6$$

and $(\lambda x_1^A.(\lambda x_2^B.x_1)^3)^4$. Their interpretations are as follows.

1	$A \to A \to B$	
2	A	
3	$A \to B$	$\to$E, 1, 2
4	B	$\to$E, 3, 2
5	$A \to B$	$\to$I, 2, 4
6	$(A \to A \to B) \to (A \to B)$	$\to$I, 1, 5

1	A	
2	B	
3	$B \to A$	$\to$I, 2, 1
4	$A \to B \to A$	$\to$I, 1, 3

Observe that the (repeat) rule does not occur anymore in the resulting deductions. In the first case, we get the same deduction as the 'original' one of 3.4. In the second deduction, we obtain a different deduction from the one of 3.4: the (repeat) rule has been removed. So in general, it is *not* the case that $(\!|-|\!) \circ [\![-]\!]$ is the identity: if we start from a flag deduction, map it to a λ-term and back, we will sometimes arrive at a different flag deduction then the one we started from. Apart from the repeat rule, there are more interesting cases sources for the non-isomorphism. This will be discussed in detail in Section 4. The other way around, it *is* the case that $[\![-]\!] \circ (\!|-|\!)$ is the identity on simply typed λ-terms. First we give the soundness theorem for the second interpretation of flag deductions as typed λ-terms.

Theorem 3.8 (Soundness of $(\!|-|\!)$). *If M is a labeled term of type A, then $(\!|M|\!)$ (following Definition 3.6) is a flag deduction with conclusion A. Moreover, if $FV(M) = \{x_{i_1}^{B_1}, \ldots, x_{i_n}^{B_n}\}$ then the assumptions of $(\!|M|\!)$ are $B_1, \ldots, B_n$, with labels $i_1, \ldots, i_n$, respectively.*

Proof. By induction on M. □

Theorem 3.9. *Given a simply typed λ-term M,*

$$[\![(\!|M|\!)]\!] \equiv M$$

Proof. By induction on M. □

4 The Fine Structure of Flag Deductions

We have already pointed out that there is no isomorphism between simply typed terms and flag deductions, because $(\!|-|\!) \circ [\![-]\!]$, is not the identity. There are various origins for this non-isomorphism and we categorize them.

Given a term M, we refer to $(\!|M|\!)$ as the *canonical flag deduction* for M. If Σ is a flag deduction, we will also call $(\!|[\![\Sigma]\!]|\!)$ the *canonical form of Σ*, because it is the canonical flag deduction for $[\![\Sigma]\!]$. This yields the class of *canonical flag* *deductions.* There is an obvious isomorphism between the typed λ-terms and the canonical flag deductions (modulo relabeling). We want the equivalence relation on flag deductions, that we alluded to before, to be defined in such a way that the class of canonical flag deductions forms a complete set of representatives. This characterizes the canonical forms from a different point of view, purely in terms of flag deductions.

R: Repeat Rule In flag deductions, the (repeat) rule can be applied everywhere. We don't want to distinguish two deductions that only differ in the applications of the (repeat) rule. The interpretation $[\![-]\!]$ maps such deductions to the same λ-term.

G: Garbage (dead ends) In a flag deduction, there may be parts that do not contribute to the final result. We can call these parts 'garbage' or 'dead ends'. Garbage can be detected by looking at the set of lines that the conclusion depends on (following the 'motivations', collecting all lines starting from the conclusion); all lines that are not encountered in this way are garbage. Note that one can not just remove one "garbage line", because there may be other (garbage) lines depending on it. (So one has to start removal with the last garbage line.)

I: Permutation of independent steps The precise order of the deduction steps is somewhat arbitrary. See the following two examples.

1	A (flag)	
2	$A{\to}B$ (flag)	
3	$A{\to}C$ (flag)	
4	$B{\to}C{\to}D$ (flag)	
5	C	$\to$E, 3, 1
6	B	$\to$E, 2, 1
7	$C{\to}D$	$\to$E, 4, 6
8	D	$\to$E, 7, 5

1	A (flag)	
2	$A{\to}C$ (flag)	
3	C	$\to$E, 2, 1
4	$A{\to}B$ (flag)	
5	B	$\to$E, 4, 1
6	$B{\to}C{\to}D$ (flag)	
7	$C{\to}D$	$\to$E, 6, 5
8	D	$\to$E, 7, 3

The conclusion C, now given in line 5, can be located anywhere between lines 3 and 8. Two consecutive flags can also be permuted, as long as they are not closed. A flag can also move over a formula, if the derivability of the formula does not depend on the flag. This is shown in the second deduction. We view all these changes in a deduction as a *permutation of independent steps.*

S: Sharing of subproofs Consider the following two flag deductions

1	B (flag)	
2	$(A{\to}B){\to}(A{\to}B){\to}C$ (flag)	
3	A (flag)	
4	B	R, 1
5	$A{\to}B$	$\to$I, 3, 4
6	$(A{\to}B){\to}C$	$\to$E, 2, 5
7	C	$\to$E, 6, 5

1	B (flag)	
2	$(A{\to}B){\to}(A{\to}B){\to}C$ (flag)	
3	A (flag)	
4	$A{\to}B$	$\to$I, 3, 1
5	$(A{\to}B){\to}C$	$\to$E, 2, 4
6	A (flag)	
7	$A{\to}B$	$\to$I, 6, 1
8	C	$\to$E, 5, 7

In the left deduction, the conclusion $A \to B$ is used twice (in lines 6 and 7), to do an $\to$-elimination. The subproof of $A \to B$ (lines 3–5) is 'shared' by the two applications of $\to$-E on lines 6 and 7. Although the example above is obviously quite trivial (for reasons of exposition), this is clearly a great advantage of flag deductions over ordinary natural deduction (and simply typed λ-terms): a result that has been derived in a certain context can be reused (i.e. as a source it can be 'shared' by consecutive other rules). In ordinary natural deduction, the result $A \to B$ would have to be derived twice. In terms of simply typed λ-calculus, this means that one and the same sub-term will occur several times. We illustrate this by computing the λ-term for this flag deduction, which is

$$y_2(\lambda z^A.x_1)(\lambda z^A.x_1) : C,$$

where x_1^B is the variable associated to line 1 and $y_2^{(A\to B)\to(A\to B)\to C}$ is the variable associated to line 2. The flag deduction associated with this λ-term is the deduction to the right above. Observe that the subproof of $A \to B$ has been copied and occurs twice in this flag deduction. Also the repeat rule has been removed.

We now define an equivalence relation $\simeq_i$ on flag deductions that equates flag deductions that only differ as a consequence of permutation of independent steps. We also define 3 reduction relations, $\longrightarrow_r$ that removes repeat rules, $\longrightarrow_g$ that removes garbage and $\longrightarrow_s$ that unshares deductions.

Definition 4.1. *Define the equivalence relation $\simeq_i$ on flag deductions as the reflexive, symmetric, transitive closure of the following relations.*

1. *(Interchange of lines) If $\Sigma \neq \emptyset$, then*

$$\begin{array}{c|ll} \vdots & \Theta & \\ l & A & m \\ l' & B & m' \\ \vdots & \Sigma & \end{array} \;\simeq_i\; \begin{array}{c|ll} \vdots & \Theta & \\ l' & B & m' \\ l & A & m \\ \vdots & \Sigma & \end{array}$$

Note that, as we assume both sides of the $\simeq_i$ to be well-formed flag deductions, it must be the case that $l \notin m'$, and $l' \notin m$.

2. *(Interchange of blocks) If $\Sigma \neq \emptyset$, then*

$$\begin{array}{c|ll} \vdots & \Theta & \\ n & \begin{array}{|l} A \\ \hline \end{array} & \\ \vdots & \begin{array}{|l} \Pi \end{array} & \\ m & A \to C & \to\text{-}I,n,p \\ l & B & m' \\ \vdots & \Sigma & \end{array} \;\simeq_i\; \begin{array}{c|ll} \vdots & \Theta & \\ l & B & m' \\ n & \begin{array}{|l} A \\ \hline \end{array} & \\ \vdots & \begin{array}{|l} \Pi \end{array} & \\ m & A \to C & \to\text{-}I,n,p \\ \vdots & \Sigma & \end{array}$$

Note that, as we assume both sides of the $\simeq_i$ to be well-formed flag deductions, it must be the case that $l \notin m'$, and $l' \notin \Pi$.

Remark 4.2 (to the Definition). Note that the well-formedness of the right hand side is not automatically implied by the well-formedness of the left hand side. We assume $\Sigma \neq \emptyset$, because if $\Sigma = \emptyset$, then the deduction on the left of the $\simeq_i$ may have a different conclusion from the one to the right.

Definition 4.3. *The reduction relations $\longrightarrow_g$, $\longrightarrow_r$ and $\longrightarrow_s$ on flag deductions are defined as follows.*

1. *If $\Sigma \neq \emptyset$ and $l \notin \Sigma$, then*

$$\begin{array}{l|ll} \vdots & \Theta & \\ n & \begin{array}{|l} A \\ \hline \Pi \end{array} & \\ \vdots & & \\ l & A \to C & \to\text{-}I,n,p \\ \vdots & \Sigma & \end{array} \quad \longrightarrow_g \quad \begin{array}{l|l} \vdots & \Theta \\ \vdots & \Sigma \end{array}$$

2. *If $\Sigma \neq \emptyset$, $l \notin \Sigma$ and $m \neq [I,-,-]$, then*

$$\begin{array}{l|ll} \vdots & \Theta & \\ l & A & m \\ \vdots & \Sigma & \end{array} \quad \longrightarrow_g \quad \begin{array}{l|l} \vdots & \Theta \\ \vdots & \Sigma \end{array}$$

3. *If $\Sigma \neq \emptyset$, then*

$$\begin{array}{l|ll} \vdots & \Theta & \\ l & A & R,m \\ \vdots & \Sigma & \end{array} \quad \longrightarrow_r \quad \begin{array}{l|l} \vdots & \Theta \\ \vdots & \Sigma[l/m] \end{array}$$

4.

$$\begin{array}{l|ll} \vdots & \Theta & \\ m & \begin{array}{|l} A \\ \hline A \end{array} & \\ l & & R,\ m \end{array} \quad \longrightarrow_r \quad \begin{array}{l|l} \vdots & \Theta \\ m & \begin{array}{|l} A \\ \hline \end{array} \end{array}$$

5. *If $m = [R,k]$ or $m = [\to\text{-}E, k_1, k_2]$, then*

$$\begin{array}{l|ll} \vdots & \Theta_1 & \\ l & A & m \\ \vdots & \Theta_2 & \\ l_0 & B_1 & \ldots,l\ldots \\ \vdots & \Theta_3 & \\ l_1 & B_2 & \ldots,l\ldots \\ \vdots & \Sigma & \end{array} \quad \longrightarrow_s \quad \begin{array}{l|ll} \vdots & \Theta_1 & \\ l & A & m \\ l' & A & m \\ \vdots & \Theta_2 & \\ l_0 & B_1 & \ldots,l\ldots \\ \vdots & \Theta_3 & \\ l_1 & B_2 & \ldots,l'\ldots \\ \vdots & \Sigma & \end{array}$$

6.

$$
\begin{array}{r|lll}
\vdots & \Theta_1 & \\
n & \begin{array}{|l} A \\ \hline \Pi \end{array} & \\
\vdots & & \\
l & A{\to}C & {\to}\text{-}I,n,p \\
\vdots & \Theta_2 & \\
l_0 & B_1 & \ldots,l\ldots \\
\vdots & \Theta_3 & \\
l_1 & B_2 & \ldots,l\ldots \\
\vdots & \Sigma &
\end{array}
\quad \longrightarrow_s \quad
\begin{array}{r|lll}
\vdots & \Theta_1 & \\
n & \begin{array}{|l} A \\ \hline \Pi \end{array} & \\
\vdots & & \\
l & A{\to}C & {\to}\text{-}I,n,p \\
n' & \begin{array}{|l} A \\ \hline \Pi' \end{array} & \\
\vdots & & \\
l' & A{\to}C & {\to}\text{-}I,n',p' \\
\vdots & \Theta_2 & \\
l_0 & B_1 & \ldots,l\ldots \\
\vdots & \Theta_3 & \\
l_1 & B_2 & \ldots,l'\ldots \\
\vdots & \Sigma &
\end{array}
$$

where Π' is just Π with fresh line numbers.

The transitive reflexive symmetric closure of $\longrightarrow_g, \longrightarrow_r, \longrightarrow_s$ and $\simeq_i$ will be denoted by $\simeq_{gris}$.

Remark 4.4. In the reduction rules, apart from *lines* being repeated, garbage or interchangeable, we also have to take into accounts *blocks*: parts of a proof that are 'guarded' by an $\to$-I-rule.

Theorem 4.5 (Preservation of equality under $[\![-]\!]^L$ and $[\![-]\!]$). *Let the two flag deductions Σ and Θ be given. If $\Sigma \simeq_i \Theta$ or $\Sigma \longrightarrow_{gr} \Theta$, then $[\![\Sigma]\!]^L \equiv [\![\Theta]\!]^L$. If $\Sigma \longrightarrow_s \Theta$, then $[\![\Sigma]\!] \equiv [\![\Theta]\!]$.*

Proof. For every base step in the definition of $\longrightarrow_{gr}$ or $\simeq_i$, we show that $[\![-]\!]^L$ maps $\simeq$ flag deductions to $\equiv$ λ-terms. Similarly for $\longrightarrow_s$. □

Lemma 4.6 (Closure under $\longrightarrow_r$, $\longrightarrow_g$, $\longrightarrow_s$). *If Σ is a flag deduction and $\Sigma \longrightarrow_g \Sigma'$ or $\Sigma \longrightarrow_r \Sigma'$ or $\Sigma \longrightarrow_g \Sigma'$, then Σ' is a flag deduction with the same conclusion.*

Proof. For every reduction step of Definition 4.3, we easily verify the statement of the Lemma. □

Lemma 4.7. *The reductions $\longrightarrow_g$, $\longrightarrow_r$ and $\longrightarrow_{gr}$ (the union of $\longrightarrow_g$ and $\longrightarrow_r$) are strongly normalizing.*

Proof. The $\longrightarrow_r$ and $\longrightarrow_g$ rules return a shorter flag deduction. □

It can also be proved that $\longrightarrow_{gr}$ is confluent. We will obtain this property only as a consequence of uniqueness of normal forms, which we will prove by defining the *gr*-normal form of a flag deduction directly (without reducing).

The end goal of this section is to prove the reverse of Theorem 4.5. To do that we consider the $\longrightarrow_{grs}$ normal forms, so we first have to show that these exist, which we do by proving that $\longrightarrow_s$ terminates on $\longrightarrow_{gr}$-normal forms. We first introduce some useful notions and state some auxiliary Lemmas.

For l a line in Σ, we want to define $\Sigma \lceil l$ as the flag deduction that arises from Σ by restricting to all those lines that are 'needed' by l. This is the transitive closure of the 'refers to' relation, where a line l refers to l' if l' is used in the motivation (the m) of line l. In taking this transitive closure we skip the applications of the repeat rule. The set of needed lines is inductively defined as follows.

Definition 4.8. *Given a flag deduction Σ and a line $\langle l, A, m\rangle$ in Σ, we define the set of* needed lines *for $\langle l, A, m\rangle$, $lines_\Sigma(\langle l, A, m\rangle)$, by*

$$\begin{aligned}
lines_\Sigma(\langle l, A, -\rangle) &= \{l\}\\
lines_\Sigma(\langle l, A, [R, l_0]\rangle) &= lines_\Sigma(l_0)\\
lines_\Sigma(\langle l, A{\to}B, [{\to}\text{-}I, l_1, l_2]\rangle) &= \{l, l_1\} \cup lines_\Sigma(l_2)\\
lines_\Sigma(\langle l, B, [{\to}\text{-}E, l_1, l_2]\rangle) &= \{l\} \cup lines_\Sigma(l_1) \cup lines_\Sigma(l_2)
\end{aligned}$$

We will usually omit the formula A and the motivation m from this notation, writing $lines_\Sigma(l)$ instead. We write $\Sigma \lceil l$ for $\Sigma \lceil lines_\Sigma(l)$, the restriction of Σ to the lines that are in $lines_\Sigma(l)$.

Lemma 4.9. *If Σ is in gr-normal form, then $\Sigma \lceil l = \Sigma$ (with l the last line of Σ).*

Proof. Obviously, $l' \in \Sigma \lceil l \Rightarrow l' \in \Sigma$, so we only have to prove that all lines that occur in Σ also appear in $\Sigma \lceil l$. This is an easy consequence of the Definition of needed lines. □

Lemma 4.10. *The set of 'needed lines' of a flag deduction is preserved under gr-reduction. That is, for $\Sigma \longrightarrow_{gr} \Sigma'$, with last lines l and l' respectively,*

$$lines_\Sigma(l) = lines_{\Sigma'}(l').$$

Which implies immediately that $\Sigma \lceil l = \Sigma' \lceil l'$.

Proof. By distinguishing cases according to the reduction step $\Sigma \longrightarrow_{gr} \Sigma'$.

Corollary 4.11. *For Σ a flag deduction with last line l,*

$$gr\text{-}normal\ form(\Sigma) = \Sigma \lceil l.$$

Proof. Σ has a gr-normal form, say $\Sigma \longrightarrow_{gr} \Sigma_1 \longrightarrow_{gr} \ldots \longrightarrow_{gr} \Sigma_n$ with Σ_n in gr-normal form. From Lemma 4.10 it follows that $\Sigma \lceil l = \ldots = \Sigma_i \lceil l_i = \ldots = \Sigma_n \lceil l_n$ where l_i is the last line of Σ_i. From Lemma 4.9, it follows that $\Sigma_n \lceil l_n = \Sigma_n$, so $\Sigma \lceil l$ is Σ_n, the gr-normal form of Σ.

Lemma 4.12. *If Σ in gr-normal form, then lab(Σ) = lab($[\![\Sigma]\!]^L$).*

Proof. For lab(Σ) $\subseteq$ lab($[\![\Sigma]\!]^L$), we prove that all labels of $\Sigma \lceil l$ (l the last line of Σ) occur in $[\![\Sigma]\!]^L$. (Then we are done, because, by Lemma 4.9, $\Sigma \lceil l = \Sigma$ for Σ in *gr*-normal form.) Let $l' \in \Sigma \lceil l$, i.e. $l' \in \text{lines}_\Sigma(l)$. We now prove $l' \in \text{lab}([\![\Sigma]\!]^L)$ by an immediate induction on $\text{lines}_\Sigma(l)$. For the reverse, lab($[\![\Sigma]\!]^L$) $\subseteq$ lab(Σ), we prove lab($[\![\Sigma]\!]^L$) $\subseteq$ lab($\Sigma \lceil l$) by induction on $[\![\Sigma]\!]^L$.

Lemma 4.13. *If Σ in gr-normal form, and $\Sigma \longrightarrow_s \Sigma'$, then Σ' is also in gr-normal form.*

Proof. Let Σ be in *gr*-normal form with last line l, so $l' \in \text{lines}_\Sigma(l)$ for all $l' \in \text{lab}(\Sigma)$. In $\Sigma \longrightarrow_s \Sigma'$ new lines are introduced, but they are also in $\text{lines}_\Sigma(l)$, as is easily checked by analyzing the two possible *s*-reduction steps. □

Lemma 4.14. *If Σ in gr-normal form and $\Sigma \longrightarrow_s \Sigma'$, then $\#$lab($[\![\Sigma]\!]^L$) < $\#$lab($[\![\Sigma']\!]^L$)*

Proof. In an *s*-reduction step, new labels are added. Now the Lemma follows immediately from Lemmas 4.12 and 4.13. □

Corollary 4.15 (Termination of $\longrightarrow_s$ on *gr*-normal forms). *If Σ is in gr-normal form, then unsharing ($\longrightarrow_s$) terminates on Σ.*

Proof. The number of labels in $[\![\Sigma]\!]^L$ strictly increases under $\longrightarrow_s$ (Lemmas 4.14 and 4.13). On the other hand, the λ-term $[\![\Sigma]\!]$ (without labels) does not change under $\longrightarrow_s$. There is a maximum to the number of labels that can occur in a labeled version of $[\![\Sigma]\!]$, so $\longrightarrow_s$ terminates. □

We now study the *i*-equality on *grs*-normal forms. The main result is that, for Σ and Θ in *grs*-normal form, if $[\![\Sigma]\!]^L = [\![\Theta]\!]^L$, then $\Sigma \simeq_i \Theta$. The main technique for establishing this result is 'merging' two flag deductions into one. The main property about merging is that, if Σ is a *grs*-normal form containing the two *independent* lines l_1 and l_2, then merging $\Sigma \lceil l_1$ and $\Sigma \lceil l_2$ yields a well-formed flag-deduction Σ'. This only works for Σ in *grs*-normal form.

Definition 4.16. *Given the flag deduction Σ, with $l_1 l_2 \in$ lab(Σ), l_1 and l_2 are* Σ-independent, *notation $l_1 \bot_\Sigma l_2$, if $l_1 \notin \text{lines}_\Sigma(l_2)$ and $l_2 \notin \text{lines}_\Sigma(l_1)$. We omit Σ when it is clear from the context.*

An important property of $\bot$ is the following.

Lemma 4.17. *If Σ is in s-nf and contains the line $\langle l, A, [\to\text{-}E, l_1, l_2]\rangle$, then $l_1 \bot l_2$.*

In the following we denote by oflag(Σ) the set of *open flags* of Σ.

Definition 4.18. *The flag deduction Σ and Θ are* compatible, *notation comp(Σ, Θ), if the following hold.*

1. *$lab(\Sigma) \cap lab(\Theta) \subset oflag(\Sigma) \cap oflag(\Theta)$, i.e. a label that occurs in both Σ and Θ must occur as an open flag in both.*
2. *If $i \in oflag(\Sigma) \cap oflag(\Theta)$ then it occurs as* the same flag *in both Σ and Θ (i,e. with the same formula).*

To define the merging of two flag deductions Σ and Θ, we view them as sequences. The goal is to prove that if both Σ and Θ are in *grs*-normal form, then the merging is a well-formed flag deduction. But this is only the case if we treat a part of a deduction that is 'under a flag' as a 'block' (one part of the sequence Σ), thus disallowing lines from Θ to be moved under a flag of Θ.

Definition 4.19. *If $comp(\Sigma, \Theta)$, we define the* merging *of Σ and Θ, notation $\Sigma||\Theta$, as the following flag deduction.*

1. *First remove $oflag(\Sigma)$ from Σ and $oflag(\Theta)$ from Θ, obtaining Σ', resp. Θ'.*
2. *Now interleave the sequences Σ' and Θ', starting at the end with an element of Σ'. In doing so, we consider a part $\langle i, A, F\rangle, \ldots, \langle l, A{\to}B, [I, i, l_2]\rangle$ as one element of the sequence.*
3. *Finally, put all elements of $oflag(\Sigma) \cup oflag(\Theta)$ on top of the sequence Δ in a canonical way (following a fixed ordering of $\mathcal{I}$).*

Lemma 4.20. *If $comp(\Sigma, \Theta)$ then $\Sigma||\Theta$ is a well-formed flag deduction.*

Lemma 4.21. *If $comp(\Sigma, \Theta)$ and $\Sigma \simeq_i \Sigma'$, then $\Sigma||\Theta \simeq_i \Sigma'||\Theta$ and $\Theta||\Sigma \simeq_i \Theta||\Sigma'$.*

Proposition 4.22. *Given two flag deductions Σ and Θ in grs-normal form, if $[\![\Sigma]\!]^L \equiv [\![\Theta]\!]^L$, then $\Sigma \simeq_i \Theta$.*

Proof. By induction on the structure of $[\![\Sigma]\!]^L$.

var $[\![\Sigma]\!]^L = x_i$. Then Σ and Θ are both $\langle i, A, [F]\rangle$.

app $[\![\Sigma]\!]^L = (M^{l_1} N^{l_2})^l$. Then Σ and Θ end with an $\to$-E rule. By induction hypothesis, $\Sigma \lceil l_1 \simeq_i \Theta \lceil l_1$ and $\Sigma \lceil l_2 \simeq_i \Theta \lceil l_2$ and so $\Sigma \simeq_i ((\Sigma \lceil l_1)||(\Sigma \lceil l_2))\langle l, B, [E, l_1, l_2]\rangle \simeq_i ((\Theta \lceil l_1)||(\Theta \lceil l_2))\langle l, B, [E, l_1, l_2]\rangle \simeq_i \Theta$

abs $[\![\Sigma]\!]^L = (\lambda x_i{:}A.M^{l_2})^l$. Then Σ and Θ end with an $\to$-I rule. They can be of one of the following shapes.

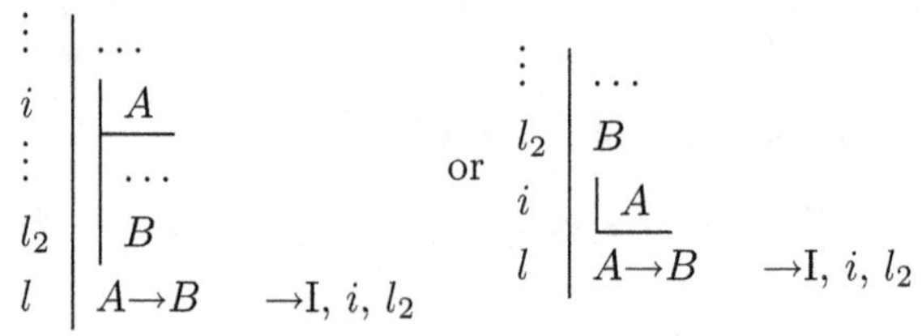

We distinguish cases according to whether $x_i \in \mathrm{FV}(M)$ (and then $i \in \mathrm{lines}(l_2)$) or $x_i \notin \mathrm{FV}(M)$ (and then $i \notin \mathrm{lines}(l_2)$). In the first case, Σ and Θ must have the first shape. By induction hypothesis, $\Sigma \lceil l_2 \simeq_i \Theta \lceil l_2$.

Then also $\Sigma \lceil l_2 \simeq_i \Theta \lceil l_2$, where the last open flag is preserved and hence $\Sigma \simeq_i \Theta$. In the second case, $\Sigma \lceil l_2$ and $\Theta \lceil l_2$ do not contain i. Hence both Σ and Θ are $\simeq_i$ equal to a deduction of the second shape. By induction hypothesis, $\Sigma \lceil l_2 \simeq_i \Theta \lceil l_2$ and we can safely add the line $\langle i, A, F\rangle$ at the end and also the line $\langle l, A{\to}B, [I, i, l_2]\rangle$, and hence $\Sigma \simeq_i \Theta$. □

To prove the final theorem, we need to more Lemmas. They could have been proved before already, but were not yet needed. Therefore we state them only now. Both are proved by induction on the structure of Σ, using the fact that if l is the label of a line that is not a flag, then l occurs at most once in a motivation of Σ.

Lemma 4.23. *If Σ is a flag deduction in s-normal form, then $[\![\Sigma]\!]^L$ is a uniquely labelled simply typed term. (That is, every label occurs at most once in $[\![\Sigma]\!]^L$.)*

Lemma 4.24. *For r a relabelling, Σ a flag deduction and M a labelled simply typed term, if $[\![\Sigma]\!]^L = M$, then $[\![r(\Sigma)]\!]^L = r(M)$.*

Theorem 4.25. *If $[\![\Sigma]\!] \equiv [\![\Theta]\!]$, then $\Sigma \simeq_{gris} \Theta$.*

Proof. Suppose $[\![\Sigma]\!] \equiv [\![\Theta]\!]$. Consider the grs-normal forms of Σ and Θ: Σ' and Θ'. Then $[\![\Sigma]\!] \equiv [\![\Sigma']\!] \equiv [\![\Theta']\!] \equiv [\![\Theta]\!]$. This implies that $[\![\Sigma']\!]^L \equiv M$, $[\![\Theta']\!] \equiv N$ with $|M| \equiv |N|$. Moreover (by Lemma 4.23), all labels in M and N are unique, so we can find a relabelling r such that $r(M)$ is N. By Lemma 4.24, this means that $[\![r(\Sigma')]\!]^L \equiv r(M) \equiv N \equiv [\![\Theta']\!]^L$. From Proposition 4.22, it now follows that $r(\Sigma') \simeq_i \Theta'$. Hence (as we work modulo relabelling), $\Sigma \simeq_{grs} \Sigma' \simeq_i \Theta' \simeq_{grs} \Theta$. □

The following follows immediately from the Theorem and Theorem 3.9.

Corollary 4.26. *Given a flag deduction Σ, $(\![[\![\Sigma]\!]]\!) \simeq_{gris} \Sigma$.*

4.1 Defining Cut-Elimination

We can now define cut-elimination on flag deductions by first taking the $\longrightarrow_s$ normal form and then eliminating cuts as discussed in Section 2.1.

Definition 4.27. *We define cut-elimination on flag deductions as follows.*

$$
\begin{array}{l|l l}
1 & \quad\begin{array}{|l} A \\ \hline \Sigma \end{array} & \\
\vdots & & \\
l & A{\to}B & \to\text{-}I\ 1,n \\
\vdots & \Theta & \\
k & A & \\
\vdots & \Pi & \\
l' & B & \to E,\ l,\ k
\end{array}
\quad \Longrightarrow_c \quad
\begin{array}{l|l l}
\vdots & \Theta & \\
k & A & \\
\vdots & \Sigma[k/1] & \\
\vdots & \Pi & \\
l' & B & R,n
\end{array}
$$

k	A	
$\vdots$	Π_0	
k'	$\quad A$ (flag)	
$\vdots$	$\quad \Sigma$	
l	$A{\rightarrow}B$	$\rightarrow$-I k',n
$\vdots$	Π_1	
l'	B	$\rightarrow E$, l, k

$\Longrightarrow_c$

k	A	
$\vdots$	Π_0	
$\vdots$	$\Sigma[k/k']$	
$\vdots$	Π_1	
l'	B	R,n

As usual, these reduction rules can also be applied in a context. In the definition, we introduce repeat rules to make sure that B remains on the last line. These can again be removed via $\longrightarrow_r$ steps.

Remark 4.28. Different from cut-elimination in Gentzen natural deduction, a $\Longrightarrow_c$-step does not involve any duplication of subderivations, which may seem odd. However, a $\Longrightarrow_c$ step can introduce sharing of subproofs and the unsharing (via $\longrightarrow_s$) involves duplication of subderivations. This also implies that, to apply another cut-elimination step we first have to take the $\longrightarrow_s$-normal form of the result.

An example where a $\Longrightarrow_c$ step creates sharing is the following. (On the right hand side, line l is shared by lines 4 and 6.)

1	$A{\rightarrow}(B{\rightarrow}C){\rightarrow}D$	
2	$A{\rightarrow}C$	
3	$\quad A$ (flag)	
4	$\quad (B{\rightarrow}C){\rightarrow}D$	→E, 1, 3
5	$\quad\quad B$ (flag)	
6	$\quad\quad C$	→E, 2, 3
7	$\quad B{\rightarrow}C$	→I, 5, 6
8	$\quad D$	→E, 4, 7
9	$A{\rightarrow}D$	→I, 3, 8
$\vdots$	Θ	
l	A	
$l+1$	D	→E, 9, l

$\Longrightarrow_c$

1	$A{\rightarrow}(B{\rightarrow}C){\rightarrow}D$	
2	$A{\rightarrow}C$	
$\vdots$	Θ	
l	A	
4	$(B{\rightarrow}C){\rightarrow}D$	→E, 1, l
5	$\quad B$ (flag)	
6	$\quad C$	→E, 2, l
7	$B{\rightarrow}C$	→I, 5, 6
8	D	→E, 4, 7
$l+1$	D	R, 8

Lemma 4.29. *If Σ is a well-formed flag deduction in $\longrightarrow_s$-normal form and $\Sigma \Longrightarrow_c \Sigma'$, then Σ' is a well-formed flag deduction with the same conclusion.*

Proof. As all the lines (except for the flags) are referred to at most once in a $\longrightarrow_s$-normal form,we can safely move around the subparts (and remove some of them) as indicated above.

Theorem 4.30. *For Σ a well-formed flag deduction in $\longrightarrow_{gs}$-normal form and for M a uniquely labelled simply typed term,*

$$\Sigma \Longrightarrow_c \Sigma_1 \longrightarrow_{sr} \Sigma_2 \simeq_i \Theta \;\; \textit{iff} \;\; [\![\Sigma]\!] \longrightarrow_\beta [\![\Theta]\!]$$
$$M \longrightarrow_\beta N \;\; \textit{iff} \;\; (\![M]\!) \Longrightarrow_c \Sigma_1 \longrightarrow_{sr} \Sigma_2 \simeq_i (\![N]\!)$$

where the Σ_1 and Σ_2 are existentially quantified.

5 Future Work

In this paper we restrict to the simplest fragment of logic: minimal proposition logic. But already there important aspects of flag deductions become visible, showing that in some way their structure (e.g. the order of the steps) is quite arbitrary but that in another way (e.g. the reusability of proven results, 'sharing'), their structure is quite useful and interesting. The Curry-Howard interpretation to simply typed λ calculus that we define here and the analysis of cut-elimination brings about this structure quite nicely.

We believe that the results of this paper can be extended to full first order predicate logic. This will be presented in forthcoming work which is a more detailed exposition of the results in this paper. A more intersting aspect is the definition of a term calculus for flag deductions directly. Simply typed λ-calculus ignores part of the structure of a flag deduction, extracting its 'computational content' and removing 'bureaucratic details'. But it also removes sharing and we don't consider that to be only a 'bureaucratic detail', but sometimes computationally relevant. It was suggested by the referees to use a λ-calculus with let-expressions to encode flag deductions faithfully. Then the reductions $\longrightarrow_{grs}$ and the congruence $\simeq_i$ can be described on these terms directly, giving a more perspicuous presentation. The 'sharing' example deduction in Section 4 then is interpreted as

$$\text{let}\, x_5 = (\lambda x_3.\text{let}\, x_4 = x_1 \,\text{in}\, x_4) \,\text{in}\, (\text{let}\, x_6 = x_2 x_5 \,\text{in}\, (\text{let}\, x_7 = x_6 x_5 \,\text{in}\, x_7))$$

This gives connections with the monadic presentation of λ-calculus, the (operational) CPS-translation, the (logical) A-translation.

Similarly, one can define a slightly different Curry-Howard embedding to simply typed terms and then $\simeq_i$ becomes σ-equivalence on λ-terms, as in the work of [Regnier 1994]. This gives a connection with proof nets. We will exploit these connections further and we thank the referees for their comments.

We note that the other suggested interpretations do not really follow the inductive structure of the flag deductions. It might be interesting to find a term-calculus for flag deductions where the basic constructors for flag deductions are the same as for the term calculus.

Acknowledgments

We greatly acknowledge the very insightful comments and suggestions of the referees on the first version of this paper. We are sorry that we don't have the space to discuss all their comments here.

References

[Barendregt 1992] H.P. Barendregt, Lambda calculi with Types. In *Handbook of Logic in Computer Science*, eds. Abramski et al., Oxford Univ. Press, pp. 117 – 309.

[Bornat and Sufrin 1999] R. Bornat and B. Sufrin, Animating Formal Proof at the Surface: The Jape Proof Calculator; *The Computer Journal*, Vol. 42, no. 3, pp. 177-192, 1999.

[Regnier 1994] L. Regnier, Une équivalence sur les lambda-termes, *TCS* 126(2), pp. 281–292, 1994.

[Fitch 1952] F. B. Fitch, *Symbolic Logic*, the Ronald Press Company, New York, 1952.

[Gentzen 1969] G. Gentzen. *Collected Works*. Edited by M.E. Szabo. North-Holland, Amsterdam, 1969.

[Howard 1980] W.H. Howard, The formulas-as-types notion of construction, in *To H.B. Curry: Essays on Combinatory Logic, Lambda Calculus and Formalism*, eds. J.P. Seldin and J.R. Hindley, Academic Press 1980, pp. 479–490.

[Prawitz 1965] D. Prawitz. *Natural Deduction.* Almquist & Wiksell, Stockholm, 1965.

Efficient λ-Evaluation with Interaction Nets

Ian Mackie

Department of Computer Science
King's College London
Strand, London WC2R 2LS, UK
ian@dcs.kcl.ac.uk

Abstract. This paper presents an efficient implementation of the λ-calculus using the graph rewriting formalism of interaction nets. Building upon a series of previous works, we obtain one of the most efficient implementations of this kind to date: out performing existing interaction net implementations, as well as other approaches. We conclude the paper with extensive testing to demonstrate the capabilities of this evaluator.

1 Introduction

One of the first algorithms to implement Lévy's [9] notion of optimal reduction for the λ-calculus was presented by Lamping [7]. With the help of linear logic [4], this algorithm was tidied up and lead to the well-known algorithm of Gonthier, Abadi and Lévy [5]. Empirical and theoretical studies of this algorithm have revealed several causes of inefficiency (accumulation of certain nodes in the graph rewriting formalism). Asperti et al. [1] devised BOHM (Bologna Optimal Higher-Order Machine) to overcome some of these issues, which has stood until now as not only the most efficient (in terms of rewriting steps) implementation of optimal reduction, but also the most efficient implementation of the λ-calculus.

Interaction nets [6] (a particular form of graph rewriting) has played a role in the above algorithms. However, the focus has been on optimality, rather than on using the interaction net framework. A parallel thread of work takes interaction nets as a focus point rather than optimality. An important reason for this choice is that, in addition to offering insight into issues such as sharing computation, they provide an operational model which captures *all* the computation: in other words, counting the rewrite steps is sufficient to measure the cost of a computation. The key observation here is that in the previously mentioned implementations of the λ-calculus, β-reduction (not including substitution) is just another graph rewrite. Our aim therefore is to use a more pragmatic approach to optimal reduction where we aim to find the *minimum number of rewrite steps* (β included). Historically, this notion of "practical" optimality began in [11], based on an interaction net encoding of linear logic due to Abramsky. Although this first λ-evaluator based on interaction nets performed fewer interactions (rewrite steps) for specific λ-terms than Lamping's algorithm, it was never a match for BOHM. A further attempt, YALE [12], provided a substantial improvement

V. van Oostrom (Ed.): RTA 2004, LNCS 3091, pp. 155–169, 2004.

which can systematically perform better than Lamping's algorithm, and approximate BOHM on specific classes of terms. However, when the need for optimality kicks in, YALE is a very poor second best.

A question therefore remained: is there an efficient interaction net implementation of the λ-calculus which does less work than BOHM? The purpose of the present paper is to answer this question in the positive. Specifically, we give a new λ-evaluator: KCLE (King's College Lambda Evaluator) which has the following features:

- It is efficient: although KCLE performs more β-reduction steps than optimal reducers, the overall number of graph rewrite steps is smaller.
- It evaluates λ-terms to full normal form, even for open terms (as a side effect, this offers a relatively simple notion of read-back, as normal forms are images of the translation function).
- It is an interaction net, so we can take advantage of many results and implementations, specifically parallel, where almost linear speedup has been achieved. We discuss other advantages of interaction nets later.

Relation to Previous Work. The present paper is a continuation of a programme of research by the author to use interaction nets as an efficient mechanism for the encoding of the λ-calculus. Specifically, it builds upon two pieces of work: [11] and [12]. It is also related to the work on interaction nets for Linear Logic [13].

Overview. The rest of this paper is structured as follows. In the next section we recall interaction nets, and motivate why we use them. In Section 3 we give the translation of the λ-calculus into interaction nets. Section 4 studies the reduction system. In Section 5 we examine properties of the encoding. Section 6 gives experimental evidence of our results, where we compare with other systems. Finally, we conclude the paper in Section 7.

2 Interaction Nets

An interaction net system [6] is specified by giving a set Σ of symbols, and a set $\mathcal{R}$ of interaction rules. Each symbol $\alpha \in \Sigma$ has an associated (fixed) *arity*. An occurrence of a symbol $\alpha \in \Sigma$ will be called an *agent*. If the arity of α is n, then the agent has $n + 1$ *ports*: a distinguished one called the *principal port* depicted by an arrow, and n *auxiliary ports* labelled $x_1, \ldots, x_n$ corresponding to the arity of the symbol. Such an agent will be drawn in the following way:

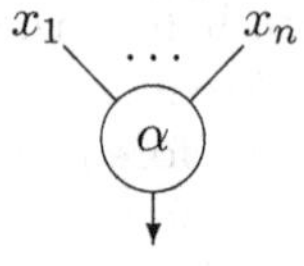

A net N built on Σ is a graph (not necessarily connected) with agents at the vertices. The edges of the graph connect agents together at the ports such that there is only one edge at every port. The ports of an agent that are not connected to another agent are called the free ports of the net. There are two special instances of a net: a wiring (no agents) and the empty net.

A pair of agents $(\alpha, \beta) \in \Sigma \times \Sigma$ connected together on their principal ports is called an *active pair*; the interaction net analog of a redex. An interaction rule $((\alpha, \beta) \Longrightarrow N) \in \mathcal{R}$ replaces an occurrence of the active pair (α, β) by a net N. The rule must satisfy two conditions: all free ports are preserved during reduction (reduction is local, *i.e.*, only the part of the net involved in the rewrite is modified), and there is at most one rule for each pair of agents. The following diagram illustrates the idea, where N is any net built from Σ.

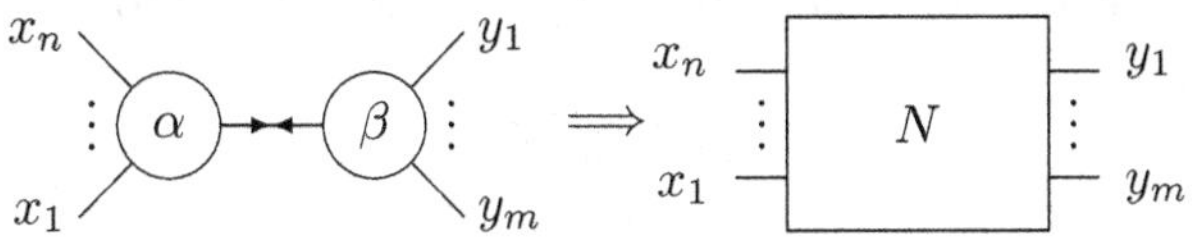

We use the notation $\Longrightarrow$ for the one step reduction relation and $\Longrightarrow^*$ for its transitive and reflexive closure. If a net does not contain any active pairs then we say that it is in normal form. One-step reduction ($\Longrightarrow$) satisfies the diamond property, and thus we obtain a very strong notion of confluence. Indeed, all reduction sequences are permutation equivalent and standard results from rewriting theory tell us that all notions of termination coincide (if one reduction sequence terminates, then all reduction sequences terminate).

We choose to base this work on interaction nets rather than general graph rewriting for several reasons. First, one perspective on interaction nets is that they are a user defined instruction set (assembly language) for an object language. We define this object language, and a compilation of a high-level language (in this case the λ-calculus) and we directly obtain an implementation. The most important aspect of this instruction set, as a consequence of the definition of interaction rules, is that it expresses *all* the elements of a computation: there is no external copying or erasing machinery for instance. Interaction nets can therefore be seen as offering a low-level operational semantics. For this reason it is one of the best formalisms for studying implementations and estimating the cost of evaluation. For a given implementation, each interaction is a known, constant time operation and therefore we are able to estimate costs easier. It is also the case that it is easy to identify the next rewrite rule (which is not always the case in traditional λ-graph rewriting).

Another reason for choosing interaction nets is that we can take advantage of its properties (such as strong confluence) to provide very simple and direct proofs of correctness of encodings. In addition to these properties, we can also take advantage of existing implementations, specifically parallel (see for instance [16]) where almost linear speedup has been demonstrated. Any system of interaction nets that is written can therefore be executed on parallel hardware: no explicit processor allocation is required.

The final reason is that interaction nets capture sharing in a very natural way—indeed, one has to work quite hard to duplicate work. They are therefore a natural fit when studying efficient implementations of any programming language, and especially the λ-calculus.

3 Translation

In this section we give a translation $\mathcal{T}(\cdot)$ of the λ-calculus into interaction nets. The agents required for the translation will be introduced when needed, and the interaction rules for these agents will be given in the following section. We remark that the translation given here is very similar to that used by YALE [12], with an essential difference that we identify closed abstractions in the translation (the rewrite rules of the next section are quite different however).

A λ-term t with $\mathsf{fv}(t) = \{x_1, \ldots, x_n\}$ will be translated as a net $\mathcal{T}(t)$ with the root edge at the top, and n free edges corresponding to the free variables:

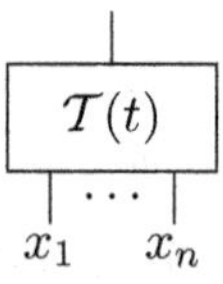

The labelling of free edges is just for the translation (and convenience), and is not part of the system. The first case of the translation function is when t is a variable, say x, then $\mathcal{T}(t)$ is translated into an edge:

Abstraction. If t is an abstraction, say $\lambda x.t'$, then we first require that $x \in \mathsf{fv}(t')$. If this condition is not satisfied, then we can add the following agent to the translation of the body:

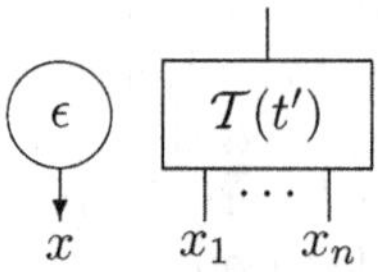

Having assured this condition, there are two alternative translations of the abstraction, which are both given in the following diagram:

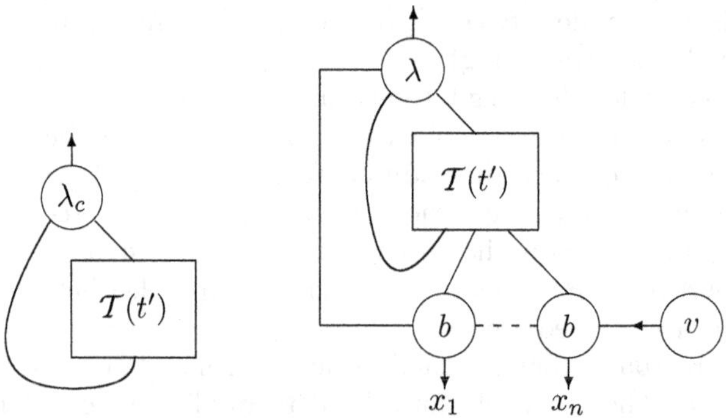

The first case, shown on the left in the above diagram, is when $\mathsf{fv}(\lambda x.t') = \emptyset$. Here we use one agent λ_c to represent a *closed abstraction*. This net corresponds closely to usual graph representations of the λ-calculus (for instance [15]), except that we explicitly connect the occurrence of the variable to the binding λ.

The second case, shown on the right, is when $\mathsf{fv}(\lambda x.t') = \{x_1, \ldots, x_n\}$. Here we introduce three different kinds of agent: λ of arity 3, for abstraction, and two kinds of agent representing a list of free variables. An agent b is used for each free variable, and we end the list with an agent v. The idea is that there is a pointer to the free variables of an abstraction; the body of the abstraction is encapsulated in a box structure. We assume, without loss of generality, that the (unique) occurrence of the variable x is in the leftmost position of $\mathcal{T}(t')$.

We remark that a closed term will never become open during reduction (although of course terms may become closed, and indeed there are interaction rules which will create a λ_c agent from a λ agent when needed). The use of the λ_c agent identifies the case where there are no free variables, and plays a crucial role in the efficient dynamics of this system.

Application. If t is an application, say uv, then $\mathcal{T}(uv)$ is given by the following net, where we have introduced an agent @ of arity 2 corresponding to an application. In addition, if u and v share common free variables, then c agents (representing copy) collect these together pairwise so that a single occurrence of each free variable occurs amongst the free edges.

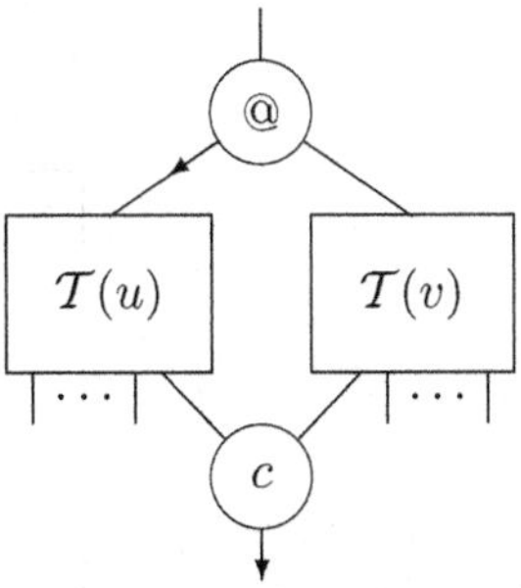

That completes the cases for the translation, which is the same whether we talk about typed or untyped terms. For typed terms, it is convenient to have a constant at base type, say $\star : I$, which can be thought of as similar to an integer for instance. For $\star$ we can choose to represent this as the λ-term $\lambda x.x$, and thus we do not need to add anything to the translation nor to the interaction rules of the next section.

We state one important static result about this translation, which is a direct consequence of the fact that no active pairs are created for the translation of normal forms.

Lemma 1. *If t is a λ-term in normal form, then $\mathcal{T}(t)$ is a net in normal form.*

Example 1. In Figure 1 we give three example nets corresponding to the term $\mathbf{2}' = \lambda xy.(\lambda z.(z(zy)))(\lambda w.xw)$, $\mathbf{K} = \lambda xy.x$, and $\mathbf{2} = \lambda fx.f(fx)$, which give a flavour of the kinds of structures that we are dealing with.

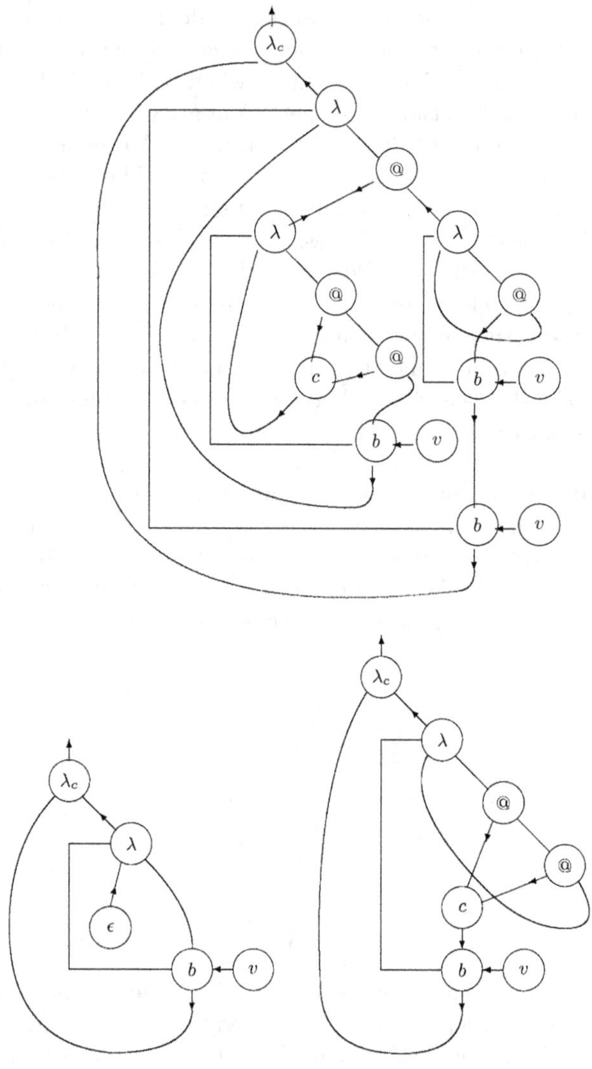

Fig. 1. Example nets

4 Reduction

In this section we give the heart of the work by defining the interaction rules for the evaluator. We back these rules up with some intuitions, and in the following section we state several properties of the rewrite system.

We begin by giving, in Figure 2, the interaction rules for KCLE. When talking about these, we shall use the notation $\alpha \bowtie \beta$ to identify the rule where the agents α and β make up the left-hand side. The final two rules in the figure ($\delta \bowtie \alpha$, and $\epsilon \bowtie \alpha$) are rule schemes, and correspond to the general pattern

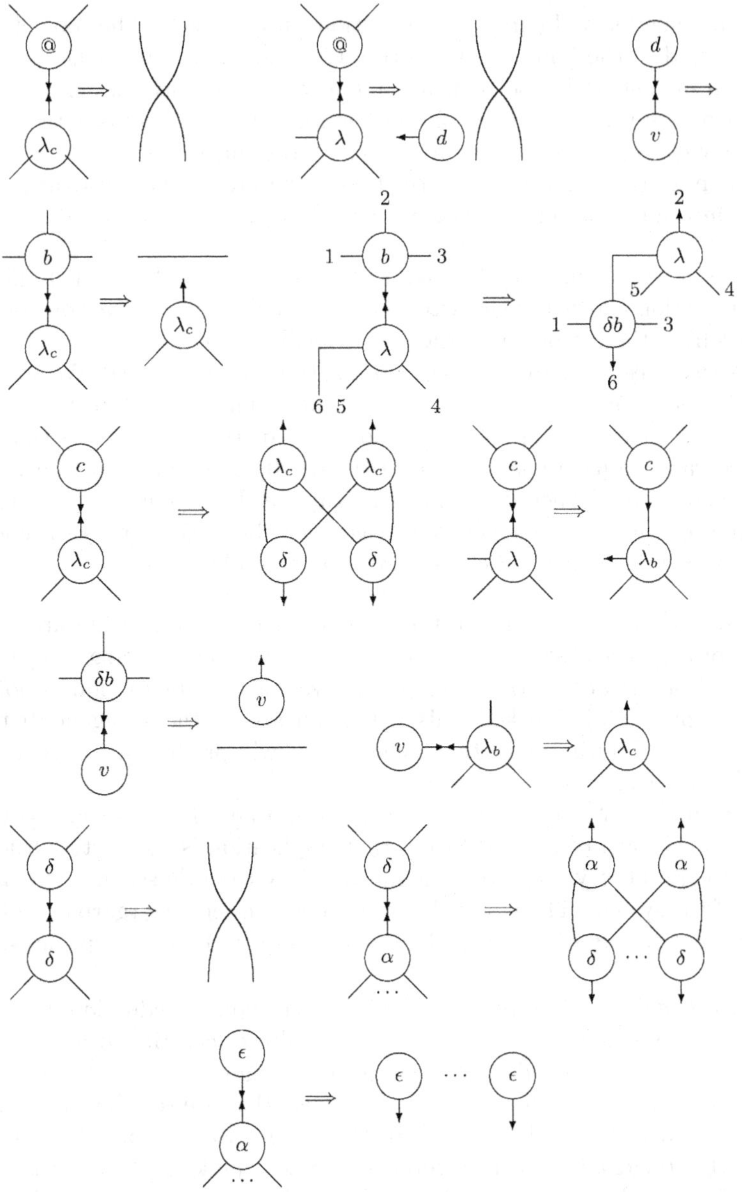

Fig. 2. KCLE Interaction Rules

of rule for these agents, where α can be any of the agents. Where necessary, to avoid ambiguity, we have labelled the free edges of the left and right-hand sides of the rules.

Intuitions. Here we give some insight into the reduction process. The first two rules $\lambda_c \bowtie @$ and $\lambda \bowtie @$ are performing β-reduction, $(\lambda x.t)u \to t[u/x]$, where

$t[u/x]$ is some form of explicit substitution. In both cases, the root of the term is connected to the body of the abstraction, and the variable edge is connected to the argument. In the second case a d agent is introduced which tries to erase the list of free variables. The only interaction rule for d here is with v, in which case they eliminate each other, as shown by the third rule.

The next two rules are concerned with the process of substituting one abstraction inside another: $(\lambda x.t)[u/y] \rightarrow \lambda x.t[u/y]$.

1. $\lambda_c \bowtie b$ corresponds to the case when u is a closed abstraction. This single interaction allows the substitution to be made, and removes one occurrence of b from the list of free variables of the abstracted term.
2. $\lambda \bowtie b$ corresponds to the case when the term u is not closed. We do not block this operation, but the substitution process may not complete (specifically, the net under the abstraction is moved inside the other abstraction, but the free variable list remains unsubstituted). If, by other reductions, the term becomes closed, then it will indeed complete. The agent δb is used to wait to interact with a v agent, in which case the substitution process is completed. The example below indicates what is going on here.

The next two rules concern the copy agent c, which can initiate the duplication of a *closed* abstraction. This is achieved by introducing δ agents inside the body of the abstraction for λ_c. However, if the abstraction is not closed, then the progress is blocked: this is the purpose of the λ_b agent. If the term becomes closed during reduction, then $v \bowtie \lambda_b$ will produce a λ_c agent in which case copying can progress.

The rules for δ concern the duplication of a net. The first one shows $\delta \bowtie \delta$ which cancel each other, indicating that duplication is complete. Otherwise, a δ agent copies everything which is indicated by the rule scheme: α can be any agent of the system. The final rule scheme concerns erasing (garbage collection). The ϵ agent simply erases everything it interacts with, and propagates.

Example Reduction. Figure 3 gives several snapshots of reduction, with the aim of illustrating some of the intuitions given above. Starting from the example term $\mathbf{2}' = \lambda xy.(\lambda z.(z(zy)))(\lambda w.xw)$, given in Figure 1, we perform the $\lambda \bowtie @$ and $c \bowtie \lambda$ interactions to obtain the first net. Here we see that a λ_b agent has been created to block the copying of the (open) abstraction. We also identify where the d agent has been introduced creates "junk" in the net: the sub-net consisting of d, b and v could all be eliminated, but cannot be erased due to the configuration of principal ports. This net is now in normal form.

However, if we construct the application of $\mathbf{2}'I$, then a number of further reductions are possible. After the $\lambda_c \bowtie @$ rule, we have the net for the identity connected to the lowest b agent. Using $\lambda_c \bowtie b$ twice allows this substitution to be made. In the process, we create a $\lambda_b \bowtie v$ interaction which converts λ_b into λ_c. The redex created by the substitution can be contracted, and copied giving us the second net in the figure. Four further interactions give the last net in the figure. If we apply this to a further I, then we get the translation of

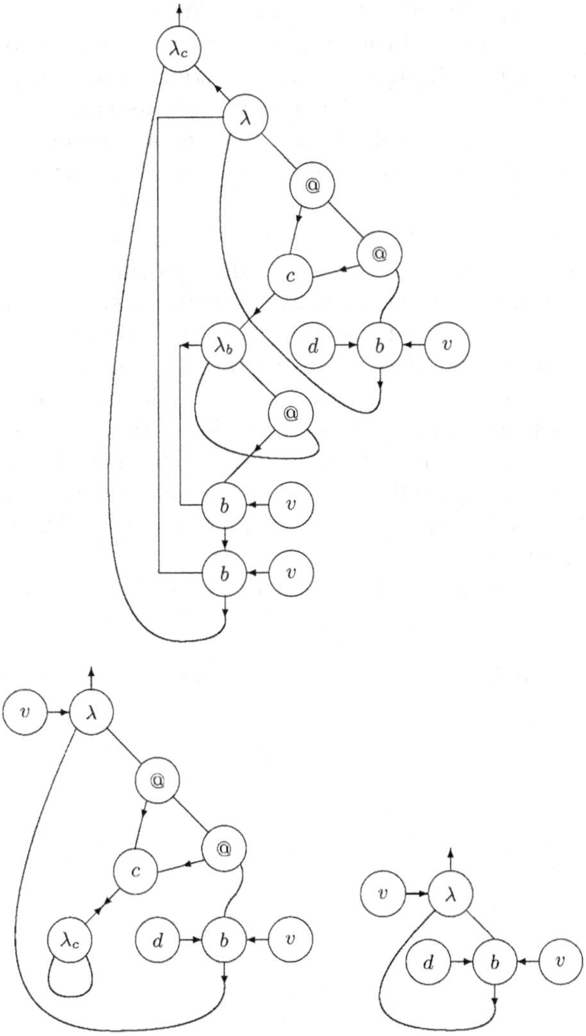

Fig. 3. Example reduction sequence (snapshots)

I as the final net after 4 interactions. The point here is that junk is removed during reduction; computation can complete to a recognisable term if enough arguments are provided.

5 Properties

As we have seen in Figure 3, reduction is weak: a net in normal form does not necessarily correspond to the translation of a λ-term in normal form. However, we define an extension to the system, called the normal form extension, which allow us to simulate full reduction in the λ-calculus. We then show that this extension does nothing if the term is already in normal form. All proofs are inductive, and use the interaction rules from the previous section. A key component in all the proofs is the confluence of interaction nets: we only need to show that there exists a specific reduction sequence of interactions, then we obtain the same result for any permutation of this sequence. We begin by defining this extension.

Normal Form Extension. We extend KCLE with additional agents and rules which will serve for two purposes. First, they provide us with an operational read-back procedure. Additional agents can be connected to all free edges of a net, which will allow the net to be reduced to the representation of the normal form of a translated term. Second, they can also be part of the system: whenever we need to force reduction to normal form (for efficiency reasons for instance), then we can introduce agents from this extension of the basic system. Since these interaction rules are to some extent less important than the previous ones we relegate them to the appendix. There are four agents $\phi_1, \phi_2, \phi_3, \phi_\lambda$ of arity 1, one agent $@'$ of arity 2, and two agents λ_u, cb of arity 3. The interaction rules for these are given in Appendix A.

Next we define a general net construction, which is needed for the correctness result.

Definition 1 (Enclosure). *For any net N, we define $\mathcal{E}_\alpha(N)$, the α-enclosure of the net N as:*

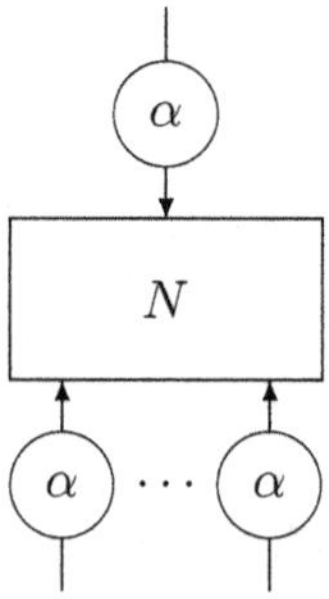

In words, all the free edges are connected to α agents. The diagram shows the case for $\mathsf{ar}(\alpha) = 1$, but any arity is possible.

Definition 2 (Normal Form Enclosure). *Define $\mathcal{T}_{\rm nf}(t)$ of a λ-term t as $\mathcal{E}_{\phi_3}(\mathcal{E}_{\phi_2}(\mathcal{E}_{\phi_1}(\mathcal{T}(t))))$ which we call the* normal form enclosure *of a term t.*

Lemma 2. *For any λ-term t in normal-form there is a terminating reduction sequence: $\mathcal{T}_{\rm nf}(t) \Longrightarrow^* \mathcal{T}(t)$ (i.e., the normal form extension has the global effect of doing nothing with λ-normal-forms).*

We can now state the main results of this section.

Theorem 1. *Let t be a λI-term (i.e., a net built without the ϵ agent).*

1. *If $t \to_\beta^* t'$ (t' a β-normal-form) then there exists a terminating sequence of interactions: $\mathcal{T}_{\mathrm{nf}}(t) \Longrightarrow^* \mathcal{T}(t')$. If t does not have a normal form, then neither does $\mathcal{T}_{\mathrm{nf}}(t)$.*
2. *If t is a closed simply typable term of base type, then there is a terminating sequence of interactions: $\mathcal{T}(t) \Longrightarrow^* \star$ (i.e., using only the rules in Figure 2).*

Remark 1. In the above theorem we can relax the condition of t being a λI-term if we do not evaluate disconnected nets. Such a strategy for interaction nets does indeed exist [3], and evaluators have been built that implement this strategy, see for instance [17]. In this extended abstract we simplify matters by simply stating the results for the λI-calculus. Also note that the first part of the theorem above does not require that the term t is closed, thus applies equally to open terms.

Proof. (Sketch.) The first part requires meticulous use of the normal form extension to show that a particular reduction sequence in the λ-calculus is simulated.

The second is less complex, and we give some additional details here. The key to the proof is a substitution lemma: it is well-known that for closed terms of ground type that during reduction there is always a closed substitution being propagated [3]. Now using this property, we can simulate substitutions. The only additional property that we require is that the encapsulation $\mathcal{E}_\delta(N)$ duplicates a net N, when N is a net in normal form.

6 Testing

There are a number of standard benchmark results that are used to test the performance of these kinds of evaluators. We base ours on those used to demonstrate BOHM [2], which is the system that we use as the basis of the comparison. These terms generate vast computations, and in particular include terms where sharing plays a significant role for the evaluation. Table 1 shows some evaluations of Church numerals. Results of the form 8(4) should be read as 8 interactions, of which 4 were $\lambda \bowtie @$ (thus giving a count of the number of β-reductions performed). The number of β-reduction steps performed is only given for curiosity: it is the number of interactions which gives a measure of actual work. To give some indication as to what some of these numbers mean, 2 2 2 2 2 II evaluates in around 5 seconds using KCLE, whereas BOHM takes approximately 10 minutes. Implementations of functional languages, such as Haskell, OCaml, SML, etc. are well known to not be able to cope with such terms, and indeed fail on this input. The next table below gives the computation of factorial using only the pure λ-calculus (using the encoding given in [2]).

Term	KCLE	BOHM
Fact 2 II	292(41)	212(41)
Fact 3 II	521(55)	373(54)
Fact 4 II	813(70)	594(68)
Fact 5 II	1180(86)	903(83)
Fact 10 II	4560(181)	5168(173)
Fact 20 II	23845(446)	53498(428)
Fact 30 II	69830(811)	241978(783)

Table 1. Benchmark: Church numerals

Term	KCLE	BOHM
2 II	8(4)	12(4)
2 2 II	32(9)	40(9)
2 2 2 II	88(18)	93(16)
2 2 2 2 II	328(51)	356(27)
2 2 2 2 2 II	983392(131124)	1074037060(61)
3 2 II	52(12)	47(12)
4 2 II	72(15)	63(15)
5 2 II	92(18)	79(18)
10 2 II	192(33)	159(33)
15 2 II	292(48)	239(48)
3 2 2 II	170(29)	157(21)
4 2 2 II	312(48)	330(26)
5 2 2II	574(83)	847(31)
10 2 2 II	15564(2082)	531672(56)
15 2 2 II	491834(65585)	537100609(81)
17 2 2 II	1966438(262199)	-
3 2 2 2 II	4018(542)	34740(40)
4 2 2 2 II	983376(131121)	1074037034(60)
2 2 2 10 II	1072(179)	10307(67)
2 2 2 20 II	2002(339)	23812(92)
2 2 2 30 II	2932(499)	97927(167)
2 2 2 2 10 II	4129096(655412)	1073933204(162)
2 2 2 2 20 II	8061226(1310772)	-
2 2 2 2 30 II	11993356(1966132)	-

The last two tables below show: a term $(\lambda z.(\lambda x.zxxx)(\lambda y.2(\lambda x.y(xI))n))II$, where n is a Church numeral that we vary, and using terms taken from [2]: $n2'2'II$, where $2' = \lambda xy.(\lambda z.(z(zy)))(\lambda w.xw)$ (cf. Figure 1).

n	KCLE	BOHM
2	178(38)	149(22)
3	199(40)	185(26)
4	220(44)	230(30)
5	241(48)	275(34)
10	346(68)	500(54)
20	556(108)	950(94)
30	766(148)	1400(134)

Term	KCLE	BOHM
1 $2'2'II$	102(17)	41(11)
2 $2'2'II$	154(24)	90(14)
3 $2'2'II$	236(35)	172(17)
4 $2'2'II$	378(54)	379(20)
5 $2'2'II$	640(89)	1016(23)
10 $2'2'II$	15630(2088)	664501(38)
15 $2'2'II$	491900(65591)	671375626(53)

A brief analysis of these tables (and other data accumulated) show a general pattern from which two key observations can be made:

1. As the terms become larger: BOHM improves with respect to the number of β-interactions, but KCLE becomes better with respect to the total number of interactions (*i.e.*, total cost of evaluation).
2. For Church numeral calculations, the ratio of the total number of interactions to β-interactions is almost constant, indicating that the cost of implementing a β-reduction step is around 7 interactions—a very small overhead for these values.

The comparisons given here only touch on the testing done, and in particular we do not include results for other systems of interaction nets, or related systems of rewriting (see for instance [8, 10, 12, 14]). We hope to provide a more complete comparison at some future occasion.

7 Conclusions

In this paper we have demonstrated that there are still new and exciting ways to implement the λ-calculus. As our experimental results have confirmed, these can be greatly more efficient than extant systems, even asymptotically better.

Our evaluator is capable of producing full normal forms of typed or untyped λ-terms (either closed or open). This therefore opens up additional applications, such as the use in proof assistants, where normal forms of such terms are required.

As we have mentioned, the additional agents and rules which form the normal form extension can also be used in the compilation. This may achieve better performance when an argument would benefit from being a normal form before copying for instance. For the moment this is done in an ad-hoc way, and we anticipate that there should be a more systematic approach to this from analysing the term during compilation.

References

1. A. Asperti, C. Giovannetti, and A. Naletto. The Bologna Optimal Higher-Order Machine. *Journal of Functional Programming*, 6(6):763–810, Nov. 1996.
2. A. Asperti and S. Guerrini. *The Optimal Implementation of Functional Programming Langua ges*, volume 45 of *Cambridge Tracts in Theoretical Computer Science*. Cambridge University Press, 1998.
3. M. Fernández and I. Mackie. A calculus for interaction nets. In G. Nadathur, editor, *Proceedings of the International Conference on Principles and Practice of Declarative Programming (PPDP'99)*, volume 1702 of *Lecture Notes in Computer Science*, pages 170–187. Springer-Verlag, September 1999.
4. J.-Y. Girard. Linear Logic. *Theoretical Computer Science*, 50(1):1–102, 1987.
5. G. Gonthier, M. Abadi, and J.-J. Lévy. The geometry of optimal lambda reduction. In *Proceedings of the 19th ACM Symposium on Principles of Programming Languages (POPL'92)*, pages 15–26. ACM Press, Jan. 1992.
6. Y. Lafont. Interaction nets. In *Proceedings of the 17th ACM Symposium on Principles of Programming Languages (POPL'90)*, pages 95–108. ACM Press, Jan. 1990.
7. J. Lamping. An algorithm for optimal lambda calculus reduction. In *Proceedings of the 17th ACM Symposium on Principles of Programming Languages (POPL'90)*, pages 16–30. ACM Press, Jan. 1990.
8. F. Lang. *Modèles de la beta-réduction pour les implantations.* PhD thesis, ENS Lyon, 1998.
9. J.-J. Lévy. Optimal reductions in the lambda calculus. In J. P. Hindley and J. R. Seldin, editors, *To H.B. Curry: Essays on Combinatory Logic, Lambda Calculus and Formalism*, pages 159–191. Academic Press, 1980.
10. S. Lippi. Encoding left reduction in the lambda-calculus with interaction nets. *Mathematical Structures in Computer Science*, 12(6):797–822, 2002.
11. I. Mackie. *The Geometry of Implementation.* PhD thesis, Department of Computing, Imperial College of Science, Technology and Medicine, September 1994.

12. I. Mackie. YALE: Yet another lambda evaluator based on interaction nets. In *Proceedings of the 3rd International Conference on Functional Programming (ICFP'98)*, pages 117–128. ACM Press, 1998.
13. I. Mackie. Interaction nets for linear logic. *Theoretical Computer Science*, 247(1):83–140, September 2000.
14. I. Mackie and J. S. Pinto. Encoding linear logic with interaction combinators. *Information and Computation*, 176(2):153–186, August 2002.
15. S. L. Peyton Jones. *The Implementation of Functional Programming Languages.* Prentice Hall International, 1987.
16. J. S. Pinto. Sequential and concurrent abstract machines for interaction nets. In J. Tiuryn, editor, *Proceedings of Foundations of Software Science and Computation Structures (FOSSACS)*, volume 1784 of *Lecture Notes in Computer Science*, pages 267–282. Springer-Verlag, 2000.
17. J. S. Pinto. Weak reduction and garbage collection in interaction nets. In *Proceedings of the 3rd International Workshop on Reduction Strategies in Rewriting and Programming*, Valencia, Spain, 2003.

A Computing Normal Forms

In this appendix we give the rules for the additional agents that are needed to obtain full normal forms. Since there are many, we adopt the textual notation for rules [6], which are in the table below. There are four agents $\phi_1, \phi_2, \phi_3, \phi_\lambda$ of arity 1, one agent $@'$ of arity 2, and two agents λ_u, cb of arity 3. Variables in this notation are bold and underlined to distinguish them from the names of agents.

Table 2. Agents used to compute normal forms

$$\phi_1(\underline{\mathbf{a}}) \bowtie \phi_1(\underline{\mathbf{a}}) \qquad \phi_1(v) \bowtie v \qquad \phi_1(\epsilon) \bowtie \epsilon$$
$$\phi_1(\lambda_c(\underline{\mathbf{a}},\underline{\mathbf{b}})) \bowtie \lambda_c(\phi_1(\underline{\mathbf{a}}),\phi_1(\underline{\mathbf{b}})) \qquad \phi_1(\delta(\underline{\mathbf{a}},\underline{\mathbf{b}})) \bowtie \delta(\phi_1(\underline{\mathbf{a}}),\phi_1(\underline{\mathbf{b}}))$$
$$\phi_1(@(\underline{\mathbf{a}},\underline{\mathbf{b}})) \bowtie @(\phi_1(\underline{\mathbf{a}}),\phi_1(\underline{\mathbf{b}})) \qquad \phi_1(c(\underline{\mathbf{a}},\underline{\mathbf{b}})) \bowtie c(\phi_1(\underline{\mathbf{a}}),\phi_1(\underline{\mathbf{b}}))$$
$$\phi_1(\lambda(\underline{\mathbf{a}},\underline{\mathbf{b}},\underline{\mathbf{c}})) \bowtie \lambda(\phi_1(\underline{\mathbf{a}}),\phi_1(\underline{\mathbf{b}}),\phi_1(\underline{\mathbf{c}})) \qquad \phi_1(\underline{\mathbf{c}}) \bowtie cb(\phi_1(\underline{\mathbf{a}}),\phi_1(\underline{\mathbf{b}}),\phi_1(b(\underline{\mathbf{c}},\underline{\mathbf{a}},\underline{\mathbf{b}})))$$
$$\phi_1(\underline{\mathbf{a}}) \bowtie b(\phi_\lambda(cb(\underline{\mathbf{b}},\underline{\mathbf{c}},\phi_1(\underline{\mathbf{a}}))),\underline{\mathbf{b}},\underline{\mathbf{c}}) \qquad \phi_1(\underline{\mathbf{a}}) \bowtie \phi_\lambda(\phi_1(\underline{\mathbf{a}}))$$
$$\phi_\lambda(d) \bowtie d \qquad \phi_\lambda(\underline{\mathbf{c}}) \bowtie \lambda_b(\lambda_u(\underline{\mathbf{a}},\underline{\mathbf{b}},\underline{\mathbf{c}}),\underline{\mathbf{a}},\underline{\mathbf{b}})$$
$$\phi_\lambda(\delta b(\underline{\mathbf{a}},\underline{\mathbf{b}},\underline{\mathbf{c}})) \bowtie \delta b(\phi_\lambda(\underline{\mathbf{a}}),\phi_\lambda(\underline{\mathbf{b}}),\underline{\mathbf{c}}) \qquad c(v,v) \bowtie v$$
$$c(\lambda_u(\underline{\mathbf{a}},\underline{\mathbf{b}},\underline{\mathbf{c}}),\lambda_u(\underline{\mathbf{d}},\underline{\mathbf{e}},\underline{\mathbf{f}})) \bowtie \lambda_u(\delta(\underline{\mathbf{a}},\underline{\mathbf{d}}),\delta(\underline{\mathbf{b}},\underline{\mathbf{e}}),C(\underline{\mathbf{c}},\underline{\mathbf{f}}))$$
$$c(cb(\underline{\mathbf{a}},\underline{\mathbf{b}},\underline{\mathbf{c}}),cb(\underline{\mathbf{d}},\underline{\mathbf{e}},\underline{\mathbf{f}})) \bowtie cb(\delta(\underline{\mathbf{a}},\underline{\mathbf{d}}),C(\underline{\mathbf{b}},\underline{\mathbf{e}}),C(\underline{\mathbf{c}},\underline{\mathbf{f}}))$$
$$\delta b(cb(\underline{\mathbf{d}},\underline{\mathbf{e}},\underline{\mathbf{c}}),cb(\underline{\mathbf{b}},\underline{\mathbf{f}},\underline{\mathbf{d}}),\underline{\mathbf{a}}) \bowtie cb(\underline{\mathbf{b}},\delta b(\underline{\mathbf{e}},\underline{\mathbf{f}},\underline{\mathbf{a}}),\underline{\mathbf{c}})$$
$$cb(\underline{\mathbf{a}},d,\underline{\mathbf{a}}) \bowtie d \qquad \lambda_u(\underline{\mathbf{a}},\underline{\mathbf{b}},d) \bowtie @(\underline{\mathbf{a}},\underline{\mathbf{b}})$$

$$\phi_2(\underline{\mathbf{a}}) \bowtie \phi_2(\underline{\mathbf{a}}) \qquad \phi_2(\epsilon) \bowtie \epsilon$$
$$\phi_2(\underline{\mathbf{a}}) \bowtie @(\phi_2(@'(\underline{\mathbf{b}},\underline{\mathbf{a}})),\phi_2(\underline{\mathbf{b}})) \qquad \phi_2(\lambda_c(\underline{\mathbf{a}},\underline{\mathbf{b}})) \bowtie \lambda_c(\phi_2(\underline{\mathbf{a}}),\phi_2(\underline{\mathbf{b}}))$$
$$\phi_2(c(\underline{\mathbf{a}},\underline{\mathbf{b}})) \bowtie c(\phi_2(\underline{\mathbf{a}}),\phi_2(\underline{\mathbf{b}})) \qquad \phi_2(b(\underline{\mathbf{a}},\underline{\mathbf{b}},\underline{\mathbf{c}})) \bowtie b(\underline{\mathbf{a}},\phi_2(\underline{\mathbf{b}}),\underline{\mathbf{c}})$$
$$\phi_2(\underline{\mathbf{c}}) \bowtie \lambda(\phi_2(\underline{\mathbf{a}}),\phi_2(\underline{\mathbf{b}}),\lambda_b(\underline{\mathbf{c}},\underline{\mathbf{a}},\underline{\mathbf{b}})) \qquad @'(b(\underline{\mathbf{a}},\underline{\mathbf{b}},\underline{\mathbf{c}}),b(\underline{\mathbf{d}},\underline{\mathbf{e}},\underline{\mathbf{a}})) \bowtie b(\underline{\mathbf{d}},@'(\underline{\mathbf{b}},\underline{\mathbf{e}}),\underline{\mathbf{c}})$$
$$@'(c(\underline{\mathbf{a}},\underline{\mathbf{b}}),c(\underline{\mathbf{c}},\underline{\mathbf{d}})) \bowtie c(@'(\underline{\mathbf{a}},\underline{\mathbf{c}}),@'(\underline{\mathbf{b}},\underline{\mathbf{d}}))$$

$$\phi_3(\underline{\mathbf{a}}) \bowtie \phi_3(\underline{\mathbf{a}}) \qquad \phi_3(v) \bowtie v \qquad \phi_3(\epsilon) \bowtie \epsilon$$
$$\phi_3(\lambda_c(\underline{\mathbf{a}},\underline{\mathbf{b}})) \bowtie \lambda_c(\phi_3(\underline{\mathbf{a}}),\phi_3(\underline{\mathbf{b}})) \qquad \phi_3(c(\underline{\mathbf{a}},\underline{\mathbf{b}})) \bowtie c(\phi_3(\underline{\mathbf{a}}),\phi_3(\underline{\mathbf{b}}))$$
$$\phi_3(\lambda(\underline{\mathbf{a}},\underline{\mathbf{b}},\underline{\mathbf{c}})) \bowtie \lambda(\phi_3(\underline{\mathbf{a}}),\phi_3(\underline{\mathbf{b}}),\underline{\mathbf{c}}) \qquad \phi_3(b(\underline{\mathbf{a}},\underline{\mathbf{b}},\underline{\mathbf{c}})) \bowtie b(\phi_3(\underline{\mathbf{a}}),\phi_3(\underline{\mathbf{b}}),\phi_3(\underline{\mathbf{c}}))$$
$$\phi_3(\underline{\mathbf{a}}) \bowtie @'(\phi_3(\underline{\mathbf{b}}),\phi_3(@(\underline{\mathbf{a}},\underline{\mathbf{b}}))) \qquad \phi_3(\underline{\mathbf{c}}) \bowtie \lambda_b(\lambda(\underline{\mathbf{a}},\underline{\mathbf{b}},\underline{\mathbf{c}}),\underline{\mathbf{a}},\underline{\mathbf{b}})$$

We will not enter into the details of any of these rules, except to remark that there are three kinds of ϕ agents. ϕ_1 is the main one, which allows the main reduction steps to proceed. ϕ_2 allows certain substitutions to complete, and ϕ_3 cleans up the representation to that it corresponds to the translation of a λ-term. ϕ_1 can be used as part of the reduction system, but the other two should only be used for the read-back procedure. All the remaining rules are simply completing the system.

Proving Properties of Term Rewrite Systems via Logic Programs

Sébastien Limet[1] and Gernot Salzer[2]

[1] Université d'Orléans, France
limet@lifo.univ-orleans.fr
[2] Technische Universität Wien, Austria
salzer@logic.at

Abstract. We present a general translation of term rewrite systems (TRS) to logic programs such that basic rewriting derivations become logic deductions. Certain TRS result in so-called cs-programs, which were originally studied in the context of constraint systems and tree tuple languages. By applying decidability and computability results of cs-programs we obtain new classes of TRS that have nice properties like decidability of unification, regular sets of descendants or finite representations of R-unifiers. Our findings generalize former results in the field of term rewriting.

1 Introduction

Term rewrite systems (TRS) are fundamental to fields like theorem proving, system verification, or functional-logic programming. Applications there require decision procedures e.g. for R-unifiability (for terms t and t', is there a substitution σ such that $t\sigma \rightarrow_R^* u \,{}_R^*\!\leftarrow t'\sigma$?) or for reachability (is term t' reachable from term t by a rewriting derivation?). Most desired properties depend on the ability to compute $R^*(E)$ (the set of terms reachable from elements in E) by means of tree languages which have a decidable emptiness test and are closed under intersection. Authors like [3, 11, 10] studied classes of TRS which effectively preserve recognizability, i.e. $R^*(E)$ is regular if E is regular.

Recognizability is usually preserved by encoding TRS derivations as tree automata and by exploiting the properties of the class of TRS under consideration. Another method encodes first the rewriting relation by means of tree tuple languages [4, 8], i.e. it computes a tree tuple language $\{\,(t, t') \mid t \rightarrow_R^* t'\,\}$, and then obtains recognizability by projection.

This paper applies the second method using a restricted class of logic programs to handle tree tuple languages. We first define a translation of rewrite systems to logic programs that maps a restricted form of rewriting, called basic rewriting, to deductions. It preserves essential structural properties such that classes of TRS correspond naturally to certain classes of logic programs, allowing to transfer results between the two formalisms.

We restrict our attention to basic rewriting, i.e., to rewriting that never modifies parts of terms considered as data. This fits the logic programming paradigm

V. van Oostrom (Ed.): RTA 2004, LNCS 3091, pp. 170–184, 2004.

where there is a clear distinction between data (function symbols) and operations on them (predicate symbols). This allows to obtain logic programs that preserve the structure of rewrite systems well. Note that basic rewriting coincides with rewriting for large classes of term rewrite systems such as right-linear ones. Right-linearity is required for most classes that preserve recognizability.

The classes of TRS we study in this paper are defined via the notion of *possible redexes* (i.e. parts of terms that may be rewritten). This notion generalizes constructor-based TRS where some symbols, called constructors, cannot be rewritten. The required properties are right-linearity and no nesting of possible redexes. The two classes differ in the form of possible redexes in the right-hand-sides. The first class, called quasi-cs TRS, requires that the depth of a variable under a possible redex is less than or equal to the depth in the right-hand-side. The second one, called instance-based TRS, requires that possible redexes are less instantiated than left-hand-sides. Instance-based TRS extend the class of TRS described in [4] and [10], and quasi-cs TRS is a new class not studied before. As in [10] we require that each term of the input language E contains a bounded number of basic positions. Instance-based TRS extend also the class of layered transducing TRS of [11]. The resulting logic program has the additional property that the projection to the last argument of each predicate is a regular language. This implies that recognizability for the basic rewriting relation is preserved.

The main interest of the framework presented in this paper compared to previous work is the replacement of very technical ad-hoc encodings of rewriting relations as tree automata by a very general translation to a high-level language like logic programming. The preservation of recognizability then reduces to a termination proof of the algorithm presented in [5] that transforms logic programs to so-called cs-programs.

Section 2 recalls some basic definitions of TRS, Section 3 describes cs-programs and their properties, and Section 4 presents the translation of basic rewriting to logic programming. Section 5 uses this translation to define two classes of TRS that preserve recognizability. The last section gives an outlook on future work. Due to space restrictions not all proofs are included in the paper; they can be found in [6].

2 Preliminaries

We recall some basic notions and notations concerning term rewrite systems; for details see [1].

Let Σ be a finite set of symbols with arity, Var be an infinite set of variables, and $\mathcal{T}(\Sigma, Var)$ be the first-order term algebra over Σ and Var. Σ consists of two disjoint subsets: the set $\mathcal{F}$ of defined function symbols (or function symbols), and the set $\mathcal{C}$ of constructor symbols. The terms of $\mathcal{T}(\mathcal{C}, Var)$ are called data-terms. A term is linear if no variable occurs more than once in it.

For a term t, $Pos(t)$ denotes the set of positions in t, $|t| = |Pos(t)|$ the size of t, and $t|_u$ the subterm of t at position u. The term $t[u \leftarrow s]$ is obtained from t by replacing the subterm at position u by s. $Var(t)$ is the set of variables occurring in t. The set $\Sigma Pos(t) \subseteq Pos(t)$ denotes the set of non-variable positions, i.e., $t|_u \notin Var$ for $u \in \Sigma Pos(t)$ and $t|_u \in Var$ for $u \in Pos(t) \setminus \Sigma Pos(t)$. A substitution is a mapping from Var to $\mathcal{T}(\Sigma, Var)$, which extends trivially to a mapping from $\mathcal{T}(\Sigma, Var)$ to $\mathcal{T}(\Sigma, Var)$. The domain of a substitution σ, $Dom(\sigma)$, is the set $\{ x \in Var \mid x\sigma \neq \sigma \}$. For $V \subseteq Var$, $\sigma|_V$ denotes the restriction of σ to the variables in V, i.e., $x\sigma|_V = x\sigma$ for $x \in V$ and $x\sigma|_V = x$ otherwise. If term t is an instance of term s, i.e. $t = s\sigma$, we say that t matches s and s subsumes t.

Let $CVar = \{ \Box_i \mid i \geq 1 \}$ be the set of context variables distinct from Var, where a context is a term in $\mathcal{T}(\Sigma, Var \cup CVar)$ such that the ith occurrence of a context variable counted from left to right is labelled by $\Box_i$. $\Box_1$ (also denoted $\Box$) is called the trivial context. A context is called n-context if it contains n context variables. For an n-context C, the expression $C[t_1, \dots, t_n]$ denotes the term $\{ \Box_i \mapsto t_i \mid 1 \leq i \leq n \} C$.

A term rewrite system (TRS) is a finite set of oriented equations called rewrite rules. Lhs and rhs are shorthand for the left-hand and right-hand side of a rule, respectively. For a TRS R, the rewrite relation is denoted by $\rightarrow_R$ and is defined by $t \rightarrow_R s$ if there exists a rule $l \rightarrow r$ in R, a non-variable position u in t, and a substitution σ, such that $t|_u = l\sigma$ and $s = t[u \leftarrow r\sigma]$. Such a step is written as $t \rightarrow_{[u, l \rightarrow r, \sigma]} s$. If a term t cannot be reduced by any rewriting rule, it is said to be irreducible. The reflexive-transitive closure of $\rightarrow_R$ is denoted by $\rightarrow_R^*$, and the symmetric closure of $\rightarrow_R^*$ by $=_R$. The relation $\rightarrow_R^n$ denotes n steps of the rewrite relation. By $t \downarrow_R s$ we denote the derivation $t \rightarrow_R^* s$ such that s is irreducible. For a set of terms, E, we define $R^*(E) = \{ t \mid t' \rightarrow_R^* t \text{ for some } t' \in E \}$ and $R^{\downarrow}(E) = \{ t \mid t' \downarrow_R t \text{ for some } t' \in E \}$. If the lhs (rhs) of every rule is linear the TRS is said to be left-(right-)linear. If it is both left- and right-linear the TRS is called linear. A TRS is constructor based if every rule is of the form $f(t_1, \dots, t_n) \rightarrow r$ where all t_i's are data-terms.

3 Cs-Programs

We use techniques from logic programming to deal with certain types of rewriting relations. Term rewrite systems are transformed to logic programs preserving their characteristic properties. The programs are manipulated by standard folding/unfolding techniques to obtain certain normal forms like *cs-programs*. Results about the latter lead directly to conclusions about rewrite systems. Therefore this section presents some results about logic programs. While some notions and theorems are just quoted from [5] (and in fact appear in a similar form already in [9]), other results are new. We presuppose basic knowledge about logic programming (see e.g. [7]).

Definition 1. *A program clause is a* cs-clause *if its body is linear and contains no function symbols, i.e., if all arguments of the body atoms are variables occurring nowhere else in the body. A cs-clause is linear if the head atom is linear. A logic program is a (linear)* cs-program *iff all its clauses are (linear) cs-clauses.*

Every logic program $\mathcal{P}$ can be transformed to an equivalent cs-program by applying two rules, unfolding and definition introduction[1]. The rules transform states $\langle \mathcal{P}, \mathcal{D}_{\text{new}}, \mathcal{D}_{\text{done}}, \mathcal{C}_{\text{new}}, \mathcal{C}_{\text{out}} \rangle$ where $\mathcal{D}_{\text{new}}$ are definitions not yet unfolded, $\mathcal{D}_{\text{done}}$ are definitions already processed but still used for simplifying clauses, $\mathcal{C}_{\text{new}}$ are clauses generated from definitions by unfolding, and $\mathcal{C}_{\text{out}}$ is the cs-program generated so far. Syntactically, definitions are written as clauses, but from the semantic point of view they are equivalences. We require the head of a definition to contain all variables occurring in the body[2]. A set of definitions, $\mathcal{D}$, is *compatible with* $\mathcal{P}$, if all predicate symbols occurring in the heads of the definitions occur just there and nowhere else in $\mathcal{D}$ and $\mathcal{P}$; the only exception are tautological definitions of the form $P(\vec{x}) \leftarrow P(\vec{x})$ where P may occur without restrictions throughout $\mathcal{D}$ and $\mathcal{P}$. The predicate symbols in the heads of $\mathcal{D}$ are called the *predicates defined by* $\mathcal{D}$.

We write $S \Rightarrow S'$ if S' is a state obtained from state S by applying one of the rules *unfolding* or *definition introduction* defined below. The reflexive and transitive closure of $\Rightarrow$ is denoted by $\stackrel{*}{\Rightarrow}$. An *initial state* is of the form $\langle \mathcal{P}, \mathcal{D}, \emptyset, \emptyset, \emptyset \rangle$ where $\mathcal{D}$ is compatible with $\mathcal{P}$. A *final state* is of the form $\langle \mathcal{P}, \emptyset, \mathcal{D}', \emptyset, \mathcal{P}' \rangle$. $\mathcal{P}$ and $\mathcal{D}$ are called the input of a derivation, $\mathcal{P}'$ its output. A derivation is *complete* if its last state is final.

Unfolding. Pick a definition not yet processed, select one or more of its body atoms according to some selection rule, and unfold them with all matching clauses from the input program. Formally:

$$\frac{\langle \mathcal{P},\, \mathcal{D}_{\text{new}} \,\dot{\cup}\, \{L \leftarrow \mathcal{R} \dot{\cup} \{A_1, \ldots, A_k\}\},\, \mathcal{D}_{\text{done}},\, \mathcal{C}_{\text{new}},\, \mathcal{C}_{\text{out}} \rangle}{\langle \mathcal{P},\, \mathcal{D}_{\text{new}},\, \mathcal{D}_{\text{done}} \cup \{L \leftarrow \mathcal{R} \cup \{A_1, \ldots, A_k\}\},\, \mathcal{C}_{\text{new}} \cup \mathcal{C},\, \mathcal{C}_{\text{out}} \rangle}$$

where $\mathcal{C}$ is the set of all clauses $L \leftarrow (\mathcal{R} \cup \mathcal{B}_1 \cup \cdots \cup \mathcal{B}_k)\mu$ such that $H_i \leftarrow \mathcal{B}_i$ is a clause in $\mathcal{P}$ for $i = 1, \ldots, k$, and such that the simultaneous most general unifier μ of $(A_1, \ldots, A_k)$ and $(H_1, \ldots, H_k)$ exists. Note that the clauses from $\mathcal{P}$ have to be renamed properly such that they share variables neither with each other nor with $L \leftarrow \mathcal{R} \cup \{A_1, \ldots, A_k\}$.

Definition Introduction. Pick a clause not yet processed, decompose its body into minimal variable-disjoint components, and replace every component that is not yet a single linear atom without function symbols by an atom that is either looked up in the set of old definitions, or if this fails is built of a new predicate symbol and the component variables. For every failed lookup introduce a new definition associating the new predicate symbol with the replaced component.

[1] Our version of definition introduction is a combination of definition introduction and folding in the traditional sense of e.g. [13].

[2] We could get rid of the restriction by introducing the notion of *linking variables* like in [9]. Since the restriction is always satisfied in our context we avoid this complication.

Formally:

$$\frac{\langle \mathcal{P},\, \mathcal{D}_{\text{new}},\, \mathcal{D}_{\text{done}},\, \mathcal{C}_{\text{new}} \,\dot{\cup}\, \{H \leftarrow \mathcal{B}_1 \dot{\cup} \cdots \dot{\cup} \mathcal{B}_k\},\, \mathcal{C}_{\text{out}}\rangle}{\langle \mathcal{P},\, \mathcal{D}_{\text{new}} \cup \mathcal{D},\, \mathcal{D}_{\text{done}},\, \mathcal{C}_{\text{new}},\, \mathcal{C}_{\text{out}} \cup \{H \leftarrow L_1, \ldots, L_k\}\rangle}$$

where $\mathcal{B}_1, \ldots, \mathcal{B}_k$ is a maximal decomposition of $\mathcal{B}_1 \cup \cdots \cup \mathcal{B}_k$ into non-empty variable-disjoint subsets,

$$L_i = \begin{cases} L\eta^{-1} & \text{if } L \leftarrow \mathcal{B}_i\eta \in \mathcal{D}_{\text{done}} \text{ for some var. renaming } \eta \\ P_i(x_1, \ldots, x_n) & \text{otherwise, } \{x_1, \ldots, x_n\} \text{ being the vars. of } \mathcal{B}_i. \end{cases}$$

for $1 \leq i \leq k$ and new predicate symbols P_i, and where $\mathcal{D}$ is the set of all $L_i \leftarrow \mathcal{B}_i$ such that L_i contains a new predicate symbol[3].

Theorem 1. *Let $\mathcal{P}$ be a logic program and $\mathcal{D}$ be a set of definitions compatible with $\mathcal{P}$. If $\langle \mathcal{P}, \mathcal{D}, \emptyset, \emptyset, \emptyset\rangle \overset{*}{\Rightarrow} \langle \mathcal{P}, \emptyset, \mathcal{D}', \emptyset, \mathcal{P}'\rangle$, then $\mathcal{P}'$ is a cs-program whose least Herbrand model semantics coincides with the one of $\mathcal{P}$ for all predicates defined by $\mathcal{D}$.*

In the following we discuss properties of logic programs that can be transformed to finite cs-programs.

Definition 2. *A clause is* quasi-cs, *if the body is linear and for every variable that occurs both in the body and the head, the depth of its occurrence in the body is smaller than or equal to the depth of all occurrences in the head. A program is quasi-cs if all its clauses are.*

Theorem 2. *Let $\mathcal{P}$ be a quasi-cs program, and let $\mathcal{D}_\mathcal{P}$ be the set of all tautologies $P(\vec{x}) \leftarrow P(\vec{x})$ such that P occurs in $\mathcal{P}$. Every $\Rightarrow$-derivation with input $\mathcal{P}$ and $\mathcal{D}_\mathcal{P}$ is finite.*

Another, new class of programs that also have finite cs-counterparts are instance-based programs.

Definition 3. *A clause $H \leftarrow \mathcal{B}$ of a logic program $\mathcal{P}$ is* instance-based *(is an ib-clause) if $\mathcal{B}$ is linear and for every pair of atoms H', A where H' is the head of a clause of $\mathcal{P}$ properly renamed and A is an atom in $\mathcal{B}$ such that H' and A unify, H' is an instance of A. A program is instance-based (is an ib-program) if it consists of ib-clauses.*

Theorem 3. *Let $\mathcal{P}$ be an ib-program, and let $\mathcal{D}_\mathcal{P}$ be the set of all tautologies $P(\vec{x}) \leftarrow P(\vec{x})$ such that P occurs in $\mathcal{P}$. Every $\Rightarrow$-derivation with input $\mathcal{P}$ and $\mathcal{D}_\mathcal{P}$ is finite.*

Proof. We show that for ib-clauses all definitions occurring in a derivation are of the form $L \leftarrow A$, where A is a single linear atom which subsumes any head clause it unifies with. This implies that only finitely many definitions are generated up

[3] A substitution η is a variable renaming for a set of atoms $\mathcal{R}$, if there exists a substitution η^{-1} such that $\mathcal{R}\eta\eta^{-1} = \mathcal{R}$.

to variable renaming, i.e., complete derivations are finite. Note that the initial definitions in $\mathcal{D}_{\mathcal{P}}$ are of this particular form.

Let $H \leftarrow \mathcal{B}$ be a ib-clause of $\mathcal{P}$, and let $\mu = \text{mgu}(A, H)$ be the most general unifier involved in unfolding a definition $L \leftarrow A$. We may divide μ into two parts, μ_A and μ_H, such that $A\mu_A = H\mu_H$. Since A is linear, $x\mu_H$ is a linear term for all x, and the variables in $x\mu_H$ do not occur in $\mathcal{B}$ (definitions and clauses are renamed apart prior to unfolding). Therefore $\mathcal{B}\mu_H$, the body of the new clause $(L \leftarrow \mathcal{B})\mu = L\mu_A \leftarrow \mathcal{B}\mu_H$, consists of linear atoms not sharing any variables with each other. Hence a maximal decomposition of $\mathcal{B}$ consists of singletons only.

It remains to show that each atom of $\mathcal{B}$ and therefore of the new definitions subsumes all the head clauses it unifies with. A subsumes H therefore $H\mu = H$ which means that $\mathcal{B}\mu = \mathcal{B}$ and therefore atoms of $\mathcal{B}$ satisfies ib property, so it subsumes all the head clauses it unifies with. □

A logic program is called monadic if it contains only unary predicate symbols. A unary predicate is monadic if it is defined by a monadic program.

Lemma 1. *The set of ground terms S is a regular tree language iff there exists a linear monadic cs-predicate P such that $P(t)$ is true for all terms t in S.*

Definition 4. *Let P be a predicate defined by a program $\mathcal{P}$. The ith argument of P is said to be* free *iff for every horn clause $P(t_1, \dots, t_n) \leftarrow A_1, \dots, A_k$ of $\mathcal{P}$, t_i is linear, $\forall x \in Var(t_i)$, x appears once in the head and once in the body and $\forall j \in 1, k$, either $Var(t_i) \cap Var(A_j) = \emptyset$ or $Var(t_i) \cap Var(A_j)$ is a single variable which occurs in a free argument of A_j.*

Lemma 2. *Let P be a predicate which i^{th} argument is free, of a cs-program $\mathcal{P}$. The set $\{t \mid \mathcal{P} \models P(t_1, \dots, t_{i-1}, t, t_{i+1}, \dots, t_n)\}$ is defined by the set of clauses constructed as follows:*

- $C_0 = \{\}$, $\mathcal{P}_0 = \{P'\}$
- $C_i = \{Q'(s) \leftarrow \{R'(s') | Q' \in \mathcal{P}_{i-1}, Q(\dots, s, \dots) \leftarrow \mathcal{B} \in \mathcal{P}$ *and* $\mathcal{B}$ *has at least one model,* $R(\dots, s', \dots) \in \mathcal{B}$ *and* $Var(s) \cap Var(s') \neq \emptyset\}$,
  $\mathcal{P}_i = \{Q'$ *occurring in a body of a clause of* $C_i\} \setminus \bigcup_{j<i} \mathcal{P}_j\}$

Proof. It is obvious that if $\mathcal{P} \models P(t_1, \dots, t_{i-1}, t, t_{i+1}, \dots, t_n)$ then $\mathcal{P}' \models P'(t)$. Now, let us prove that if $\mathcal{P}' \models P'(t)$ then there is some $t_1, \dots, t_n$ such that $\mathcal{P} \models P(t_1, \dots, t_{i-1}, t, t_{i+1}, \dots, t_n)$. It is done by induction on the height of the proof tree.

- $h = 0$ obvious by construction
- $h > 0$ Let $P'(s) \leftarrow \mathcal{B}'$ be the top clause of the proof tree $P(\dots, s, \dots) \leftarrow \mathcal{B}$ the corresponding clause of $\mathcal{P}$ and $\sigma = \text{mgu}(t, s)$. By construction, we have $\mathcal{P} \models \mathcal{B}\mu$, this means that $\forall Q(\vec{t})$ s.t. $Var(Q(\vec{t})) \cap Var(s) = \emptyset$, we have $\mathcal{P} \models Q(\vec{t})\mu$. $\forall Q'(s')\sigma \in \mathcal{B}'\sigma$, $\mathcal{P}' \models Q'(s')\sigma$. Since s appears at a free position of P, s' occurs at a free position of Q, so by induction hypothesis $\mathcal{P} \models Q(\dots, s', \dots)\sigma'$ with $s'\sigma' = s'\sigma$. Since all atoms of $\mathcal{B}$ are variable disjoint, the substitution $\mu' = \sigma' \uplus \mu|_{\{x \notin Dom(\sigma')\}}$ is such that $Q(\dots, s', \dots)\mu' = Q(\dots, s', \dots)\sigma'$ if $Q'(s') \in \mathcal{B}'$ and $Q(\vec{t})\mu' = Q(\vec{t})\mu$ if $Var(Q(\vec{t})) \cap Var(t) = \emptyset$. So $\mathcal{P} \models \mathcal{B}\mu'$ this implies $\mathcal{P} \models P(\dots, s, \dots)\mu'$ where $s\mu' = t$.

In the following, $Reg(\mathcal{P})$ denotes the set of clauses constructed in Lemma 2.

Definition 5. *Let $\mathcal{P}$ be a cs-program. A clause $P(\vec{x}) \leftarrow P'(\vec{t}) \cup \mathcal{B}_{reg}$ is called a* regular join definition *compatible with $\mathcal{P}$ if*

- *$P(\vec{x})$ is linear and $\vec{x} \subseteq Var(P'(\vec{t})) \cup Var(\mathcal{B}_{reg})$*
- *$P'(\vec{t})$ is linear*
- *the elements of $\mathcal{B}_{reg}$ are linear monadic predicates.*

$P'(\vec{t})$ is called the key atom. If all atoms of $\mathcal{B}_{reg}$ are of depth 0 the definition is said to be a flat regular join definition.

The strategy which chooses atoms of maximal depth terminates but does not preserve freeness of arguments. For example, consider the clause
$$A(s(c(x,y_1)), s(c(x,y_2)), c(x_3,y_3)) \leftarrow A(x,y_1,x_3), B(y_2,y_3)$$
where the last arguments of A and B are free. When unfolding the regular join definition $P(z_1,z_2,z_3) \leftarrow A(s(z_1),s(z_2),z_3), C(z_1), C(z_2)$ where C is linear monadic, we get
$$P(c(x,y_1), c(x,y_2), c(x_3,y_3)) \leftarrow A(x,y_1,x_3), B(y_2,y_3), C(c(x,y_1)), C(c(x,y_2)).$$
The body of this clause cannot be decomposed, so the new clause in $\mathcal{C}_{\text{new}}$ is $P(c(x,y_1), c(x,y_2), c(x_3,y_3)) \leftarrow Q(x,y_1,y_2,x_3,y_3)$. But now the variables x_3 and y_3 are no longer "independent". This could have been avoided if $C(c(x,y_1))$ and $C(c(x,y_2))$ had been unfolded before introducing the new clause in $\mathcal{C}_{\text{new}}$. Indeed suppose that $C(c(x',x'')) \leftarrow C(x'), C(x'')$ is in the definition of C, then the next unfolding step yields
$$P(c(x,y_1), c(x,y_2), c(x_3,y_3)) \leftarrow A(x,y_1,x_3), B(y_2,y_3), C(x), C(y_1), C(y_2)$$
where the two free variable are separated into two different components. This observation leads to the following restricted rule that unfolds atoms without introducing clauses in $\mathcal{C}_{\text{new}}$. Together with a suitable strategy it preserves freeness of arguments.

Restricted Unfolding.

$$\frac{\langle \mathcal{P},\ \mathcal{D}_{\text{new}} \mathbin{\dot\cup} \{L \leftarrow \mathcal{R} \dot\cup \{A_1,\ldots,A_k\}\},\ \mathcal{D}_{\text{done}},\ \mathcal{C}_{\text{new}},\ \mathcal{C}_{\text{out}} \rangle}{\langle \mathcal{P},\ \mathcal{D}_{\text{new}} \cup \mathcal{C},\ \mathcal{D}_{\text{done}},\ \mathcal{C}_{\text{new}},\ \mathcal{C}_{\text{out}} \rangle}$$

Let A be an atom and $\mathcal{P}$ be a logic program. A is said to be *sufficiently instantiated* if it is an instance of all clause heads of $\mathcal{P}$ that unify with it (if there is no such head, the condition is vacuously satisfied). A variable in an atom is said to be not sufficiently instantiated if the variable gets instantiated by the most general unifier of the atom with some clause head.

A freeness preserving strategy for $\Rightarrow$-derivations starting from a regular join definition consists in unfolding atoms of $\mathcal{B}_{reg}$ by Restricted Unfolding and the key atom by general Unfolding. The atom for the unfolding operation is selected in the following way: choose an atom of $\mathcal{B}_{reg}$ with a depth greater than 0; if there is none, choose one whose variables are not sufficiently instantiated; otherwise choose the key atom.

Before giving the result, we make some observation concerning the unification problems occurring in the discussion.

Let t be a term and a variable $x \in Var(t)$, we denote by $MinDepth(x,t)$ the minimal depth at which x occurs in t and by $\Delta Depth(t)$ the maximal depth difference between two occurrences of the same variable in t. Note that for linear, variable disjoint terms s and t unifiable by $\mu = \mathrm{mgu}(s,t)$ we have $Depth(s\mu) \leq \max(Depth(s), Depth(t))$.

Property 1. Let t be a term and s be a linear term such that $Var(t) \cap Var(s) = \emptyset$. Let $\mu = \mathrm{mgu}(s,t)$. Then $x\mu$ is linear and $Depth(x\mu) \leq Depth(s)$ for all $x \in Var(t)$, and $Depth(x\mu) \leq \max(Depth(t), Depth(s) + \Delta Depth(t))$ for all $x \in Var(s)$. Moreover, $\Delta Depth(t\mu) = \Delta Depth(t)$ and the maximal number of occurrences of a variable in $t\mu$ is the same as in t.

Lemma 3. *Any regular join definition can be transform into a flat one using restricted unfolding.*

Theorem 4. *Any $\Rightarrow$-derivation using a freeness preserving strategy starting with a cs-program and a flat regular join definition is terminating.*

Proof. Let τ be the maximal depth of an atom occurring in the whole program and in the first definition. According to the freeness preserving strategy, the key atom K will be unfolded only if all the variables it shares with $\mathcal{B}_{reg}$ are sufficiently instantiated. Since all predicates of $\mathcal{B}_{reg}$ are non-copying monadic, they define each a regular tree language. Let us call intersection problem $\bigcup_{0 \leq i \leq n} C_i[P_1^i(x_1^i), \ldots, P_n^i(x_{n_i}^i)]$ where C_i is a context which depth is inferior to $2 \times \tau$ and P_j^i and predicate symbols and x_j^i are pairwise different variables. The solution of such a problem is $\{t | \forall i \in 0,n, t = C^i[t_1^i, \ldots, t_n^i], \mathcal{P} \models \bigcup_{1 \leq i \leq n_i} P_j^i(t_j)\}$. since $\mathcal{P}$ is finite there is only a finite number of intersection problems. For a unification problem UP $MinSol = Min(\{Depth(t) | t \text{ solution of } UP\})$. Let us call h the maximum of all the $MinSol$.

We prove that $Depth(K)$ is less or equal to h. We first have to remark that all subterms of K which occurs at a depth greater than τ, has been generated by one or more non-copying monadic predicates. A variable x of K is not sufficiently instantiated if either it occurs at an occurrence of K which is also an occurrence of one clause head it unifies with or it occurs at an occurrence greater than one duplicated variable of one clause head. The first case can be solved by unfolding this atom and its "descendants" to make this branch growing. The depth of the result is less than $2 \times \tau$. For the second case let $PB_{H,x} = \{C_i[\vec{t_i}] = K|_u | H|_u = x\}$ where $Depth(C_i) \leq 2 \times \tau$ and each terms of the $\vec{t_i}$ have been generated by one non-copying monadic predicate. $PB_{H,x}$ is an intersection problem, therefore its minimal solution has a depth inferior or equal to h. This means that unfolding atoms involved in this problem leads to an instance of depth inferior to $\tau + h$ (x is a variable of H so it occurs at a depth inferior to τ).

Let $H' \leftarrow \mathcal{B}'$ the clause used for unfolding the key atom and $\mu = \mathrm{mgu}(K, H')$. Since all variables shared by the key atom are sufficiently instantiated $x\mu = x$ for $x \in Var(\mathcal{B}_{reg})$ so $\mathcal{B}_{reg}\mu = B'_{reg}$ and is of depth 0. The variables of H' are instantiated by terms of depth less than h, which means that $\mathcal{B}'\mu$ is of depth less than h. Since K is linear $x\mu$ is linear for $x \in Var(H')$ and so $\mathcal{B}'\mu$ is linear. Finally since since atoms of $\mathcal{B}_{reg}$ are unary and of depth 0 they share variables with at most one argument of atoms of $\mathcal{B}'\mu$. The new maximal decomposition of $\mathcal{B}_{reg}\mathcal{B}'\mu$ in variable disjoint components contains at most one atom of $\mathcal{B}'\mu$ so they are all body of regular join definition compatible with $\mathcal{P}$.

Since the depth of the key atom is bounded and linear, number of body of regular join definition compatible with $\mathcal{P}$ is bounded so the algorithm terminates.

Theorem 5. *Let $P(\vec{x}, x) \leftarrow P'(\vec{t}, x), \mathcal{B}_{reg}$ be a regular join definition, such that x occurs at a free argument of P' and $x \notin Var(\mathcal{B}_{reg})$. The argument where x occurs in P is a free argument of P in the resulting cs-program.*

Proof. Since x occurs once in the body it will be instantiated only when unfolding the key atom and in this case each variable of the instance will occur in a different key atom at free argument. So all clauses introduced in $\mathcal{C}_{\text{new}}$ are of the form $P(\vec{t}, t) \leftarrow P_1(\vec{x_1}, x_1), \ldots, P_n(\vec{x_n}, x_n)$ where $x_1, \ldots, x_n$ are variables of t and t do not share variables with $\vec{t}$. So t and the x_is occur at a free argument.

4 Encoding Basic Rewriting by Logic Programs

In this section we present the way we encode rewriting relation by a logic program. This translation intends to obtain logic programs for which it is possible to deduce recognizability preservation, namely cs-programs for which one argument is free. The translation presented here works for any TRS but does not lead to a cs-program in general.

One of the main differences between term rewriting and logic programming formalisms is the clear distinction in logic programming between the predicate symbols and the function symbols i.e. between the data and the operations applied on them. This distinction is usually not made in term rewriting system. For example, considering the TRS $f(x) \to g(x, x), a \to b, a \to c, g(b, c) \to c$ and the term $f(a)$, we have the following derivation $f(a) \to g(a, a) \to g(a, b) \to c$. In the first rewriting step a is considered as a "data" of f and in the second and third steps a is an "operation". In pure logic programming such a symbol which is sometimes a predicate symbol and sometimes a function symbol does not exist. Since our aim is to prove TRS properties using cs-programs, we intend to encode the TRS derivations by a logic program which is as close as possible to the original TRS. This is why we define a transformation procedure which tends to preserve the structure of the terms. The price to pay is to encode only a restricted form of rewriting relation which fits well to logic programming formalism, namely basic rewriting. Roughly speaking, in basic rewriting it is forbidden to rewrite a subterm which has been considered as data in a former step. Fortunately basic rewriting and rewriting relations coincides for large classes of TRS.

4.1 Basic Rewriting

A rewriting derivation $t_0 \to_R t_1 \ldots \to_R t_n$ is said to be P-basic if the set of basic position of t_0 (denoted $BasPos(t_0)$) is P and for each step $t_i \to_{[u_i, l_i \to r_i, \sigma_i]}$

t_{i+1}, u_i belongs to the set $BasPos(t_i)$ and the set of basic positions of t_{i+1} is $(BasPos(t_i) \setminus \{ v \in Pos(t_i) \mid u_i \leq v \}) \cup \{ u_i.w \mid w \in \Sigma Pos(r_i) \}$. Each step of a basic derivation is denoted $\rightarrow_{bas}$ and we write $t \rightarrow_{bas}^{P*} s$ if t rewrites into s with a P-basic derivation. $\rightarrow_{bas}^{*}$ denotes the relation $\{(t, t') | t \rightarrow_{bas}^{Pos(t)*} t'\}$. For a set of terms E, $R_{bas}^{*}(E) = \{ t \mid \exists t' \in E, t' \rightarrow_{bas}^{*} t \}$. Most of the time P is abusively ommited in the following.

Basic rewriting and rewriting coincides for large classes of TRS, in particular for right-linear TRS.

Lemma 4. *Let R be right-linear TRS. Then $\stackrel{R}{\rightarrow}^{*}$ and $\stackrel{R}{\rightarrow}_{bas}^{*}$ are the same relation.*

Note that for the TRS given in introduction of this section, $\stackrel{R}{\rightarrow}^{*}$ and $\stackrel{R}{\rightarrow}_{bas}^{*}$ are different since $f(a)$ cannot be rewritten in c with a basic derivation.

The following definitions will be needed further down. They allow to point out positions in a term that may be rewritten. Let R be a TRS and t a term. A position u of t is called a *possible redex position* if $t|_u$ is of the form $C[t_1, \ldots, t_n]$ where C is neither trivial nor a variable, does not contain any possible redex position and C unifies with at least one lhs of R and u_i, the position of t_i, for $1 \leq i \leq n$ is a redex position. C is called the *possible redex* at occurrence u of t. The set of all possible redex positions of t is denoted $PRedPos_R(t)$ and the set of all possible redexes of t is denoted $PRed_R(t)$ (R may be ommited if clear from context). $PRedVar_R(t) = Var(t) \cap (\bigcup_{C \in PRed_R(t)} Var(C))$ is the set of variables of t occurring in one of its possible redexes. For a variable x of $PRedVar_R(t)$ $PRedDepth_R(x)$ is the maximal depth of x in a possible redex of t. The context C that does not contain any possible redex and that is such that $t = C[t_1, \ldots, t_n]$ where u_i for $1 \leq i \leq n$ is a redex position, is called the *irreducible part* of t and is denoted $Irr_R(t)$. If $u < v$ are two possible redex postions of t then v is said to be *nested.*

Example 1. For the TRS $R = \{f(s(x)) \rightarrow c(f(p(x)), f(f(x)))\}$ and the term $t = c(f(p(x)), f(f(x)))$ we have $PRedPos_R(t) = \{2, 2.1\}$, $PRed_R(t) = \{f(\square), f(x)\}$, $Irr_R(t) = c(f(p(x)), \square_1)$, $PRedVar(t) = \{x\}$, and $PRedDepth(x) = 1$. t contains a nested redex.

Notice that for any term t, $Pos(t)$-basic derivations and $PRedPos(t)$-basic derivations are the same since positions of t which are not in $PRedPos(t)$ cannot be rewritten.

4.2 Translating TRS to Logic Programs

Table 1 specifies the rules for transforming terms and rewrite rules to clause logic. For a TRS R, let $\mathcal{LP}(R)$ denote the logic program consisting of the clauses obtained by applying the fourth rule to all rewrite rules in R. For sake of simplicity, we will denote by x_u the fresh variable introduced in the third rule for the subterm $f(s_1, \ldots, s_n)$ at occurrence u of a rhs s and A_u the atom produced by this rule.

Table 1. Converting rewrite rules to clause logic

$$\frac{\top}{v \leadsto \langle v, \emptyset \rangle} \quad \text{if } v \in Var$$

$$\frac{s_1 \leadsto \langle t_1, \mathcal{G}_1 \rangle \dots s_n \leadsto \langle t_n, \mathcal{G}_n \rangle}{f(s_1, \dots, s_n) \leadsto \langle f(t_1, \dots, t_n), \bigcup_i \mathcal{G}_i \rangle} \quad \text{if } \varepsilon \notin PRedPos_R(f(s_1, \dots, s_n))$$

$$\frac{s_1 \leadsto \langle t_1, \mathcal{G}_1 \rangle \dots s_n \leadsto \langle t_n, \mathcal{G}_n \rangle}{f(s_1, \dots, s_n) \leadsto \langle x, \bigcup_i \mathcal{G}_i \bigcup \{P_f(t_1, \dots, t_n, x)\} \rangle} \quad \text{if } \varepsilon \in PRedPos_R(f(s_1, \dots, s_n))$$

$$\frac{s \leadsto \langle t, \mathcal{G} \rangle}{f(s_1, \dots, s_n) \to s \leadsto P_f(s_1, \dots, s_n, t) \leftarrow \mathcal{G}} \quad \text{if } f(s_1, \dots, s_n) \to s \in \mathcal{R}$$

Example 2. The following rewrite rules and clauses specify multiplication and addition.

$$\begin{array}{ll} *(0,x) \to 0 & \leadsto P_*(0,x,0) \leftarrow \\ *(s(x),y) \to +(y,*(x,y)) & \leadsto P_*(s(x),y,x_\varepsilon) \leftarrow P_+(y,x_2,x_\varepsilon), P_*(x,y,x_2) \\ +(0,x) \to x & \leadsto P_+(0,x,x) \leftarrow \\ +(s(x),y) \to s(+(x,y)) & \leadsto P_+(s(x),y,s(x_1)) \leftarrow P_+(x,y,x_1) \end{array}$$

Let $\mathcal{P}_{Id} = \{\, P_f(x_1, \dots, x_n, f(x_1, \dots, x_n)) \leftarrow \ \mid f \in \mathcal{F}\,\}$. $\mathcal{P}_{Id}$ allows to stop any derivation any time.

Theorem 6. *Let R be a TRS, s a term such that $s \leadsto \langle s', \mathcal{G} \rangle$. $s \to^*_{bas} t$ iff $\mathcal{LP}(R) \cup \mathcal{P}_{Id} \models \mathcal{G}\mu$ and $t = s'\mu$.*

Unfortunately – but this is not a surprise – the transformation of any term rewriting system does not usually lead to a cs-program. This is mainly due to the non linearity of the bodies of resulting clauses as well as their non flatness. Non-linearity has itself two causes. The first one is the non-linearity of the rhs of the rewriting rule, the second is due to nested redexes. Therefore term rewriting systems for which basic derivations can be expressed by a cs, should have linear rhs with no nested possible redexes. In Section 5, we present two classes of TRS where non-flatness have been weaken thanks to quasi-cs-programs and ib-programs.

5 Term Rewrite Systems Preserving Recognizability

In this section, we give two classes of TRS for which the encoding presented section 4 leads to a cs-program which has the additionnal property that the last argument of each predicate is free (i.e. it is a regular language). This argument encodes the resulting term of the rewrite derivation, this allows to deduce recognizability preservation for certain kind of input languages.

Definition 6. *A quasi-cs-TRS is a TRS with the following properties:*

- *it is right linear*
- *no rhs contains nested possible redexes*
- *For $l \to r \in R$, $x \in PRedVar(r)$ implies that either $x \notin Var(l)$ or $PRedDepth(x)$ is less or equal to the minimal depth at which x occurs in l.*

Definition 7. *An ib-TRS is a TRS with the following properties:*

- *it is right linear*
- *each rhs does not contain nested possible redexes*
- *if C is a possible redex of a right-hand-side and l a left-hand-side then l is an instance of C*

Theorem 7. *For both ib-TRS and quasi-cs TRS, $\to^*$ can be represented by a cs-program.*

Proof. Since both ib-TRS and quasi-cs-TRS are right linear, $\to^*$ and $\to^*_{bas}$ are equal for this two classes. Nowt is sufficient to prove that the logic program obtained from these classes of TRS are ib- or quasi-cs-programs. Since ib- and quasi-cs-TRS are right linear and do not contain nested possible redexes in the rhs, the logic program obtained contains only clauses with linear bodies. Moreover, for quasi-cs-TRS, the depth of the variable in the possible redexes of the rhs is less or equal to its minimal depth in the lhs, the depth of the variable in the bodies is less or equal to the minimal depth of this variable in the head. For ib-TRS, the fact that possible redexes of the rhs subsume lhs, ensures the ib property of the resulting logic program.

Lemma 5. *The resulting program of a quasi-cs-TRS or an ib-TRS is such that the last component of each predicate is free.*

In fact, this lemma is not true for the rules as given in Table 1. A rule like $+(x, 0) \to x$ is transformed to the clause $P_+(x, 0, x) \leftarrow$; obviuosly, the 3rd position of P_+ is not free because x occurs twice in the head of the clause. This problem appears for each rewrite rule $l \to r$ such that $Irr_R(r) \setminus PRedVar_R(r) \neq \emptyset$. Notice that since $l \to r$ is right linear, $x \in PRedVar(r)$ occurs once in the last argument of the head and once in the body. So for all $f(s_1, \ldots, s_n) \to r$ such that the resulting clause is $P_f(s_1, \ldots, s_n, s) \leftarrow \mathcal{B}$, we define $\sigma_r = \{\, x \mapsto x_r \mid x \in Var(s) \setminus \mathcal{B} \,\}$ and $\mathcal{B}_r = \{\, P_{Id}(x, x_r) \mid x \in Dom(\sigma_r) \,\}$ and we transform the clause $P_f(s_1, \ldots, s_n, s) \leftarrow \mathcal{B}$ to $P_f(s_1, \ldots, s_n, s\sigma') \leftarrow \mathcal{B}_r \cup \mathcal{B}$. This new clause is still a quasi or ib-clause and the last argument is linear and does not share variables with other arguments. If P_{Id} is defined by the set of clauses (which are both ib- and quasi-cs-clause) $P_{Id}(f(\vec{x}), f(\vec{x'})) \leftarrow P_{Id}(\vec{x}, x')$ then the semantics of P_f remains unchanged with regard to the Herbrand Models.

For example, considering the rewriting rule $+(x, 0) \to x$, we have $\sigma_r = \{x \mapsto x_r\}$, $\mathcal{B}_r = \{P_{Id}(x, x_r)\}$, and the transformed clause is $P(x, 0, x_r) \leftarrow P_{Id}(x, x_r)$.

Now we show how to use these results to compute $R^*_{bas}(E)$ for R which is either a quasi-cs or an ib TRS and E a regular tree language. In [10] it has been shown that $R^*(E)$ is not regular for the TRS $R = \{f(g(x)) \to g(f(x))\}$ which is both ib and quasi-cs, and the regular tree language $E = (fg) * 0$. $R^*(E)$ is the set of terms which contains as much f as g, in particular $R^{\downarrow}(E) = g^n f^n 0$, which is not a regular language. One can remark that terms of E contains unbounded number of possible redexes. So E is to be restricted to a regular tree language with a bounded number of possible redexes.

Definition 8. *A regular tree languages E is* possible-redex-bounded *if there exists an integer k such that every term t in E contains at most k possible redexes.*

Theorem 8. *Let E be a possible redex bounded regular tree language and R be a right-linear TRS which is either instance-based or quasi-cs. Then $R^*_{bas}(E)$ is a regular tree language.*

Proof. A possible redex bounded regular tree language can be described by a logic program with two kinds of clauses:
$P(f(x_1, \ldots, x_n)) \leftarrow P_1(x_1), \ldots, P_n(x_n)$ for positions that are not rewritten and $P(y) \leftarrow P_f(x_1, \ldots, x_n, y), P_1(x_1), \ldots, P_n(x_n)$ for possible redex positions (notice that this clause is almost a regular join definition) in this case body of clauses defining P do not contain P and no other clauses is headed by P.

The set of non-regular definitions L_* can be stratified in the following way:
$L_0 = \{P(y) \leftarrow P_f(x_1, \ldots, x_n, y), P_1(x_1), \ldots, P_n(x_n)$ s.t. the definition of the P_i does not contain any join definition $\}$.
$L_i = \{P(y) \leftarrow P_f(x_1, \ldots, x_n, y), P_1(x_1), \ldots, P_n(x_n) \notin \bigcup_{j<i} L_j$ s.t. the definition of the P_i contains no non regular definitions but those of $\bigcup_{j<i} L_j\}$.

Then for a right-linear irreducible-based TRS R which is either ib or quasi-cs and tree language defined as above by a logic program *Input*, we first compute $\mathcal{LP}(R)$ and then transform it into the cs-program $cs(\mathcal{LP}(R))$, then we apply the following algorithm:

$\mathcal{P}_0 = (\mathit{Input} \setminus L_*)$, $\mathit{Done} = \emptyset$, $k = 1$
while $\mathit{Done} \neq L_*$
  Let $i = \mathit{Min}(\{ j \mid L_j \setminus \mathit{Done} \neq \emptyset \})$
  Let $C \in L_i \setminus \mathit{Done}$
  Let $\mathcal{P}'$ **be the cs-program computed from** C **and** $P_{k-1} \cup cs(\mathcal{LP}(R))$
  $P_k = \mathit{Reg}(\mathcal{P}')$
  $k = k + 1$
end while

By the definition of *Input*, $\mathcal{P}_0$ is a linear monadic program. So the first iteration of the algorithm is a regular join definition transformation. If for a step k C is a regular join definition compatible with $P_k \cup cs(\mathcal{LP}(R))$, we have from freeness property the lonely argument of each predicate of L_i will be free, therefore $\mathit{Reg}(\mathcal{P}')$ produces only linear monadic clauses, so $\mathcal{P}_{k+1}$ linear monadic.

As a consequence $\mathcal{P}_*$ which defines the language $R^*_{bas}(E)$, describes a regular tree language.

This result can be used for example to compute the set of descendants of ground constructor instances of a linear term, for a constructor based TRS.

Indeed, let the term $f(s(g(x))$ where constructor symbols are s and 0. Its ground constructor instances contains only two possible redexes. The program is

$P(x) \leftarrow P_f(x_1, x), P_s(x_1)$ $P_s(s(x)) \leftarrow P'(x)$ $P'(x) \leftarrow P_g(x_1, x), P_*(x_1)$
$P_*(0) \leftarrow$ $P_*(s(x)) \leftarrow P_*(x)$

But it is also possible to compute the set of descendants of the regular language $s^*(f(s^*(g(x))))$ where f and g may occur at any depth in the terms of the language. This weaken the restriction of [10] on the kind of regular language allowed for computing $R^*(E)$, and can be useful for reachability problems issued from infinite state system verification for example.

Our result on ib-TRS extends those of [10] because all constructor based TRS satisfying Réty restrictions are ib-TRS but the TRS containing the single rule $f(s(x)) \rightarrow f(f(p(x)))$ does not satisfy the condition on nested function symbols. Quasi cs-TRS is neither included nor includes Réty's TRS. Indeed the former TRS is an instance-based one. On the other hand $\{f(x) \rightarrow g(s(x)), g(s(p(x))) \rightarrow g(x)\}$ respects Réty's restrictions but is not an ib-TRS. Another main class of TRS which has been studied is right-linear finite path overlapping TRS defined in [12]. This class allows nested possible redexes which is forbidden for cs-TRS. On the other hand rewriting rules like $f(s(x)) \rightarrow f(x)$ are not allowed in [12], but as the name of the class indicates, right-linearity is required, therefore basic rewriting relation is equivalent to rewrite relation, so it should be possible to handle this kind of TRS in our framework. More recently [11] defines the class of layered transducing TRS (LT-TRS for short) which are preserving recognizability under some conditions. A LT-TRS is a linear TRS working over a signature where some unary function symboles are distinguished as markers. The rule of LT-TRS are of one of the following two forms: $f(q_1(x_1), \ldots, q_n(x_n)) \rightarrow q(t)$ or $q(x) \rightarrow q'(t)$ where $q, q', q_1, \ldots, q_n$ are markers and t, t' do not contain any markers. The class of LT-TRS is a strict subclass of ib-cs TRSs since t and t' do not contain any possible redexes. The authors of this article define two conditions on LT-TRS to obtain the recognizability preservation. The first one corresponds to the conditions of theorem 8 on the input language. The second condition defining the IO separated LT-TRS, is on the marker symbols and allows to get the preservation for any input languages which is not the case for ib-cs TRSs.

6 Conclusion and Future Work

The translation of basic rewriting to logic programming presented in this paper provides a simple way to obtain finite presentations for derivations that allow to study properties of term rewrite systems. Its generality allows to extend already known results as well as to get new ones. Other classes of logic programs also transform to finite cs-programs. Pseudo-regular programs, for example, extend regular relations of [2] by weakening restriction on copying clauses. This should lead to further classes of term rewrite systems preserving recognizability. Pseudo-regular tuple languages are closed under intersection, therefore we expect that the corresponding class of term rewrite systems is less restrictive on the right-hand-sides. Our final aim is to give a complete characterization of term rewrite systems preserving regularity that correspond to finite cs-programs.

A nice side effect of translating everything to logic programs is that we obtain without much effort prototype implementations in Prolog which allow to experiment with the results[4].

Acknowledgements

We would like to thank the four referees for their substantial comments and valuable suggestions. We are particularly grateful for the reference to [9]: while this work predates some of our results in [5] and makes them obsolete, it supports our conviction that well-established notions and techniques from logic programming are of advantage also to other term-oriented fields like term rewriting or tree tuple languages.

References

1. F. Baader and T. Nipkow. *Term Rewriting and All That.* Cambridge University Press, United Kingdom, 1998.
2. H. Comon, M. Dauchet, R. Gilleron, D. Lugiez, S. Tison, and M. Tommasi. *Tree Automata Techniques and Applications (TATA).* `http://www.grappa.univ-lille3.fr/tata`, 1997.
3. P. Gyenizse and S. Vágvölgyi. Linear generalized semi-monadic rewrite systems effectively preserve recognizability. *Theoretical Computer Science*, 194:87–122, 1998.
4. S. Limet and P. Réty. E-unification by means of tree tuple synchronized grammars. *Discrete Mathematics and Theoretical Computer Science*, 1:69–98, 1997.
5. S. Limet and G. Salzer. Manipulating tree tuple languages by transforming logic programs. In Ingo Dahn and Laurent Vigneron, editors, *Electronic Notes in Theoretical Computer Science*, volume 86. Elsevier, 2003.
6. S. Limet and G. Salzer. Proving properties of term rewrite systems via logic programs. Technical report, RR-2004-5, LIFO Université d'Orléans, www.univ-orleans.fr/SCIENCES/LIFO/prodsci/publications/lifo2004.htm.en, 2004.
7. J.W. Lloyd. *Foundations of Logic Programming.* Springer, 1984.
8. D. Lugiez and Ph. Schnoebelen. The regular viewpoint on pa-processes. *Theoretical Computer Science*, 274(1-2):89–115, 2002.
9. M. Proietti and A. Pettorossi. Unfolding-definition-folding, in this order, for avoiding unnecessary variables in logic programs. *Theoretical Computer Science*, 142(1):89–124, 1995.
10. P. Réty. Regular sets of descendants for constructor-based rewrite systems. In *Proc. of the 6th conference LPAR*, number 1705 in LNAI. Springer, 1999.
11. H. Seki, T. Takai, Y. Fujinaka, and Y. Kaji. Layered transducing term rewriting system and its recognizability preserving property. In S. Tison, editor, *Proc. of 13th Conference RTA*, volume 2378 of *LNCS*. Springer, 2002.
12. T. Takai, Y. Kaji, and H. Seki. Right-linear finite path overlapping term rewriting systems effectively preserve recognizability. In *11th International Conference RTA*, volume 1833 of *LNCS*, pages 270–273. Springer, 2000.
13. H. Tamaki and T. Sato. Unfold/fold transformation of logic programs. In S. Tärnlund, editor, *Proc. 2nd Int. Logic Programming Conf. (ICLP)*, pages 127–138. University of Uppsala, Sweden, 1984.

[4] See e.g. `www.logic.at/css` for a Prolog program for computing cs-programs.

On the Modularity of Confluence in Infinitary Term Rewriting

Jakob Grue Simonsen

Department of Computer Science, University of Copenhagen (DIKU),
Universitetsparken 1, DK-2100 Copenhagen Ø, Denmark
simonsen@diku.dk

Abstract. We show that, unlike the case in finitary term rewriting, confluence is not a modular property of infinitary term rewriting systems, even when these are non-collapsing. We also give a positive result: two sufficient conditions for the modularity of confluence in the infinitary setting.

1 Introduction

Modularity is the study of properties of rewriting systems that are, or are not, preserved when combining different systems. In finitary term rewriting, a number of properties, e.g., confluence [9, 14], are known to be modular whereas others, e.g. termination [13], are known not to be; see Ch. 8 of [11] for an overview. Modularity has, however, been left completely uninvestigated in the setting of *infinitary term rewriting*, a formalism developed in a series of landmark papers [3, 4, 6, 7]. In this paper, we take the first steps to investigate modularity in the setting of *strongly convergent* infinitary rewriting.

1.1 Contributions

We show that:

- Confluence is *not* a modular property of infinitary term rewriting systems, even for non-collapsing systems.
- Confluence is preserved under disjoint union of a set of *left-linear* iTRSs iff the set has the property of being *essentially non-collapsing*, i.e. at most one system contains collapsing rules.
- Confluence is preserved under disjoint union of a set of arbitrary (i.e. not necessarily left-linear), non-collapsing iTRSs if only terms of *finite rank* are considered.

1.2 Organization of the Paper

Section 2 introduces basic concepts from infinitary rewriting and defines what it means for a property to be modular in this setting. Section 3 contains the

V. van Oostrom (Ed.): RTA 2004, LNCS 3091, pp. 185–199, 2004.

counterexample to modularity of confluence. Sections 4 and 5 presents the two sufficient conditions for confluence to be modular, whereas Section 6 briefly discusses the difficulties in extending the results to the setting of *weakly* convergent rewriting.

2 Preliminaries

We assume familiarity with finitary term rewriting (ample introductions are [2, 8, 1] and Chapter 2 of [12]) and basic ordinal theory (see e.g. [10]). The successor of an ordinal α is denoted by $\alpha + 1$, and the least infinite ordinal by ω. If α is a limit ordinal, we indicate this by writing $Lim(\alpha)$. We assume a countable set of variables and a "Hilbert-hotel" style renaming for all terms considered so that fresh variables are always available. Positions in (finite) terms are elements of $\{1, 2, \ldots\}^*$ defined in the usual way. The subterm of term s at position p is denoted $s|_p$. The root symbol of a term is the symbol at position ϵ. If $\mathbf{f}$ is a unary function symbol and $k \in \omega$, we denote by $\mathbf{f}^k(s)$ k successive applications of $\mathbf{f}$ to the term s; we extend the notation to include $\mathbf{f}^\omega$ with the obvious meaning. Let $\Box \notin \Sigma \cup \mathcal{X}$. A term with *holes* is a term over Σ with variable set $\mathcal{X} \cup \{\Box\}$. A term with a hole at position p will be written as $s[]_p$, a term where the holes are at positions $i \in I$ where I is some (possibly infinite) set is written as $s[]_{i \in \mathcal{I}}$ and is called a (many-hole) *context*. Observe that a context may have no holes. Substituting terms from an $\mathcal{I}$-indexed sequence of terms $(s_i)_{i \in \mathcal{I}}$ into a many-hole context is defined in the obvious way. In the following, we recall a number of concepts from infinitary rewriting; our definitions are as in [5], with a few differences in nomenclature.

Definition 1. *Let $Ter(\Sigma)$ be the set of finite terms over the (not necessarily finite) signature Σ with alphabet $\mathcal{X}$. Define the metric $d : Ter(\Sigma) \times Ter(\Sigma) \longrightarrow [0; 1]$ by $d(t, t') \triangleq 0$ if $t = t'$ and $d(t,' t) = 2^{-k}$ otherwise, where k is the length of the shortest position at which t and t' differ. The completion of the metric space $(Ter(\Sigma), d)$, denoted $Ter^\infty(\Sigma)$ is called the* set of finite and infinite terms *(or simply* terms*) over Σ. The* depth *of a position, u, in a term is the length, $|u|$, of u.*

Definition 2. *An* infinitary rewrite rule *is a pair $\mathbf{l} \longrightarrow \mathbf{r}$ where $\mathbf{l} \in Ter(\Sigma)$ and $\mathbf{r} \in Ter^\infty(\Sigma)$ such that $\mathbf{l}$ is not a variable and every variable in $\mathbf{r}$ occurs in $\mathbf{l}$. An* infinitary term rewriting system, *denoted* iTRS, *is a pair $\mathcal{R} \triangleq (\Sigma, R)$, consisting of a signature Σ and a set of infinitary rewrite rules R.*

Definition 3. *A term s is* linear *if every variable of $\mathcal{X}$ occurs at most once in s. A rule $\mathbf{l} \longrightarrow \mathbf{r}$ is* left-linear *if $\mathbf{l}$ is linear. $\mathbf{l} \longrightarrow \mathbf{r}$ is* collapsing *if $\mathbf{r} \in \mathcal{X}$.*

Definition 4. *Let α be any ordinal. A* derivation of length α *is a sequence of rewrite steps $(s_\beta \longrightarrow s_{\beta+1})_{\beta<\alpha}$. In the step $s_\beta \longrightarrow s_{\beta+1}$, assume that the redex contracted is at position u_β of s_β; the* depth, *denoted d_β, of the redex is the depth of u_β. The derivation is called* weakly convergent *(aka.* Cauchy convergent*) if,* *for every limit ordinal* $\lambda \leq \alpha$*, the distance* $d(s_\beta, s_\lambda)$ *tends to* 0 *as* β *approaches* λ *from below. It is called* strongly convergent *if it is Cauchy convergent and, in addition,* d_β *tends to infinity as* β *approaches* λ *from below. If* $(s_\beta \longrightarrow s_{\beta+1})_{\beta<\alpha}$ *is convergent with limit* t *and* $s_0 = s$*, we write* $s \longrightarrow^\alpha t$*, and say that* t *is a* derivative *of* s*. When the length of the derivation is bounded above by* γ*, we shall occasionally write* $s \longrightarrow^{\leq\gamma} t$*. When the length of a* strongly *convergent derivation is unimportant, we write* $s \twoheadrightarrow t$*.*

Observe that concatenating any finite number of strongly convergent derivations yields a strongly convergent derivation. The following lemma concerning strongly convergent rewriting is due to Kennaway et al. [5, 6]:

Lemma 1 (Compression). *In every left-linear iTRS, if* $s \twoheadrightarrow t$*, then* $s \longrightarrow^{\leq\omega} t$*.*

Definition 5. *A* peak *of an iTRS R is a triple* $t \twoheadleftarrow s \twoheadrightarrow t'$ *of terms such that* $s \twoheadrightarrow t$ *and* $s \twoheadrightarrow t'$*. A* valley *of an iTRS R is a triple* $t \twoheadrightarrow s' \twoheadleftarrow t'$ *of terms such that* $t \twoheadrightarrow s'$ *and* $t' \twoheadrightarrow s'$*. If there exists a valley* $t \twoheadrightarrow s' \twoheadleftarrow t'$*, then* t *and* t' *are said to be* joinable*, written* $t \sim t'$*, and* s' *is said to be their* join*. A term* s *of R is said to be* confluent *(aka.* transfinitely Church-Rosser*) if every peak* $t \twoheadleftarrow s \twoheadrightarrow t'$ *has a corresponding valley* $t \twoheadrightarrow s' \twoheadleftarrow t'$*. The iTRS R is said to be confluent if all of its terms are confluent.*

The next auxiliary lemma, the *Dovetailing Lemma*, will prove to be useful in Sections 4 and 5.

Lemma 2 (Dovetailing). *If* $\{s_i\}_{i\in\mathcal{I}}$ *is a set of parallel subterms of some term* $s = C[s_i]_{i\in\mathcal{I}}$ *such that there are terms* t_i *with* $s_i \twoheadrightarrow t_i$ *for all* $i \in \mathcal{I}$*, then* $s \twoheadrightarrow C[t_i]_{i\in\mathcal{I}}$*.*

Proof. Since function symbols have finite arity, there are finitely many of s_i with root symbol at any given depth k. Since concatenation of a finite number of strongly convergent derivations yields a strongly convergent derivation, there exists, for every $k \in \omega$, a strongly convergent derivation S_k turning all s_i with root symbol at depth k into the respective t_i. Clearly, the derivation $S_0 \cdot S_1 \cdot S_2 \cdots$ is strongly convergent with limit $C[t_i]_{i\in\mathcal{I}}$. □

Definition 6. *The* direct sum *of a set* $\mathcal{R} = \{R_k\}_{k\in\mathcal{K}}$ *of iTRSs over signatures* $\{\Sigma_k\}_{k\in\mathcal{K}}$*, denoted by* $\oplus\mathcal{R}$*, is the iTRS* $\bigcup_{k\in\mathcal{K}} R_k$ *over signature* $\biguplus_{k\in\mathcal{K}} \Sigma_k$*, where* $\uplus$ *is disjoint union of sets. If* $\{R_k\}_{k\in\mathcal{K}} = \{R_0, R_1\}$*, we write* $R_0 \oplus R_1$*. A term over* $\biguplus_{k\in\mathcal{K}} \Sigma_k$ *is called* monochrome *if all of its function symbols are elements of a single* Σ_k*.*

When the involved signatures are disjoint, we refer to the iTRSs of $\mathcal{R}$ as being disjoint. Observe that, unlike the finitary case, we allow sums of any (finite or infinite) number of iTRSs. As we shall see, this has no impact on the modularity of confluence.

Definition 7. *The* rank *of a term s over $\{\Sigma_k\}_{k\in\mathcal{K}}$, denoted by* $\mathbf{rank}(s)$, *is the maximal number of signature changes in maximal paths starting from the root, if such a number exists, and ∞ otherwise. If* $\mathbf{rank}(s) \neq \infty$, *we say that s* is of finite rank.

Note that $\mathbf{rank}(s) = \infty$ does not imply the existence of a maximal path encountering infinitely many signature changes. All maximal paths could encounter only finitely many signature changes, but an upper bound on the number of such changes may not exist.

Definition 8. *A predicate P on the class of iTRSs is said to be* modular *if, given an arbitrary set $\{R_k\}_{k\in\mathcal{K}}$, the direct sum $\oplus\mathcal{R}$ has property P iff all elements of $\{R_k\}_{k\in\mathcal{K}}$ have property P.*

In (finitary) TRSs, the number of elements of $\{\Sigma_k\}_{k\in\mathcal{K}}$ contributing function symbols to any term s is finite. Hence, it is sufficient to consider finite sets $\mathcal{K}$ in this setting, showing that our definition reduces to the usual one in the finitary case.

Definition 9. *Let the root symbol of the term s over $\biguplus_{k\in\mathcal{K}}\Sigma_k$ belong to the signature Σ_r. The* cap *of s, denoted* $\mathbf{cap}(s)$, *is the maximal monochrome, linear term $C[x_i]_{i\in\mathcal{I}}$ containing the root symbol of s such that there is a sequence $(s_i)_{i\in\mathcal{I}}$ of terms with root symbols in $\biguplus_{k\in\mathcal{K}\setminus\{r\}}\Sigma_k$ and $s = C[s_i]_{i\in\mathcal{I}}$, in which case we write $C[\![s_i]\!]_{i\in\mathcal{I}}$ for clarity. The s_i are called the* principal subterms *of s.*

Observe that a term s may have an infinite number of principal subterms.

Definition 10. *The set of* blocks *of a term s, denoted $Bl(s)$ over $\biguplus_{k\in\mathcal{K}}\Sigma_k$ is defined by the following coinduction:*

1. $\mathbf{cap}(s) \in Bl(s)$.
2. *If $s = C[\![s_i]\!]_{i\in\mathcal{I}}$, then, for all $i \in \mathcal{I}$, $Bl(s_i) \subseteq Bl(s)$.*

A block, b, is collapsing *if $b \twoheadrightarrow x$ for some $x \in \mathcal{X}$, and we say that b* collapses to x. *If the underlying iTRS is confluent, each collapsing block b can collapse to at most one x and we call that x the* collapsing variable *of that block.*

Definition 11. *A rewrite step $s \longrightarrow t$ is* outer *if the redex contracted is in the cap of s, otherwise it is* inner. *An outer step is indicated by $\stackrel{o}{\longrightarrow}$, an inner step by $\stackrel{i}{\longrightarrow}$.*

In the remainder of the paper, we make essential use of the *descendant relation* well-known from strongly convergent rewriting [5, Def. 12.5.1], that tracks positions across derivations. Observe that since we do not in general track *residuals* (i.e. what "happens" to redexes), we do not require the iTRS involved to be left-linear (certain residuals *will* be tracked in Section 4 where all systems are assumed to be left-linear).

Definition 12 (The Descendant Relation). *Let R be an iTRS, let s be a term of R, and let $s \twoheadrightarrow t$. The set of* descendants *of any position $u \in Pos(s)$ across $s \twoheadrightarrow t$, denoted $u/(s \twoheadrightarrow t)$, is defined by induction on the length α of $s \twoheadrightarrow t$:*

- *$\alpha = 0$. Then, $u/(s \twoheadrightarrow t) = \{u\}$.*
- *$\alpha = \beta+1$. Let q be any position of s_β and assume that the redex r contracted in $s_\beta \longrightarrow s_{\beta+1}$ is of the rule $\mathbf{l} \longrightarrow \mathbf{r}$ and situated at position u. If $q \preceq u$, then $q/(s_\beta \longrightarrow s_{\beta+1}) = \{q\}$. If $u \prec q$, then there is exactly one variable occurrence x in $\mathbf{l}$ at position p_x such that $u \cdot p_x \cdot p' = q$ for some position p'. Let $\{p_x^k\}_{k \in \mathcal{K}}$ be the set of positions of occurrences of x in $\mathbf{r}$. Then, $q/(s \twoheadrightarrow s_{\beta+1}) = \{u \cdot p_x^k \cdot p'\}$ We then define $u/(s \longrightarrow s_{\beta+1})$ to be $\bigcup_{q \in u/(s \longrightarrow s_\beta)} (u/(s_\beta \longrightarrow s_{\beta+1}))$.*
- *$Lim(\alpha)$. Here, a position q of s_α is a descendant of a position u of s iff q is a descendant of u in s_β for all sufficiently large $\beta < \alpha$.*

We shall speak of descendants of variable occurrences and principal subterms, meaning "the position of a variable occurrence" and "position of the root symbol of a principal subterm". Note that the definition of descendant entails that if $C[\![s_i]\!]_{i \in \mathcal{I}} \twoheadrightarrow C'[\![t_j]\!]_{j \in \mathcal{J}}$ and some t_j is a descendant of s_i, then $s_i \twoheadrightarrow t_j$. Strong convergence is crucial in this respect (cf. Section 6).

3 A General Counterexample to Modularity of Confluence

We now turn to the modularity of confluence. As in the case of orthogonal systems, there is a trivial counterexample based on the presence of two collapsing rules: If $R_0 = \{\mathbf{f}(x) \longrightarrow x\}$ and $R_1 = \{\mathbf{g}(x) \longrightarrow x\}$, then both R_0 and R_1 are confluent, but in $R_0 \oplus R_1$ there is a peak $\mathbf{f}^\omega \twoheadleftarrow \mathbf{f}(\mathbf{g}(\mathbf{f}(\mathbf{g}(\cdots)))) \twoheadrightarrow \mathbf{g}^\omega$, and $\mathbf{f}^\omega$ and $\mathbf{g}^\omega$ are obviously not joinable. We will therefore restrict our attention to non-collapsing systems.

In infinitary rewriting, we may need "balancing" rules to make non-left-linear rules applicable if we desire confluence. To appreciate this, consider $S \triangleq \{\mathbf{f}(x, x) \longrightarrow \mathbf{a}\}$ which is (finitarily) confluent by Newman's Lemma; but when considering S as an iTRS, we lose (infinitary) confluence:

Example 1. Consider S. From the term $h \triangleq \mathbf{f}(h, h)$ we get the following two derivatives $k \triangleq \mathbf{f}(\mathbf{a}, k)$ and $p \triangleq \mathbf{f}(p, \mathbf{a})$, both of which are normal forms of S, i.e. S cannot be confluent. □

Suitably extending S yields a confluent iTRS; consider the following right-ground system:

$$R \triangleq \begin{cases} \mathbf{f}(x, x) \longrightarrow \mathbf{a} \\ \mathbf{f}(\mathbf{a}, x) \longrightarrow \mathbf{a} \\ \mathbf{f}(x, \mathbf{a}) \longrightarrow \mathbf{a} \\ \mathbf{f}(\mathbf{f}(x, y), z) \longrightarrow \mathbf{a} \\ \mathbf{f}(x, \mathbf{f}(y, z)) \longrightarrow \mathbf{a} \end{cases}$$

We have the following:

Proposition 1. *R is confluent.*

Proof. We claim that if $\mathbf{f}(s, s')$ is a term and if $\mathbf{f}(s, s') \twoheadrightarrow t$ is a strongly convergent derivation of length at least 1, then $t \longrightarrow \mathbf{a}$ (observe that $t \notin \mathcal{X}$). We reason as follows: If $t = \mathbf{a}$, we are done. Otherwise, write $t = \mathbf{f}(w, w')$ and split on cases according to w and w':

1. $w = \mathbf{a}$ or $w' = \mathbf{a}$. Here, $t \longrightarrow \mathbf{a}$ by an application of either the rule $\mathbf{f}(\mathbf{a}, x) \longrightarrow \mathbf{a}$ or $\mathbf{f}(x, \mathbf{a}) \longrightarrow \mathbf{a}$.
2. $w = \mathbf{f}(r, r')$ or $w' = \mathbf{f}(r, r')$. In this case, $t \longrightarrow \mathbf{a}$ by an application of either the rule $\mathbf{f}(\mathbf{f}(x, y), z) \longrightarrow \mathbf{a}$ or the rule $\mathbf{f}(x, \mathbf{f}(y, z)) \longrightarrow \mathbf{a}$.
3. $w = x$ and $w' = y$ for $x, y \in \mathcal{X}$. Since there are no collapsing rules, this is only possible if $s = x$ and $s' = y$. If $x \neq y$, $\mathbf{f}(x, y)$ is a normal form, which contradicts the assumption that $\mathbf{f}(s, s') \twoheadrightarrow t$ has length at least 1. Thus, we must have $x = y$, i.e. $w = w'$ and the rule $\mathbf{f}(x, x) \longrightarrow \mathbf{a}$ yields $\mathbf{f}(w, w') \longrightarrow \mathbf{a}$. □

Make a "copy", R', of R, renaming $\mathbf{f}$ to $\mathbf{g}$ and $\mathbf{a}$ to $\mathbf{b}$, and copying the rules *mutatis mutandis.* The resulting system is clearly confluent, but $R \oplus R'$ is not confluent:

Proposition 2. *The term $s \triangleq \mathbf{f}(\mathbf{g}(s, s), \mathbf{g}(s, s))$ is not confluent (in $R \oplus R'$).*

Proof. It is clear that $s \longrightarrow \mathbf{a}$ and that $\mathbf{g}(s, s) \longrightarrow \mathbf{b}$. There is a strongly convergent derivation of the "right" subterm $\mathbf{g}(s, s)$ with limit $s'' \triangleq \mathbf{g}(\mathbf{a}, \mathbf{f}(\mathbf{b}, s''))$. Since the "left" subterm $\mathbf{g}(s, s)$ rewrites in one step to $\mathbf{b}$, s can in ω steps be rewritten to $s' \triangleq \mathbf{f}(\mathbf{b}, \mathbf{g}(\mathbf{a}, s'))$, which is a normal form. Thus, there is a peak $\mathbf{a} \twoheadleftarrow s \twoheadrightarrow s'$ for which no corresponding valley exists. □

Corollary 1. *$R \oplus R'$ is not confluent.*

Corollary 2. *Confluence is not a modular property of iTRSs.*

The counterexample to confluence crucially employs two facts:

1. One of the considered systems has a rule that is not left-linear.
2. The specific term considered does not have finite rank.

The further main results of this paper are that if restrictions are imposed on one of the two facts above, modularity of confluence may be recovered.

4 Modularity of Confluence for Left-Linear Systems

In this section we consider combinations of confluent, left-linear, pairwise disjoint systems, and subsequently derive necessary and sufficient conditions for modularity of confluence. We begin by proving our results for non-collapsing iTRSs and later extend them to sets of iTRSs $\mathcal{R}$ such that $\oplus\mathcal{R}$ is essentially non-collapsing.

Definition 13. *The term s is said to be* insulated *if it contains no collapsing blocks.*

Proposition 3. *If R is a left-linear iTRS, s is insulated and $s \twoheadrightarrow t$, then t is insulated.*

Proof. By left-linearity and insulation of s. □

Thus, for *left-linear* terms, insulation of s corresponds to the notion of *preservation* well known from the study of modularity in finitary rewriting [9].

Proposition 4 (Outer and Inner Derivations Commute). *Let $\mathcal{R}$ be a set of left-linear, pairwise disjoint iTRSs and let s be an insulated term with a peak $t \overset{i}{\twoheadleftarrow} s \overset{o}{\twoheadrightarrow} t'$. Then, there exists a term s' and a valley $t' \overset{i}{\twoheadrightarrow} s' \overset{o}{\twoheadleftarrow} t$.*

Proof. Straightforward induction on the length of the longest of the two derivations in the peak (in case of equal length, pick any of them). □

Proposition 5 (Postponement of Inner Derivation). *Let $\mathcal{R}$ be a set of left-linear, pairwise disjoint iTRSs and let s be an insulated term with $s \twoheadrightarrow t$ (in $\oplus\mathcal{R}$). Then, there is a term t' such that $s \overset{o}{\twoheadrightarrow} t' \overset{i}{\twoheadrightarrow} t$.*

Proof. By left-linearity and insulation, inner rewrite steps can neither destroy nor create outer redexes, and there is thus a term t' and a strongly convergent *outer* derivation $s \twoheadrightarrow t'$ such that $\mathbf{cap}(t') = \mathbf{cap}(t)$, and such that the set of descendants of any position of a variable occurrence in $\mathbf{cap}(s)$ is identical under $s \twoheadrightarrow t$ and $s \twoheadrightarrow t'$. Every principal subterm t_j of t is a descendant of some subterm s_i of s, and by disjointness of the iTRSs and strong convergence, we have $s_i \twoheadrightarrow t_j$. By strong convergence of $s \twoheadrightarrow t$ and the definition of descendant, t_j is eventually "fixed" at a single position p_j. The principal subterm of t' at p_j is a descendant of s_i, and since there were no inner steps in $s \twoheadrightarrow t'$ must be identical to s_i. Hence, $s_i \twoheadrightarrow t_j$. Since t_j was arbitrary, the same argument holds for all principal subterms of t, and an application of the Dovetailing Lemma concludes the proof. □

Proposition 6. *Let $\mathcal{R}$ be a set of left-linear, pairwise disjoint, confluent iTRSs, and let s be an insulated term. Then the following diagram commutes for any peak $t \twoheadleftarrow s \twoheadrightarrow t'$:*

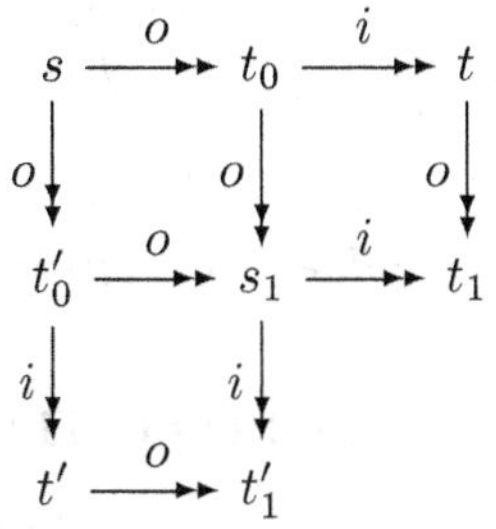

where all rewrite steps in the peak $t_1 \overset{i}{\twoheadleftarrow} s_1 \overset{i}{\twoheadrightarrow} t'_1$ take place at depth ≥ 1.

Proof. Use Proposition 5 twice to erect the leftmost and uppermost sides of the diagram. Since the systems were assumed to be confluent and left-linear, outer derivation is confluent, whence we get commutativity of the upper-left square. Two applications of Proposition 4 furnish commutativity of the two remaining squares. All rewrite steps in the peak $t'_1 \twoheadleftarrow s_1 \twoheadrightarrow t_1$ are inner, and so by insulation take place at depth ≥ 1. □

Lemma 3. *Let $\mathcal{R}$ be a set of left-linear, confluent, pairwise disjoint iTRSs. Then every insulated term is confluent (in $\oplus\mathcal{R}$).*

Proof. Let $t \twoheadleftarrow s \twoheadrightarrow t'$ be a peak of $\oplus\mathcal{R}$ with s insulated. By Proposition 6, we can erect a diagram as in that proposition. Consider the peak $t_1 \twoheadleftarrow s_1 \twoheadrightarrow t'_1$ and observe that $\mathbf{cap}(s_1) = \mathbf{cap}(t_1) = \mathbf{cap}(t'_1)$, since s was insulated. Write $s_1 = C[\![s'_i]\!]_{i \in \mathcal{I}}$. Then the inner derivations in $t_1 \twoheadleftarrow s_1 \twoheadrightarrow t'_1$ occur in the s_i. Applying the proposition coinductively (*viz.* the below diagram) to the inner derivations in the s_i – using the Dovetailing Lemma to order arrange derivations in parallel subterms – yields strongly convergent derivations $t \twoheadrightarrow s' \twoheadleftarrow t'$ for some term s', as all redex contractions at the "kth application" of Proposition 6 take place at depth $\geq k$.

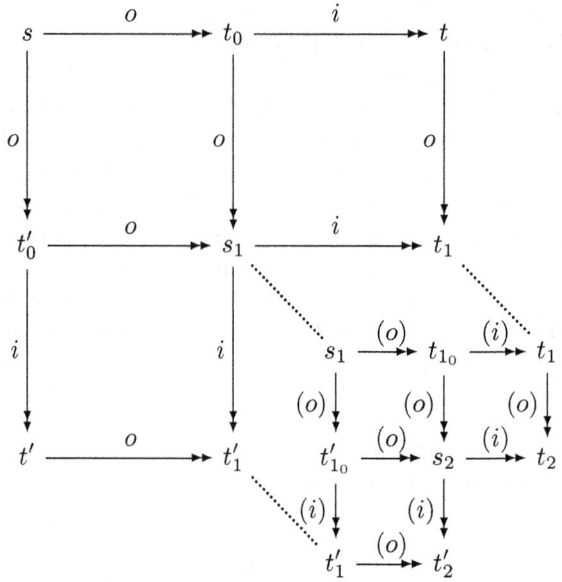

□

Corollary 3. *Modularity is confluent for left-linear, non-collapsing, iTRSs.*

4.1 Essentially Non-collapsing Sets of iTRSs

We now give a simple condition on sets of iTRSs that will turn out to be a necessary and sufficient condition for the modularity of confluence.

Definition 14. *A set, $\mathcal{R}$, of pairwise disjoint iTRSs is said to be* essentially non-collapsing *if at most one iTRS of $\mathcal{R}$ contains a collapsing rule. If there exists an R such that R is the unique iTRS among $\mathcal{R}$ that contains a collapsing rule, we call R the* collapsing colour *of $\mathcal{R}$.*

The definition is similar to the notion of *almost-non-collapsing* iTRS well-known from the study of *orthogonal* iTRSs [5]; observe however, that the term is used here as a property of a *set* of iTRSs, not the individual iTRSs.

Proposition 7. *Let R be a left-linear iTRS. If there is a strongly convergent derivation $s \longrightarrow^{\alpha} x$ for some $x \in \mathcal{X}$, then $s \longrightarrow^{k} x$ for some $k \in \omega$.*

Proof. By the Compression Lemma, we may assume that $\alpha \leq \omega$. The fact that $s \longrightarrow^{\leq\omega} x$ is *convergent* now furnishes the desideratum. □

Lemma 4. *If $\mathcal{R}$ is a set of left-linear, pairwise disjoint, confluent iTRSs such that $\oplus\mathcal{R}$ is confluent, then $\mathcal{R}$ is essentially non-collapsing.*

Proof. By contraposition. If $\mathcal{R}$ were *not* essentially non-collapsing, there would be at least two iTRSs, (Σ_1, R_1) and (Σ_2, R_2), each containing a collapsing rule. We may write as $C_1[x] \longrightarrow x$, resp. $C_2[x] \longrightarrow x$ where the first rule is from R_1, and the second from R_2. The term $C_1[C_2[C_1[C_2[\cdots]]]]$ has the two derivatives $C_1[C_1[\cdots]]$ and $C_2[C_2[\cdots]]$ which are terms over disjoint alphabets and are hence joinable only if both terms can be rewritten to a variable, y. By Proposition 7, this can only happen if $C_i[C_i[\cdots]] \longrightarrow^{*} y$. Since this derivation is finite, there exists an n such that a stack, $C_i[\cdots[C_i[x]]]$ of n copies of $C_i[x]$ rewrites to y. But, clearly, $C_i[\cdots[C_i[x]]] \longrightarrow^{*} x$, whence confluence of the underlying iTRS yields $x = y$. By left-linearity, we may assume that there are no copies of x in $C_i[C_i[\cdots]]$. Thus, $C_1[C_1[\cdots]]$ and $C_2[C_2[\cdots]]$ have no common join, contradicting confluence of $\oplus\mathcal{R}$. □

Thus, essential non-collapsingness is a necessary condition for modularity of confluence. To see that it is also sufficient, we proceed as follows:

Definition 15. *Let $\mathcal{R}$ be a set of left-linear, confluent, pairwise disjoint iTRSs. Let s be a term and write $s = C[\![s_i]\!]_{i\in\mathcal{I}}$ (observe that if s is monochrome, we have $\mathcal{I} = \emptyset$). We define the term $\tilde{s}$ as follows:*

$$\begin{array}{ll} C[\tilde{s}_i]_{i\in\mathcal{I}} & \textit{if } C[x_i]_{i\in\mathcal{I}} \textit{ is not collapsing} \\ \tilde{s}_m & \textit{if } C[x_i]_{i\in\mathcal{I}} \twoheadrightarrow x_m \end{array}$$

That is, $\tilde{s}$ is the term obtained from s by collapsing all collapsing blocks in a top-down fashion; by essential non-collapsingness, only blocks of a single colour will be collapsed. Observe that by confluence of the elements of $\mathcal{R}$, each block can collapse in at most one way, whence $\tilde{s}$ is well-defined. Note also that $s \twoheadrightarrow \tilde{s}$.

Proposition 8. *Let $\mathcal{R}$ be an essentially non-collapsing set of left-linear, pairwise disjoint, confluent iTRSs. Then $\tilde{s}$ is insulated, and the set of descendants of any $u \in Pos(s)$ is the same for any strongly convergent derivation $s \twoheadrightarrow \tilde{s}$ and satisfies $|u/(s \twoheadrightarrow \tilde{s})| \leq 1$.*

Proof. By left-linearity, contraction of redexes in one block can only create redexes in another if the block collapses. Since the considered systems are left-linear and there is at most one collapsing colour, $\tilde{s}$ must be insulated. By confluence of the systems, each collapsing block $C[x_1, \ldots, x_m]$ has a unique collapsing variable x_i, and we hence have $|u/(s \twoheadrightarrow \tilde{s})| \leq 1$ for any such derivation. $u/(s \twoheadrightarrow \tilde{s})$ is clearly independent of the choice of derivation. □

Definition 16. *Let $\mathcal{R}$ be an essentially non-collapsing set of left-linear, confluent, pairwise disjoint iTRSs, For any term s, we define $P_s(u)$ as the predicate on $Pos(s)$ that is true iff u has a descendant across $s \twoheadrightarrow \tilde{s}$. Furthermore, we set $U_s \triangleq \{u \in Pos(s) : P_s(u)\}$.*

By the previous proposition, we see that $P_s(u)$, and hence also U_s, is well-defined.

Proposition 9. *Let $\mathcal{R}$ be an essentially non-collapsing set of left-linear, confluent, pairwise disjoint iTRSs, and s be a term. Then U_s is partially ordered by $\prec$ and the graph of $(U_s, \prec)$ is a (possibly infinite) directed tree $\mathcal{T}_s$. The number of children of $\mathcal{T}_s$ at any vertex u is the arity of the function symbol at position u in s.*

Proof. U_s is partially ordered by $\prec$ since $Pos(s)$ is. The graph of $(Pos(s), \prec)$ is a directed tree, and clearly U_s is connected, hence also a directed tree. If a block collapses, at least one position below it has a descendant across $s \twoheadrightarrow \tilde{s}$, whence a position $u \in Pos(s)$ has exactly as many children in $(U_s, \prec)$ as it has in $(Pos(s), \prec)$. □

Proposition 10. *Let $\mathcal{R}$ be an essentially non-collapsing set of left-linear, confluent, pairwise disjoint iTRSs, and let $s \twoheadrightarrow s'$ have length at most ω (by the Compression Lemma, if need be). Let d be any non-negative integer. There is a non-negative integer d' such that, for all $u \in U_{s'}$ with $|u| \geq d'$, the depth, d'', of the single element in $u/(s' \twoheadrightarrow \tilde{s}')$ satisfies $d'' \geq d$. Furthermore, there is a $k \in \omega$ such that for $k' > k$, if the redex, r, contracted in $s_{k'} \xrightarrow{r} s_{k'+1}$ is at position u, then the unique residual of r by $s_{k'} \twoheadrightarrow \tilde{s}_{k'}$ is at depth $\geq d$.*

Proof. By Proposition 9 and the pigeon hole principle, the number of vertices at each depth in $\mathcal{T}_{s'}$ is finite. Write $u = p_1^{nc} \cdot p_1^{c} \cdot p_2^{nc} \cdot p_2^{c} \cdots p_m^{nc}$ where the p_i^{nc} and p_i^{c} are the positions of variables in non-collapsing and collapsing blocks of s', respectively, and where p_1^{nc} is possibly the empty position. Clearly, the depth of the single element in $u/(s' \twoheadrightarrow \tilde{s}')$ is $cl(u) \triangleq |p_1^{nc} \cdot p_2^{nc} \cdots p_m^{nc}|$, and $cl(u)$ is thus the depth of u in $\mathcal{T}_{s'}$. Let $\{u_1, \ldots, u_m\}$ be the set of vertices in $\mathcal{T}_{s'}$ at depth d, and set $d' \triangleq \max\{|u_1|, \ldots, |u_m|\}$; then $P_{s'}(u)$ and $|u| \geq d'$ implies $cl(u) \geq d$. Strong convergence of $s \twoheadrightarrow s'$ and the fact that the length is at most ω yield existence of a $k \in \omega$ such that all redexes contracted in $s_k \twoheadrightarrow s'$ are at depths $\geq d'$. Hence, for any $k' > k$, if $s_{k'} \xrightarrow{r} s_{k'+1}$, the unique residual of r by $s_{k'} \twoheadrightarrow \tilde{s}_{k'}$ will be at depth $\geq d$ in $\tilde{s}_{k'}$. □

Proposition 11. *Let $\mathcal{R}$ be an essentially non-collapsing set of left-linear, confluent, pairwise disjoint iTRSs, and let $s \twoheadrightarrow t$. Then $\tilde{s} \twoheadrightarrow \tilde{t}$.*

Proof. By the Compression Lemma, we may assume that $s \longrightarrow^{\leq\omega} t$, and proceed by induction on the length, α, of the derivation:

- $\alpha = 0$. Trivial.
- $\alpha = j + 1$.
  Consider $s_j \xrightarrow{r} s_{j+1}$; if the redex r is at position u and $u \notin U_{s_j}$, we have $\tilde{s}_j = \tilde{s}_{j+1}$, and we are done. If $u \in U_{s_j}$, contracting $r/(s_j \twoheadrightarrow \tilde{s}_j)$ clearly yields $\tilde{s}_{j+1}$ in one step.
- $\alpha = \omega$. By Proposition 10, for each depth $d \in \omega$, there is a $k \in \omega$ such that all steps in $\tilde{s}_{k'} \longrightarrow \tilde{s}_{k'+1}$ are below depth d for $k' > k$, showing that the resulting derivation is strongly convergent with limit $\tilde{t}$. □

Proposition 12. *Let R be the collapsing colour of an essentially non-collapsing set $\mathcal{R}$ of left-linear, confluent iTRSs. If $\tilde{s} \twoheadrightarrow t$, then $s \twoheadrightarrow t$.*

Proof. $s \twoheadrightarrow \tilde{s} \twoheadrightarrow t$. □

We can now prove the first positive result of the paper:

Theorem 1. *Let $\mathcal{R}$ be a set of confluent, left-linear, pairwise disjoint iTRS. Then, $\oplus\mathcal{R}$ is confluent iff $\mathcal{R}$ is essentially non-collapsing.*

Proof. If $\oplus\mathcal{R}$ is confluent, it follows from Lemma 4 that $\mathcal{R}$ must be essentially non-collapsing. Conversely, if $\mathcal{R}$ is essentially non-collapsing, let $t \twoheadleftarrow s \twoheadrightarrow t'$ be a peak of $\oplus\mathcal{R}$. By Proposition 11, there exists a peak $\tilde{t} \twoheadleftarrow \tilde{s} \twoheadrightarrow \tilde{t}'$. Lemma 3 now yields existence of a term s' and strongly convergent sequences $\tilde{t} \longrightarrow s'$ and $\tilde{t}' \longrightarrow s'$. An application of Proposition 12 concludes the proof. □

4.2 Mutually Orthogonal Systems

Confluence of left-linear systems in finitary rewriting can be ensured by less strict demands than that of disjointness. In both first- and higher-order finitary rewriting, *mutual orthogonality* (and the more lax *mutual weak orthogonality*) is sufficient for confluent systems to be confluent under direct sum [15]. The techniques of [6, 5] for proving confluence results in orthogonal (strongly convergent) transfinite rewriting use reasoning about residuals and the depths of redexes contracted in valleys as their linchpin; this does not generalize to arbitrary confluent iTRSs, hence not to the setting of modularity, since we cannot necessarily track residuals in non-orthogonal systems. Unlike the case with disjoint systems, contraction of a redex in one system can create redexes in others without being the application of a collapsing rule; as we cannot properly gauge the effect of such creations without tracking residuals, there appears to be no easy way of extending our results for left-linear systems to the setting of mutual orthogonality.

5 Confluence of Terms of Finite Rank

In this section, we show that when only terms of *finite rank* are considered, confluence is modular for non-collapsing, not necessarily left-linear, systems. The methods employed are akin to Toyama's original proof of (finitary) confluence of TRSs [14] and the initial part of the later, more elegant proof [9]. The parts of these papers dealing with collapsing rules do not appear to be applicable when working with *strongly* convergent derivations.

Proposition 13. *If the confluent terms s and s' satisfy $s \sim s'$ (i.e. s and t are joinable), then any derivative, t, of s is joinable with any derivative, t', of s'.*

Proof. Straightforward. □

Definition 17. *For sequences of terms $(s_k)_{k\in\mathcal{K}}$ and $(t_k)_{k\in\mathcal{K}}$, we write $(s_k)_{k\in\mathcal{K}} \propto (t_k)_{k\in\mathcal{K}}$ when it is the case that $t_{k'} = t_{k''}$ if $s_{k'} \sim s_{k''}$ for all $k', k'' \in \mathcal{K}$.*

Proposition 14. *Let $\mathcal{R}$ be a set of non-collapsing, pairwise disjoint iTRSs, let $s = C[\![s_i]\!]_{i\in\mathcal{I}}$, and assume that $s \twoheadrightarrow t$ with $t = C'[\![t_j]\!]_{j\in\mathcal{J}}$. Choose variables $(x_i)_{i\in\mathcal{I}}$ such that $(s_i)_{i\in\mathcal{I}} \propto (x_i)_{i\in\mathcal{I}}$. Then $C[x_i]_{i\in\mathcal{I}} \twoheadrightarrow C'[y_j]_{j\in\mathcal{J}}$ such that y_j is a descendant of x_i across $C[x_i]_{i\in\mathcal{I}} \twoheadrightarrow C'[y_j]_{j\in\mathcal{J}}$ iff t_j is a descendant of s_i across $s \twoheadrightarrow t$ for all $i \in \mathcal{I}$, $j \in \mathcal{J}$.*

Proof. By induction on the length, α, of $s \twoheadrightarrow t$.

- $\alpha = 0$. Straightforward.
- $\alpha = \beta+1$. Write $s_\beta = D[\![t'_k]\!]_{k\in\mathcal{K}}$; by the induction hypothesis we may assume that there exists a strongly convergent derivation $C[x_i]_{i\in\mathcal{I}} \twoheadrightarrow D[z_k]_{k\in\mathcal{K}}$ such that z_k is a descendant of x_i iff t'_k is a descendant of s_i, for all $k \in \mathcal{K}$, $i \in \mathcal{I}$ Consider the single rewrite step $s_\beta \longrightarrow s_{\beta+1}$. Assume that the redex contracted is of the rule $\mathbf{l} \longrightarrow \mathbf{r}$ in s_β and at position u. If the redex is not outer, or the rule is left-linear, the desideratum follows immediately. Assume, then that the redex is outer and that the rule is not left-linear. Since the induction hypothesis furnishes that z_k is a descendant of x_i iff t'_k is a descendant of s_i, for all $k \in \mathcal{K}$, $i \in \mathcal{I}$, whence $(t'_k)_{k\in\mathcal{K}} \propto (z_k)_{k\in\mathcal{K}}$, and the rule $\mathbf{l} \longrightarrow \mathbf{r}$ is applicable at position u in $D[z_k]_{k\in\mathcal{K}}$. The demand on the descendants is clearly fulfilled.
- *Lim*(α). Observe that the rewrite steps of $C[x_i]_{i\in\mathcal{I}} \twoheadrightarrow C'[y_j]_{j\in\mathcal{J}}$ correspond exactly to the outer steps of $s \twoheadrightarrow t$. If $C[x_i]_{i\in\mathcal{I}} \twoheadrightarrow C'[y_j]_{j\in\mathcal{J}}$ were not strongly convergent, neither would $s \twoheadrightarrow t$ be. It is clear by the definition of the descendant relation that the demand on the descendants is fulfilled. □

Proposition 15. *Let R be a non-collapsing iTRS, let $s = C[\![s_i]\!]_{i\in\mathcal{I}}$ such that the s_i are all confluent, and choose variables $(x_i)_{i\in\mathcal{I}}$ such that $(s_i)_{i\in\mathcal{I}} \propto (x_i)_{i\in\mathcal{I}}$. If $C[x_i]_{i\in\mathcal{I}} \twoheadrightarrow C'[z_k]_{k\in\mathcal{K}}$, then $C[\![s_i]\!]_{i\in\mathcal{I}} \twoheadrightarrow C'[\![t_k]\!]_{k\in\mathcal{K}}$ such that t_k is a descendant of s_i across $C[\![s_i]\!]_{i\in\mathcal{I}} \twoheadrightarrow C'[\![t_k]\!]_{k\in\mathcal{K}}$ iff z_k is a descendant of x_i across $C[x_i]_{i\in\mathcal{I}} \twoheadrightarrow C'[z_k]_{k\in\mathcal{K}}$.*

Proof. By induction on the length, α, of $C[x_i]_{i\in\mathcal{I}} \twoheadrightarrow C'[z_k]_{k\in\mathcal{K}}$.

- $\alpha = 0$. Straightforward
- $\alpha = \beta + 1$. Write $s_\beta = D[r_l]_{l\in\mathcal{L}}$. We have $C[x_i]_{i\in\mathcal{I}} \longrightarrow^\beta D[z_l]_{l\in\mathcal{L}}$. By the induction hypothesis, we have $C[s_i]_{i\in\mathcal{I}} \twoheadrightarrow D[r_l]_{l\in\mathcal{L}}$ for suitable $(r_l)_{l\in\mathcal{L}}$ such that the demand on the descendant relation is satisfied. Assume that the redex contracted is of the rule $\mathbf{l} \longrightarrow \mathbf{r}$ in s_β and at position u in $D[z_l]_{l\in\mathcal{L}}$. If the rule is left-linear, the desideratum follows immediately. If the rule is not left-linear, applicability of the rule in $D[z_l]_{l\in\mathcal{L}}$, $(s_i)_{i\in\mathcal{I}} \propto (x_i)_{i\in\mathcal{I}}$ and the descendants part of the induction hypothesis furnishes that if $z_j = z_{j'}$, r_j and $r_{j'}$ are joinable. Since $\mathbf{l}$ is a finite term, only a finite number of principal subterms need to be reduced to a common term in order for the rule to be applicable in $D[r_l]_{l\in\mathcal{L}}$. Thus, by Proposition 13, there exists a strongly convergent derivation $D[r_l]_{l\in\mathcal{L}} \twoheadrightarrow D[r'_l]_{l\in\mathcal{L}}$ (with all steps performed at depth $\geq |u|$) such that $\mathbf{l} \longrightarrow \mathbf{r}$ is applicable at position u in $D[r'_l]_{l\in\mathcal{L}}$ and the demand on the descendant relation is satisfied.
- $Lim(\alpha)$. There are two kinds of rewrite steps performed in $C[\![s_i]\!]_{i\in\mathcal{I}} \twoheadrightarrow C'[\![t_k]\!]_{k\in\mathcal{K}}$: "outer" steps corresponding to (and of the same depth as) the steps in $s \twoheadrightarrow C'[z_k]_{k\in\mathcal{K}}$, and "inner" steps performed to make non-left-linear rules applicable in the successor case above. The inner steps are all performed at a depth greater than that of the non-left-linear outer step that prompted them. Hence, the resulting derivation is strongly convergent; the demand on the descendant relation is clearly satisfied. □

Lemma 5. *Let $\mathcal{R}$ be a set of non-collapsing iTRSs and let $s = C[\![s_i]\!]_{i\in\mathcal{I}}$. Assume that outer derivation and the s_i are confluent for all $i \in \mathcal{I}$, and let $D[\![t_l]\!]_{l\in\mathcal{L}} \twoheadleftarrow s \twoheadrightarrow D'[\![t_{l'}]\!]_{l'\in\mathcal{L}'}$ be a peak. Then there exists a valley $t \twoheadrightarrow s' \twoheadleftarrow t'$.*

Proof. Apply Propositions 14 and 15 twice:

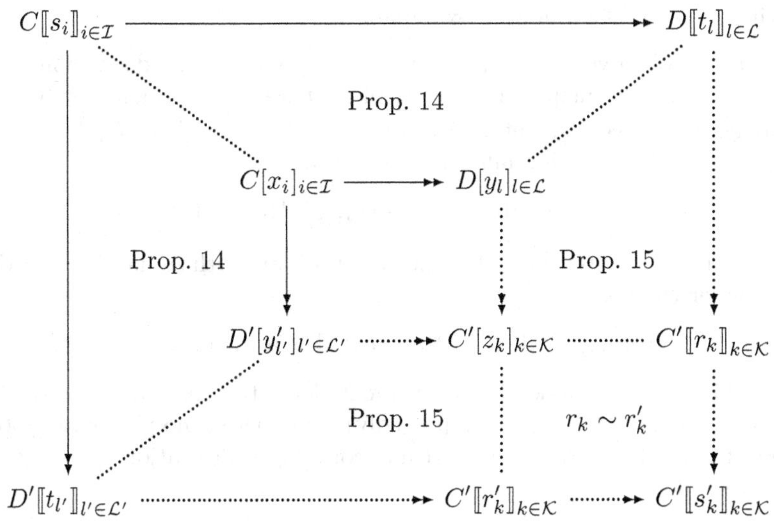

For the lower right rectangle, observe that the demand on the descendant relations in Propositions 14 and 15 ensure that r_k and r'_k are descendants of the same s_i for all $k \in \mathcal{K}$. Since the s_i were confluent, we get $r_k \sim r'_k$ for all $k \in \mathcal{K}$. An application of the Dovetailing Lemma yields that performing the $|\mathcal{K}|$ derivations needed to obtain $C'[\![s'_k]\!]$ from $C'[r_k]$ (resp. $C'[r'_k]$) can be done in a strongly convergent fashion. □

We now have the second positive result of this paper:

Theorem 2. *Let $\mathcal{R}$ be a set of non-collapsing, confluent iTRSs. Then, every term s over $\biguplus_{i \in \mathcal{I}} \Sigma_i$ with finite rank is confluent.*

Proof. By induction on $\mathbf{rank}(s)$. If $\mathbf{rank}(s) = 0$, the result follows immediately, since monochrome terms were assumed to be confluent. If $\mathbf{rank}(s) > 0$, note that outer derivation is confluent, as are all principal subterms of s, since they have rank strictly less than $\mathbf{rank}(s)$. The result follows by an application of Lemma 5. □

Systems containing collapsing rules exhibit severe technical complications that appear to be solvable neither with the techniques presented herein, nor with the standard techniques from finitary rewriting [9, 14].

6 Weakly Convergent Rewriting

In the previous sections, we have considered *strongly* convergent derivations. The more general setting of *weak* convergence is not very well understood, and sports far fewer auxiliary results. A major hurdle in this setting is that it is not clear how to define a suitable descendant relation. To appreciate the impact of this on the study of modularity, observe that the techniques of finitary rewriting, as well as those in this paper, depend crucially on the property of non-collapsing rewriting that, in a derivation $s \twoheadrightarrow t$, we can identify principal subterms of t with descendants of principal subterms of s. This is lost in weakly convergent rewriting, to wit the following example:

Example 2. We give an example of a weakly convergent derivation $s \longrightarrow^{\alpha} t$ such that, for a principal subterm t_j of t, there is *no* principal subterm s_i of s satisfying $s_i \twoheadrightarrow t_j$. Let $R_0 \triangleq \{\mathbf{a}(x) \longrightarrow \mathbf{b}(x)\}$ and let R_1 by the system consisting of the following infinite set of rules:

$$\mathbf{f}(x, \mathbf{g}^k(\mathbf{c}), \mathbf{d}(y, z)) \longrightarrow \mathbf{f}(y, \mathbf{g}^{k+1}(\mathbf{c}), z) \text{ for } k \in \omega$$

Clearly, the two systems are disjoint, and both are orthogonal. Let $s \triangleq \mathbf{d}(\mathbf{a}^{\omega}, s)$ and ponder the term:

$$\mathbf{f}(\mathbf{a}(\mathbf{c}), \mathbf{g}(\mathbf{c}), \mathbf{d}(\mathbf{a}(\mathbf{c}), \mathbf{d}(\mathbf{a}(\mathbf{a}(\mathbf{c})), \mathbf{d}(\mathbf{a}(\mathbf{a}(\mathbf{a}(\mathbf{c}))), \mathbf{d}(\cdots)))))$$

from which there is a weakly convergent derivation having limit $\mathbf{f}(\mathbf{a}^{\omega}, \mathbf{g}^{\omega}, s)$ (contract redexes at position ϵ repeatedly). But there is no principal subterm, s_i of the starting term such that a weakly convergent derivation $s_i \longrightarrow^{\beta} \mathbf{a}^{\omega}$ exists for any ordinal β. □

Acknowledgement

The author is grateful to the anonymous referees for thorough comments that have significantly improved the presentation of the paper.

References

1. F. Baader and T. Nipkow. *Term Rewriting and All That.* Cambridge University Press, Cambridge, United Kingdom, 1998.
2. N. Dershowitz and J.-P. Jouannaud. Rewrite systems. In Jan van Leeuwen, editor, *Handbook of Theoretical Computer Science*, volume B, pages 243 – 320. Elsevier Science, Amsterdam, Holland, 1990.
3. N. Dershowitz, S. Kaplan, and D. Plaisted. Rewrite, Rewrite, Rewrite, Rewrite, Rewrite .... *Theoretical Computer Science*, 83(1):71–96, 1991.
4. J.R. Kennaway, J.W. Klop, M.R. Sleep, and F.J. de Vries. Infinitary Lambda Calculus. *Theoretical Computer Science*, 175:93–125, 1997.
5. R. Kennaway and F.-J. de Vries. Infinitary Rewriting. In *Term Rewriting Systems* [12], chapter 12, pages 668–711.
6. R. Kennaway, J.W. Klop, R. Sleep, and F.J. de Vries. Transfinite Reductions in Orthogonal Term Rewriting. *Information and Computation*, 119(1):18 – 38, 1995.
7. R. Kennaway, V. van Oostrom, and F.J. de Vries. Meaningless Terms in Rewriting. *The Journal of Functional and Logic Programming*, 1999-1:1–34, 1999.
8. J.W. Klop. Term Rewriting Systems. In S. Abramsky, D. Gabbay, and T. Maibaum, editors, *Handbook of Logic in Computer Science*, volume 2, pages 1–116. Oxford University Press, 1992.
9. J.W. Klop, A. Middeldorp, Y. Toyama, and R. de Vrijer. Modularity of Confluence: A Simplified Proof. *Information Processing Letters*, 49:101 – 109, 1994.
10. E. Mendelson. *Introduction to Mathematical Logic.* Chapman and Hall, 4. edition, 1997.
11. E. Ohlebusch. *Advanced Topics in Term Rewriting.* Springer-Verlag, 2002.
12. TeReSe. *Term Rewriting Systems*, volume 55 of *Cambridge Tracts in Theoretical Computer Science.* Cambridge University Press, 2003.
13. Y. Toyama. Counterexamples to Termination for the Direct Sum of Term Rewriting Systems. *Information Processing Letters*, 25:141 – 143, 1987.
14. Y. Toyama. On the Church-Rosser Property for the Direct Sum of Term Rewriting Systems. *Journal of the American Society for Computing Machinery*, 34:128 – 143, 1987.
15. V. van Oostrom and F. van Raamsdonk. Weak Orthogonality Implies Confluence: the Higher-Order Case. In *Proceedings of the Third International Symposium on Logical Foundations of Computer Science (LFCS '94)*, volume 813 of *Lecture Notes in Computer Science*, pages 379–392. Springer-Verlag, 1994.

MU-TERM: A Tool for Proving Termination of Context-Sensitive Rewriting*

Salvador Lucas

DSIC, Universidad Politécnica de Valencia
Camino de Vera s/n, E-46022 Valencia, Spain
slucas@dsic.upv.es

Abstract. Restrictions of rewriting can eventually achieve termination by pruning all infinite rewrite sequences issued from every term. Context-sensitive rewriting (*CSR*) is an example of such a restriction. In *CSR*, the replacements in some arguments of the function symbols are permanently forbidden. This paper describes MU-TERM, a tool which can be used to automatically prove termination of *CSR*. The tool implements the generation of the appropriate orderings for proving termination of *CSR* by means of polynomial interpretations over the rational numbers. In fact, MU-TERM is the first termination tool which generates term orderings based on such polynomial interpretations. These orderings can also be used, in a number of different ways, for proving termination of ordinary rewriting. Proofs of termination of *CSR* are also possible via existing transformations to TRSs (without any replacement restriction) which are also implemented in MU-TERM.

1 Introduction

Context-sensitive rewriting (*CSR* [12]) is useful for describing semantic aspects of a number of programming languages (e.g., Maude, OBJ2, OBJ3, or CafeOBJ) and analyzing the computational properties of the corresponding programs, in particular termination (see [11]). Termination of *CSR* can also be used to prove top-termination of TRSs and to easily define normalizing reduction strategies [12]. Termination of *CSR* has also been related to termination of some (auxiliary) evaluation modes of functional languages like Haskell (see [9]).

In *CSR*, a *replacement map* μ discriminates, for each symbol of the signature, the argument positions $\mu(f)$ on which the replacements are allowed.

Example 1. Consider the TRS $\mathcal{R}$:

```
nats  → cons(0,incr(nats))    incr(cons(x,xs)) → cons(s(x),incr(xs))
pairs → cons(0,incr(odds))    head(cons(x,xs)) → x
odds  → incr(pairs)           tail(cons(x,xs)) → xs
```

* Work partially supported by MCYT project STREAM TIC2001-2705-C03-01, MCyT Acción Integrada HU 2003-0003 and Agencia Valenciana de Ciencia y Tecnología GR03/025.

V. van Oostrom (Ed.): RTA 2004, LNCS 3091, pp. 200–209, 2004.

with $\mu(\texttt{cons}) = \{1\}$ and $\mu(f) = \{1, \ldots, ar(f)\}$, for any other symbols f in the signature. The infinite sequence $\underline{\texttt{nats}} \to \texttt{cons(0,incr(}\underline{\texttt{nats}}\texttt{))} \to \cdots$ is not possible with *CSR* because of $\mu(\texttt{cons}) = \{1\}$.

Automatic proofs of termination are always desirable although difficult. Several methods have been developed for proving termination of *CSR* under μ for a given TRS $\mathcal{R}$ (i.e., for proving the *μ-termination* of $\mathcal{R}$). However, no tool for proving termination of *CSR* has been reported to date. Our tool, MU-TERM, is intended to fill this gap. Two main approaches to prove termination of *CSR* have been investigated in the literature so far:

1. *Indirect proofs* which are based on transforming the problem of proving termination of *CSR* into a proof of termination of rewriting. For instance, [5,8,9,10,16] describe a number of transformations Θ from TRSs $\mathcal{R}$ and replacement maps μ that produce TRSs $\mathcal{R}^\mu_\Theta$. If we are able to prove *termination* of $\mathcal{R}^\mu_\Theta$ (using the standard methods), then the μ-termination of $\mathcal{R}$ is ensured. Our tool implements all these transformations and also provides interfaces for the use of external tools for proving termination (of $\mathcal{R}^\mu_\Theta$): AProVE[1], CiME[2], Termptation[3], and the Tyrolean/Tsukuba Termination Tool[4] (TTT).
2. *Direct proofs*, which are based on using μ-reduction orderings (see [16]) such as the (context-sensitive) recursive path orderings [3], polynomial orderings [7,14], semantic path orderings [4], and Knuth-Bendix orderings [4]. These are orderings $>$ on terms which can be used to directly compare the left- and right-hand sides of the rules in order to conclude the μ-termination of the TRS. The MU-TERM tool implements automatic proofs of termination of *CSR* by using the polynomial interpretations over the rational numbers of [14].

The modular analysis of termination of *CSR* described in [6] can also be used: MU-TERM attempts a *safe* decomposition of the TRS in such a way that the components satisfy the modularity requirements described in [6]. If it succeeds in performing a non-trivial decomposition (i.e., MU-TERM obtains more than one component), then individual proofs of termination are attempted for each component.

The tool can also be used for proving *termination* of rewriting in a number of ways. This is because term rewriting is a particular case of *CSR* where the replacement map $\mu_\top(f) = \{1, \ldots, ar(f)\}$, for all symbols f in the signature is used. Polynomials over the rationals [14] are used to generate appropriate reduction orderings. As far as the author knows, MU-TERM is currently the only termination tool which uses such kind of polynomials. On the other hand, 'proper' μ-reduction orderings (where μ can be different from $\mu_\top$) based on such polyno-

[1] `http://www-i2.informatik.rwth-aachen.de/AProVE`
[2] `http://cime.lri.fr`
[3] `http://www.lsi.upc.es/~albert`
[4] `http://cl2-informatik.uibk.ac.at`

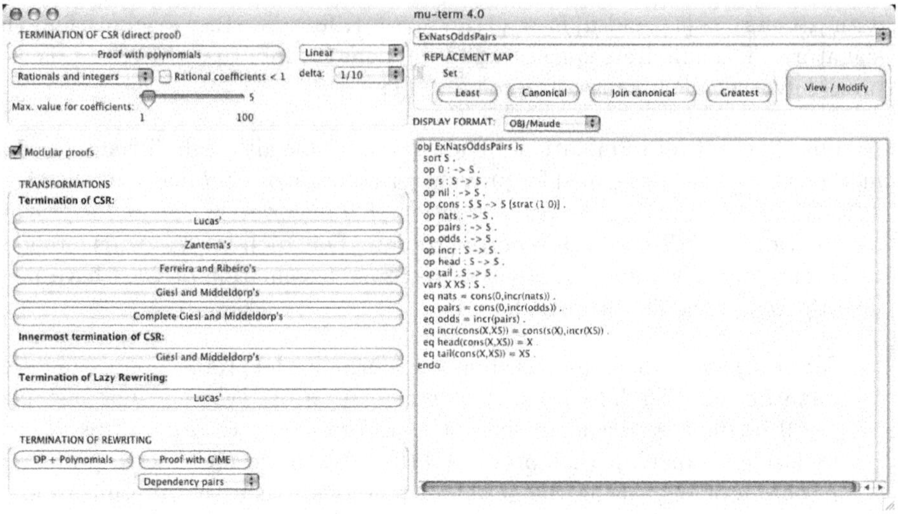

Fig. 1. Screenshot of the main window of MU-TERM

mial interpretations are also used together with the dependency pairs approach [1] to prove termination of rewriting.

The MU-TERM system is available at

`http://www.dsic.upv.es/~slucas/csr/termination/muterm`

2 Interface and Functionality

MU-TERM has a graphical user interface (see Figure 1). In the following, we explain the funcionalities of the tool.

Menu File. MU-TERM holds a list of TRSs (possibly with the corresponding replacement maps) which can be modified and transformed. The TRS which is displayed in the main window is called the *current TRS*. MU-TERM uses the *current TRS* to perform most actions selected by the user: prove termination, transform, etc.

TRSs can be loaded from files containing the rules in the 'simple format' l `->` r, where l and r are terms in the usual prefix syntax (infix operators are not allowed) with arguments enclosed between '(' and ')' and separated by ','. In this format, *variable identifiers begin with a capital letter*. TRSs are introduced in the system from text files via menu `File`; after successfully reading the file, the TRS becomes the *current TRS*.

The system is also able to deal with modules following a subset of the full OBJ / Maude grammar. The advantage is that, in contrast to the simple format (which needs a second phase for the introduction of the replacement map, see below), we are able to specify the *replacement map* at once, by means of the

OBJ / Maude *strategy annotations.* For instance, the TRS $\mathcal{R}$ of Example 1 was introduced as the OBJ / Maude module showed in Figure 1, where the strategy annotation (1 0) for `cons` is interpreted by $\mu(\texttt{cons}) = \{1\}$ (i.e., zeros in strategy annotations are just removed).

On the other hand, the user is allowed to display and save the *current TRS* in(to) a number of different formats which permit to export TRSs to AProVE, C*i*ME, Termptation, TTT, and to the format of the *Termination Problems Data Base*[5] (TPDB).

Panel Termination of CSR (direct proof). The tool implements the techniques described in [14]. A proof of μ-termination of a TRS $\mathcal{R}$ is transformed into the problem of solving a set of constraints over the (unknown) coefficients of a polynomial interpretation for the signature of $\mathcal{R}$. An interesting feature of our technique is that we generate polynomial interpretations with *rational* coefficients. For instance, such rational coefficients permit to deal with the lack of monotonicity in the non-replacing arguments of symbols which can be necessary to prove termination of *CSR* (see [14] for further details and below for an example). This can be achieved by using rational coefficients q with $0 \leq q < 1$ in the monomials which correspond to non-replacing arguments (see [14, Section 5]).

In order to try a direct proof of the μ-termination of a TRS $\mathcal{R}$, MU-TERM builds polynomial interpretations with undefined (rational) coefficients for each symbol of the signature. We are able to deal with four different kinds of polynomial interpretations: linear, simple, simple-mixed (see [15]) and quadratic. Then, according to the technique described in [14, Section 5], MU-TERM automatically generates a set of Diophantine constraints on the undeterminate coefficients of the polynomial interpretations. Solving such constraint implies that the system is μ-terminating and the solution yields the polynomial interpretation which induces the corresponding polynomial ordering.

We use C*i*ME as an *auxiliary* tool to solve the constraints generated by the system. Our choice of C*i*ME is motivated by the availability of a language for expressing Diophantine constraints, and commands for solving them. This is present in C*i*ME but currently missing (or unavailable) in other termination tools which (internally) may use similar constraint solvers for dealing with polynomial orderings (e.g., AProVE). C*i*ME, however, is only able to solve Diophantine inequations yielding nonnegative *integers* as solutions. MU-TERM deals with the task of making the use of rational numbers compatible with this limitation: given a Diophantine constraint $e_1 \geq e_2$ containing an occurrence of $\frac{p}{q}$ in e_1 (or e_2), we obtain an equivalent constraint $q \cdot e_1 \geq q \cdot e_2$. Then, we propagate the multiplication of q to all coefficients in e_1 and e_2, thus removing the occurrences of q in the denominator of any rational coefficient involving it. We repeat this simplification process until no rational coefficient is present. In fact, each 'rational' coefficient is internally handled as a pair of integers. When a polynomial interpretation involving such coefficients is displayed, MU-TERM uses the notation p/q. We provide two ways to implement the connection with C*i*ME:

[5] See `http://www.lri.fr/ marche/wst2004-competition/format.html`

1. *Automatic* connection. This only works with systems where C*i*ME is directly available (Windows, MacOS X, ...). The constraints generated by MU-TERM are sent to C*i*ME; the answer is automatically processed by MU-TERM to show the polynomial interpretation (with *rational*, nonnegative coefficients) which proves the μ-termination of $\mathcal{R}$. For instance, the μ-termination of $\mathcal{R}$ in Example 1 can be automatically proved in this way (see Figure 2).

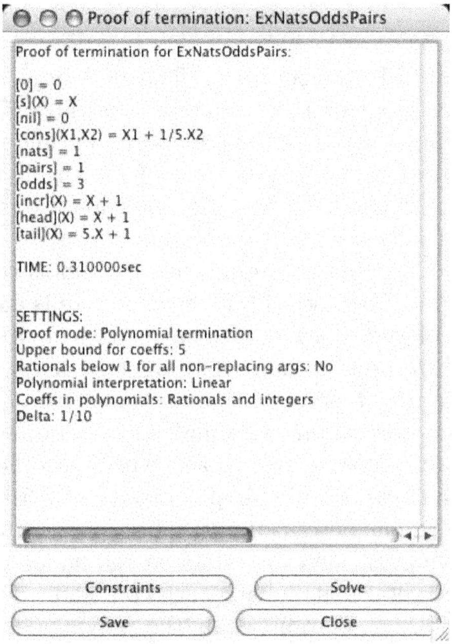

Fig. 2. Polynomial proof of μ-termination of $\mathcal{R}$ in Example 1

2. *User-assisted* connection. If C*i*ME is not available on the computer that runs MU-TERM, it is still possible to remotely use C*i*ME. For this purpose, the tool can provide a C*i*ME version of the appropriate constraint. The user would, then, deal with the necessary communication with C*i*ME to achieve any possible proof of termination. Again, as far as we know, other available tools for proving termination of rewriting (e.g., AProVE, Termptation, TTT, ...) do not provide any means to direcly solve constraints generated by an external system like MU-TERM.

If the *modular proofs* are activated, then MU-TERM computes a maximal *safe* decomposition of the current system (according to the results in [6], which concern disjoint and constructor sharing systems) and separately proves termination of each module by computing the appropriate polynomial interpretations.

Panel Transformations. Here we can apply different transformations for proving termination of *CSR*. Each of them transforms the *current* TRS; the current

TRS, however, remains unchanged (the transformed system is added to the list of TRSs). This permits to easily apply many transformations to the same current TRS. Available transformations correspond to [5,9,10,16] for proving termination of *CSR*. As an example, the μ-termination of the following TRS `Ex1_Zan97` [16, Example 1]:

```
g(x) → h(x)        c → d
h(d) → g(c)
```

(with $\mu(\mathtt{g}) = \mu(\mathtt{h}) = \varnothing$) cannot be directly proved by using a linear, simple, simple-mixed, or quadratic polynomial interpretation with nonnegative coefficients. However, it is possible to prove the μ-termination of $\mathcal{R}$ by using Giesl and Middeldorp's transformation in [9] to obtain a TRS `Ex1_Zan97_GM`

```
a__g(X)       → a__h(X)      mark(h(X)) → a__h(X)     a__g(X) → g(X)
a__c          → d            mark(c)    → a__c        a__h(X) → h(X)
a__h(d)       → a__g(c)      mark(d)    → d           a__c    → c
mark(g(X))    → a__g(X)
```

which can be proved terminating with MU-TERM also (see below).

Again, if the *modular proofs* are activated, then MU-TERM uses the computed maximal decomposition of the current system to individually apply the transformations. Then, if the transformed system becomes the current TRS and the user tries a proof of termination, the proof will be individually attempted for each transformed component. Note that this does not mean that $\mathcal{R}^{\mu}_{\Theta}$ itself admits a safe modular decomposition (normally, this is not the case, see [6] for a deeper discussion) but only that we take indirect benefit from the possibility of decomposing the original system $\mathcal{R}$.

Regarding termination of *innermost CSR* [11], we implement the correct and complete transformation of [8]. Some of the aforementioned transformations are also correct for proving termination of innermost *CSR* (see [8,11]), i.e., if the innermost termination of the transformed system can be proved, this implies the innermost termination of *CSR* for the original one.

We also include a transformation between pairs of TRSs and replacement maps which permits to prove termination of *lazy rewriting* as termination of *CSR* (see [13]). This transformation, however, does not benefit from any modular decomposition, since (as far as the author knows) modularity of termination of lazy rewriting has not been investigated yet.

Panel Termination of rewriting. Term rewriting is a particular case of *CSR* where the (top) replacement map $\mu_\top(f) = \{1, \ldots, ar(f)\}$, for all $f \in \mathcal{F}$ is used. Thus, the polynomial $\mu_\top$-reduction orderings generated by MU-TERM as explained above can also be used to prove *termination* of rewriting. We can use this fact, for instance, to prove termination of the TRSs which are obtained from the aforementioned transformations. For instance, `Ex1_Zan97_GM` above can be proved polynomially terminating with MU-TERM (see Figure 3).

On the other hand, in [1,2], Arts and Giesl discuss the use of weakly monotonic and non-monotonic orderings for proving termination of TRSs in combination with the dependency pairs approach. We have implemented the use of

Proof of termination: Ex1_Zan97_GM

Proof of termination for Ex1_Zan97_GM:

[g](X) = 2.X
[h](X) = 2.X
[c] = 0
[d] = 1
[a__g](X) = 2.X + 2
[a__h](X) = 2.X + 1
[a__c] = 2
[mark](X) = X + 3

TIME: 0.010000sec

SETTINGS:
Proof mode: Polynomial termination
Upper bound for coeffs: 5
Rationals below 1 for all non-replacing args: No
Polynomial interpretation: Linear
Coeffs in polynomials: Rationals and integers
Delta: 1/10

Constraints   Solve

Save   Close

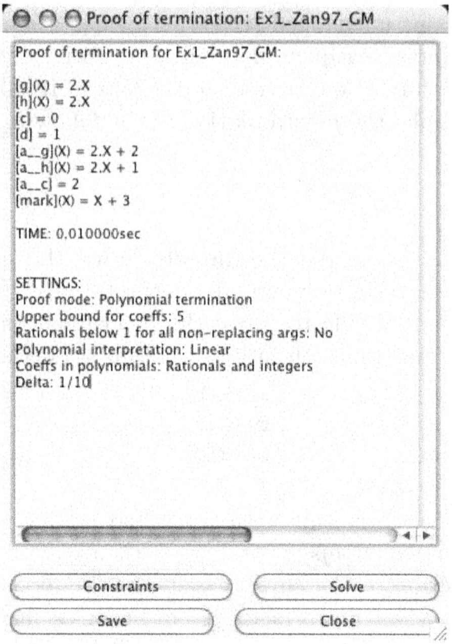

Fig. 3. Polynomial termination of rewriting with MU-TERM

polynomial interpretations over the rationals to generate weak and μ-monotonic orderings (where monotonicity is required only for the arguments of f belonging to $\mu(f)$, see [16]) which are suitable to be used together with the dependency pairs approach for proving termination of rewriting. For instance, in Figure 4 we show a proof of termination of the TRS ExNatsOddsPairs_GM obtained from the CSRS $\mathcal{R}$ (labelled ExNatsOddsPairs when loaded into MU-TERM) of Example 1. Note the new symbols with prefix nF_ which correspond to the 'tuple' symbols introduced by the computation of the dependency pairs (see [1]).

Once again, if the *modular proofs* are activated, then MU-TERM uses the computed maximal decomposition of the current system to perform the corresponding (polynomial or dependency-pairs based) proofs. This is because the modularity results for *CSR* [6] boils down to valid (and known) results for rewriting when the top replacement map μ_T is considered.

Additionally, we also provide a connection to direcly use the termination expert of CiME (rather than its constraint solver) in its different modalities.

Panel Replacement map. The controls in this panel permit to initialize / display / modify the replacement map associated to the current TRS. We provide a number of different ways to initialize a replacement map: the top replacement map μ_T and the *least* replacement map $\mu_\perp$ (with $\mu_\perp(f) = \varnothing$ for every symbol f of the signature) are the simplest ones. MU-TERM also permits to specify the *canonical* replacement map $\mu_{\mathcal{R}}^{can}$ of $\mathcal{R}$ which is *the most restrictive replacement map ensuring that the non-variable subterms of the left-hand sides of the rules*

The TRS ExNatsOddsPairs_GM	Proof of termination with MU-TERM

```
a__nats → cons(0,incr(nats))
a__pairs → cons(0,incr(odds))
a__odds → a__incr(a__pairs)
a__incr(cons(X,XS))
  → cons(s(mark(X)),incr(XS))
a__head(cons(X,XS)) → mark(X)
a__tail(cons(X,XS)) → mark(XS)
mark(nats) → a__nats
mark(pairs) → a__pairs
mark(odds) → a__odds
mark(incr(X)) → a__incr(mark(X))
mark(head(X)) → a__head(mark(X))
mark(tail(X)) → a__tail(mark(X))
mark(0) → 0
mark(s(X)) → s(mark(X))
mark(nil) → nil
mark(cons(X1,X2))
  → cons(mark(X1),X2)
a__nats → nats
a__pairs → pairs
a__odds → odds
a__incr(X) → incr(X)
a__head(X) → head(X)
a__tail(X) → tail(X)
```

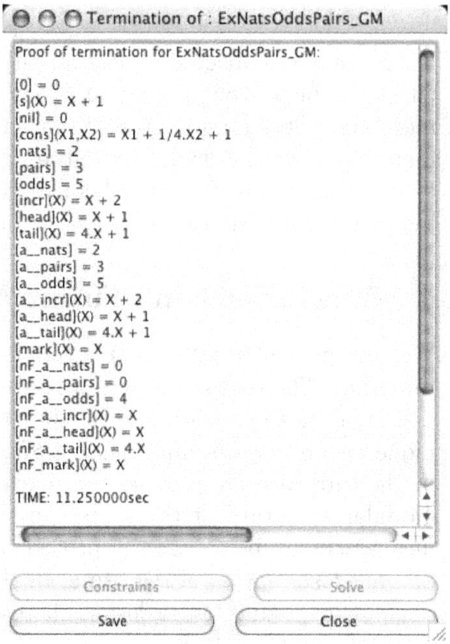
Termination of : ExNatsOddsPairs_GM
Proof of termination for ExNatsOddsPairs_GM:
[0] = 0
[s](X) = X + 1
[nil] = 0
[cons](X1,X2) = X1 + 1/4.X2 + 1
[nats] = 2
[pairs] = 3
[odds] = 5
[incr](X) = X + 2
[head](X) = X + 1
[tail](X) = 4.X + 1
[a__nats] = 2
[a__pairs] = 3
[a__odds] = 5
[a__incr](X) = X + 2
[a__head](X) = X + 1
[a__tail](X) = 4.X + 1
[mark](X) = X
[nF_a__nats] = 0
[nF_a__pairs] = 0
[nF_a__odds] = 4
[nF_a__incr](X) = X
[nF_a__head](X) = X
[nF_a__tail](X) = 4.X
[nF_mark](X) = X
TIME: 11.250000sec
Constraints
Solve
Save
Close

Fig. 4. Termination with dependency pairs and polynomials over the rationals

of $\mathcal{R}$ are replacing [12]. The class of replacement map which is currently being used is indicated with a colour code: red for the least replacement map, blue for the top one, green if the replacement map is greather than or equal to the canonical replacement map, and gray in any other case.

Once a replacement map has been initialized (or was already given, for instance, by loading an OBJ/Maude module), it is possible to modify it by double clicking on the corresponding symbol and appropriately setting the arguments as replacing or non-replacing.

3 Use of MU-TERM

MU-TERM is written in Haskell[6], and wxHaskell[7] has been used to develop the graphical user interface. The system consists of around 30 Haskell modules containing more than 4000 lines of code. Compiled versions and instructions for the installation are available on the MU-TERM WWW site.

[6] See http://haskell.org/

[7] See http://wxhaskell.sourceforge.net

We have used MU-TERM for developing the experiments reported in the following WWW page:

`http://www.dsic.upv.es/~slucas/csr/termination/examples`

where we have collected almost all published examples of non-terminating TRSs which can be proved μ-terminating for concrete replacement maps μ. We consider more than 40 examples (out from around 20 different references) which were managed with MU-TERM for either attempting a direct proof of termination or eventually transforming them to try a proof with either MU-TERM or other external termination tool.

4 Conclusions and Future Work

We have presented MU-TERM, a tool for proving termination of context-sensitive rewriting. The tool is written in Haskell and has a graphical user interface.

As far as the author knows, MU-TERM is the first implementation of a technique for proving termination of *CSR*; moreover, MU-TERM implements the first method for directly proving termination of *CSR* and also can benefit from the modular structure of the system by automatically decomposing it into modules which can be separately proved terminating. The tool can also be used for proving *termination* of rewriting. In this sense, MU-TERM is the first tool which implements reduction orderings based on polynomial interpretations over the rational numbers. It is also the first tool which makes use of such polynomials for the automatic generation of reduction pairs which can be used to prove termination of rewriting by using the dependency pairs approach.

Future extensions of the tool will address the problem of efficiently using negative coefficients in polynomial interpretations (see [14, Section 5.3] for further motivations and discussion). We also plan to investigate new families of real functions which are well-suited for automatization purposes. We will focus on those functions which can also be used to prove termination of *CSR*, by introducing mechanisms for loosing monotonicity in some arguments. In particular, polynomial fractions were proposed in [14] as suitable candidates for this purpose. In all these cases, however, more research is necessary before being able to provide reasonable techniques for dealing with these kind of functions. Another promising line of work is the implementation of the CS-RPO.

We plan to improve the generation of reports and the inclusion of new, richer formats for input systems (e.g., Conditional TRSs, Many sorted TRSs, TRSs with AC symbols, etc.). We also plan to directly connect MU-TERM with AProVE.

Acknowledgements

I thank Albert Rubio and Cristina Borralleras for providing most of the proofs (and disproofs) of termination of *CSR* with CS-RPO summarized in the aforementioned WWW site. I also thank Claude Marché and Xavier Urbain for their technical support regarding C*i*ME. Finally, I thank the anonymous referees for their comments.

References

1. T. Arts and J. Giesl. Termination of Term Rewriting Using Dependency Pairs *Theoretical Computer Science*, 236:133-178, 2000.
2. T. Arts and J. Giesl. A collection of examples for termination of term rewriting using dependency pairs. Technical report, AIB-2001-09, RWTH Aachen, Germany, 2001.
3. C. Borralleras, S. Lucas, and A. Rubio. Recursive Path Orderings can be Context-Sensitive. In *Proc. of CADE'02*, LNAI 2392:314-331, Springer-Verlag, Berlin, 2002.
4. C. Borralleras. Ordering-based methods for proving termination automatically. PhD Thesis, Departament de Llenguatges i Sistemes Informàtics, Universitat Politècnica de Catalunya, May 2003.
5. M.C.F. Ferreira and A.L. Ribeiro. Context-Sensitive AC-Rewriting. In *Proc. of RTA'99*, LNCS 1631:286-300, Springer-Verlag, Berlin, 1999.
6. B. Gramlich and S. Lucas. Modular termination of context-sensitive rewriting. In *Proc. of PPDP'02*, pages 50-61, ACM Press, New York, 2002.
7. B. Gramlich and S. Lucas. Simple termination of context-sensitive rewriting. In *Proc. of RULE'02*, pages 29-41, ACM Press, New York, 2002.
8. J. Giesl and A. Middeldorp. Termination of Innermost Context-Sensitive Rewriting. In *Proc. DLT'02*, LNCS 2450:231-244, Springer-Verlag, Berlin, 2003.
9. J. Giesl and A. Middeldorp. Transformation Techniques for Context-Sensitive Rewrite Systems. *Journal of Functional Programming*, to appear, 2004.
10. S. Lucas. Termination of context-sensitive rewriting by rewriting. In *Proc. of ICALP'96*, LNCS 1099:122-133, Springer-Verlag, Berlin, 1996.
11. S. Lucas. Termination of Rewriting With Strategy Annotations. In *Proc. of LPAR'01*, LNAI 2250:669-684, Springer-Verlag, Berlin, 2001.
12. S. Lucas. Context-sensitive rewriting strategies. *Information and Computation*, 178(1):293-343, 2002.
13. S. Lucas. Lazy Rewriting and Context-Sensitive Rewriting. Electronic Notes in Theoretical Computer Science, volume 64. Elsevier Sciences, 2002.
14. S. Lucas. Polynomials for proving termination of context-sensitive rewriting. In *Proc. of FOSSACS'04*, LNCS 2987:318-332, Springer-Verlag, Berlin 2004.
15. J. Steinbach. Generating Polynomial Orderings. *Information Processing Letters* 49:85-93, 1994.
16. H. Zantema. Termination of Context-Sensitive Rewriting. In *Proc. of RTA'97*, LNCS 1232:172-186, Springer-Verlag, Berlin, 1997.

Automated Termination Proofs with AProVE

Jürgen Giesl, René Thiemann, Peter Schneider-Kamp, and Stephan Falke

LuFG Informatik II, RWTH Aachen, Ahornstr. 55, 52074 Aachen, Germany
{giesl,thiemann,psk}@informatik.rwth-aachen.de
spf@i2.informatik.rwth-aachen.de

Abstract. We describe the system ProVE, an automated prover to verify (innermost) termination of term rewrite systems (TRSs). For this system, we have developed and implemented efficient algorithms based on classical simplification orders, dependency pairs, and the size-change principle. In particular, it contains many new improvements of the dependency pair approach that make automated termination proving more powerful and efficient. In ProVE, termination proofs can be performed with a user-friendly graphical interface and the system is currently among the most powerful termination provers available.

1 Introduction

The system AProVE (Automated Program Verification Environment) offers a variety of techniques for automated termination proofs of TRSs: First, it provides efficient implementations of classical simplification orders to prove termination "directly" (*recursive path orders* possibly with status [6, 19], *Knuth-Bendix orders* [20], and *polynomial orders* [22]), cf. Sect. 2. To increase the power of automated termination proofs, we implemented the *dependency pair* technique [2, 13] in AProVE which allows the application of classical orders to examples where automated termination analysis would fail otherwise (Sect. 3). In contrast to most other implementations, we integrated numerous refinements such as *narrowing*, *rewriting*, and *instantiation* of dependency pairs [2, 12, 14, 15], recent improvements to reduce the constraints generated by the dependency pair technique [14, 15, 28], etc. Therefore, AProVE succeeds on many examples where all other automated termination provers fail. Thus, the principles used in AProVE's implementation may also be very helpful for other tools based on dependency pairs (Arts [1], CiME [5], TTT [18]) or on other related approaches for termination of TRSs (Termptation [4], Cariboo [10]). Apart from direct termination proofs and dependency pairs, as a third termination technique, AProVE offers the *size-change principle* [23] and it is also possible to combine this principle with dependency pairs [27] (Sect. 4). The tool is written in Java and proofs can be performed both in a fully automated or in an interactive mode via a graphical user interface. The modular design of AProVE's implementation is explained in Sect. 5. In Sect. 6 we show how to run the system and compare AProVE with related tools.

V. van Oostrom (Ed.): RTA 2004, LNCS 3091, pp. 210–220, 2004.

2 Direct Termination Proofs

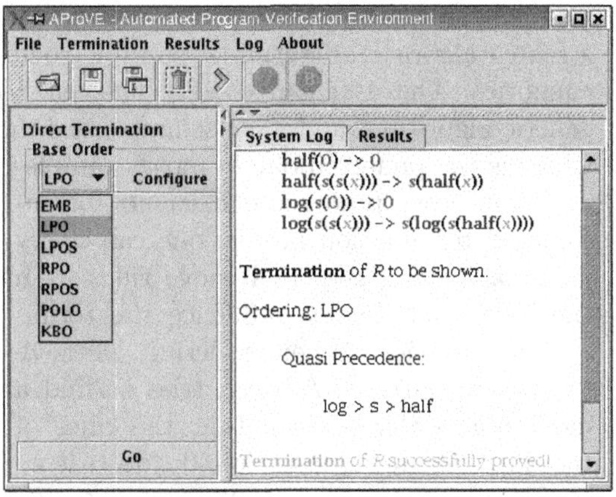

This section describes the base orders of AProVE which can be used for direct termination proofs, but also for proofs with constraint generation techniques like dependency pairs or the size-change principle.

In direct termination proofs, the system tries to find a reduction order where all rules are decreasing. The following *path orders* are available: the embedding order (EMB), the lexicographic path order (LPO, [19]), the LPO with status which compares subterms lexicographically w.r.t. arbitrary permutations (LPOS), the recursive path order comparing subterms as multisets (RPO, [6]), and the RPO with status which combines LPOS and RPO (RPOS).

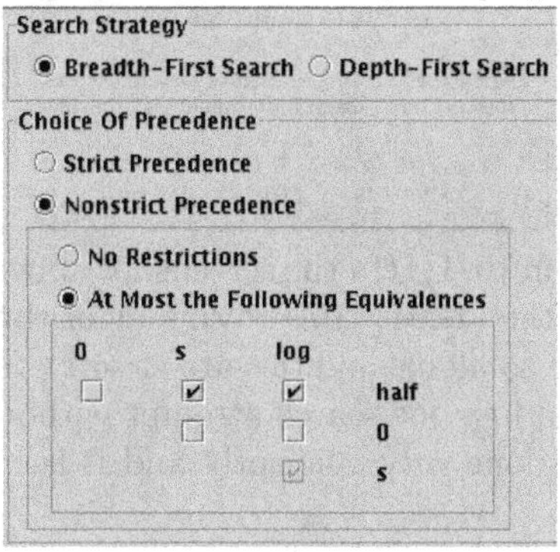

Path orders may be parameterized by a precedence on function symbols and a status which determines how arguments of function symbols are compared. To explore the search space for these parameters, the system leaves them as unspecified (or "*minimal*") as possible. The user can decide between depth-first or breadth-first search and one can configure path orders by deciding whether different function symbols may be equivalent w.r.t. the precedence ("Nonstrict Precedence"). It is also possible to restrict potential equivalences to certain pairs of function symbols.

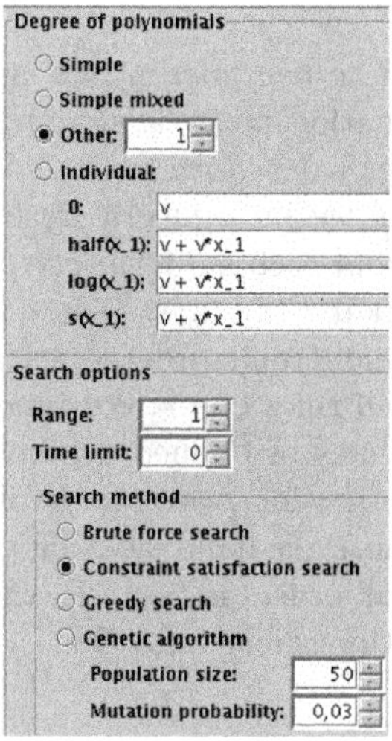

AProVE also offers *Knuth-Bendix orders* (KBO, [20]) using the polynomial-time algorithm of [21] and the technique of [9] to compute the degenerate subsystem of homogeneous linear inequalities.

The last class of orders in AProVE are *polynomial orders* (POLO, [22]) where every function symbol is associated with a polynomial with natural coefficients. The user can specify the degree of the polynomials and the range of the coefficients. One can also provide individual polynomials for some function symbols manually. To prove termination, AProVE generates a set of polynomial inequalities stating that left-hand sides of rules should be greater than the corresponding right-hand sides. By the method of partial derivation [11, 22], these inequalities are transformed into inequalities only containing coefficients, but no variables anymore. Finally, a search algorithm determines suitable coefficients that satisfy the resulting inequalities. The user can choose between brute force search, greedy search, a genetic algorithm, and a constraint-based method based on interval arithmetic, which is the preferred one in most examples.

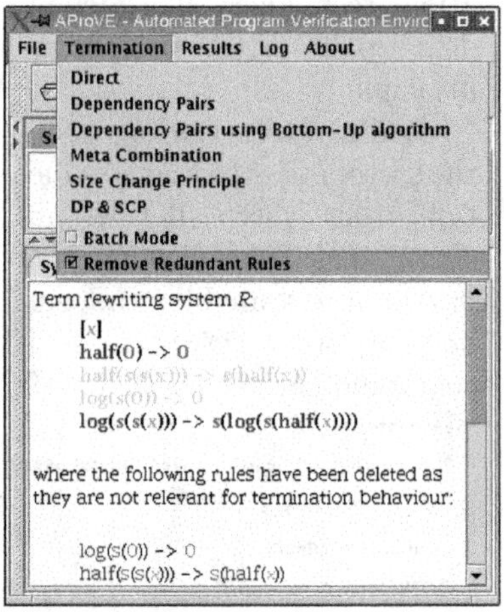

To improve power and efficiency of automated termination proofs, one can apply a pre-processing step to remove rules from the TRS that do not influence the termination behavior. When selecting "`Remove Redundant Rules`", AProVE tries to find a monotonic order $\succ$ such that the rules of the TRS $\mathcal{R}$ are at least weakly decreasing (i.e., at least $l \succsim r$ for all $l \to r \in \mathcal{R}$). Then rules which are strictly decreasing are removed, i.e., it suffices to prove termination of $\mathcal{R} \setminus \{l \to r \mid l \succ r\}$. This extends existing related approaches to remove rules [16, 22, 30].

For this pre-processing, we use linear polynomial interpretations with coefficients from $\{0, 1\}$. (In the screenshot above, we mapped $\mathsf{s}(x)$ to $x + 1$ and half and log to the identity.) AProVE's algorithm for polynomial orders solves constraints where some inequalities are strictly decreasing and all others are weakly decreasing in just one search attempt without backtracking [15]. So removal of rules can be done very efficiently and it is repeated until no rule can be removed anymore.

3 Termination Proofs with Dependency Pairs

The *dependency pair* approach [2, 13] increases the power of automated termination analysis significantly. The root symbols of left-hand sides of rules are called *defined* and all other symbols are *constructors*. For each defined symbol f we introduce a fresh *tuple symbol* F. Then for each rule $f(s_1, ..., s_n) \to r$ and each subterm $g(t_1, ..., t_m)$ of r with defined root g, we build a dependency pair $F(s_1, ..., s_n) \to G(t_1, ..., t_m)$. To prove termination one has to find a weakly monotonic order $\succ$ such that $s \succ t$ for all dependency pairs $s \to t$ and $l \succsim r$ for all rules $l \to r$. For innermost termination, $l \succsim r$ is only required for the usable rules of defined symbols in right-hand sides of dependency pairs. The *usable rules* for f are the f-rules together with the usable rules for all defined symbols in right-hand sides of f-rules. Moreover, if $\succsim$ is $\mathcal{C}_\varepsilon$-compatible (which holds for all orders in Sect. 2), then even for termination one only has to require $l \succsim r$ for the usable rules [28].

General Options and Base Order

In AProVE, one can select whether to use dependency pairs for termination or for innermost termination proofs. The system also checks if a TRS is non-overlapping (then innermost termination implies termination). AProVE contains recent improvements which combine different modularity criteria and reduce the set of usable rules [14, 28].[1] They can be switched off for experimental purposes. To search for orders $\succ$, one can select any base order from Sect. 2.

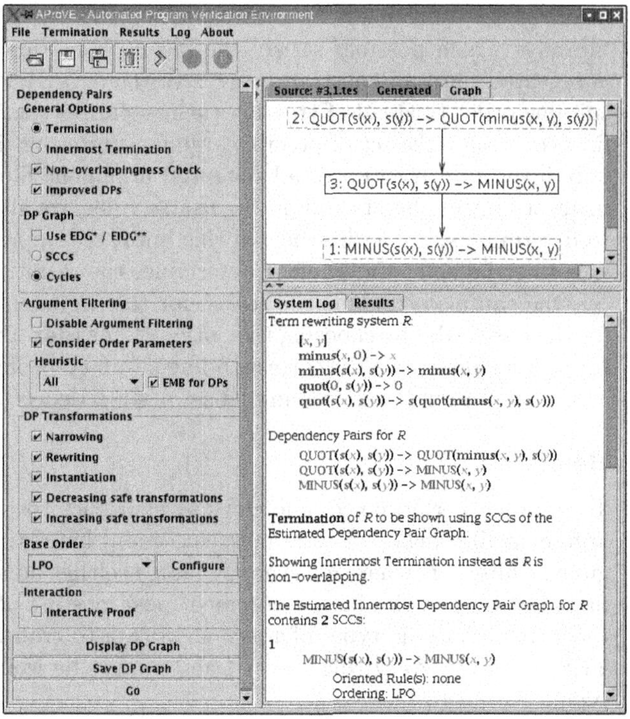

Argument Filter

However, most of these orders are *strongly* monotonic, while the dependency pair approach only requires *weak* monotonicity. (For polynomial orders, a weakly monotonic variant is obtained by permitting the coefficient 0. But `LPO(S)`, `RPO(S)`, and `KBO` are always strongly monotonic.) Thus, before searching for an order, some of the arguments of the function symbols in the constraints can be eliminated by an *argument filtering* π [2]. For example, a binary function symbol f can be turned into a unary symbol by eliminating f's first argument. Then π replaces all terms $f(t_1, t_2)$ in the constraints by $f(t_2)$. Hence, we can obtain a weakly monotonic order $\succ_\pi$ from a strongly monotonic order $\succ$ and an argument filtering π by defining $s \succ_\pi t$ iff $\pi(s) \succ \pi(t)$. Moreover, we developed an improvement by first applying the argument filtering and determining the usable rules afterwards [14, 28]. The advantage is that the argument filtering may eliminate some symbols f from the right-hand sides of dependency pairs and rules. Then, one does not have to require $l \succsim_\pi r$ for the f-rules anymore. For this improvement, one has to select "`Improved DPs`" in the `General Options`.

As there are exponentially many argument filterings, this search space must be explored efficiently. AProVE uses a depth-first algorithm [14] which treats the

[1] Currently, the results of [28] are only available in AProVE 1.1-beta, which does not yet contain all options of AProVE 1.0. AProVE 1.1 will combine their features.

constraints one after another. We start with the set of argument filterings possibly satisfying the first constraint. Here we use the idea of [17] to keep argument filterings as "undefined" as possible. Then this set is reduced further to those filterings which possibly satisfy the second constraint as well. This procedure is repeated until all constraints are investigated. By inspecting constraints in a suitable order (instead of treating them separately as in [17]), already after the first constraint the set of possible argument filterings is rather small. Thus, one only inspects a fraction of all potential argument filterings. To use our refinement of filtering before computing usable rules, we also developed an algorithm to determine suitable filterings in this improved approach automatically, which is non-trivial since the filtering determines the resulting constraints.

One can also combine the search for the argument filtering with the search for the base order by choosing the option "`Consider Order Parameters`". Then the system also stores the corresponding parameters of the order for each possible argument filtering (e.g., a minimal set of precedences and stati, cf. Sect. 2).

Heuristics

To improve performance, one can use heuristics to restrict the set of possible argument filterings [14]. The most successful heuristic "`Type`" only regards argument filterings where for every symbol f, either no argument position is eliminated or all non-eliminated argument positions are of the same type. Here, we use a monomorphic type inference algorithm to transform a TRS into a sorted TRS (where in every rule $l \to r$, l and r must be well-typed terms of the same type).

When selecting the heuristic "`EMB for DPs`", only the very simple embedding order is used for orienting constraints like $s \succ_\pi t$ which come from dependency pairs $s \to t$. Only for constraints $l \succsim_\pi r$ from rules $l \to r$, one may apply more complicated orders like `LPO`, `RPO(S)`, etc. Since our depth-first algorithm to determine argument filterings starts with the dependency pairs, this reduces the search space significantly without compromising power very much.

This depth-first algorithm uses a top-down approach where constraints from f-rules are considered before g-rules, if f calls g. As an alternative heuristic, we also offer a "`Bottom-Up algorithm`" which starts with determining an argument filtering for constructors and then moves upwards through the recursion hierarchy where g is treated before f, if f calls g. To obtain an efficient technique, here the system only determines one single argument filtering at each choice point, even if several ones are possible and it does not perform any backtracking. This algorithm reduces the search space enormously, but is also restricts the power, since the proof can fail if one selects the "wrong" argument filtering at some point. Thus, this heuristic is suitable as a fast pre-processing step and if it fails, one should still apply the full dependency pair approach afterwards, cf. Sect. 5.

DP Graph

For TRSs with many rules, (innermost) termination proofs should be performed in a modular way. To this end, one constructs an estimated (innermost) dependency graph and regards its cycles separately [2, 13]. One can select between standard [2] and more powerful recent estimations (`EDG*` / `EIDG**`) [15, 17].

For each cycle, only one dependency pair must be strictly decreasing and the others just have to be weakly decreasing. As shown in [17], one should not compute all cycles, but only maximal cycles (*strongly connected components (SCCs)*). The reason is that the chosen argument filtering and base order may make several dependency pairs in an SCC strictly decreasing. In that case, subcycles of the SCC containing such a strictly decreasing dependency pair do not have to be considered anymore. So after solving the constraints for the initial SCCs, all strictly decreasing dependency pairs are removed and one now builds SCCs from the remaining dependency pairs, etc. This algorithm is chosen by selecting "`Cycles`". The algorithm "`SCCs`" requires a strict decrease for all dependency pairs in an SCC and is only intended for experimental purposes.

In order to benefit from all existing refinements on modularity of dependency pairs, we developed and implemented an improved technique which permits the combination of recent results on modularity of $\mathcal{C}_\varepsilon$-terminating TRSs [29] with arbitrary estimations of dependency graphs, cf. [14, 28]. This improvement is only available if one selects "`Improved DPs`" in the `General Options`.

DP Transformations

To increase power, a dependency pair can be transformed into several new pairs by *narrowing*, *rewriting*, and *instantiation* [2, 12, 14, 15]. In contrast to [12, 14], AProVE can instantiate dependency pairs both w.r.t. the pairs before and behind it in chains (the latter is called *forward* instantiation) [15]. The user can select which of these transformations should be applicable. Usually, all transformations should be enabled, since they are often crucial for the success of the proof and they can never "harm": if the termination proof succeeds without transformations, then it also succeeds when performing transformations [15], but not vice versa. However, the problem is when to use these transformations, since they may be applicable infinitely often. Moreover, transformations may increase runtime by producing a large number of similar constraints. AProVE performs transformations in "safe" cases where their application is guaranteed to terminate [14]. We distinguish between *increasing* and *decreasing* safe transformations. Decreasing transformations delete dependency pairs or SCCs and therefore, they do not have a negative impact on the efficiency. The user can disable both kinds of safe transformations. If turned on, decreasing transformations are applied before trying

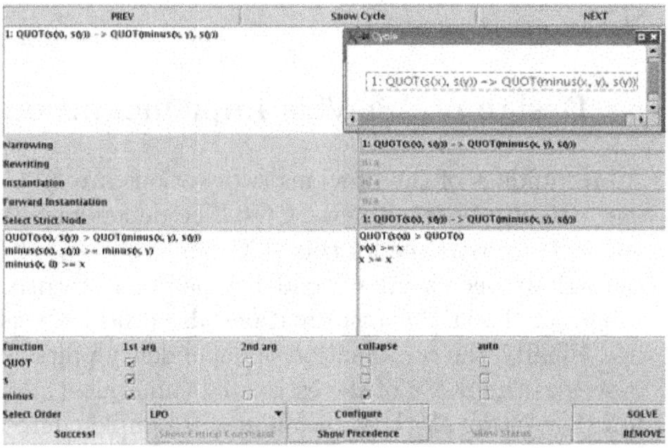

to solve the constraints for an SCC. Increasing transformations are only used a limited number of times when a proof attempt fails, and then the proof is re-attempted again.

Interaction
In addition to the fully automated mode, (innermost) termination proofs with dependency pairs can be performed interactively. Here, the user can specify which transformation steps should be performed and for any cycle or SCC, one can determine (parts of) the argument filtering, the base order, and the dependency pair which should be strictly decreasing. The constraints resulting from such selections are immediately displayed, such that interactive proofs are supported in a very comfortable way. This mode is intended for the development of new heuristics and for machine-assisted proofs of particularly challenging examples.

4 Termination Proofs with the Size-Change Principle

A new *size-change principle* for termination of functional programs was presented in [23] and extended to TRSs in [27]. A similar principle is also known for logic programs [8]. AProVE offers the technique of [27, Thm. 11] for size-change termination of TRSs using the embedding order as underlying base order.[2]

AProVE also contains the combination of the size-change principle with dependency pairs from [27], which often succeeds with much simpler argument filterings and base orders than the pure dependency pair approach. Again, each SCC of the estimated (innermost) dependency graph is treated separately. In case of failure for some SCC, the dependency pairs are transformed by narrowing, rewriting, or instantiation and the proof attempt is re-started. If the user has selected the "`hybrid`" algorithm, then the pure dependency pair method is tried as soon as the limits for the transformations are reached. Thus, then the combined method is used as a fast technique which is checked first for every SCC and only if it fails, one uses the ordinary dependency pair approach on this SCC.

5 Design of AProVE's Implementation

All techniques of the previous two sections are *SCC-processors* which transform one SCC into a set of new SCCs: The dependency pair approach takes an SCC and if the constraints for this SCC can be solved using some base order, it deletes the strictly decreasing dependency pairs and returns the SCCs of the remaining subgraph. The DP transformations also produce a set of new SCCs out of a given one. Finally, the combination of dependency pairs with the size-change principle processes the SCCs of the estimated (innermost) dependency graph one by one, too. Therefore, all these techniques are implemented as modules which take one

[2] As shown in [27], only very restricted base orders are sound in connection with the size-change principle. In addition to the results in [27], the full embedding order may be used, where $f(\dots, x_i, \dots) \succ x_i$ also holds for defined function symbols f.

SCC as input and return a set of SCCs. So AProVE uses the following main algorithm, where one may choose different SCC-processors in Step 4 (b).

1. `Remove redundant rules of the TRS which do not influence termination.`
2. `Check whether the TRS is non-overlapping. Then it is sufficient to prove innermost termination instead of termination.`
3. `Compute initial SCCs of the estimated (innermost) dependency graph.`
4. `While there are SCCs left and there is no failure:`
   (a) `Remove one SCC` $\mathcal{P}$ `from the set of SCCs.`
   (b) `Choose an SCC-processor.`
   (c) `Transform` $\mathcal{P}$ `with the chosen SCC-processor.`
   (d) `Add the resulting new set of SCCs to the remaining SCCs.`

Due to this modular structure, procedures which combine different termination techniques can easily be implemented in AProVE. One just has to configure which SCC-processors should be taken in Step 4 (b). It is advantageous if one first tries to use fast SCC-processors which benefit from successful heuristics. In this way, SCCs that are easy to handle can be treated efficiently. Only for SCCs where these fast SCC-processors fail, one should use slower but more powerful SCC-processors afterwards. Examples for such termination procedures offered in AProVE are the `hybrid` algorithm described in Sect. 4 or the following "`Meta Combination`" algorithm. This algorithm is particularly useful if one does not want to get involved with the details of termination proving, but wants to use AProVE in a "black box"-mode. In Step 4 (b), it always takes the first processor from the following list that is applicable (i.e., that can transform the SCC $\mathcal{P}$ into a new set of SCCs different from $\mathcal{P}$). Here, we use linear polynomial interpretations with coefficients from $\{0,1\}$ and LPOs with "`Nonstrict Precedence`".

- `Decreasing safe transformations`
- "`DPs using Bottom-Up algorithm`" `with POLO and LPO as base orders`
- `Dependency pairs with the heuristic` "`EMB for DPs`" `and LPO`
- `Full dependency pair approach with POLO as base order`
- `Increasing safe transformations`

6 Running AProVE and Comparison with Other Tools

AProVE accepts four input languages: logic and (first-order) functional programs, conditional and unconditional TRSs. Functional and logic programs are translated into conditional TRSs and conditional TRSs are transformed further into unconditional TRSs [12, 24]. For logic programs, these transformations correspond to the approach of the termination prover TALP [25].

The results of the termination proof are displayed in `html`-format and can be stored in `html`- or LaTeX-format. Moreover, a "`System Log`" describes all (possibly failed) proof attempts. Any termination proof attempt may be interrupted by a stop-button. Instead of running the system on only one TRS or program, one can also run it on collections and directories of examples in a "`Batch Mode`".

Compared with other recent automated termination provers for TRSs (Arts [1], Cariboo [10], CiME [5], Termptation [4], TTT [18]), AProVE is the only system incorporating improvements like automated dependency pair transformations, applying argument filterings before usable rules, and combining modularity results based on $\mathcal{C}_\varepsilon$-termination with recent dependency graph estimations. Moreover, it offers more base orders than any other system, it can also handle conditional TRSs, and integrates the size-change principle. Finally, AProVE's design permits the combination of powerful heuristics and different termination techniques as in the "Meta Combination" algorithm of Sect. 5. In addition, the system has a graphical user interface and a comfortable interactive component.

The next version of AProVE will also feature AC-rewriting and we try to improve its performance on string rewrite systems and logic programs. Our future work is also concerned with extensions to handle imperative programs, higher-order functional programs, and context-sensitive rewriting. Moreover, we plan to add a component to detect programs and systems that are *not* terminating.

Due to the numerous improvements developed and integrated in AProVE, it succeeded on more examples than any other system at the exhibition/competition of automated termination provers at the International Termination Workshop 2003. These results are confirmed by the following experiments, where we give an empirical comparison of AProVE 1.0 (using the "Meta Combination" algorithm) with the only two other tools currently available on the web (CiME and Termptation). The tools were tested on the collections of [3, 7, 26] (130 TRSs for termination, 151 TRSs for innermost termination). To show that the techniques described in [18] are a substantial restriction, in the last row we ran AProVE in a mode where we switched off all improvements and only used the methods available in [18]. Since [18] contains several base orders and argument filtering heuristics, we took the ones which gave the best overall result on this collection.

	Termination		Innermost Term.	
System	Power	Time	Power	Time
AProVE	95.4 %	26.2 s	98.0 %	34.3 s
CiME	71.5 %	660.7 s	—	—
Termptation	65.4 %	521.8 s	72.8 %	681.7 s
AProVE with techniques of [18]	52.3 %	679.1 s	—	—

The "Power" column contains the percentage of those examples in the collection where the proof attempt was successful. The "Time" column gives the overall time for running the system on all examples of the collection (also on the ones where the proof attempt failed). For each example we used a time-out of 60 seconds on a Pentium IV with 2.4 GHz and 1 GB memory. For more details on the above experiments and to download AProVE, the reader is referred to `http://www-i2.informatik.rwth-aachen.de/AProVE`.

References

1. T. Arts. System description: The dependency pair method. In *Proc. 11th RTA*, LNCS 1833, pages 261–264, 2000.
2. T. Arts and J. Giesl. Termination of term rewriting using dependency pairs. *Theoretical Computer Science*, 236:133–178, 2000.
3. T. Arts and J. Giesl. A collection of examples for termination of term rewriting using dependency pairs. Technical Report AIB-2001-09[3], RWTH Aachen, 2001.
4. C. Borralleras, M. Ferreira, and A. Rubio. Complete monotonic semantic path orderings. In *Proc. 17th CADE*, LNAI 1831, pages 346–364, 2000.
5. E. Contejean, C. Marché, B. Monate, and X. Urbain. CiME. `http://cime.lri.fr`.
6. N. Dershowitz. Termination of rewriting. *J. Symb. Comp.*, 3:69–116, 1987.
7. N. Dershowitz. 33 examples of termination. In *Proc. French Spring School of Theoretical Computer Science*, LNCS 909, pages 16–26, 1995.
8. N. Dershowitz, N. Lindenstrauss, Y. Sagiv, and A. Serebrenik. A general framework for automatic termination analysis of logic programs. *Applicable Algebra in Engineering, Communication and Computing*, 12(1,2):117–156, 2001.
9. J. Dick, J. Kalmus, and U. Martin. Automating the Knuth-Bendix ordering. *Acta Informatica*, 28:95–119, 1990.
10. O. Fissore, I. Gnaedig, and H. Kirchner. Cariboo: An induction based proof tool for termination with strategies. In *Proc. 4th PPDP*, pages 62–73. ACM, 2002.
11. J. Giesl. Generating polynomial orderings for termination proofs. In *Proc. 6th RTA*, LNCS 914, pages 426–431, 1995.
12. J. Giesl and T. Arts. Verification of Erlang processes by dependency pairs. *Appl. Algebra in Engineering, Communication and Computing*, 12(1,2):39–72, 2001.
13. J. Giesl, T. Arts, and E. Ohlebusch. Modular termination proofs for rewriting using dependency pairs. *Journal of Symbolic Computation*, 34(1):21–58, 2002.
14. J. Giesl, R. Thiemann, P. Schneider-Kamp, and S. Falke. Improving dependency pairs. In *Proc. 10th LPAR*, LNAI 2850, pages 165–179, 2003.
15. J. Giesl, R. Thiemann, P. Schneider-Kamp, and S. Falke. Mechanizing dependency pairs. Technical Report AIB-2003-08[3], RWTH Aachen, Germany, 2003.
16. J. Giesl and H. Zantema. Liveness in rewriting. In *Proc. 14th RTA*, LNCS 2706, pages 321–336, 2003.
17. N. Hirokawa and A. Middeldorp. Automating the dependency pair method. In *Proc. 19th CADE*, LNAI 2741, 2003.
18. N. Hirokawa and A. Middeldorp. Tsukuba termination tool. In *Proc. 14th RTA*, LNCS 2706, pages 311–320, 2003.
19. S. Kamin and J. J. Lévy. Two generalizations of the recursive path ordering. Unpublished Manuscript, University of Illinois, IL, USA, 1980.
20. D. Knuth and P. Bendix. Simple word problems in universal algebras. In J. Leech, editor, *Comp. Problems in Abstract Algebra*, pages 263–297. Pergamon, 1970.
21. K. Korovin and A. Voronkov. Verifying orientability of rewrite rules using the Knuth-Bendix order. In *Proc. 10th RTA*, LNCS 2051, pages 137–153, 2001.
22. D. Lankford. On proving term rewriting systems are Noetherian. Technical Report MTP-3, Louisiana Technical University, Ruston, LA, USA, 1979.
23. C. S. Lee, N. D. Jones, and A. M. Ben-Amram. The size-change principle for program termination. In *Proc. POPL '01*, pages 81–92, 2001.
24. E. Ohlebusch. Termination of logic programs: Transformational approaches revisited. *Appl. Algebra in Engineering, Comm. and Comp.*, 12(1,2):73–116, 2001.

[3] Available from `http://aib.informatik.rwth-aachen.de`

25. E. Ohlebusch, C. Claves, and C. Marché. TALP: A tool for the termination analysis of logic programs. In *Proc. 11th RTA*, LNCS 1833, pages 270–273, 2000.
26. J. Steinbach. Automatic termination proofs with transformation orderings. In *Proc. 6th RTA*, LNCS 914, pages 11–25, 1995. Full version appeared as Technical Report SR-92-23, Universität Kaiserslautern, Germany.
27. R. Thiemann and J. Giesl. Size-change termination for term rewriting. In *Proc. 14th RTA*, LNCS 2706, pages 264–278, 2003.
28. R. Thiemann, J. Giesl, and P. Schneider-Kamp. Improved modular termination proofs using dependency pairs. In *Proc. 2nd IJCAR*, LNAI, 2004. To appear.
29. X. Urbain. Automated incremental termination proofs for hierarchically defined term rewriting systems. In *Proc. 1st IJCAR*, LNAI 2083, pages 485–498, 2001.
30. H. Zantema. TORPA: Termination of rewriting proved automatically. In *Proc. 15th RTA*, LNCS, 2004.

An Approximation Based Approach to Infinitary Lambda Calculi

Stefan Blom

CWI, P.O.-box 94.079, 1090 GB Amsterdam, The Netherlands
sccblom@cwi.nl

Abstract. We explore an alternative for metric limits in the context of infinitary lambda calculus with transfinite reduction sequences. We will show how to use the new approach to get calculi that correspond to the 111, 101 and 001 infinitary lambda calculi of Kennaway et al, which have been proved to correspond to Berarducci Trees, Levy-Longo Trees and Böhm Trees respectively. We will identify subsets of the sets of meaningless terms of the metric calculi and prove that the approximation based calculi are equivalent to their metric counterparts up to these subsets.

Keywords: lambda calculus, infinitary rewriting.

1 Introduction

There exist two approaches to infinitary rewriting: complete developments of infinite sets of redexes on infinite terms (see [Cor93]) and transfinite reduction sequences where one assigns a limit to an infinite reduction sequence and then starts again (see [KdV03] for an overview). We follow the latter approach.

Transfinite reduction sequences have been studied in both the context of term rewriting systems ([DK89,KKSdV93,KKSdV95b]) and in the context of the lambda calculus ([KKSdV95a,KKSdV97]). In this line of work, the basic means of assigning a result to an infinite sequence is to view the set of infinite terms as a metric space and use the metric limit. On top of this basic principle a lot of fine tuning can be done. The approach works well, but it is not very elegant in the sense that we cannot assign a result to *every* infinite reduction sequence. Therefore, we have explored an alternative means of assigning a result to an infinite sequence: the limes inferior (liminf). We have also explored how to fine tune the concept.

Important notions for this paper are the depth of a sub-term in a term and prefix of a term. The default definition of depth of a sub-term in a term is the number of symbols between the top symbol of the sub-term and the top of the term. A prefix of a term t is a finite term smaller than t in the $\leq_\Omega$ order induced by $\Omega \leq_\Omega s$, for any term s. Note that we use Ω as a constant rather than as a term. Also note that where we use Ω the constant $\bot$ is often used instead.

V. van Oostrom (Ed.): RTA 2004, LNCS 3091, pp. 221–232, 2004.

To illustrate some of the concepts in infinitary rewriting, we will use the classical ABC term rewriting example:

$$\begin{array}{ll} A(X) & \to X \\ B(X) & \to X \\ C & \to A(B(C)) \end{array}$$

Note that we can reproduce the behavior of this TRS in the lambda calculus by interpreting A, B and C as follows:

$$A(x) = (\lambda u.x)\mathbf{A} \quad B(x) = (\lambda u.x)\mathbf{B} \quad C = (\lambda u.A(B(u\,u)))\,(\lambda u.A(B(u\,u))) \ .$$

As a simple example, consider the infinite sequence

$$C \to A(B(C)) \to A(B(A(B(C)))) \to \cdots$$

The result of this sequence is $AB^\omega \equiv A(B(A(B(\cdots))))$. This may be denoted $C \to^\omega AB^\omega$. Likewise, the result of

$$C \to A(B(C)) \to A(C) \to A(A(B(C))) \to A(A(C)) \to \cdots$$

is $A^\omega \equiv A(A(\cdots))$. We also have $C \to^\omega B^\omega$. All of these results are metric limits.

The sequence

$$C \to A(B(C)) \to B(C) \to C \to A(B(C)) \to B(C) \to C \to \cdots \tag{1}$$

does not have a metric limit. To get its liminf, we look at the sets of prefixes of these terms:

$$\{\Omega, C\}, \{\Omega, A(\Omega), A(B(\Omega)), A(B(C))\}, \{\Omega, B(\Omega), B(C)\}, \{\Omega, C\}, \cdots$$

The only prefix which is present in all sets in a tail of the sequence is Ω. Thus, the liminf of this sequence is Ω. Likewise, the sequence

$$A(C) \to A(A(B(C))) \to A(A(C)) \to A(C) \to \cdots$$

has no metric limit, but its liminf is $A(\Omega)$.

To deal with sequences that do not have a limit, several metric calculi use the notion of 0-active term or root active term. A term is 0-active is there exists an infinite reduction which contracts a redex at the root of the term infinitely often. For example, in (1) every redex is contracted at the root so C is 0-active. These metric calculi add a rule that 0-active terms can be rewritten to Ω.

Adding such a rule creates some confusion over what is the correct result of some sequences. For example, the term A^ω is also 0-active, as shown by the sequence

$$A^\omega \equiv \underline{A}(A^\omega) \to A^\omega \equiv \underline{A}(A^\omega) \to A^\omega \equiv \underline{A}(A^\omega) \to \cdots \tag{2}$$

However, the limit of the terms in the sequence is A^ω. This means that we have two candidates for the correct result of this sequence: Ω and A^∞. The more advanced work on metric calculi avoids this confusion by distinguishing between weakly and strongly converging reduction sequences. A weakly converging reduction sequence is any reduction sequence whose limit exists. A strongly converging reduction sequence is a sequence whose limit exists and in which the depth of the redexes tends to infinity. As all redexes in (2) are contracted at the root, it is weakly converging sequence but not a strongly converging sequence. In the sequence

$$A^\omega \equiv \underline{A}(A^\omega) \to A^\omega \equiv A(\underline{A}(A^\omega)) \to A(A^\omega) \equiv A(A(\underline{A}(A^\omega))) \to \cdots \qquad (3)$$

the redexes are contracted at depths $0, 1, 2, \cdots$. So this sequence is strongly converging. We want to adhere to the strong convergence intuition, so what we want for our new limit is that the result of (2) is Ω and that the result of (3) is A^ω. This can be achieved by taking the liminf of the contexts in which the redexes are contracted (equating $\Box$ with Ω). The intuition behind this choice is that we take the liminf of the sets of preserved prefixes, where a prefix is preserved if it does not overlap with the redex being contracted. So the result of (2) is the liminf of $\Omega, \Omega, \cdots$, which is Ω. The result of (3) is the liminf of $\Omega, A(\Omega), A(A(\Omega)), \cdots$, which is$A^\omega$. Because of this intuition, we will refer to our calculi as *preservation calculi* and its limit notion as *preservation limit.*

The preservation limit of any sequence which contracts a redex at the root infinitely often becomes Ω. This means that for the case of the standard depth function, we can replace strongly converging metric limits plus 0-active to ω rule by just the preservation limit.

For non-standard depth functions the situation is slightly more complicated. The standard depth function increases the depth by one every time one passes through a function symbol. Kennaway et al. parameterized the lambda calculus depth with a string xyz, where $x, y, z \in \{0, 1\}$. When passing through a lambda the depth increases by x, when passing through the left argument of an application it increases by y and when passing through the right argument of an application it increases by z. So in the term $\lambda u.v\,w$, the depth of $v\,w$ is x, the depth of v is $x + y$ and the depth of w is $x + z$. With this alternative notion of depth, we must replaced at the root by at depth 0 in the above explanations. Most of the theory goes through in the same way as before, but we create a new class of problematic terms: the 0-undefined terms: terms which contain an Ω at depth 0. Both the metric and the liminf calculi must rewrite non-trivial 0-undefined terms to Ω. The metric calculi do so by extending the set of 0-active terms to a set of meaningless terms, the liminf calculi will need to add a rule. To see why adding a rule is necessary, consider the 001 depth notion and the term $(\lambda u.v\,u)\,((\lambda v.v\,v)\,(\lambda v.v\,v))$. We have

$$(\lambda u.v\,u)\,((\lambda v.v\,v)\,(\lambda v.v\,v)) \xrightarrow[0-\text{active}]{} (\lambda u.v\,u)\,\Omega \xrightarrow[\beta]{} v\,\Omega \xrightarrow[0-\text{undefined}]{} \Omega$$

and

$$(\lambda u.v\,u)\,((\lambda v.v\,v)\,(\lambda v.v\,v)) \xrightarrow[\beta]{} v\,((\lambda v.v\,v)\,(\lambda v.v\,v)) \xrightarrow[0-\text{active}]{} \Omega$$

Without the 0-undefined rule confluence would have been lost in this case.

There are three useful calculi: 111, 101 and 001. These calculi correspond to Berarducci Trees ([Ber94]), Levy-Longo Trees ([Lév78]) and Böhm Trees, respectively. The preservation 111 calculus and the metric 111 calculus differ only by the lengths of reduction sequences, the end point are exactly identical. The 101 and 001 versions not only differ by length, but we must also take the end points modulo 0-undefined terms.

Overview. The remainder of the paper is organized as follows. In Sect. 2, we introduce a notion of infinite term which is in between de Bruijn notation and sets of prefixes. In Sect. 3, we introduce the metric infinitary lambda calculi of Kennaway et al. using our own notation. In Sect. 4, we introduce preservation limits for all of the lambda calculi defined in the preceding section. In Sect. 5, we compare metric sequences with preservation sequences. This section is followed by the conclusion.

2 Preliminaries

When working with infinitary calculi, it is often not enough to know where in a term a symbol occurs, but also what symbols are above it. Thus, in our notion of position we also encode the symbols encountered along the way. More precisely a position will be a finite string of numbers, where a 1 stands for passing through a lambda, a 2 for the left argument of an application and a 3 for the right argument of an application. The empty string is denoted ϵ. The set of positions is given by:

$$\mathsf{Pos} = \{1, 2, 3\}^* \ .$$

If p and p' are positions then pp' is the concatenation (as strings) of the two. If p is a position and s is an arbitrary symbol then $p\,s = s$. (E.g. $1\,2\,3\,@ = @$.)

Using positions we can represent terms as suitable partial functions from the set of positions to the set of symbols. In order to avoid problems with α-conversion, we use a single λ symbol and represent bound variables by the position of the lambda they are bound to. We will often view (partial) functions as relations. That is, we will switch freely between writing $f(x) = y$ and writing $(x, y) \in f$.

Definition 1. *The set of terms T^∞ is the set of partial functions $t : \mathsf{Pos} \rightarrow \{@, \lambda\} \cup \mathcal{V} \cup \mathsf{Pos}$ such that*

- *If $t(p\,1)$ is defined then $t(p) = \lambda$.*
- *If $t(p\,2)$ is defined or $t(p\,3)$ is defined then $t(p) = @$.*
- *If $t(p) = q$ then $\exists q' : q\,1\,q' = p$.*

Two important sets of positions in a term t are the set of *defined positions* ($\mathcal{D}(t)$) and the set of *undefined edge positions* ($\Omega(t)$). The former is simply the domain of the partial function, denoted $\mathcal{D}(t)$. The latter is the set of positions which are not defined in the term, but their immediate predecessor is defined in the term.

Definition 2. *For any term* $t \in T^\infty$ *let*

$$\mathcal{D}(t) = \{p \mid (p, s) \in t\} \ ;$$
$$\Omega(t) = (\{p1 \mid (p, \lambda) \in t\} \cup \{p2, p3 \mid (p, @) \in t\}) \setminus \{p \mid (p, s) \in t\} \ .$$

The binary operator $\uparrow$ removes a sub-term at a position. That is, it replaces the sub-term by Ω:

Definition 3. *For any term* $t \in T^\infty$ *and position* $p \in \mathsf{Pos}$ *let*

$$t \uparrow p = \{(p', s') \mid (p', s') \in t \wedge \neg \exists p'' : pp'' = p'\}$$

We will now define how normal lambda term and context syntax in terms of our definition of term. The term syntax definitions are straight forward:

Definition 4.

$$\begin{aligned}
[\![\Omega]\!] &= \emptyset \\
[\![x]\!] &= \{(\epsilon, x)\} \\
[\![M\,N]\!] &= [\![M]\!]\,[\![N]\!] \\
t_1\,t_2 &= \{(\epsilon, @)\} \cup \{(2\,p, 2\,s) \mid (p, s) \in t_1\} \cup \{(3\,p, 3\,s) \mid (p, s) \in t_2\} \\
[\![\lambda x.M]\!] &= \lambda x.[\![M]\!] \\
\lambda x.t &= \{(\epsilon, \lambda)\} \cup \{(1\,p, 1\,s) \mid (p, s) \in t \wedge s \neq x\} \cup \{(1\,p, \epsilon) \mid (p, s) \in t \wedge s = x\}
\end{aligned}$$

The trick to defining contexts is to use a mapping from abstraction positions to variable names to control which variables are bound by which position[1].

Definition 5. *A context is a tuple* $(t, p, \Box)$*, where* t *is a term,* $p \in \Omega(t)$ *is the position of the hole in* C *and* $\Box : \{p \mid (p, \lambda) \in t\} \to \mathcal{V}$ *is a surjective function.*

$$(t_1, p, \Box)[t_2] = t_1 \cup \{(p\,p', b(s)) \mid (p', s) \in t_2\}$$

$$b(s) = \begin{cases} p'' \ , \ \textit{if } \Box(p'') = s \\ p\,s \ , \ \textit{if } s \in \mathsf{Pos} \\ s \quad , \ \textit{otherwise} \end{cases}$$

We also need substitutions.

Definition 6. *A substitution is a function* $\sigma : \mathcal{V} \to T^\infty$*. The application of the substitution* σ *applied to a term* t *is given by:*

$$t\sigma = \{(p, s) \mid (p, s) \in t \wedge t \notin \mathcal{V}\} \cup (\cup\{\{(p\,p', p\,s') \mid (p', s') \in \sigma(x)\} \mid x \in \mathcal{V} \wedge (p, x) \in t\})$$

And finally we can define β-reduction at a position in the usual way:

Definition 7. *If* $C \equiv (t, p, \Box)$ *is a context,* x *is a variable and* M, N *are terms then*

$$C[(\lambda x.M)\,N] \xrightarrow[\beta]{p} C[M[x := N]] \ .$$

[1] The idea of controlling the binding in a context is not new. For example, it was explored in [BdV01].

3 Metric Infinitary Rewriting

In this section, we introduce the metric infinitary lambda calculi of Kennaway et al. (see [KKSdV97]). Note that we have made a few minor modification to the presentation to allow better comparison with the preservation infinitary lambda calculi in the next chapter.

The depth of a symbol in a term is derived from the length of the position.

Definition 8. *For $x, y, z \in \{0,1\}$, the length $|.|_{xyz} : \mathsf{Pos} \to \mathbb{N}$ of a position in an xyz calculus is given by*

$$\begin{aligned} |1\,p|_{xyz} &= x + |p|_{xyz} \\ |2\,p|_{xyz} &= y + |p|_{xyz} \\ |3\,p|_{xyz} &= z + |p|_{xyz} \end{aligned}$$

Terms with infinitely many symbols at the same depth are excluded in the metric calculi. In other words, we restrict to terms with finite levels.

Definition 9. *The set of finite level xyz terms is*

$$T^{xyz} = \{t \in T^{\infty} \mid \neg\exists p, p_1, p_2, \cdots : \forall n : |p_n|_{xyz} = 0 \wedge p\,p_1 \cdots p_n \in \mathcal{D}(t)\} \ .$$

Based on the notion of length and these sets of terms, we can define a metric space

Definition 10. *The xyz metric is defined as*

$$d_{xyz}(t_1, t_2) = \begin{cases} 0 & , \textit{if } t_1 = t_2 \\ 2^{-\min\{|p|_{xyz} | t_1(p) \neq t_2(p)\}} & , \textit{otherwise} \end{cases}$$

In the metric calculi terms which are meaningless because they have an Ω at depth 0 and terms which are meaningless because they allow an infinite reduction with contracts infinitely many redexes at depth 0 can be treated as the same thing. However, for the preservation calculi in the next chapter we need to be able to distinguish between them. Thus, we introduce the following 2 rewrite rules to deal with meaningless terms.

Definition 11. *The rewrite rules $\xrightarrow[\Omega_{xyz}]{}$ and $\xrightarrow[\infty_{xyz}]{}$ are given by*

$$\begin{aligned} t &\xrightarrow[\Omega_{xyz}]{} \Omega, \textit{ if } \exists p \in \Omega(t) : |p|_{111} > 0 \wedge |p|_{xyz} = 0 \\ t &\xrightarrow[\infty_{xyz}]{} \Omega, \textit{ if } \exists t_i, p_i : t = t_0 \xrightarrow[\beta]{}\!\!\!\!\!\rightarrow \xrightarrow[\beta]{p_0} t_1 \xrightarrow[\beta]{}\!\!\!\!\!\rightarrow \xrightarrow[\beta]{p_1} t_2 \cdots \wedge |p_i|_{xyz} = 0 \end{aligned}$$

The common part in the definition of reduction sequence for the metric and preservation calculi is the notion of pre-reduction sequence.

Definition 12. *Given a rewrite relation R with positions, a pre-reduction sequence of length α is a tuple $((t_i)_{i \in \alpha+1}, (p_i)_{i \in \alpha})$, such that for all $i \in \alpha$ $t_i \xrightarrow[R]{p_i} t_{i+1}$. We say that the* depths of the redexes tend to infinity *if for all limit ordinals $\alpha' \leq \alpha$ $\lim_{i \in \alpha'} |p_i|_{xyz} = \infty$.*

In a pre-reduction sequence every terms must rewrite to its successor if it exists and optionally the depths of redexes must tend to infinity. A reduction sequence is a pre-reduction sequence where at every limit ordinal, the term is the "limit" of the sequence preceding it.

Definition 13. *A metric reduction sequence of length* α *is a pre-reduction sequence* $((t_i)_{i\in\alpha+1}, (p_i)_{i\in\alpha})$ *in which the depths of redexes tend to infinity, such that for all limit ordinals* $\alpha' \leq \alpha$ $t_{\alpha'} = \lim_{i\in\alpha'} t_i$.

We now have the components necessary to define the infinitary lambda calculi of Kennaway et al. We have done so in Table 1. Next, we define the new bits needed for our preservation calculi.

Table 1. The xyz infinitary lambda calculus Λ^{xyz}.

set of terms:	T^{xyz}
rewrite rules:	$\xrightarrow[\beta]{}, \xrightarrow[\infty_{xyz}]{}, \xrightarrow[\Omega_{xyz}]{}$
transfinite sequences:	strongly converging in d_{xyz} metric

4 Preserved Approximation Infinitary Rewriting

In this section we introduce our new preservation infinitary lambda calculi. Its limit definition uses the standard *limes inferior* on sequences of sets and the notion of preserved part of a term.

Definition 14. *Given a limit ordinal* α *and sequence of sets* $(S_i)_{i\in\alpha}$ *the limes inferior is defined as*

$$\liminf_{i\in\alpha} S_i = \bigcup_{j\in\alpha} \bigcap_{j\leq k<\alpha} S_k \ .$$

The preservation limit of a reduction sequence cuts the terms a position above the redex (see Def. 3), which is determined by applying the so called preservation function to the position of the redex, and then takes the liminf of the resulting preserved parts of the terms:

Definition 15. *Let* $\nabla : \mathsf{Pos} \to \mathsf{Pos}$ *be a function such that* $\forall p \in \mathsf{Pos} : \ \nabla(p)$ *is a prefix of* p. *A preservation reduction sequence of length* α *using* ∇ *is a pre-reduction sequence* $((t_i)_{i\in\alpha+1}, (p_i)_{i\in\alpha})$, *such that for all limit ordinals* $\alpha' \leq \alpha$ $t_{\alpha'} = \liminf_{i\in\alpha'}(t_i \uparrow \nabla_{xyz}(p_i))$.

The preservation functions that are needed to get behavior similar to the xyz infinitary lambda calculi are:

Definition 16. *Given a position p, the preserved part of p (denoted $\nabla_{xyz}(p)$) is given by*

$$\begin{aligned}
\nabla_{xyz}(\epsilon) &= \epsilon \\
\nabla_{xyz}(p\,1) &= \begin{cases} p\,1 & , \textit{if } x = 1 \\ \nabla_{xyz}(p) & , \textit{otherwise} \end{cases} \\
\nabla_{xyz}(p\,2) &= \begin{cases} p\,2 & , \textit{if } y = 1 \\ \nabla_{xyz}(p) & , \textit{otherwise} \end{cases} \\
\nabla_{xyz}(p\,3) &= \begin{cases} p\,3 & , \textit{if } z = 1 \\ \nabla_{xyz}(p) & , \textit{otherwise} \end{cases}
\end{aligned}$$

The preservation calculi are defined in Table 2. In the next section, we will study the relation between the metric and the preservation calculi. In the remainder of this section, we will discuss briefly the Böhm tree definition of Lévy.

Table 2. The xyz preservation lambda calculus.

set of terms:	T^{xyz}
rewrite rules:	$\xrightarrow[\beta]{}$, $\xrightarrow[\Omega_{xyz}]{}$
transfinite sequences:	preservation sequences using ∇_{xyz}

In his thesis ([Lév78]) Lévy defines the Böhm tree as follows

$$\mathrm{BT}(M) = \{a \in \Lambda \mid M \twoheadrightarrow_{\beta} N, a \leq_{\Omega} \omega(N)\} \ ,$$

where $\omega(N)$ is the normal form of N with respect to

$$\begin{aligned}
(\lambda x.M)\,N &\to \Omega \\
\Omega\,M &\to \Omega \\
\lambda x.\Omega &\to \Omega
\end{aligned}$$

Infinite terms are represented as sets of finite approximations. (The order $\leq_{\Omega}$ is generated by $\Omega \leq_{\Omega} M$.) He also defined what we refer to as the Lévy-Longo tree as:

$$\mathrm{LL}(M) = \{a \in \Lambda \mid M \twoheadrightarrow_{\beta} N, a \leq_{\Omega} \omega(N)\} \ ,$$

where $\omega(N)$ is the normal form of N with respect to

$$\begin{aligned}
(\lambda x.M)\,N &\to \Omega \\
\Omega\,M &\to \Omega
\end{aligned}$$

As mentioned before, these trees correspond to the 001 and 101 calculi respectively. The connection between that work and this is that we could in each case have defined $\omega(N)$ as the normal from of N with respect to

$$\begin{aligned}
(\lambda x.M)\,N &\to \Omega \\
M &\to \Omega, \text{ if } M \xrightarrow[\Omega_{xyz}]{} \Omega
\end{aligned}$$

5 Equivalences

In this section we will prove that every metric reduction sequence is a preservation $\beta, \Omega_{xyz}, \infty_{xyz}$ reduction sequence with depths of redexes tending to infinity and vice versa. We will also show how to construct preservation β, Ω_{xyz} sequences from a preservation $\beta, \Omega_{xyz}, \infty_{xyz}$ sequences and vice versa.

The equivalence of metric reduction sequences and preservation $\beta, \Omega_{xyz}, \infty_{xyz}$ reduction sequences with depths of redexes tending to infinity is really due to the fact that when the depth of redexes tends to infinity the limit notions are the same:

Lemma 17. *If $((t_i)_{i\in\alpha+1}, (p_i)_{i\in\alpha})$ is a pre-reduction sequence over R where the depths of the redexes tend to infinity then for any limit ordinal $\alpha' \leq \alpha$:*

$$\lim_{i\to\alpha'} t_i = \liminf_{i\to\alpha'} t_i \uparrow \nabla_{xyz}(p_i) \ .$$

Proof. Let $s_1 = \lim_{i\to\alpha'} t_i$, $s_2 = \liminf_{i\to\alpha'} t_i$ and $s_3 = \liminf_{i\to\alpha'} t_i \uparrow \nabla_{xyz}(p_i)$. We will prove $s_1 = s_2$ by distinguishing two cases:

"$\subseteq$" Given $(p, s) \in s_1$. By definition of limit, we can find $k < \alpha'$ such that for all $j, k \leq j < \alpha'$, we have $d_{xyz}(s_1, t_j) < 2^{|p|_{xyz}}$. From $d_{xyz}(s_1, t_j) < 2^{|p|_{xyz}}$ it follows that $t_j(p) = s$, which implies that $(p, s) \in \cap_{k\leq j<\alpha'} t_j \subseteq s_2$.

"$\supseteq$" Given $(p, s) \in s_2$. By definition of lim inf, we can find $k < \alpha'$ such that $(p, s) \in \cap_{k\leq j<\alpha'} t_j$. If $(p, s) \notin s_1$ then $d_{xyz}(t_j, s_1) \geq 2^{|p|_{xyz}} > 0$, which contradicts the fact that $\lim_{i\to\alpha'} t_i = s_1$.

We will prove $s_2 = s_3$ by case distinction also:

"$\subseteq$" Given $(p, s) \in s_2$. By definition of lim inf, we can find $k_1 < \alpha'$ such that for all $j, k_1 \leq j < \alpha'$ $(p, s) \in t_j$. Because $\lim_{i\to\alpha'} |p_i|_{xyz} = \infty$, we can find $k_2 < \alpha'$ such that for all $j, k_2 \leq j < \alpha'$ $|p_j|_{xyz} > |p|_{xyz}$. Let $k = \max(k1, k_2)$ then for all $j, k \leq j < \alpha'$ we have $(p, s) \in t_j \uparrow \nabla_{xyz}(p_j)$ and thus $(p, s) \in s_3$.

"$\supseteq$" Trivial. □

If the limits are the same then the reduction sequences must be the same as well:

Theorem 18. *1. Every metric xyz reduction sequence is a preservation xyz sequence with ∞_{xyz} steps.*
2. Every preservation xyz reduction sequence with ∞_{xyz} steps in which the depths of the redexes tend to infinity is a metric xyz reduction sequence.

Proof. Corollary from Lemma 17.

The key to the translation of preservation sequences with ∞_{xyz} steps to sequences without them is to replace these steps with subsequences of length ω, which do the same job. However, the sequence removes the undefinedness more thoroughly than the step. For example, in the 001-calculi:

$$(\lambda x.x\,x\,x)(\lambda x.x\,x\,x) \xrightarrow[\beta]{\epsilon} (\lambda x.x\,x\,x)(\lambda x.x\,x\,x)(\lambda x.x\,x\,x) \xrightarrow[\infty_{001}]{2} \Omega(\lambda x.x\,x\,x)$$

translates to

$$\begin{aligned}(\lambda x.x\,x\,x)(\lambda x.x\,x\,x) &\xrightarrow[\beta]{\epsilon} (\lambda x.x\,x\,x)(\lambda x.x\,x\,x)(\lambda x.x\,x\,x)\\ &\xrightarrow[\beta]{2} (\lambda x.x\,x\,x)(\lambda x.x\,x\,x)(\lambda x.x\,x\,x)(\lambda x.x\,x\,x)\\ &\xrightarrow[\beta]{22} \cdots\end{aligned}$$

The limit of that sequence is Ω instead of $\Omega(\lambda x.x\,x\,x)$ but we do have that $\Omega(\lambda x.x\,x\,x) \xrightarrow[\Omega_{001}]{\epsilon} \Omega$ and we also have $(\lambda x.x\,x\,x)(\lambda x.x\,x\,x) \xrightarrow[\infty_{001}]{\epsilon} \Omega$.

We say that two terms are equivalent up to 0-undefined sub-terms if there exists a third term to which both terms rewrite with an infinitary sequence, which uses only Ω_{xyz} steps. Thus, the following theorem implies that any sequence with ∞_{xyz} steps can be translated to a sequence without those steps and which ends in a term which is equivalent up to 0-undefined sub-terms:

Theorem 19. *If* $t_1 \xrightarrow[\beta,\Omega_{xyz},\infty_{xyz}]{}{}^{\alpha} t_2$ *is a preservation sequence then there exist* α', α'', t_3, *such that* $t_1 \xrightarrow[\beta,\Omega_{xyz}]{}{}^{\alpha'} t_3$ *and* $t_2 \xrightarrow[\Omega_{xyz}]{}{}^{\alpha''} t_3$ *are preservation sequences.*

Proof. Given $((s_i)_{i\in(\alpha+1)}, (p_i)_{i\in\alpha})$ with $s_0 = t_1$ and $s_\alpha = t_2$. The construction consists of two steps. The first step is projection of the sequence to Ω_{xyz} (infinitary) normal forms. The second step is the replacement of ∞_{xyz} step by β sequences.

By $\xrightarrow[\Omega_{xyz}]{}{}^{\mathrm{nf}}$ we denote reduction to Ω_{xyz} (infinitary) normal form of a term. Such a reduction always exists, ends in a unique term and can have length at most ω. (The set of all redexes in a term is countable and no redex can duplicate any other (erasure is possible).) Thus, we can find $(s'_i)_{i\in(\alpha+1)}$ such that $s_i \xrightarrow[\Omega_{xyz}]{}{}^{\mathrm{nf}} s'_i$. Let $t_3 = s'_\alpha$.

For every step $s_i \xrightarrow{p_i} s_{i+1}$ in the original reduction, we construct a sequence between s'_o and s'_{i+1} by distinguishing the type of the step:

"$s_i \xrightarrow[\beta]{p_i} s_{i+1}$" If the redex is erased then $s'_i = s'_{i+1}$. Otherwise $s'_i \xrightarrow[\beta]{p_i} \xrightarrow[\Omega_{xyz}]{}{}^{\mathrm{nf}} s'_{i+1}$.

"$s_i \xrightarrow[\Omega_{xyz}]{p_i} s_{i+1}$" This redex is certainly erased so $s'_i = s'_{i+1}$.

"$s_i \xrightarrow[\infty_{xyz}]{p_i} s_{i+1}$" It is possible that the redex is erased and that we have $s'_i = s'_{i+1}$. Otherwise the step projects down to $s'_i \xrightarrow[\infty_{xyz}]{\nabla_{xyz}(p_i)} s'_{i+1}$.
In the latter case, we can find a context C with the hole at $\nabla_{xyz}(p_i)$, such that $s'_i = C[u_0]$, $u_0 \xrightarrow[\infty_{xyz}]{\epsilon} \Omega$ and $C[\Omega] = s'_{i+1}$. Thus, there exists a sequence $u_0 \xrightarrow[\beta]{pp_0} u_1 \xrightarrow[\beta]{pp_1} u_2 \cdots$ where infinitely often $|pp_i|_{xyz} = 0$. Hence the lim inf of this sequence is Ω. We also have

$$C[u_0] \xrightarrow[\beta]{\nabla_{xyz}(p_i)\,pp_0} C[u_1] \xrightarrow[\beta]{\nabla_{xyz}(p_i)\,pp_1} C[u_2] \cdots \to^{\omega} C[\Omega]$$

This construction yields a pre-reduction sequence from t_1 to t_3. To prove that this is also a reduction sequence, let us consider the limit points. These limit points case be of two types:

- A limit point which was introduced by projecting β or translating an ∞_{xyz} step. In this case the limit condition is satisfied because the subsequence is by construction a reduction sequence.
- A limit point, which survived projection. That is, a limit point in the given sequence such that infinitely many steps in front of it were translated into non-empty subsequences. In this case the results follow from the observation that if $s'_i \xrightarrow{pp_0} \xrightarrow{pp_1} \cdots s'_{i+1}$ in $\beta > 0$ steps then

$$s_i \uparrow \nabla_{xyz}(p_i) \xrightarrow[\Omega_{xyz}]{}^{\mathrm{nf}} \cap_{j \in \beta} s'_{i\,j} \uparrow \nabla_{xyz}(pp_j) \ .$$

In other words, the approximations preserved by a step in the original sequence and preserve by rewriting to Ω_{xyz} normal form are precisely those approximations preserved by the sub-sequence. Hence, first projecting and then taking the limit has the same result as first taking the limit and then projecting.

Theorem 20. *If $t_1 \xrightarrow[\beta,\Omega_{xyz}]{}^{\alpha} t_2$ is a preservation sequence then there exists α' such there exists a preservation sequence $t_1 \xrightarrow[\beta,\infty_{xyz},\Omega_{xyz}]{}^{\alpha'} t_2$ in which the depths of the redexes tends to infinity.*

Proof. By induction on the length of the sequence.

Given the original α. If α is not a limit ordinal we can apply the induction hypothesis on the predecessor of α. We will now construct ordinals $\alpha_0 \leq \alpha_1 \ldots < \alpha$ such that all redex between α_i and α are at depth at least i. Let $\alpha_0 = 0$. Given α_i, apply the induction hypothesis to the sequence up to α_i. Let

$$P = \{p \mid (p, s) \in t_{\alpha_i}, p = \nabla_{xyz}(p), |p|_{xyz} = i\} \ .$$

Not that P is a finite set. Every redex constructed after α_i has a prefix which is in P because every redex contracted after α_i occurs at at least depth i. Let

$$P_\infty = \{p \in P \mid \neg\exists\alpha' < \alpha : \forall\alpha' \leq \beta < \alpha : p \text{ is not a prefix of } p_\beta \vee |p_\beta|_{xyz} > i\} \ .$$

For each element p of P we replace the first redex contracted below p by an ∞_{xyz} redex contracted at position p. We can now find α_{i+1} such that

$$\forall \alpha_{i+1} \leq \beta < \alpha : |p_\beta|_{xyz} > i \ .$$

6 Conclusion

We have introduced a new notion of limit for infinitary rewriting: the preservation limit. The new notion is more elegant in two ways. First, the limit of any sequence is defined. Second, the new notion captures the notion of 0-active term in the limit definition so it needs only 0-undefined terms in its set of meaningless terms.

For the infinitary lambda calculi $\Lambda^{111}, \Lambda^{101}$ and Λ^{001}, we have proved that metric reduction sequences can be transformed into preservation reduction sequences with the same begin and end terms. We have also show that preservation sequences can be transformed into metric sequences with the same begin terms and the same end terms up to 0-undefined sub-terms.

In [Ken92] one can find an abstraction of metric infinitary rewriting. It would be interesting future work to develop an abstract version of preservation infinitary rewriting and compare the two, or even to develop a framework of which both are specific examples.

References

[BdV01] Mirna Bognar and Roel de Vrijer. A calculus of lambda calculus contexts. *Journal of Automated Reasoning*, 27(1):29–59, 2001.

[Ber94] A. Berarducci. Infinite λ-calculus and non-sensible λ-models. In *Proceedings of the conference in honour of Roberto Magari.* To appear in the series on Logic and Algebra, Marcel & Decker, New York, 1994.

[Cor93] A. Corradini. Term rewriting in CT_Σ. In M.-C. Gaudel and J.-P. Jouannaud, editors, *Proc. Colloquium on Trees in Algebra and Programming (CAAP '93), Springer-Verlag LNCS 668*, pages 468–484, 1993.

[DK89] N. Dershowitz and S. Kaplan. Rewrite, rewrite, rewrite, rewrite, rewrite. In *Proc. ACM Conference on Principles of Programming Languages, Austin, Texas*, pages 250–259, 1989.

[KdV03] Richard Kennaway and Fer-Jan de Vries. Infinitary rewriting. In Marc Bezem, Jan Willem Klop, and Roel de Vrijer, editors, *Term Rewriting Systems*, volume 55 of *Cambridge Tracts in Theoretical Computer Science*, chapter 12, pages 668–711. Cambridge university press, 2003.

[Ken92] J.R. Kennaway. On transfinite abstract reduction systems. Technical Report CS-R9205, CWI, 1992.

[KKSdV93] J. R. Kennaway, J. W. Klop, M. R. Sleep, and F. J. de Vries. An infinitary church-rosser property for non-collapsing orthogonal term rewriting systems. In M. R. Sleep, M. J. Plasmeijer, and M. C. D. J. van Eekelen, editors, *Term Graph Rewriting: Theory and Practice*, pages 47–59. John Wiley & Sons, 1993.

[KKSdV95a] J.R. Kennaway, J.W. Klop, M.R. Sleep, and F.J. de Vries. Infinitary lambda calculus and bohm models. In Jieh Hsiang, editor, *Rewriting Techniques and Applications*, volume 914 of *Lecture Notes in Computer Science*, pages 257–270. Springer Verlag, April 1995.

[KKSdV95b] J.R. Kennaway, J.W. Klop, M.R. Sleep, and F.J. de Vries. Transfinite reduction in orthogonal term rewriting systems. *Information and Computation*, 119(1):18–38, 1995.

[KKSdV97] J.R. Kennaway, J.W. Klop, M.R. Sleep, and F.J. de Vries. Infinitary lambda calculus. *TCS*, 175:93–125, 1997.

[Lév78] J.-J. Lévy. *Réductions Correctes et Optimales dans le Lambda-Calcul.* PhD thesis, Universite Paris VII, October 1978.

Böhm-Like Trees for Term Rewriting Systems

Jeroen Ketema

Department of Computer Science
Faculty of Sciences, Vrije Universiteit Amsterdam
De Boelelaan 1081a, 1081 HV Amsterdam, The Netherlands

Abstract. In this paper we define Böhm-like trees for term rewriting systems (TRSs). The definition is based on the similarities between the Böhm trees, the Lévy-Longo trees, and the Berarducci trees. That is, the similarities between the Böhm-like trees of the λ-calculus. Given a term t a tree partially represents the root-stable part of t as created in each maximal fair reduction of t. In addition to defining Böhm-like trees for TRSs we define a subclass of Böhm-like trees whose members are monotone and continuous.

1 Introduction

In the theory of the λ-calculus there occur three very similar trees. These are the Böhm trees [1], the Lévy-Longo trees or lazy trees [2], and the Berarducci trees [3]. We call these trees the *Böhm-like trees.* In this paper we define Böhm-like trees for term rewriting system (TRSs). We also define a subclass of Böhm-like trees whose members are monotone and continuous.

The definition of Böhm-like trees for TRSs is based on the similarities between the Böhm-like trees of the λ-calculus. Given a term t a tree partially represents the root-stable part of t as created in each maximal fair reduction of t. Maximal means it is either a reduction to normal form or an infinite reduction. Fair means that every redex occurring in the reduction is eventually contracted.

The actual part as represented by a particular Böhm-like tree depends on the definition of that tree. In the λ-calculus, Böhm trees represent subterms in head normal form, Lévy-Longo trees represent subterms in weak head normal form, and Berarducci trees represent all root-stable subterms.

A root-stable part and a Böhm-like tree can become infinitely large in a maximal reduction. For example, if Y denotes a λ-term that behaves as a fixed-point combinator, then

$$Y(\lambda xy.x) \to_\beta^* \lambda y_1.Y(\lambda xy.x) \to^* \lambda y_1.\lambda y_2.Y(\lambda xy.x) \to_\beta^* \dots$$

and $\lambda y_1.\lambda y_2.\lambda y_3.\dots$ is the Lévy-Longo tree of $Y(\lambda xy.x)$. It is also the Berarducci tree of $Y(\lambda xy.x)$.

Construction. To obtain a partial representation of the root-stable part of a term t, as created in each maximal fair reduction, we construct partial representations of the root-stable parts as created in each finite reduction of t. That is,

V. van Oostrom (Ed.): RTA 2004, LNCS 3091, pp. 233–248, 2004.

we construct partial representations of the root-stable parts of the final terms. If we construct representations for final terms of increasingly longer reductions, then in the limit we get a partial representation of the root-stable part of t as created in each maximal fair reduction.

Approaches. There are three approaches to formalising the above limit process. We discuss each of these in turn. The differences between the approaches originate from the different ways in which they represent trees.

Ideal Completion. In this approach unspecified subterms and a partial order on terms are defined first. Then, employing the partial order, trees are defined by means of ideal completion. That is, trees are represented by ideals. The finite and infinite ideals represent respectively the finite and infinite trees. Constructing the partial representation of the root-stable part as created in a finite reduction is done with the help of functions. These functions are called a direct approximant functions. Given a final term of a finite reduction a direct approximant function strips out subterms, leaving them unspecified. At least the non-root-stable subterms are stripped out. The exact definition of a direct approximant function depends on the particular Böhm-like tree [2, 4–6].

Partial Functions. In this approach trees are represented as partial functions from the set of positions to the union of the signature and the variables. The partial functions with a finite and infinite domain represent respectively the finite and infinite trees. Given a term t, the symbol that occurs at a certain position in a Böhm-like tree of t is acquired by recursively reducing t and the subterms of the reduct of t until they are in head normal form, in weak head normal form, or root-stable, depending on the particular tree [1].

Metric Completion. In this approach a metric on terms is defined first. Then, trees are defined by means of metric completion of the set of terms. The terms and the elements created by metric completion represent respectively the finite and infinite trees. The Böhm-like tree of a term is obtained by means of infinitary rewriting in a transfinitely confluent version of the λ-calculus. Rewrite rules of the form $t \to \bot$ are used to obtain transfinite confluence. The actual terms t that occur in the rewrite rules $t \to \bot$ depend on the particular Böhm-like tree [3, 7].

Current Approach. In this paper we use ideal completion to define Böhm-like trees for TRSs. However, to keep the discussion simple

we consider only confluent left-linear TRSs.

Considering non-confluent TRSs at least requires additional clauses in Definition 5.1, as Blom [8] and Ariola and Blom [9] show.

Related Work. The related work can be divided into three categories. First, using ideal completion Boudol [10] and Ariola [11] already defined one particular Böhm-like tree. We discuss this tree in Example 6.11.

Second, Kennaway, Van Oostrom, and De Vries [12] define Böhm-like trees for TRSs on a similar level of abstraction as we do. They use metric completion and infinitary rewriting. To obtain transfinite confluence they formulate sufficient conditions on the terms that may occur in the rewrite rules $t \to \bot$. A comparison of their sufficient conditions and our approach is non-trivial and outside the scope of this paper.

Third, Boudol [10], Blom [8], and Ariola and Blom [9] use ideal completion to define Böhm-like trees that are more abstract than the ones defined here. In their approaches, as we further explain in Sect. 5, the range of the direct approximant functions no longer need to be terms. Their approaches offer excellent frameworks for studying the most abstract properties shared between Böhm-like trees. However, their trees no longer represent the root-stable part of a term as created in each maximal fair reduction. In addition, their direct approximant functions cannot be restricted by relating the domain and range of the functions with the help of partial order on terms. We use such relations when defining the subclass of Böhm-like trees that is monotone and continuous.

Overview. In the rest of this paper we proceed as follows. In Sect. 2 we give some preliminary definitions. Then, in Sect. 3, we define unspecified subterms and the related partial order. In Sect. 4 we define trees, and in Sect. 5 we give a definition of Böhm-like trees. After this we consider a subclass of computable direct approximant functions. The Böhm-like trees based on these direct approximant functions are monotone and continuous. We give the definition of the subclass in Sect. 6, and in Sect. 7 we prove that the trees are monotone and continuous. In Sect. 8, the final section, we give some possible directions for further research.

2 Preliminaries

Most of the notation and concepts we use in this paper correspond to that in the books by Baader and Nipkow [13] and Stoltenberg-Hansen, Lindström, and Griffor [14]. In this section we summarise the most relevant notation and concepts.

Given a signature Σ and a set of variables X, we denote by Σ_n the subset of Σ whose elements have arity n. By $\mathcal{T}er(\Sigma, X)$ we denote the set of terms over Σ and X. We call a term $t \in \mathcal{T}er(\Sigma, X)$ *linear* if each variable from X occurs at most once in t.

If $t \in \mathcal{T}er(\Sigma, X)$, then $\mathcal{P}os(t)$ denotes the set of positions of t. The positions have an associated *prefix order*. We say that p is a *prefix* of q, denoted $p \leq q$, if there exists a position r such that $p \cdot r = q$. Here, the symbol $\cdot$ denotes the *concatenation* of positions and r may be the *empty position* ϵ. We call the positions p and q *parallel* if neither $p \leq q$ nor $q \leq p$.

We denote the subterm of a term t at position $p \in \mathcal{P}os(t)$ by $t|_p$. The replacement of a subterm at position p in t by a term s is denoted $t[s]_p$.

Given a term t and a substitution σ, we denote the application of σ to t by $\sigma(t)$. We also use notation like $t[x := t_1; y := t_2]$. In this case we have a substitution that replaces x by t_1 and y by t_2.

By $\mathcal{R} = (\Sigma, R)$ we denote a TRS over a signature Σ and with the set of rewrite rules R. The elements of R are denoted $l \to r$, where $l, r \in \mathcal{T}er(\Sigma, X)$. We call $\mathcal{R}$ *left-linear*, if the left-hand sides of all its rewrite rules are linear. The rewrite relation defined by R is denoted $\to$. Its reflexive and transitive-reflexive closures are respectively denoted $\to^=$ and $\to^*$.

A TRS $\mathcal{R}$ is *subcommutative*, if for every $s \to t_1$ and $s \to t_2$ there exists a u such that $t_1 \to^= u$ and $t_2 \to^= u$. Moreover, $\mathcal{R}$ is *confluent*, if for every $s \to^* t_1$ and $s \to^* t_2$ there exists a u such that $t_1 \to^* u$ and $t_2 \to^* u$. A term t is in *normal form* with respect to $\mathcal{R}$ if no redex occurs in t. The TRS $\mathcal{R}$ is *terminating* if all reductions are finite.

We call a term t in a TRS *root-stable* if we cannot rewrite t to a term which is a redex. We call a subterm $t|_p$ with $p \in \mathcal{P}os(t)$ root-stable if for all $q \leq p$ the term $t|_q$ is root-stable.

By $\mathcal{P} = (P, \sqsubseteq)$ we denote a partial order $\sqsubseteq$ over a set P. If $Q \subseteq P$, then Q is *consistent* if there exists a $p \in P$ such that for all $q \in Q$ we have $q \sqsubseteq p$. If Q has a least upper bound, then we denote it by $\bigsqcup Q$.

Given a partial order $\mathcal{P} = (P, \sqsubseteq)$, we call a non-empty set $D \subseteq P$ *directed*, if for all $p, q \in D$ there exists an $r \in D$ such that $p \sqsubseteq r$ and $q \sqsubseteq r$. Moreover, we call a non-empty set $D \subseteq P$ *downward closed*, if for all $p \sqsubseteq q$ with $p \in P$ and $q \in D$ we have $p \in D$.

A partial order $\mathcal{P} = (P, \sqsubseteq)$ is a *conditional upper semi-lattice with least element (cusl)*, if P has a least element and if every consistent subset of P has a least upper bound. A set $I \subseteq P$ is an *ideal*, if it is downward closed and if every $\{p, q\} \subseteq I$ is consistent and has a least upper bound in I. An ideal is called finite if it has finite cardinality, otherwise it is called infinite.

For every directed set $D \subseteq P$ in a cusl $\mathcal{P} = (P, \sqsubseteq)$ we can define an ideal, denoted $\downarrow D$, called the *downward closure* of D

$$\downarrow D = \{p \in P \mid p \sqsubseteq q \text{ for some } q \in D\} .$$

Moreover, we have that $\mathcal{P}^\infty = (P^\infty, \subseteq)$ is a partial order. Here, P^∞ denotes $\{I \subseteq P \mid I \text{ is an ideal of } \mathcal{P}\}$ and $\subseteq$ denotes subset inclusion. The partial order $\mathcal{P}^\infty$ is called the *ideal completion* of $\mathcal{P}$.

3 Partial Terms

Let Σ be signature and X a set of variables. To represent unspecified subterms we extend the signature with a constant $\bot$ which neither occurs in Σ nor in X. The *unspecified subterms* are defined as those subterms that are equal to $\bot$.

We call the set of terms over the signature $\Sigma \cup \{\bot\}$ the set of *partial terms*. We denote the set by $\mathcal{T}er(\Sigma_\bot, X)$. We leave out the adjective partial when it is obvious from the context.

Given a TRS $\mathcal{R} = (\Sigma, R)$ we can define the TRS $\mathcal{S} = (\Sigma \cup \{\bot\}, R)$. The definition of $\mathcal{S}$ is sound, as $\Sigma \subseteq \Sigma \cup \{\bot\}$. Moreover, $\mathcal{S}$ has the same confluence and termination properties as $\mathcal{R}$, as we can consider $\bot$ to be a variable which we have singled out.

With the help of $\bot$ we can define a partial order on terms, called the prefix order. We can also define a strict partial order, called the strict prefix order.

Definition 3.1. *Let Σ be a signature and X a set of variables.*

1. *The* prefix order *on $\mathcal{T}er(\Sigma_\bot, X)$, denoted $\preccurlyeq$, is the smallest binary relation such that*
   (a) $x \preccurlyeq x$ for all $x \in X$,
   (b) $\bot \preccurlyeq t$ for all $t \in \mathcal{T}er(\Sigma_\bot, X)$, and
   (c) $f(s_1, \ldots, s_n) \preccurlyeq f(t_1, \ldots, t_n)$ for all $f \in \Sigma_n$ and $s_i \preccurlyeq t_i$ with $1 \leq i \leq n$.
2. *The* strict prefix order *on $\mathcal{T}er(\Sigma_\bot, X)$, denoted $\prec$, is the smallest binary relation such that for all $s, t \in \mathcal{T}er(\Sigma_\bot, X)$*

$$s \prec t \text{ iff } s \preccurlyeq t \text{ and } s \neq t\,.$$

If $s \preccurlyeq t$, then we call s a *prefix* of t. Moreover, if $s \prec t$, then we call s a *strict prefix* of t. The term s is a prefix of the term t if either s and t are equal or if there exist unspecified subterms in s which are specified in t but not the other way around. See Fig. 1 for a graphical representation.

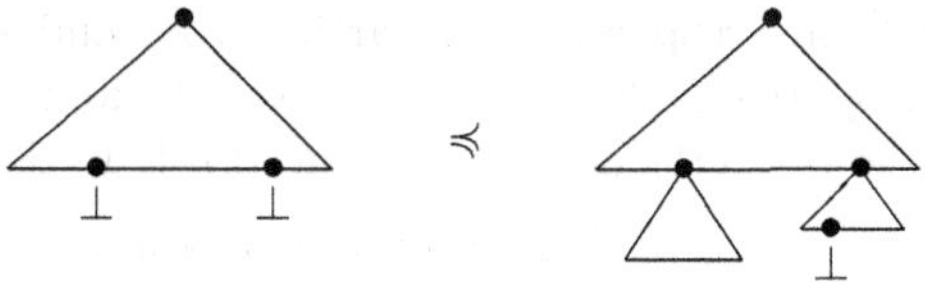

Fig. 1. The prefix order on $\mathcal{T}er(\Sigma_\bot, X)$

By induction on the structure of terms it follows that $\mathcal{PT} = (\mathcal{T}er(\Sigma_\bot, X), \preccurlyeq)$ and $\mathcal{SPT} = (\mathcal{T}er(\Sigma_\bot, X), \prec)$ are respectively a partial order and a strict partial order. The pair $\mathcal{PT}$ is in fact a cusl. The existence of a least element, the constant $\bot$, follows by the second clause of the prefix order and the anti-symmetry of partial orders. By the same facts and by induction on the structure of terms it follows that every consistent set of terms has a least upper bound.

We have the following relations between the prefix orders and the positions of terms.

Lemma 3.2. *Let $s, t \in \mathcal{T}er(\Sigma_\bot, X)$.*

1. *For all $s \preccurlyeq t$*
   - *$\mathcal{P}os(s) \subseteq \mathcal{P}os(t)$, and*
   - *for all $p \in \mathcal{P}os(s)$ such that $s|_p \neq \bot$, the root symbol of $s|_p$ is equal to the root symbol of $t|_p$.*
2. *For all $s \prec t$, there exist $p \in \mathcal{P}os(s)$ such that $s|_p = \bot$ and $t|_p \neq \bot$.*

Proof. By induction on the structure of terms. □

Using the previous lemma we prove well-foundedness of the strict prefix order.

Proposition 3.3. *The strict prefix order on $\mathcal{T}er(\Sigma_\perp, X)$ is well-founded.*

Proof. Let $s, t \in \mathcal{T}er(\Sigma_\perp, X)$ with $s \prec t$. From Lemma 3.2 it follows that

$$\#\{p \mid p \in \mathcal{P}os(s),\ s|_p \neq \perp\} < \#\{p \mid p \in \mathcal{P}os(t),\ t|_p \neq \perp\},$$

where $\#S$ denotes the cardinality of S. Hence, as $<$ is a well-founded order on the natural numbers, the result follows. □

We can extend the prefix order to substitutions by means of a point-wise definition. That is, given substitutions σ and τ

$$\sigma \preccurlyeq \tau \text{ iff } \sigma(x) \preccurlyeq \tau(x) \text{ for all } x \in X .$$

Using this definition we can also extend the strict prefix order to substitutions

$$\sigma \prec \tau \text{ iff } \sigma \preccurlyeq \tau \text{ and } \sigma(x) \prec \tau(x) \text{ for some } x \in X .$$

Thus, for all variables we must have $\sigma(x) \preccurlyeq \tau(x)$ and for at least one variable we must also have $\sigma(x) \prec \tau(x)$.

The extensions of the prefix order and the strict prefix order to substitutions are again respectively a partial order and a strict partial order. This follows easily from their definitions and the fact that the prefix order and the strict prefix order on terms are respectively a partial order and a strict partial order.

The following property holds with respect to the extension of the prefix order to substitutions. The property plays an essential rôle in Sect. 6.

Lemma 3.4. *Let $s, t \in \mathcal{T}er(\Sigma_\perp, X)$ such that t is linear. If $s \preccurlyeq \tau(t)$ for some substitution τ, then there exists an $s' \in \mathcal{T}er(\Sigma_\perp, X)$ and a substitution σ' such that $s = \sigma'(s')$, $s' \preccurlyeq t$, $\sigma' \preccurlyeq \tau$, and s' linear.*

Proof. Suppose $s \preccurlyeq \tau(t)$ for some substitution τ. We prove the result by induction on the number of positions $p \in \mathcal{P}os(s)$ such that $s|_p = \perp$ and $t|_p \neq \perp$.

Base Case. There are no positions p such that $s|_p = \perp$ and $t|_p \neq \perp$. Hence, $s = \tau(t)$ and the result follows by defining $s' = t$ and $\sigma' = \tau$.

Induction Step. Suppose the result holds for some number of positions $n \geq 0$. Let us prove it holds for $n + 1$ positions.

As $n + 1 > 0$, there exists a position $p \in \mathcal{P}os(s)$ such that $s|_p = \perp$ and $\tau(t)|_p \neq \perp$. With respect to p there are two possibilities

1. p is a non-variable position of t, or
2. there exists a variable position q of t such that $p = q \cdot r$.

In the first case define

$$t' = t[\perp]_p$$
$$\tau'(x) = \tau(x) \text{ for all } x \in X.$$

In the second case define

$$t' = t$$
$$\tau'(x) = \begin{cases} \tau(x)[\perp]_r & \text{if } x = t|_q \\ \tau(x) & \text{otherwise.} \end{cases}$$

In both cases $t' \preccurlyeq t$, $\tau' \preccurlyeq \tau$ and t' linear. Moreover, as $s|_p = \bot$ and $\tau(t)|_p \neq \bot$, it follows that $s \preccurlyeq \tau'(t') \prec \tau(t)$ and that p is the only position such that $\tau'(t')|_p = \bot$ and $\tau(t)|_p \neq \bot$. Consequently, the number of positions p with $s|_p = \bot$ and $\tau'(t')|_p \neq \bot$ is n, and by the induction hypothesis it follows that there exist an s' and σ' such that $s = \sigma'(s')$, $s' \preccurlyeq t'$, $\sigma' \preccurlyeq \tau'$, and s' linear. The actual result follows by transitivity of the prefix orders on terms and substitutions. □

We conclude this section with two remarks regarding the previous lemma.

Remark 3.5. If the position p as used in the induction step is a variable position of the term t, then there is in fact more than one way to construct t' and τ'. Consider, for example, $s = f(\bot, a)$, $t = f(x, y)$, and $\tau = [x := a; y := a]$. Following the proof of the lemma we have

$$f(x,y)[x := \bot; y := a] = f(\bot, a) \preccurlyeq f(a,a) = f(x,y)[x := a; y := a]\,.$$

However, we also have

$$f(\bot,y)[x := a; y := a] = f(\bot, a) \preccurlyeq f(a,a) = f(x,y)[x := a; y := a]\,.$$

That is, in the first case $t' = f(x, y)$ and $\tau' = [x := \bot; y := a]$ and in the second case $t' = f(\bot, y)$ and $\tau' = [x := a; y := a]$.

Remark 3.6. If t is not assumed to be linear, it is in general not possible to prove the lemma. Consider, for example, $s = f(g(\bot), g(a))$, $t = f(x, x)$, and $\tau = [x := g(a)]$. Although we have

$$f(g(\bot), g(a)) \preccurlyeq f(g(a), g(a)) = f(x,x)[x := g(a)]\,,$$

there does not exist a substitution σ' such that $\sigma'(f(x,x)) = f(g(\bot), g(a))$. The first argument of s is not equal to its second argument.

4 Trees

We define the set of trees by means of ideal completion.

Definition 4.1. *Let Σ be a signature and X a set of variables. The set of* trees, *denoted $\mathcal{T}^\infty(\Sigma_\bot, X)$, is defined by*

$$\mathcal{T}^\infty(\Sigma_\bot, X) = \{I \subseteq \mathcal{T}er(\Sigma_\bot, X) \mid I \textit{ is an ideal of } \mathcal{PT}\}\,.$$

In this definition the finite and infinite ideals represent respectively the *finite trees* and *infinite trees*. We do not explain ideal completion any further. This has been done elsewhere [14].

The following three concepts are related to trees.

Definition 4.2. *Let $S, T \in \mathcal{T}^\infty(\Sigma_\bot, X)$. Define*

Prefix Order *$S \preccurlyeq T$ iff for all $s \in S$ there exist $t \in T$ such that $s \preccurlyeq t$,*
Positions *$\mathcal{P}os(T) = \bigcup\{\mathcal{P}os(t) \mid t \in T\}$, and*
Subtree *$T|_p = \{t|_p \mid t \in T,\ p \in \mathcal{P}os(t)\}$ if $p \in \mathcal{P}os(T)$.*

Two remarks are in order with respect to this definition. First, as trees are ideals, the prefix order is in fact subset inclusion. Hence, the least upper bound of a consistent set of trees is its union. Second, as follows immediately from its definition, $T|_p$ is an ideal, and it is finite when T is finite.

We can clarify the chosen terminology with the help an isomorphism ι from $\mathcal{T}er(\Sigma_\bot, X)$ to the finite ideals of $\mathcal{T}^\infty(\Sigma_\bot, X)$. Given a term t, the isomorphism is defined by

$$\iota(t) = \downarrow\{t\} = \{s \mid s \preccurlyeq t\}.$$

The set $\iota(t)$ is finite. This follows from the definition of the prefix order and from the fact that t has a finite number of symbols. The set $\iota(t)$ is also an ideal. This follows by the definition of downward closure.

The inverse of ι assigns to each finite ideal I its least upper bound. That is,

$$\iota^{-1}(I) = \bigsqcup I.$$

The existence of the least upper bound of I follows by the definition of finite ideals. By this fact and the facts about $\iota(t)$ it follows easily that ι actually is an isomorphism. Hence, each term corresponds to a finite ideal and vice versa. As we can view every term as a finite tree, we also call a finite ideal a finite tree.

The following observations relate the concepts from Definition 4.2 with the prefix order, the set of positions, and the replacement of a subterm, as defined in the preliminaries. We assume that $s, t \in \mathcal{T}er(\Sigma_\bot, X)$ and that S and T are finite ideals of $\mathcal{T}^\infty(\Sigma_\bot, X)$.

$$\begin{array}{rclcrcl} s \preccurlyeq t & \text{iff} & \iota(s) \preccurlyeq \iota(t) & \qquad & S \preccurlyeq T & \text{iff} & \iota^{-1}(S) \preccurlyeq \iota^{-1}(T) \\ \mathcal{P}os(t) & = & \mathcal{P}os(\iota(t)) & & \mathcal{P}os(T) & = & \mathcal{P}os(\iota^{-1}(T)) \\ \iota(t|_p) & = & \iota(t)|_p & & \iota^{-1}(T|_p) & = & \iota^{-1}(T)|_p \end{array}$$

5 Böhm-Like Trees

A Böhm-like tree of a term t partially represents the root-stable part of t as created in each maximal fair reduction of t. To obtain a Böhm-like tree of t we construct partial representations of the root-stable parts of the final terms of all finite reductions. This is done with a direct approximant function. The definition of such a function depends on the particular Böhm-like tree. However, all direct approximant functions must satisfy the following definition. It summarises the properties shared between the direct approximants functions defined in earlier papers [2, 4–6, 11].

Definition 5.1. *Let $\mathcal{R} = (\Sigma, R)$ be a TRS. A* direct approximant function *of $\mathcal{R}$ is a function $\omega : \mathcal{T}er(\Sigma_\bot, X) \to \mathcal{T}er(\Sigma_\bot, X)$, such that for all $s, t \in \mathcal{T}er(\Sigma_\bot, X)$ and substitutions σ*

1. *$\omega(t) \preccurlyeq t$,*
2. *if $t|_p = \sigma(l)$, then $\omega(t) \preccurlyeq t[\bot]_p$ for all $p \in \mathcal{P}os(t)$ and $l \to r \in R$, and*
3. *if $s \to t$, then $\omega(s) \preccurlyeq \omega(t)$.*

In the remainder of this section we assume $\mathcal{R} = (\Sigma, R)$ is a confluent left-linear TRS and ω is a direct approximant function of $\mathcal{R}$.

Given a term t, we call $\omega(t)$ the *direct approximant* of t. Note that by the first and second clause of Definition 5.1 a direct approximant is in normal form with respect to $\mathcal{R}$.

The first clause of Definition 5.1 expresses that a direct approximant of a term is a prefix of that term. Note that the root-stable part of a term, or any of its prefixes, is such a prefix. The first clause is a consequence of the second clause for terms not in normal form.

The second and third clause of Definition 5.1 are motivated by the following lemma. It expresses that a direct approximant only provides information on the root-stable part of a term.

Lemma 5.2. *Let $t \in \mathcal{T}er(\Sigma_\bot, X)$ and $p \in \mathcal{P}os(t)$. If $t|_p$ is not a root-stable subterm of t, then there exists a $q \leq p$ such that $q \in \mathcal{P}os(\omega(t))$ and $\omega(t)|_q = \bot$.*

Proof. This follows immediately from the definition of root-stable subterms and the second and third clause of Definition 5.1. □

We are now almost ready to define Böhm-like trees. However, we first need to define the notion of auxiliary set. An auxiliary set of a term t consists of the direct approximants of all the reducts of t.

Definition 5.3. *If $t \in \mathcal{T}er(\Sigma_\bot, X)$, then its* auxiliary set, *denoted $\mathcal{A}(t)$, is defined by*

$$\mathcal{A}(t) = \{\omega(s) \mid t \to^* s\}\,.$$

Auxiliary sets have the following property.

Lemma 5.4. *Let $t \in \mathcal{T}er(\Sigma_\bot, X)$. The set $\mathcal{A}(t)$ is directed.*

Proof. The set $\mathcal{A}(t)$ is non-empty, as follows from the fact that $\omega(t) \in \mathcal{A}(t)$. Moreover, for all $s_1, s_2 \in \mathcal{A}(t)$ there exist an $r \in \mathcal{A}(t)$ such that $s_1 \preccurlyeq r$ and $s_2 \preccurlyeq r$, as follows from the third clause of Definition 5.1 and the assumption that all considered TRSs are confluent. □

The set $\mathcal{A}(t)$ is not necessarily a tree. Consider, for example, the TRS $\mathcal{R} = (\{c\}, \emptyset)$ with c a constant. Since there are no reduction rules, the identity function on $\mathcal{T}er(\{c\}_\bot, X)$ is a direct approximant function. Hence, we have $\mathcal{A}(c) = \{c\}$. This is not a tree, as $\bot \notin \{c\}$. However, as $\mathcal{A}(t)$ is directed and as trees are ideals we can obtain a tree by closing $\mathcal{A}(t)$ downward. This leads to the following definition of Böhm-like trees.

Definition 5.5. *If $t \in \mathcal{T}er(\Sigma_\bot, X)$, then its* Böhm-like tree, *denoted BLT(t), is defined by*

$$BLT(t) = \downarrow\!\mathcal{A}(t)\,.$$

We have for each t that $\downarrow\!\mathcal{A}(t)$ exists and is unique. Hence, BLT is a function from $\mathcal{T}er(\Sigma_\bot, X)$ to $\mathcal{T}^\infty(\Sigma_\bot, X)$. By Lemma 5.2 and the fact that root-stability is preserved under reduction, Böhm-like trees only provide information on root-stable parts.

We now give two examples of direct approximant functions and Böhm-like trees.

Example 5.6 (Trivial Trees). Given a term t, its trivial direct approximant is defined by $\omega_{\mathrm{T}}(t) = \bot$.

The three clauses of Definition 5.1 hold trivially. As we have for all $t \to^* s$ that $\omega_{\mathrm{T}}(s) = \bot$, it follows that $\mathrm{BLT}(t) = \mathcal{A}(t) = \{\bot\}$. Note that ω_{T} is minimal in the sense that it does not provide any information on root-stable subterms.

Example 5.7 (Berarducci-Like Trees). Given a term t, its Berarducci-like direct approximant ω_{BeL} replaces precisely all non-root-stable subterms of t by $\bot$.

Again, the three clauses of Definition 5.1 hold trivially. Note that ω_{BeL} is maximal in the sense that it preserves all root-stable subterms. Unfortunately, as root-stability is undecidable, ω_{BeL} is in general not computable.

Berarducci-like trees are modelled after the Berarducci trees from the λ-calculus [3]. The direct approximant function associated with the Berarducci trees also replaces precisely all non-root-stable subterms by $\bot$.

To make the Berarducci-like trees more concrete let us consider combinatory logic (CL) with the combinators S, K, and I and the usual reduction rules.

The following trees are Berarducci-like trees for CL.

$$\begin{aligned}\mathrm{BLT}(K\bot) &= \{\bot, \bot\bot, K\bot\}\\ \mathrm{BLT}(YK) &= \{\bot, \bot\bot, K\bot, \bot(\bot\bot), \ldots\}\\ \mathrm{BLT}(SII(SII)) &= \{\bot\}\end{aligned}$$

The subterm Y in the second tree denotes a term that behaves as a fixed-point combinator. In the case of the last tree note that for every $SII(SII) \to^* t$ we have $t \to^* SII(SII)$. Hence, no reduct of $SII(SII)$ is root-stable.

We end this section with a proof that Böhm-like trees are preserved under rewriting and by discussing some related work.

Proposition 5.8. *Let $s, t \in Ter(\Sigma_\bot, X)$. If $s \to^* t$, then $BLT(s) = BLT(t)$.*

Proof. Suppose $s \to^* t$. We prove $\mathrm{BLT}(s) \preccurlyeq \mathrm{BLT}(t)$ and $\mathrm{BLT}(t) \preccurlyeq \mathrm{BLT}(s)$. The result follows from the observation that the prefix order on trees is in fact subset inclusion.

By the definition of Böhm-like trees there exists for every $t'' \in \mathrm{BLT}(s)$ a term t' such that $s \to^* t'$ and $t'' \preccurlyeq \omega(t')$. As we assume that every TRS is confluent, there exists an r such that $t \to^* r$ and $t' \to^* r$. Thus, $\omega(r) \in \mathcal{A}(t) \subseteq \mathrm{BLT}(t)$. Moreover, by the third clause of Definition, 5.1 $\omega(t') \preccurlyeq \omega(r)$. Hence, $t'' \preccurlyeq \omega(r)$ and $\mathrm{BLT}(s) \preccurlyeq \mathrm{BLT}(t)$.

As every reduct of t is a reduct of s, we have $\mathcal{A}(t) \subseteq \mathcal{A}(s)$. By the definition of downward closure $\downarrow\mathcal{A}(t) \subseteq \downarrow\mathcal{A}(s)$. Thus, $\mathrm{BLT}(t) \preccurlyeq \mathrm{BLT}(s)$. □

In the work by Boudol [10], Blom [8], and Ariola and Blom [9] a more abstract approach is taken to defining Böhm-like trees. They use more abstract definitions of direct approximant functions.

Boudol [10] only requires of the range of the direct approximant function that its is an algebra over $\mathcal{T}er(\Sigma_\bot, X)$. The range does not need to be $\mathcal{T}er(\Sigma_\bot, X)$. In correspondence with this, Boudol drops the first clause of Definition 5.1.

Blom [8] and Ariola and Blom [9] require the domain of the direct approximant function only to be an ARS $\mathcal{A} = (A, \to)$ with a partial order on A. The ARS $\mathcal{A}$ does not need to be confluent. The range of the direct approximant function may be an arbitrary (complete) partial order. In correspondence with this, they drop the first and second clause of Definition 5.1. They also add a new clause to compensate for the fact that $\mathcal{A}$ does not need to be confluent.

6 Direct Approximant TRSs

In this section we define a class of confluent and terminating TRSs, the direct approximant TRSs (ωTRSs). We prove that the function that assigns to each term in such a TRS its unique normal form is a direct approximant function. In the next section we prove that the Böhm-like trees based on ωTRSs are monotone and continuous.

Not every direct approximant function can be defined by means of a confluent and terminating TRS. An example is the function ω_{BeL} from the previous section. This function cannot be defined by means of a TRS, as unique normal forms of confluent and terminating TRSs are always computable, while root-stability and, hence, ω_{BeL} is not.

As in the case of direct approximant functions, the definition of ωTRSs is relative to a given TRS. The definition summarises the properties shared between the TRSs used to define direct approximant functions in earlier papers [6,10,11, 15].

Definition 6.1. *Let $\mathcal{R} = (\Sigma, R)$ be a confluent left-linear TRS. A* direct approximant TRS (ωTRS) *of $\mathcal{R}$ is a left-linear TRS $\mathcal{D} = (\Sigma_\bot, D)$, whose rewrite relation, denoted $\to_\omega$, satisfies*

1. *$e = \bot$ for all $d \to_\omega e \in D$,*
2. *$\bot$ is a normal form with respect to $\to_\omega$,*
3. *$t \to_\omega^* \bot$ for all $t \preccurlyeq d$ with $d \to_\omega \bot \in D$ (see Fig. 2), and*
4. *$l \to_\omega^* \bot$ for all $l \to r \in R$.*

In the remainder of this section we assume $\mathcal{R} = (\Sigma, R)$ is a confluent left-linear TRS and $\mathcal{D} = (\Sigma_\bot, D)$ is a ωTRS of $\mathcal{R}$. We proceed as follows. First, we give an example of a ωTRS. Then, we prove ωTRSs are confluent and terminating using the first, second, and third clause of Definition 6.1. Finally, we prove that the unique normal forms define direct approximants using the third and fourth clause.

Example 6.2 (Huet-Lévy ωTRSs). The rewrite rules of the Huet-Lévy ωTRS are all rules of the form $t \to_\omega \bot$ such that $\bot \neq t \preccurlyeq l$ and $l \to r \in R$.

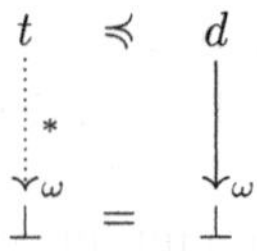

Fig. 2. Definition 6.1.(3)

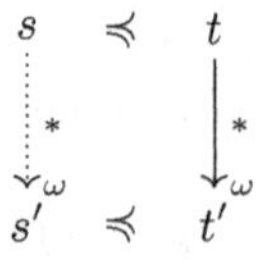

Fig. 3. Lemma 6.7

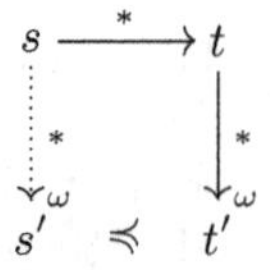

Fig. 4. Lemma 6.8

The four clauses of Definition 6.1 follow trivially from the definition of Huet-Lévy ωTRSs. The direct approximant function defined by a Huet-Lévy ωTRS originates from the work by Huet and Lévy [16]. The first formulation as a TRS is by Klop and Middeldorp [15]. The definition of Klop and Middeldorp differs slightly from ours, but equality of the transitive-reflexive closures follows easily with the help of Lemma 3.4.

The Huet-Lévy TRS for CL has no less than 28 rewrite rules. However, using the fact that the third clause of Definition 6.1 is formulated in terms of the transitive-reflexive closure of $\to_\omega$, we can define a ωTRS with the same transitive-reflexive closure but with only four rewrite rules.

$$\begin{array}{rr} Sxyz \to_\omega \bot & Kxy \to_\omega \bot \\ Ix \to_\omega \bot & \bot x \to_\omega \bot \end{array}$$

Hence, the formulation of the third clause of Definition 6.1 enables us to define more "economic" ωTRSs.

To prove confluence of $\mathcal{D}$ we first show that confluence holds for ωTRSs for which the third clause of Definition 6.1 can be strengthened to

$t \to_\omega^= \bot$ *for all* $t \preccurlyeq d$ *with* $d \to_\omega \bot \in D$.

That is, t must rewrite to $\bot$ in at most one step and not just in finitely many steps. We call ωTRSs with this strengthened third clause *single-step ωTRSs* .

Lemma 6.3. *If $\mathcal{E} = (\Sigma_\bot, E)$ is a single-step ωTRS, then $\mathcal{E}$ is confluent.*

Proof. The ωTRS $\mathcal{E} = (\Sigma_\bot, E)$ is subcommutative by the first clause of Definition 6.1 and the single-step assumption. Confluence is implied by subcommutativity [13, Lemma 2.7.4]. □

Using confluence of single-step ωTRSs we can prove confluence of $\mathcal{D}$.

Proposition 6.4. *The ωTRS $\mathcal{D}$ is confluent.*

Proof. Define a TRS $\mathcal{E} = (\Sigma_\bot, E)$, such that $t \to_\omega \bot \in E$ for all $t \in \mathcal{T}er(\Sigma_\bot, X)$ with $\bot \neq t \preccurlyeq d$ and $d \to_\omega \bot \in D$. The TRS $\mathcal{E}$ is a single-step ωTRS, as follows easily from its definition. Moreover, by the definition of $\mathcal{E}$, the transitive-reflexive closures of $\mathcal{D}$ and $\mathcal{E}$ are equal. Hence, $\mathcal{D}$ is confluent by Lemma 6.3. □

To prove termination of $\mathcal{D}$ we need the following lemma with respect to the rewrite relation of $\mathcal{D}$.

Lemma 6.5. *Let* $s, t \in \mathcal{T}er(\Sigma_\perp, X)$. *If* $s \to_\omega t$, *then* $t \prec s$.

Proof. By the first clause of Definition 6.1 a reduction step $s \to_\omega t$ is a replacement of a subterm s' at a position p in s by $\perp$. As $\perp \prec s'$, we have $t = s[\perp]_p \prec s[s']_p = s$. □

We can now prove termination.

Proposition 6.6. *The* ω*TRS* $\mathcal{D}$ *is terminating.*

Proof. By Lemma 6.5 and Proposition 3.3. □

By Propositions 6.4 and 6.6 each term t in $\mathcal{D}$ has a unique normal form. We denote this unique normal form by $\omega(t)$. We now prove that ω defines a direct approximant function. In order to do this, we first prove three lemmas.

Lemma 6.7. *Let* $s, t, t' \in \mathcal{T}er(\Sigma_\perp, X)$. *If* $s \preccurlyeq t$ *and* $t \to_\omega^* t'$, *then there exists an* $s' \in \mathcal{T}er(\Sigma_\perp, X)$ *such that* $s' \preccurlyeq t'$ *and* $s \to_\omega^* s'$ *(see Fig. 3).*

Proof. We give a proof for the case $t \to_\omega t'$. The result follows by induction on the length of $t \to_\omega^* t'$.

Suppose the redex contracted in $t \to_\omega t'$ occurs at position p. There are two cases to consider depending on the occurrence of p in s.

The position p *does not occur in* s. By the definition of the prefix order there exists a $q \leq p$ such that $s|_q = \perp$. Define $s' = s$. As $t \to_\omega t'$ replaces the subterm at position p by $\perp$ we have by $s|_q = \perp$ and $q \leq p$ that $s \preccurlyeq t'$. Moreover, $s \to_\omega^* s = s'$.

The position p *occurs in* s. In this case, $s|_p \preccurlyeq t|_p$. As $t|_p$ is a redex, we have by Lemma 3.4 and the third clause of Definition 6.1 that $s|_p \to_\omega^* \perp = t'|_p$. Define $s' = s[\perp]_p$. As $t' = t[\perp]_p$, we have $s' \preccurlyeq t'$. Moreover, as $s|_p \to_\omega^* \perp$, we have $s \to_\omega^* s'$. □

Lemma 6.8. *Let* $s, t, t' \in \mathcal{T}er(\Sigma_\perp, X)$. *If* $s \to^* t$ *and* $t \to_\omega^* t'$, *then there exists an* $s' \in \mathcal{T}er(\Sigma_\perp, X)$ *such that* $s \to_\omega^* s'$ *and* $s' \preccurlyeq t'$ *(see Fig. 4).*

Proof. We give a proof for the case $s \to t$. The result follows by induction on the length of $s \to^* t$.

Suppose the redex contracted in $s \to t$ occurs at position p. As $s[\perp]_p \preccurlyeq t$, there exists by Lemma 6.7 an s' such that $s' \preccurlyeq t'$ and $s[\perp]_p \to_\omega^* s'$. Moreover, $s \to_\omega^* s'$, because by the fourth clause of Definition 6.1 $s \to_\omega^* s[\perp]_p$. □

Lemma 6.9. *Let* $s, t \in \mathcal{T}er(\Sigma_\perp, X)$. *The following properties hold*

1. $\omega(t) \preccurlyeq t$,
2. $\omega(t) = \omega(t[\omega(t|_p)]_p)$ *for all* $p \in \mathcal{P}os(t)$,
3. $\omega(\omega(t)) = \omega(t)$,
4. $\omega(s) \preccurlyeq \omega(t)$ *if* $s \preccurlyeq t$, *and*
5. $\omega(s) \preccurlyeq \omega(t)$ *if* $s \to t$.

Proof. 1. As $\omega(t)$ is the unique normal form of t, we have $t \to_\omega^* \omega(t)$. The result follows by repeated application of Lemma 6.5.
2. For every $t|_p \to_\omega^* s$ we have $t = t[t|_p]_p \to_\omega^* t[s]_p$. Hence, as $t|_p \to_\omega^* \omega(t|_p)$, the result follows by confluence of ωTRSs.
3. By the second clause of the current lemma with $p = \epsilon$.
4. As $t \to_\omega^* \omega(t)$, there exists by Lemma 6.7 an s' such that $s' \preccurlyeq \omega(t)$. Moreover, by confluence of ωTRSs $\omega(s') = \omega(s)$ and by the first clause of the current lemma $\omega(s') \preccurlyeq s'$. Hence, by transitivity of the prefix order $\omega(s) \preccurlyeq \omega(t)$.
5. Analogous to the fourth clause of the current lemma using Lemma 6.8 instead of Lemma 6.7. □

We can now prove the following theorem.

Theorem 6.10. *The function $\omega : \mathcal{T}er(\Sigma_\perp, X) \to \mathcal{T}er(\Sigma_\perp, X)$ which assigns to each term its unique normal form with respect to $\mathcal{D}$ is a direct approximant function.*

Proof. The first clause of Definition 5.1 follows from Lemma 6.9.(1). The second clause follows from the fourth clause of Definition 6.1 and the fact that ωTRSs are confluent. The third clause follows from Lemma 6.9.(5). □

We now know that each ωTRS defines direct approximant function. Hence, it also defines a Böhm-like tree.

Example 6.11 (Huet-Lévy Trees). The Huet-Lévy ωTRS of Definition 6.2 defines the *Huet-Lévy tree.*

The Huet-Lévy tree is the Böhm-like tree already defined by Boudol [10] and Ariola [11].

Huet-Lévy trees provide more information than the trivial trees, but less than the Berarducci-like trees. For example, given the TRS with the single rewrite rule $f(a) \to b$ we have the following trees.

$$\begin{array}{rr} \mathrm{BLT}_{\mathrm{T}}(f(\perp)) = \{\perp\} & \mathrm{BLT}_{\mathrm{T}}(f(a)) = \{\perp\} \\ \mathrm{BLT}_{\mathrm{HL}}(f(\perp)) = \{\perp\} & \mathrm{BLT}_{\mathrm{HL}}(f(a)) = \{\perp, b\} \\ \mathrm{BLT}_{\mathrm{BeL}}(f(\perp)) = \{\perp, f(\perp)\} & \mathrm{BLT}_{\mathrm{BeL}}(f(a)) = \{\perp, b\} \end{array}$$

7 Monotonicity and Continuity

In this section we prove that a Böhm-like tree whose the direct approximant function can be defined by means of a ωTRS is monotone and continuous. As in the previous section, we assume $\mathcal{R} = (\Sigma, R)$ is an confluent left-linear TRS and $\mathcal{D} = (\Sigma_\perp, D)$ is a ωTRS of $\mathcal{R}$.

Proposition 7.1. *The Böhm-like tree defined by $\mathcal{D}$ is a monotone function. That is, for all $s, t \in \mathcal{T}er(\Sigma_\perp, X)$, if $s \preccurlyeq t$, then $BLT(s) \preccurlyeq BLT(t)$.*

Proof. Let $s, t \in \mathcal{T}er(\Sigma_\perp, X)$ such that $s \preccurlyeq t$. Suppose $s'' \in \text{BLT}(s)$. By the definition of $\text{BLT}(s)$ there exists an s' such that $s'' \preccurlyeq \omega(s')$ and $s \to^* s'$. As all assumed TRSs are left-linear, there exists a t' such that $t \to^* t'$ and $s' \preccurlyeq t'$. By Lemma 6.9.(4) we have $\omega(s') \preccurlyeq \omega(t')$. Hence, as $\omega(t') \in \text{BLT}(t)$, we also have $\text{BLT}(s) \preccurlyeq \text{BLT}(t)$. □

Proposition 7.2. *The Böhm-like tree defined by $\mathcal{D}$ is a continuous function. That is, if $t \in \mathcal{T}er(\Sigma_\perp, X)$, then $BLT(t) = \bigsqcup\{BLT(s) \mid s \preccurlyeq t\}$.*

Proof. Let $t \in \mathcal{T}er(\Sigma_\perp, X)$. As $t \preccurlyeq t$, we have $\text{BLT}(t) \in \{\text{BLT}(s) \mid s \preccurlyeq t\}$. Thus, $\text{BLT}(t) \preccurlyeq \bigsqcup\{\text{BLT}(s) \mid s \preccurlyeq t\}$. Moreover, by Proposition 7.1 we have for all $s \preccurlyeq t$ that $\text{BLT}(s) \preccurlyeq \text{BLT}(t)$ and, thus, $\bigsqcup\{\text{BLT}(s) \mid s \preccurlyeq t\} \preccurlyeq \text{BLT}(t)$. Combining both facts, we get the result. □

From the above two propositions we can conclude that the Huet-Lévy trees of the previous section are monotone and continuous. Note that Ariola [11] already proves this.

There exist Böhm-like trees that are not monotone and continuous. Consider, for example, the TRS with the single rewrite rule $f(a) \to b$ and its Berarducci-like tree. Given the terms $f(\perp)$ and $f(a)$ we have that $f(\perp) \preccurlyeq f(a)$, but

$$\text{BLT}(f(\perp)) = \{\perp, f(\perp)\} \not\preccurlyeq \{\perp, b\} = \text{BLT}(f(a))$$

and

$$\text{BLT}(f(a)) = \{\perp, b\} \neq \{\perp, b, f(\perp)\} = \bigcup\{\text{BLT}(s) \mid s \preccurlyeq f(a)\}\,.$$

In fact, the last set is not even a tree.

8 Further Directions

There are at least four interesting directions for further research. First, does pre-congruence hold for the presented Böhm-like trees, as it does for the Böhm-like trees of the λ-calculus [1–3] and Huet-Lévy trees [11]? That is, suppose $C[\square]$ is a context and s and t are terms, does it hold that

$$\text{BLT}(C[s]) \preccurlyeq \text{BLT}(C[t]) \text{ if } \text{BLT}(s) \preccurlyeq \text{BLT}(t)$$

Second, can we extend Böhm-like trees to higher-order rewriting systems, such that we also cover the Böhm-like trees of the λ-calculus? Third, similar to Berarducci-like trees and the Berarducci trees of the λ-calculus, do Böhm trees [1] and Lévy-Longo trees [2] have a counterpart for TRSs? Fourth, how does the current approach relate to the infinitary rewriting approach [12]?

Acknowledgements

I would like to thank Stefan Blom, Bas Luttik, Jan Willem Klop, Femke van Raamsdonk, Roel de Vrijer, and the anonymous referees for their helpful comments and remarks.

References

1. Barendregt, H.P.: The Lambda Calculus: Its Syntax and Semantics. Second edn. Elsevier Science (1985)
2. Lévy, J.J.: An algebraic interpretation of the $\lambda\beta$K-calculus and the labelled λ-calculus. In Böhm, C., ed.: λ-calculus and Computer Science Theory. Volume 37 of LNCS. Springer-Verlag (1975) 147–165
3. Berarducci, A.: Infinite λ-calculus and non-sensible models. In Ursini, A., Aglianò, P., eds.: Logic and Algebra, Marcel Dekker (1996) 339–378
4. Lévy, J.J.: Réductions correctes et optimales dans le lambda-calcul. PhD thesis, Université de Paris VII (1978)
5. Hyland, M.: A syntactic characterization of the equality in some models for the lambda calculus. Journal of the London Mathematical Society (2) **12** (1976) 361–370
6. Wadsworth, C.P.: The relation between computational and denotational properties for Scott's D_∞-models of the lambda-calculus. SIAM Journal on Computing **5** (1976) 488–521
7. Kennaway, J.R., Klop, J.W., Sleep, M., de Vries, F.J.: Infinitary lambda calculus. Theoretical Computer Science **175** (1997) 93–125
8. Blom, S.C.C.: Term Graph Rewriting: syntax and semantics. PhD thesis, Vrije Universiteit Amsterdam (2001)
9. Ariola, Z.M., Blom, S.: Skew confluence and the lambda calculus with letrec. Annals of Pure and Applied Logic **117** (2002) 95–168
10. Boudol, G.: Computational semantics of term rewriting systems. In Nivat, M., Reynolds, J.C., eds.: Algebraic methods in semantics. Cambridge University Press (1985) 169–236
11. Ariola, Z.M.: Relating graph and term rewriting via Böhm models. Applicable Algebra in Engineering, Communication and Computing **7** (1996) 401–426
12. Kennaway, R., van Oostrom, V., de Vries, F.J.: Meaningless terms in rewriting. The Journal of Functional and Logic Programming **1** (1999)
13. Baader, F., Nipkow, T.: Term Rewriting and All That. Cambridge University Press (1998)
14. Stoltenberg-Hansen, V., Lindström, I., Griffor, E.R.: Mathematical Theory of Domains. Cambridge University Press (1994)
15. Klop, J.W., Middeldorp, A.: Sequentiality in orthogonal term rewriting systems. Journal of Symbolic Computation **12** (1991) 161–195
16. Huet, G., Lévy, J.J.: Computations in orthogonal rewriting systems. In Lassez, J.L., Plotkin, G., eds.: Computational Logic. MIT Press (1991) 395–443

Dependency Pairs Revisited

Nao Hirokawa and Aart Middeldorp

Institute of Computer Science
University of Innsbruck
6020 Innsbruck, Austria

Abstract. In this paper we present some new refinements of the dependency pair method for automatically proving the termination of term rewrite systems. These refinements are very easy to implement, increase the power of the method, result in simpler termination proofs, and make the method more efficient.

1 Introduction

Since the introduction of the dependency pair method (Arts and Giesl [1]) and the monotonic semantic path order (Borralleras, Ferreira, and Rubio [5]), two powerful methods that facilitate termination proofs that can be obtained automatically, there is a renewed interest in the study of termination for term rewrite systems. Three important issues which receive a lot of attention in current research on termination are to make these methods faster, to improve the methods such that more and more (challenging) rewrite systems can be handled, and to extend the methods beyond the realm of ordinary first-order term rewriting. Especially in connection with the dependency pair method many improvements, extensions, and refinements have been published. The method forms an important ingredient in several software tools for proving terminating. To mention a few (in order of appearance): CiME [6], T$_\mathsf{T}$T [15], AProVE [12], and TORPA [24].

In this paper we go back to the foundations of the dependency pair method. Starting from scratch, we give a systematic account of the method. Along the way we derive two new refinements which are very easy to implement, increase the termination proving power[1], give rise to simpler termination proofs, and make the method much faster.

We use the following term rewrite system (TRS for short) from Dershowitz [7] to illustrate the developments in the remainder of the paper:

$$
\begin{aligned}
1:&\quad \neg\neg x \to x\\
2:&\quad \neg(x \vee y) \to \neg x \wedge \neg y\\
3:&\quad \neg(x \wedge y) \to \neg x \vee \neg y\\
4:&\quad x \wedge (y \vee z) \to (x \wedge y) \vee (x \wedge z)\\
5:&\quad (y \vee z) \wedge x \to (x \wedge y) \vee (x \wedge z)
\end{aligned}
$$

[1] Note however the discussion at the end of Section 5.

V. van Oostrom (Ed.): RTA 2004, LNCS 3091, pp. 249–268, 2004.

Termination of this TRS is easily shown by the multiset path order. This, however, does not mean that automatic termination tools easily find a termination proof. For instance, both CiME and the fully automatic "Meta Combination" algorithm in AProVE 1.0 fail to prove termination.

We assume familiarity with the basics of term rewriting ([3, 20]). We just recall some basic notation and terminology. The set of terms $\mathcal{T}(\mathcal{F}, \mathcal{V})$ constructed from a signature $\mathcal{F}$ and a disjoint set $\mathcal{V}$ of variables is abbreviated to $\mathcal{T}$ when no confusion can arise. The set of variables appearing in a term t is denoted by $\mathcal{V}\mathsf{ar}(t)$. The root symbol of a term t is denoted by $\mathrm{root}(t)$. Defined function symbols are root symbols of left-hand sides of rewrite rules. We use $\xrightarrow{\epsilon}$ to denote root rewrite steps and $\xrightarrow{>\epsilon}$ to denote rewrite steps in which the selected redex occurs below the root. A substitution is a mapping σ from variables to terms such that its domain $\mathcal{D}\mathsf{om}(\sigma) = \{x \mid x \neq \sigma(x)\}$ is finite. We write $t\sigma$ to denote the result of applying the substitution σ to the term t. A relation R on terms is closed under substitutions if $(s\sigma, t\sigma) \in R$ whenever $(s,t) \in R$, for all substitutions σ. We say that R is closed under contexts if $(u[s]_p, u[t]_p) \in R$ whenever $(s,t) \in R$, for all terms u and positions p in u. The superterm relation is denoted by $\trianglerighteq$ (i.e., $s \trianglerighteq t$ if t is a subterm of s) and $\vartriangleright$ denotes its strict part.

2 Dependency Pairs

In this section and in Section 4 we recall the basics of the dependency pair method of Arts and Giesl [1]. We provide proofs of all results.

Let us start with some easy observations. If a TRS $\mathcal{R}$ is not terminating then there must be a *minimal* non-terminating term, minimal in the sense that all its proper subterms are terminating. Let us denote the set of all minimal non-terminating terms by $\mathcal{T}_\infty$.

Lemma 1. *For every term $t \in \mathcal{T}_\infty$ there exists a rewrite rule $l \to r$, a substitution σ, and a non-variable subterm u of r such that $t \xrightarrow{>\epsilon}{}^* l\sigma \xrightarrow{\epsilon} r\sigma \trianglerighteq u\sigma$ and $u\sigma \in \mathcal{T}_\infty$.*

Proof. Let A be an infinite rewrite sequence starting at t. Since all proper subterms of t are terminating, A must contain a root rewrite step. By considering the first root rewrite step in A it follows that there exist a rewrite rule $l \to r$ and a substitution σ such that A starts with $t \xrightarrow{>\epsilon}{}^* l\sigma \xrightarrow{\epsilon} r\sigma$. Write $l = f(l_1, \ldots, l_n)$. Since the rewrite steps in $t \to^* l\sigma$ take place below the root, $t = f(t_1, \ldots, t_n)$ and $t_i \to^* l_i\sigma$ for all $1 \leqslant i \leqslant n$. By assumption the arguments $t_1, \ldots, t_n$ of t are terminating. Hence so are the terms $l_1\sigma, \ldots, l_n\sigma$. It follows that $\sigma(x)$ is terminating for every $x \in \mathcal{V}\mathsf{ar}(r) \subseteq \mathcal{V}\mathsf{ar}(l)$. As $r\sigma$ is non-terminating it has a subterm $t' \in \mathcal{T}_\infty$. Because non-terminating terms cannot occur in the substitution part, there must be a non-variable subterm u of r such that $t' = u\sigma$. □

Observe that the term $l\sigma$ in Lemma 1 belongs to $\mathcal{T}_\infty$ as well. Further note that $u\sigma$ cannot be a proper subterm of $l\sigma$ (since all arguments of $l\sigma$ are terminating).

Corollary 2. *Every term in $\mathcal{T}_\infty$ has a defined root symbol.* □

If we were to define a new TRS $\mathcal{S}$ consisting of all rewrite rules $l \to u$ for which there exist a rewrite rule $l \to r \in \mathcal{R}$ and a subterm u of r with defined function symbol, then the sequence in the conclusion of Lemma 1 is of the form $\xrightarrow{>\epsilon}{}^{*}_{\mathcal{R}} \cdot \xrightarrow{\epsilon}_{\mathcal{S}}$. The idea is now to get rid of the position constraints by marking the root symbols of the terms in the rewrite rules of $\mathcal{S}$.

Definition 3. *Let $\mathcal{R}$ be a TRS over a signature $\mathcal{F}$. Let $\mathcal{F}^\sharp$ denote the union of $\mathcal{F}$ and $\{f^\sharp \mid f$ is a defined symbol of $\mathcal{R}\}$ where $f^\sharp$ is a fresh function symbol with the same arity as f. We call these new symbols* dependency pair symbols. *Given a term $t = f(t_1, \ldots, t_n) \in \mathcal{T}(\mathcal{F}, \mathcal{V})$ with f a defined symbol, we write $t^\sharp$ for the term $f^\sharp(t_1, \ldots, t_n)$. If $l \to r \in \mathcal{R}$ and u is a subterm of r with defined root symbol such that u is not a proper subterm of l then the rewrite rule $l^\sharp \to u^\sharp$ is called a* dependency pair *of $\mathcal{R}$. The set of all dependency pairs of $\mathcal{R}$ is denoted by $\mathsf{DP}(\mathcal{R})$.*

The idea of excluding dependency pairs $l^\sharp \to u^\sharp$ where u is a proper subterm of l is due to Dershowitz [8]. Although dependency pair symbols are defined symbols of $\mathsf{DP}(\mathcal{R})$, they are not defined symbols of $\mathcal{R}$. In the following, defined symbols always refer to the original TRS $\mathcal{R}$.

Example 4. The example in the introduction admits the following 9 dependency pairs:

$$
\begin{array}{rlrl}
6: & \neg^\sharp(x \vee y) \to \neg x \wedge^\sharp \neg y & & \\
7: & \neg^\sharp(x \vee y) \to \neg^\sharp x & 11: & x \wedge^\sharp (y \vee z) \to x \wedge^\sharp y \\
8: & \neg^\sharp(x \vee y) \to \neg^\sharp y & 12: & x \wedge^\sharp (y \vee z) \to x \wedge^\sharp z \\
9: & \neg^\sharp(x \wedge y) \to \neg^\sharp x & 13: & (y \vee z) \wedge^\sharp x \to x \wedge^\sharp y \\
10: & \neg^\sharp(x \wedge y) \to \neg^\sharp y & 14: & (y \vee z) \wedge^\sharp x \to x \wedge^\sharp z
\end{array}
$$

Lemma 5. *For every term $s \in \mathcal{T}_\infty$ there exist terms $t, u \in \mathcal{T}_\infty$ such that $s^\sharp \to^*_{\mathcal{R}} t^\sharp \to_{\mathsf{DP}(\mathcal{R})} u^\sharp$.*

Proof. Immediate from Lemma 1, Corollary 2, and the preceding definition. □

Definition 6. *For any subset $T \subseteq \mathcal{T}$ consisting of terms with a defined root symbol, we denote the set $\{t^\sharp \mid t \in T\}$ by $T^\sharp$.*

An immediate consequence of the previous lemma is that for every non-terminating TRS $\mathcal{R}$ there exists an infinite rewrite sequence of the form

$$t_1 \to^*_{\mathcal{R}} t_2 \to_{\mathsf{DP}(\mathcal{R})} t_3 \to^*_{\mathcal{R}} t_4 \to_{\mathsf{DP}(\mathcal{R})} \cdots$$

with $t_i \in \mathcal{T}^\sharp_\infty$ for all $i \geqslant 1$. Hence, to prove termination of a TRS $\mathcal{R}$ it is sufficient to show that $\mathcal{R} \cup \mathsf{DP}(\mathcal{R})$ does not admit such infinite sequences. Every such sequence contains a tail in which all applied dependency pairs are used infinitely many times. For finite TRSs, the set of those dependency pairs forms a cycle in the dependency graph. *From now on, we assume that all TRSs are finite.*

As a side remark, note that all terms in $\mathcal{T}^\sharp_\infty$ are terminating with respect to $\mathcal{R}$ but admit an infinite rewrite sequence with respect to $\mathcal{R} \cup \mathsf{DP}(\mathcal{R})$.

Definition 7. *The nodes of the* dependency graph $\mathsf{DG}(\mathcal{R})$ *are the dependency pairs of* $\mathcal{R}$ *and there is an arrow from* $s \to t$ *to* $u \to v$ *if and only if there exist substitutions* σ *and* τ *such that* $t\sigma \to^*_{\mathcal{R}} u\tau$. *A* cycle *is a nonempty subset* $\mathcal{C}$ *of dependency pairs of* $\mathsf{DP}(\mathcal{R})$ *if for every two (not necessarily distinct) pairs* $s \to t$ *and* $u \to v$ *in* $\mathcal{C}$ *there exists a nonempty path in* $\mathcal{C}$ *from* $s \to t$ *to* $u \to v$.

Definition 8. *Let* $\mathcal{C} \subseteq \mathsf{DP}(\mathcal{R})$. *An infinite rewrite sequence in* $\mathcal{R} \cup \mathcal{C}$ *of the form*

$$t_1 \to^*_{\mathcal{R}} t_2 \to_{\mathcal{C}} t_3 \to^*_{\mathcal{R}} t_4 \to_{\mathcal{C}} \cdots$$

with $t_1 \in \mathcal{T}^\sharp_\infty$ *is called* $\mathcal{C}$-minimal *if all rules in* $\mathcal{C}$ *are applied infinitely often.*

Hence proving termination boils down to proving the absence of $\mathcal{C}$-minimal rewrite sequences, for any cycle $\mathcal{C}$ in the dependency graph $\mathsf{DG}(\mathcal{R})$.

Example 9. The example in the introduction has the following dependency graph:

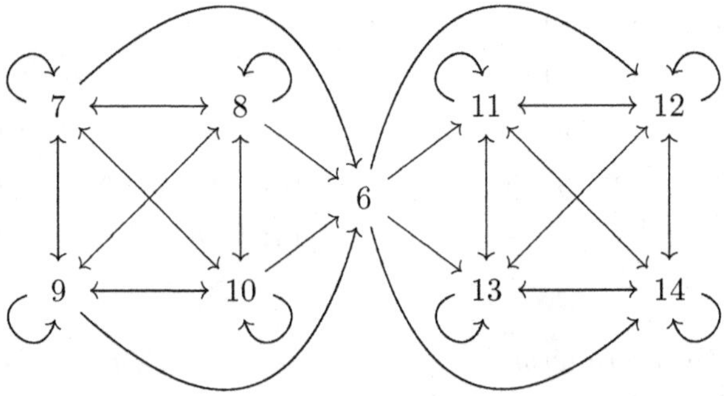

It contains 30 cycles: all nonempty subsets of both $\{7, 8, 9, 10\}$ and $\{11, 12, 13, 14\}$.

Although the dependency graph is not computable in general, sound approximations exist that can be computed efficiently (see [1, 17]). Soundness here means that every cycle in the real dependency graph is a cycle in the approximated graph. For the example TRS all known approximations compute the real dependency graph.

3 Subterm Criterion

We now present a new criterion which permits us to ignore certain cycles of the dependency graph.

Definition 10. *Let $\mathcal{R}$ be a TRS and $\mathcal{C} \subseteq \mathsf{DP}(\mathcal{R})$ such that every dependency pair symbol in $\mathcal{C}$ has positive arity. A* simple projection *for $\mathcal{C}$ is a mapping π that assigns to every n-ary dependency pair symbol $f^\sharp$ in $\mathcal{C}$ an argument position $i \in \{1, \ldots, n\}$. The mapping that assigns to every term $f^\sharp(t_1, \ldots, t_n) \in \mathcal{T}^\sharp$ with $f^\sharp$ a dependency pair symbol in $\mathcal{C}$ its argument at position $\pi(f^\sharp)$ is also denoted by π.*

Theorem 11. *Let $\mathcal{R}$ be a TRS and let $\mathcal{C}$ be a cycle in $\mathsf{DG}(\mathcal{R})$. If there exists a simple projection π for $\mathcal{C}$ such that $\pi(\mathcal{C}) \subseteq \trianglerighteq$ and $\pi(\mathcal{C}) \cap \vartriangleright \neq \varnothing$ then there are no $\mathcal{C}$-minimal rewrite sequences.*

Before presenting the proof, let us make some clarifying remarks about the notation. If R is a set of rewrite rules and O is a relation on terms then the expression $\pi(R)$ denotes the set $\{\pi(l) \to \pi(r) \mid l \to r \in R\}$, the inclusion $R \subseteq O$ abbreviates "$(l, r) \in O$ for all $l \to r \in O$", and the inequality $R \cap O \neq \varnothing$ abbreviates "$(l, r) \in O$ for at least one $l \to r \in O$". So the conditions state that after applying the simple projection π, every rule in $\mathcal{C}$ is turned into an identity or a rule whose right-hand side is a proper subterm of the left-hand side. Moreover, the latter case applies at least once.

Proof. Suppose to the contrary that there exists a $\mathcal{C}$-minimal rewrite sequence:

$$t_1 \to_{\mathcal{R}}^* u_1 \to_{\mathcal{C}} t_2 \to_{\mathcal{R}}^* u_2 \to_{\mathcal{C}} t_3 \to_{\mathcal{R}}^* \cdots \tag{1}$$

All terms in this sequence have a dependency pair symbol in $\mathcal{C}$ as root symbol. We apply the simple projection π to (1). Let $i \geqslant 1$.

- First consider the dependency pair step $u_i \to_{\mathcal{C}} t_{i+1}$. There exist a dependency pair $l \to r \in \mathcal{C}$ and a substitution σ such that $u_i = l\sigma$ and $t_{i+1} = r\sigma$. We have $\pi(u_i) = \pi(l)\sigma$ and $\pi(t_{i+1}) = \pi(r)\sigma$. We have $\pi(l) \trianglerighteq \pi(r)$ by assumption. So $\pi(l) = \pi(r)$ or $\pi(l) \vartriangleright \pi(r)$. In the former case we trivially have $\pi(u_i) = \pi(t_{i+1})$. In the latter case the closure under substitutions of $\vartriangleright$ yields $\pi(u_i) \vartriangleright \pi(t_{i+1})$. Because of the assumption $\pi(\mathcal{C}) \cap \vartriangleright \neq \varnothing$, the latter holds for infinitely many i.
- Next consider the rewrite sequence $t_i \to_{\mathcal{R}}^* u_i$. All steps in this sequence take place below the root and thus we obtain the (possibly shorter) sequence $\pi(t_i) \to_{\mathcal{R}}^* \pi(u_i)$.

So by applying the simple projection π, sequence (1) is transformed into an infinite $\to_{\mathcal{R}} \cup \vartriangleright$ sequence containing infinitely many $\vartriangleright$ steps, starting from the term $\pi(t_1)$. Since the relation $\vartriangleright$ is well-founded, the infinite sequence must also contain infinitely many $\to_{\mathcal{R}}$ steps. By making repeated use of the well-known relational inclusion $\vartriangleright \cdot \to_{\mathcal{R}} \subseteq \to_{\mathcal{R}} \cdot \vartriangleright$ ($\vartriangleright$ commutes over $\to_{\mathcal{R}}$ in the terminology of [4]), we obtain an infinite $\to_{\mathcal{R}}$ sequence starting from $\pi(t_1)$. In other words, the term $\pi(t_1)$ is non-terminating with respect to $\mathcal{R}$. Let $t_1 = f^\sharp(s_1, \ldots, s_n)$. Because $t_1 \in \mathcal{T}_\infty^\sharp$, $f(s_1, \ldots, s_n)$ is a minimal non-terminating term. Consequently, its argument $\pi(t_1) = s_{\pi(f^\sharp)}$ is terminating with respect to $\mathcal{R}$, providing the desired contradiction. □

The remarkable thing about the above theorem is that it permits us to discard cycles of the dependency graph without considering any rewrite rules. This is extremely useful. Moreover, the criterion is very simple to check.

Example 12. Consider the cycle $\mathcal{C} = \{7, 8, 9, 10\}$. The only dependency pair symbol in $\mathcal{C}$ is $\neg^\sharp$. Since $\neg^\sharp$ is a unary function symbol, there is just one simple projection for $\mathcal{C}$: $\pi(\neg^\sharp) = 1$. By applying π to $\mathcal{C}$, we obtain

$$\begin{aligned} 7\colon &\quad x \vee y \to x \\ 8\colon &\quad x \vee y \to y \\ 9\colon &\quad x \wedge y \to x \\ 10\colon &\quad x \wedge y \to y \end{aligned}$$

We clearly have $\pi(\mathcal{C}) \subseteq \rhd$. Hence we can ignore $\mathcal{C}$ (and all its subcycles). The only cycles that are not handled by the criterion of Theorem 11 are the ones that involve 13 or 14; applying the simple projection $\pi(\wedge^\sharp) = 1$ produces

$$\begin{aligned} 13\colon &\quad y \vee z \to x \\ 14\colon &\quad y \vee z \to x \end{aligned}$$

whereas $\pi(\wedge^\sharp) = 2$ gives

$$\begin{aligned} 13\colon &\quad x \to y \\ 14\colon &\quad x \to z \end{aligned}$$

None of these rules are compatible with $\trianglerighteq$.

In implementations one shouldn't compute all cycles of the dependency graph (since there can be exponentially many in the number of dependency pairs), but use the technique of Hirokawa and Middeldorp [14] to recursively solve strongly connected components (which gives rise to a linear algorithm): if all pairs in a strongly connected component (SCC for short) are compatible with $\trianglerighteq$ after applying a simple projection, the ones that are compatible with $\rhd$ are removed and new SCCs among the remaining pairs are computed. This is illustrated in the final two examples in this section. The last example furthermore shows that the subterm criterion is capable of proving the termination of TRSs that are considered to be challenging in the termination literature (cf. the remarks in [10, Example 9]).

Example 13. Consider the following TRS from [7]:

$$\begin{aligned} 1\colon &\quad \mathsf{sort}([\,]) \to [\,] \\ 2\colon &\quad \mathsf{sort}(x : y) \to \mathsf{insert}(x, \mathsf{sort}(y)) \\ 3\colon &\quad \mathsf{insert}(x, [\,]) \to x : [\,] \\ 4\colon &\quad \mathsf{insert}(x, v : w) \to \mathsf{choose}(x, v : w, x, v) \\ 5\colon &\quad \mathsf{choose}(x, v : w, y, 0) \to x : (v : w) \\ 6\colon &\quad \mathsf{choose}(x, v : w, 0, \mathsf{s}(z)) \to v : \mathsf{insert}(x, w) \\ 7\colon &\quad \mathsf{choose}(x, v : w, \mathsf{s}(y), \mathsf{s}(z)) \to \mathsf{choose}(x, v : w, y, z) \end{aligned}$$

There are 5 dependency pairs:

$$\begin{array}{rr@{\;\to\;}l}
8: & \mathsf{sort}^\sharp(x:y) & \mathsf{insert}^\sharp(x,\mathsf{sort}(y)) \\
9: & \mathsf{sort}^\sharp(x:y) & \mathsf{sort}^\sharp(y) \\
10: & \mathsf{insert}^\sharp(x,v:w) & \mathsf{choose}^\sharp(x,v:w,x,v) \\
11: & \mathsf{choose}^\sharp(x,v:w,\mathsf{0},\mathsf{s}(z)) & \mathsf{insert}^\sharp(x,w) \\
12: & \mathsf{choose}^\sharp(x,v:w,\mathsf{s}(y),\mathsf{s}(z)) & \mathsf{choose}^\sharp(x,v:w,y,z)
\end{array}$$

The dependency graph

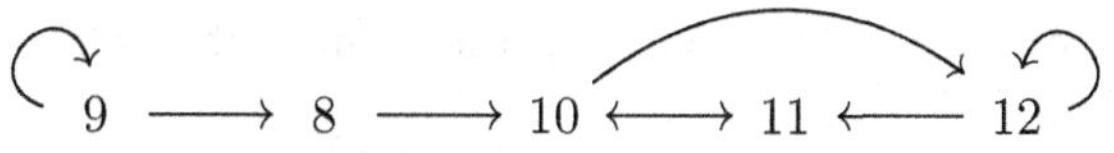

contains 2 SCCs: $\{9\}$ and $\{10,11,12\}$. The first one is handled by the simple projection $\pi(\mathsf{sort}^\sharp) = 1$:

$$9: \quad x:y \to x$$

For the other SCC we take $\pi(\mathsf{insert}^\sharp) = \pi(\mathsf{choose}^\sharp) = 2$:

$$\begin{array}{rl}
10: & v:w \to v:w \\
11: & v:w \to w \\
12: & v:w \to v:w
\end{array}$$

After removing the strictly decreasing pair 11, we are left with 10 and 12. The restriction of the dependency graph to these two pairs contains one SCC: $\{12\}$. This pair is handled by the simple projection $\pi(\mathsf{choose}^\sharp) = 3$:

$$12: \quad s(y) \to y$$

Hence the TRS is terminating.

Example 14. Consider the following TRS from [19]:

$$\begin{array}{rr@{\;\to\;}l}
1: & \mathsf{intlist}([\,]) & [\,] \\
2: & \mathsf{intlist}(x:y) & \mathsf{s}(x):\mathsf{intlist}(y) \\
3: & \mathsf{int}(\mathsf{0},\mathsf{0}) & \mathsf{0}:[\,] \\
4: & \mathsf{int}(\mathsf{0},\mathsf{s}(y)) & \mathsf{0}:\mathsf{int}(\mathsf{s}(\mathsf{0}),\mathsf{s}(y)) \\
5: & \mathsf{int}(\mathsf{s}(x),\mathsf{0}) & [\,] \\
6: & \mathsf{int}(\mathsf{s}(x),\mathsf{s}(y)) & \mathsf{intlist}(\mathsf{int}(x,y))
\end{array}$$

There are 4 dependency pairs:

$$\begin{array}{rr@{\;\to\;}l}
7: & \mathsf{intlist}^\sharp(x:y) & \mathsf{intlist}^\sharp(y) \\
8: & \mathsf{int}^\sharp(\mathsf{0},\mathsf{s}(y)) & \mathsf{int}^\sharp(\mathsf{s}(\mathsf{0}),\mathsf{s}(y)) \\
9: & \mathsf{int}^\sharp(\mathsf{s}(x),\mathsf{s}(y)) & \mathsf{intlist}^\sharp(\mathsf{int}(x,y)) \\
10: & \mathsf{int}^\sharp(\mathsf{s}(x),\mathsf{s}(y)) & \mathsf{int}^\sharp(x,y)
\end{array}$$

The dependency graph

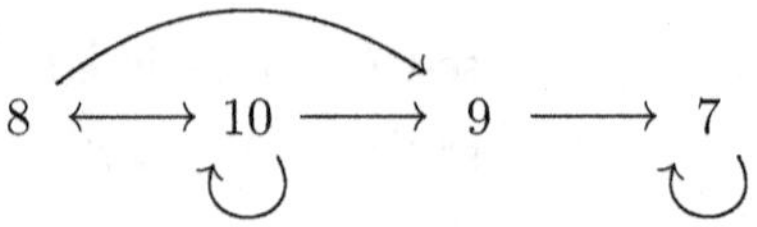

contains 2 SCCs: $\{7\}$ and $\{8, 10\}$. The first one is handled by the simple projection $\pi(\mathsf{intlist}^\sharp) = 1$:

$$7: \quad x : y \to y$$

For the second one we use the simple projection $\pi(\mathsf{int}^\sharp) = 2$:

$$\begin{aligned} 8: &\quad \mathsf{s}(y) \to \mathsf{s}(y) \\ 10: &\quad \mathsf{s}(y) \to y \end{aligned}$$

After removing the strictly decreasing pair 10, we are left with 8. Since the restriction of the dependency graph to the remaining pair 8 contains no SCCs, the TRS is terminating.

An empirical evaluation of the subterm criterion can be found in Section 6.

4 Reduction Pairs and Argument Filterings

What to do with cycles $\mathcal{C}$ of the dependency graph that cannot be handled by the criterion of the preceding section? In the dependency pair approach one uses a pair of orderings $(\gtrsim, >)$ that satisfy the properties stated below such that (1) all rules in $\mathcal{R}$ are oriented by $\gtrsim$, (2) all rules in $\mathcal{C}$ are oriented by $\gtrsim \cup >$, and (3) at least one rule in $\mathcal{C}$ is oriented by $>$.

Definition 15. *A* rewrite preorder *is a preorder (i.e., a transitive and reflexive relation) on terms which is closed under contexts and substitutions. A* reduction pair $(\gtrsim, >)$ *consists of a rewrite preorder* $\gtrsim$ *and a compatible well-founded order* $>$ *which is closed under substitutions. Compatibility means that the inclusion* $\gtrsim \cdot > \subseteq >$ *or the inclusion* $> \cdot \gtrsim \subseteq >$ *holds.*

Since we do not demand that $>$ is the strict part of the preorder $\gtrsim$, the identity $\gtrsim \cdot > = >$ need not hold, although the reduction pairs that are used in practice do satisfy this identity.

A typical example of a reduction pair is $(\geq_{\mathrm{lpo}}, >_{\mathrm{lpo}})$, where $>_{\mathrm{lpo}}$ is the lexicographic path order induced by the (strict) precedence $>$ and $\geq_{\mathrm{lpo}}$ denotes its reflexive closure. Both $\geq_{\mathrm{lpo}}$ and $>_{\mathrm{lpo}}$ are closed under contexts and the identity $\geq_{\mathrm{lpo}} \cdot >_{\mathrm{lpo}} = >_{\mathrm{lpo}}$ holds.

A general semantic construction of reduction pairs, which covers polynomial interpretations, is based on the concept of algebra. If we equip the carrier A of an algebra $\mathcal{A} = (A, \{f_{\mathcal{A}}\}_{f \in \mathcal{F}})$ with a well-founded order $>$ such that every interpretation function is weakly monotone in all arguments (i.e.,

$f_\mathcal{A}(x_1, \ldots, x_n) \geqslant f_\mathcal{A}(y_1, \ldots, y_n)$ whenever $x_i \geqslant y_i$ for all $1 \leqslant i \leqslant n$, for every n-ary function symbol $f \in \mathcal{F}$) then $(\gtrsim_\mathcal{A}, >_\mathcal{A})$ is a reduction pair. Here the relations $\gtrsim_\mathcal{A}$ and $>_\mathcal{A}$ are defined as follows: $s \gtrsim_\mathcal{A} t$ if $[\alpha]_\mathcal{A}(s) \geqslant [\alpha]_\mathcal{A}(t)$ and $s >_\mathcal{A} t$ if $[\alpha]_\mathcal{A}(s) > [\alpha]_\mathcal{A}(t)$, for all assignments α of elements of A to the variables in s and t ($[\alpha]_\mathcal{A}(\cdot)$ denotes the usual evaluation function associated with the algebra $\mathcal{A}$). In general, the relation $>_\mathcal{A}$ is not closed under contexts, $\gtrsim_\mathcal{A}$ is not a partial order, and $>_\mathcal{A}$ is not the strict part of $\gtrsim_\mathcal{A}$. Compatibility holds because of the identity $\gtrsim_\mathcal{A} \cdot >_\mathcal{A} = >_\mathcal{A}$.

In order for reduction pairs like $(\geq_{\text{lpo}}, >_{\text{lpo}})$ whose second component is closed under contexts to benefit from the fact that closure under contexts is not required, the conditions (1), (2), and (3) mentioned at the beginning of this section may be simplified by deleting certain (arguments of) function symbols occurring in $\mathcal{R}$ and $\mathcal{C}$ before testing orientability.

Definition 16. *An* argument filtering *for a signature $\mathcal{F}$ is a mapping π that assigns to every n-ary function symbol $f \in \mathcal{F}$ an argument position $i \in \{1, \ldots, n\}$ or a (possibly empty) list $[i_1, \ldots, i_m]$ of argument positions with $1 \leqslant i_1 < \cdots < i_m \leqslant n$. The signature $\mathcal{F}_\pi$ consists of all function symbols f such that $\pi(f)$ is some list $[i_1, \ldots, i_m]$, where in $\mathcal{F}_\pi$ the arity of f is m. Every argument filtering π induces a mapping from $\mathcal{T}(\mathcal{F}, \mathcal{V})$ to $\mathcal{T}(\mathcal{F}_\pi, \mathcal{V})$, also denoted by π:*

$$\pi(t) = \begin{cases} t & \text{if } t \text{ is a variable} \\ \pi(t_i) & \text{if } t = f(t_1, \ldots, t_n) \text{ and } \pi(f) = i \\ f(\pi(t_{i_1}), \ldots, \pi(t_{i_m})) & \text{if } t = f(t_1, \ldots, t_n) \text{ and } \pi(f) = [i_1, \ldots, i_m] \end{cases}$$

Note that the simple projections of the preceding sections can be viewed as special argument filterings.

Example 17. Applying the argument filtering π with $\pi(\wedge) = \pi(\vee) = [\,]$ and $\pi(\neg) = [1]$ to the rewrite rules of our leading example results in the following simplified rules:

$$\begin{aligned} 1: &\quad \neg\neg x \rightarrow x \\ 2: &\quad \neg(\vee) \rightarrow \wedge \\ 3: &\quad \neg(\wedge) \rightarrow \vee \\ 4: &\quad \wedge \rightarrow \vee \\ 5: &\quad \wedge \rightarrow \vee \end{aligned}$$

These rules are oriented from left to right by the lexicographic path order with precedence $\neg > \wedge > \vee$ (which does not imply termination of the original TRS.)

We are now ready to state and prove the standard dependency pair approach to the treatment of cycles in the dependency graph.

Theorem 18 ([9]). *Let $\mathcal{R}$ be a TRS and let $\mathcal{C}$ be a cycle in $\mathsf{DG}(\mathcal{R})$. If there exist an argument filtering π and a reduction pair $(\gtrsim, >)$ such that $\pi(\mathcal{R}) \subseteq \gtrsim$, $\pi(\mathcal{C}) \subseteq \gtrsim \cup >$, and $\pi(\mathcal{C}) \cap > \neq \varnothing$ then there are no $\mathcal{C}$-minimal rewrite sequences.*

Although the condition $\pi(\mathcal{C}) \subseteq {\gtrsim} \cup {>}$ is weaker than $\pi(\mathcal{C}) \subseteq {\gtrsim}$, in practice there is no difference since all reduction pairs that are used in automatic tools satisfy the inclusion ${>} \subseteq {\gtrsim}$.

Proof. Suppose to the contrary that there exists a $\mathcal{C}$-minimal rewrite sequence:

$$t_1 \to_{\mathcal{R}}^* u_1 \to_{\mathcal{C}} t_2 \to_{\mathcal{R}}^* u_2 \to_{\mathcal{C}} t_3 \to_{\mathcal{R}}^* \cdots \tag{2}$$

We show that after applying the argument filtering π we obtain an infinite descending sequence with respect to the well-founded order $>$. Let $i \geqslant 1$.

- First consider the dependency pair step $u_i \to_{\mathcal{C}} t_{i+1}$. Since $u_i \in \mathcal{T}^\sharp$, the step takes place at the root positions and thus there exist a dependency pair $l \to r \in \mathcal{C}$ and a substitution σ such that $u_i = l\sigma$ and $t_{i+1} = r\sigma$. Define the substitution σ_π as the composition of σ and π, i.e., $\sigma_\pi(x) = \pi(\sigma(x))$ for every variable x. A straightforward induction proof reveals that $\pi(t\sigma) = \pi(t)\sigma_\pi$ for every term t. Hence $\pi(u_i) = \pi(l)\sigma_\pi$ and $\pi(t_{i+1}) = \pi(r)\sigma_\pi$. From the assumption $\pi(\mathcal{C}) \subseteq {\gtrsim} \cup {>}$ we infer that $\pi(l) \gtrsim \pi(r)$ or $\pi(l) > \pi(r)$. Since both $\gtrsim$ and $>$ are closed under substitutions, we have $\pi(u_i) \gtrsim \pi(t_{i+1})$ or $\pi(u_i) > \pi(t_{i+1})$. As in the proof of Theorem 11, the latter holds for infinitely many i because of the assumption $\pi(\mathcal{C}) \cap {>} \neq \varnothing$.
- Next consider the rewrite sequence $t_i \to_{\mathcal{R}}^* u_i$. Using the assumption $\pi(\mathcal{R}) \subseteq {\gtrsim}$, we obtain $\pi(t_i) \gtrsim^* \pi(u_i)$ and thus $\pi(t_i) \gtrsim \pi(u_i)$ as in the preceding case.

So (2) is transformed into an infinite descending sequence consisting of $\gtrsim$ and $>$ steps, where there are an infinite number of the latter. Using the compatibility of $\gtrsim$ and $>$, we obtain an infinite descending sequence with respect to $>$, providing the desired contradiction. □

Example 19. The argument filtering of Example 17 cannot be used to handle the remaining SCC $\{11, 12, 13, 14\}$ in our leading example. This can be seen as follows. Because $\pi(\vee) = [\,]$, irrespective of the choice of $\pi(\wedge^\sharp)$, variables y and z will no longer appear in the left-hand sides of the simplified dependency pairs. Hence they cannot appear in the right-hand sides, and this is only possible if we take 1, [1], or [ ] for $\pi(\wedge^\sharp)$. The first two choices transform dependency pairs 13 and 14 into rules in which the variable x appears on the right-hand side but not on the left-hand side, whereas the third choice turns all dependency pairs into the identity $\wedge^\sharp = \wedge^\sharp$.

Since the original TRS is compatible with the multiset path order, it is no surprise that the constraints of Theorem 18 for both SCCs are satisfied by the full argument filtering π (that maps every n-ary function symbol to $[1, \ldots, n]$) and the reduction pair $(\geq_{\mathrm{mpo}}, >_{\mathrm{mpo}})$ with the precedence $\neg > \wedge > \vee$. However, it can be shown that there is no argument filtering π such that the resulting constraints are satisfied by a polynomial interpretation or the lexicographic path order.

Observe that the proof of Theorem 18 does not use the fact that $\mathcal{C}$-minimal rewrite sequences start from terms in $\mathcal{T}_\infty^\sharp$. In the next section we show that by restoring the use of minimality, we can get rid of some of the constraints originating from $\mathcal{R}$.

5 Usable Rules

More precisely, we show that the concept of *usable* rules which was introduced in [1] to optimize the dependency pair method for *innermost* termination, can also be used for termination. The resulting termination criterion is stronger than previous results in this area ([10, 23]). We start by recalling the definition of usable rules.

Definition 20. *We write $f \blacktriangleright g$ if there exists a rewrite rule $l \to r \in \mathcal{R}$ such that $f = \mathrm{root}(l)$ and g is a defined function symbol in $\mathcal{F}\mathsf{un}(r)$. For a set $\mathcal{G}$ of defined function symbols we denote by $\mathcal{R}{\restriction}\mathcal{G}$ the set of rewrite rules $l \to r \in \mathcal{R}$ with $\mathrm{root}(l) \in \mathcal{G}$. The set $\mathcal{U}(t)$ of usable rules of a term t is defined as $\mathcal{R}{\restriction}\{g \mid f \blacktriangleright^* g$ for some $f \in \mathcal{F}\mathsf{un}(t)\}$. Finally, if $\mathcal{C}$ is a set of dependency pairs then*

$$\mathcal{U}(\mathcal{C}) = \bigcup_{l \to r \in \mathcal{C}} \mathcal{U}(r)$$

Example 21. None of the dependency pairs that appear in an SCC in our leading example have defined symbols in their right-hand sides, so for both SCCs the set of usable rules is empty. The same is true for the TRSs of Examples 13 and 14.

The following definition is the key to our result. It is a variation of a similar definition in Urbain [23], which in turn is based on a definition of Gramlich [13].

Definition 22. *Let $\mathcal{R}$ be a TRS over a signature $\mathcal{F}$ and let $\mathcal{G} \subseteq \mathcal{F}$. The* interpretation *$I_\mathcal{G}$ is a mapping from terminating terms in $\mathcal{T}(\mathcal{F}^\sharp, \mathcal{V})$ to terms in $\mathcal{T}(\mathcal{F}^\sharp \cup \{\mathsf{nil}, \mathsf{cons}\}, \mathcal{V})$, where nil and cons are fresh function symbols, inductively defined as follows:*

$$I_\mathcal{G}(t) = \begin{cases} t & \text{if } t \text{ is a variable} \\ f(I_\mathcal{G}(t_1), \ldots, I_\mathcal{G}(t_n)) & \text{if } t = f(t_1, \ldots, t_n) \text{ and } f \notin \mathcal{G} \\ \mathsf{cons}(f(I_\mathcal{G}(t_1), \ldots, I_\mathcal{G}(t_n)), t') & \text{if } t = f(t_1, \ldots, t_n) \text{ and } f \in \mathcal{G} \end{cases}$$

where in the last clause t' denotes the term $\mathrm{order}(\{I_\mathcal{G}(u) \mid t \to_\mathcal{R} u\})$ with

$$\mathrm{order}(T) = \begin{cases} \mathsf{nil} & \text{if } T = \varnothing \\ \mathsf{cons}(t, \mathrm{order}(T \setminus \{t\})) & \text{if } t \text{ is the minimum element of } T \end{cases}$$

Here we assume an arbitrary but fixed total order on $\mathcal{T}(\mathcal{F}^\sharp \cup \{\mathsf{nil}, \mathsf{cons}\}, \mathcal{V})$.

Because we deal with finite TRSs, the relation is $\to_\mathcal{R}$ is finitely branching and hence the set $\{u \mid t \to_\mathcal{R} u\}$ of one-step reducts of t is finite. Moreover, every term in this set is terminating. The well-definedness of $I_\mathcal{G}$ now follows by a straightforward induction argument. The difference with Urbain's definition is that we insert $f(I_\mathcal{G}(t_1), \ldots, I_\mathcal{G}(t_n))$ in the list t' when $f \in \mathcal{G}$. This modification is crucial for obtaining Theorem 29 below.

In the following $\mathcal{C}_\mathcal{E}$ denotes the TRS consisting of the two projection rules

$$\mathsf{cons}(x, y) \to x$$
$$\mathsf{cons}(x, y) \to y$$

These rules are used to extract elements from the lists constructed by the interpretation $I_\mathcal{G}$. To improve readability, we abbreviate $\mathsf{cons}(t_1, \ldots \mathsf{cons}(t_n, \mathsf{nil}) \ldots)$ to $[t_1, \ldots, t_n]$ in the next example.

Example 23. Consider the non-terminating TRS $\mathcal{R}$ consisting of the following three rewrite rules:

$$\begin{aligned} 1:&\quad x + \mathsf{0} \to \mathsf{0} \\ 2:&\quad x \times \mathsf{0} \to \mathsf{0} \\ 3:&\quad x \times \mathsf{s}(y) \to (x + \mathsf{0}) \times \mathsf{s}(y) \end{aligned}$$

There are two dependency pairs:

$$\begin{aligned} 4:&\quad x \times^\sharp \mathsf{s}(y) \to (x + \mathsf{0}) \times^\sharp \mathsf{s}(y) \\ 5:&\quad x \times^\sharp \mathsf{s}(y) \to x +^\sharp \mathsf{0} \end{aligned}$$

The dependency graph

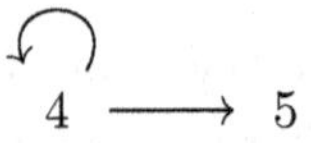

contains 1 cycle: $\mathcal{C} = \{4\}$. The following is a $\mathcal{C}$-minimal rewrite sequence:

$$\begin{aligned} ((\mathsf{0}+\mathsf{0}) \times \mathsf{0}) \times^\sharp \mathsf{s}(\mathsf{0}) &\to_\mathcal{C} (((\mathsf{0}+\mathsf{0}) \times \mathsf{0}) + \mathsf{0}) \times^\sharp \mathsf{s}(\mathsf{0}) \\ &\to_\mathcal{R} ((\mathsf{0}+\mathsf{0}) \times \mathsf{0}) \times^\sharp \mathsf{s}(\mathsf{0}) \\ &\to_\mathcal{R} \mathsf{0} \times^\sharp \mathsf{s}(\mathsf{0}) \\ &\to_\mathcal{C} (\mathsf{0}+\mathsf{0}) \times^\sharp \mathsf{s}(\mathsf{0}) \\ &\to_\mathcal{R} \mathsf{0} \times^\sharp \mathsf{s}(\mathsf{0}) \\ &\to_\mathcal{C} \cdots \end{aligned}$$

We have $\mathcal{U}(\mathcal{C}) = \{1\}$. Let $\mathcal{G}$ be the set of defined symbols of $\mathcal{R} \setminus \mathcal{U}(\mathcal{C})$, i.e., $\mathcal{G} = \{\times\}$. Applying the definition of $I_\mathcal{G}$ yields

$$\begin{aligned} I_\mathcal{G}(\mathsf{0} \times \mathsf{0}) &= \mathsf{cons}(I_\mathcal{G}(\mathsf{0}) \times I_\mathcal{G}(\mathsf{0}), \mathrm{order}(\{I_\mathcal{G}(\mathsf{0})\})) \\ &= \mathsf{cons}(\mathsf{0} \times \mathsf{0}, \mathrm{order}(\{\mathsf{0}\})) \\ &= \mathsf{cons}(\mathsf{0} \times \mathsf{0}, \mathsf{cons}(\mathsf{0}, \mathsf{nil})) \\ &= [\mathsf{0} \times \mathsf{0}, \mathsf{0}] \end{aligned}$$

and

$$\begin{aligned} I_\mathcal{G}((\mathsf{0}+\mathsf{0}) \times \mathsf{0}) &= \mathsf{cons}(I_\mathcal{G}(\mathsf{0}+\mathsf{0}) \times I_\mathcal{G}(\mathsf{0}), \mathrm{order}(\{I_\mathcal{G}(\mathsf{0} \times \mathsf{0}), I_\mathcal{G}(\mathsf{0})\})) \\ &= \mathsf{cons}((\mathsf{0}+\mathsf{0}) \times \mathsf{0}, \mathrm{order}(\{[\mathsf{0} \times \mathsf{0}, \mathsf{0}], \mathsf{0}\})) \\ &= [(\mathsf{0}+\mathsf{0}) \times \mathsf{0}, \mathsf{0}, [\mathsf{0} \times \mathsf{0}, \mathsf{0}]] \end{aligned}$$

if we assume that $\mathsf{0}$ is smaller than $[\mathsf{0} \times \mathsf{0}, \mathsf{0}]$ in the given total order. Now, by applying $I_\mathcal{G}$ to all terms in the above $\mathcal{C}$-minimal rewrite sequence, we obtain the following infinite rewrite sequence in $\mathcal{U}(\mathcal{C}) \cup \mathcal{C} \cup \mathcal{C}_\mathcal{E}$:

$$\begin{aligned}
[(\mathsf{0}+\mathsf{0}) \times \mathsf{0}, \mathsf{0}, [\mathsf{0} \times \mathsf{0}, \mathsf{0}]] \times^\sharp \mathsf{s}(\mathsf{0}) &\to_\mathcal{C} ([(\mathsf{0}+\mathsf{0}) \times \mathsf{0}, \mathsf{0}, [\mathsf{0} \times \mathsf{0}, \mathsf{0}]] + \mathsf{0}) \times^\sharp \mathsf{s}(\mathsf{0}) \\
&\to_{\mathcal{U}(\mathcal{C})} [(\mathsf{0}+\mathsf{0}) \times \mathsf{0}, \mathsf{0}, [\mathsf{0} \times \mathsf{0}, \mathsf{0}]] \times^\sharp \mathsf{s}(\mathsf{0}) \\
&\to^+_{\mathcal{C}_\mathcal{E}} \mathsf{0} \times^\sharp \mathsf{s}(\mathsf{0}) \\
&\to_\mathcal{C} (\mathsf{0}+\mathsf{0}) \times^\sharp \mathsf{s}(\mathsf{0}) \\
&\to_{\mathcal{U}(\mathcal{C})} \mathsf{0} \times^\sharp \mathsf{s}(\mathsf{0}) \\
&\to_\mathcal{C} \cdots
\end{aligned}$$

We start with some preliminary results. The first one addresses the behaviour of $I_\mathcal{G}$ on instantiated terms. The second states that $I_\mathcal{G}$ preserves any top part without $\mathcal{G}$-symbols.

Definition 24. *If σ is a substitution that assigns to every variable in its domain a terminating term then we denote the substitution that assigns to every variable x the term $I_\mathcal{G}(\sigma(x))$ by $\sigma_{I_\mathcal{G}}$.*

Lemma 25. *Let $\mathcal{R}$ be a TRS over a signature $\mathcal{F}$ and let $\mathcal{G} \subseteq \mathcal{F}$. Let t be a term and σ a substitution. If $t\sigma$ is terminating then $I_\mathcal{G}(t\sigma) \to^*_{\mathcal{C}_\mathcal{E}} t\sigma_{I_\mathcal{G}}$ and, if t does not contain $\mathcal{G}$-symbols, $I_\mathcal{G}(t\sigma) = t\sigma_{I_\mathcal{G}}$.*

Proof. We use induction on t. If t is a variable then $I_\mathcal{G}(t\sigma) = I_\mathcal{G}(\sigma(t)) = t\sigma_{I_\mathcal{G}}$. Let $t = f(t_1, \dots, t_n)$. We distinguish two cases.

1. If $f \notin \mathcal{G}$ then $I_\mathcal{G}(t\sigma) = f(I_\mathcal{G}(t_1\sigma), \dots, I_\mathcal{G}(t_n\sigma))$. The induction hypothesis yields $I_\mathcal{G}(t_i\sigma) \to^*_{\mathcal{C}_\mathcal{E}} t_i\sigma_{I_\mathcal{G}}$ for $1 \leqslant i \leqslant n$ and thus
$$I_\mathcal{G}(t\sigma) \to^*_{\mathcal{C}_\mathcal{E}} f(t_1\sigma_{I_\mathcal{G}}, \dots, t_n\sigma_{I_\mathcal{G}}) = t\sigma_{I_\mathcal{G}}$$
If there are no $\mathcal{G}$-symbols in $t_1, \dots, t_n$ then we obtain $I_\mathcal{G}(t_i\sigma) = t_i\sigma_{I_\mathcal{G}}$ for all $1 \leqslant i \leqslant n$ from the induction hypothesis and thus $I_\mathcal{G}(t\sigma) = t\sigma_{I_\mathcal{G}}$.
2. If $f \in \mathcal{G}$ then
$$I_\mathcal{G}(t\sigma) = \mathsf{cons}(f(I_\mathcal{G}(t_1\sigma), \dots, I_\mathcal{G}(t_n\sigma)), t') \to_{\mathcal{C}_\mathcal{E}} f(I_\mathcal{G}(t_1\sigma), \dots, I_\mathcal{G}(t_n\sigma))$$
for some term t'. We obtain $f(I_\mathcal{G}(t_1\sigma), \dots, I_\mathcal{G}(t_n\sigma)) \to^*_{\mathcal{C}_\mathcal{E}} t\sigma_{I_\mathcal{G}}$ as in the preceding case and thus $I_\mathcal{G}(t\sigma) \to^*_{\mathcal{C}_\mathcal{E}} t\sigma_{I_\mathcal{G}}$ as desired.

□

The preceding lemma is not true for Urbain's interpretation function.

Lemma 26. *Let $\mathcal{R}$ be a TRS over a signature $\mathcal{F}$ and let $\mathcal{G} \subseteq \mathcal{F}$. If $t = C[t_1, \dots, t_n]$ is terminating and the context C contains no $\mathcal{G}$-symbols then $I_\mathcal{G}(t) = C[I_\mathcal{G}(t_1), \dots, I_\mathcal{G}(t_n)]$.*

Proof. Let t' be the term $C[x_1, \ldots, x_n]$ where $x_1, \ldots, x_n$ are fresh variables. We have $t = t'\sigma$ for the substitution $\sigma = \{x_i \mapsto t_i \mid 1 \leqslant i \leqslant n\}$. The preceding lemma yields $I_{\mathcal{G}}(t) = t'\sigma_{I_{\mathcal{G}}}$. Clearly $t'\sigma_{I_{\mathcal{G}}} = C[I_{\mathcal{G}}(t_1), \ldots, I_{\mathcal{G}}(t_n)]$. □

The next lemma states an easy connection between usable rules and defined symbols of the other rules.

Lemma 27. *Let $\mathcal{R}$ be a TRS over a signature $\mathcal{F}$ and let $\mathcal{C} \subseteq \mathsf{DP}(\mathcal{R})$. Furthermore, let $\mathcal{G}$ be the set of defined symbols of $\mathcal{R} \setminus \mathcal{U}(\mathcal{C})$.*

1. *$\mathcal{R} = \mathcal{U}(\mathcal{C}) \cup (\mathcal{R}{\restriction}\mathcal{G})$.*
2. *If $l \to r \in \mathcal{U}(\mathcal{C})$ then r contains no $\mathcal{G}$-symbols.*

Proof. The first statement is obvious. For the second statement we reason as follows. Suppose to the contrary that r contains a function symbol $g \in \mathcal{G}$. We have $l \to r \in \mathcal{U}(t)$ for some $s \to t \in \mathcal{C}$. So there exists a function symbol $f \in \mathcal{F}\mathsf{un}(t)$ such that $f \blacktriangleright^* \mathrm{root}(l)$. We have $\mathrm{root}(l) \blacktriangleright g$ by the definition of $\blacktriangleright$ and hence also $f \blacktriangleright^* g$. Therefore $\mathcal{R}{\restriction}\{g\} \subseteq \mathcal{U}(t) \subseteq \mathcal{U}(\mathcal{C})$. So g is a defined symbol of a rule in $\mathcal{U}(\mathcal{C})$. This contradicts the assumption that $g \in \mathcal{G}$. □

The following lemma is the key result for the new termination criterion. It states that rewrite steps in $\mathcal{R}$ are transformed by $I_{\mathcal{G}}$ into rewrite sequences in $\mathcal{U}(\mathcal{C}) \cup \mathcal{C}_{\mathcal{E}}$, provided $\mathcal{G}$ is the set of defined symbols of $\mathcal{R} \setminus \mathcal{U}(\mathcal{C})$.

Lemma 28. *Let $\mathcal{R}$ be a TRS over a signature $\mathcal{F}$ and let $\mathcal{C} \subseteq \mathsf{DP}(\mathcal{R})$. Furthermore, let $\mathcal{G}$ be the set of defined symbols of $\mathcal{R} \setminus \mathcal{U}(\mathcal{C})$. If terms s and t are terminating and $s \to_{\mathcal{R}} t$ then $I_{\mathcal{G}}(s) \to^+_{\mathcal{U}(\mathcal{C}) \cup \mathcal{C}_{\mathcal{E}}} I_{\mathcal{G}}(t)$.*

Proof. Let p be the position of the rewrite step $s \to_{\mathcal{R}} t$. We distinguish two cases.

- First suppose that there is a function symbol from $\mathcal{G}$ at a position $q \leqslant p$. In this case we may write $s = C[s_1, \ldots, s_i, \ldots, s_n]$ and $t = C[s_1, \ldots, t_i, \ldots, s_n]$ with $s_i \to_{\mathcal{R}} t_i$, where $\mathrm{root}(s_i) \in \mathcal{G}$ and the context C contains no $\mathcal{G}$-symbols. We have $I_{\mathcal{G}}(s_i) \to_{\mathcal{C}_{\mathcal{E}}} \mathrm{order}(\{I_{\mathcal{G}}(u) \mid s_i \to_{\mathcal{R}} u\})$. Since $s_i \to_{\mathcal{R}} t_i$, we can extract $I_{\mathcal{G}}(t_i)$ from the term $\mathrm{order}(\{I_{\mathcal{G}}(u) \mid s_i \to_{\mathcal{R}} u\})$ by appropriate $\mathcal{C}_{\mathcal{E}}$ steps, so $I_{\mathcal{G}}(s_i) \to^+_{\mathcal{C}_{\mathcal{E}}} I_{\mathcal{G}}(t_i)$. We now obtain $I_{\mathcal{G}}(s) \to^+_{\mathcal{C}_{\mathcal{E}}} I_{\mathcal{G}}(t)$ from Lemma 26.
- In the other case $s = C[s_1, \ldots, s_i, \ldots, s_n]$ and $t = C[s_1, \ldots, t_i, \ldots, s_n]$ with $s_i \xrightarrow{\epsilon}_{\mathcal{R}} t_i$, where $\mathrm{root}(s_i) \notin \mathcal{G}$ and the context C contains no $\mathcal{G}$-symbols. Since $\mathrm{root}(s_i) \notin \mathcal{G}$ the applied rewrite rule $l \to r$ in the step $s_i \xrightarrow{\epsilon}_{\mathcal{R}} t_i$ must come from $\mathcal{U}(\mathcal{C})$ according to part 1 of Lemma 27. Let σ be the substitution with $\mathcal{D}\mathsf{om}(\sigma) \subseteq \mathcal{V}\mathsf{ar}(l)$ such that $s_i = l\sigma$ and $t_i = r\sigma$. According to part 2 of Lemma 27, r contains no $\mathcal{G}$-symbols and thus we obtain $I_{\mathcal{G}}(s_i) \to^*_{\mathcal{C}_{\mathcal{E}}} l\sigma_{I_{\mathcal{G}}}$ and $I_{\mathcal{G}}(t_i) = r\sigma_{I_{\mathcal{G}}}$ from Lemma 25. Clearly $l\sigma_{I_{\mathcal{G}}} \to_{\mathcal{U}(\mathcal{C})} r\sigma_{I_{\mathcal{G}}}$ and thus $I_{\mathcal{G}}(s_i) \to^+_{\mathcal{U}(\mathcal{C}) \cup \mathcal{C}_{\mathcal{E}}} I_{\mathcal{G}}(t_i)$. Lemma 26 now yields the desired $I_{\mathcal{G}}(s) \to^+_{\mathcal{U}(\mathcal{C}) \cup \mathcal{C}_{\mathcal{E}}} I_{\mathcal{G}}(t)$. □

After these preparations, the main result[2] of this section is now easily proved.

[2] This result has been independently obtained by Thiemann *et al.* [21].

Theorem 29. *Let $\mathcal{R}$ be a TRS and let $\mathcal{C}$ be a cycle in* $\mathsf{DG}(\mathcal{R})$. *If there exist an argument filtering π and a reduction pair $(\gtrsim, >)$ such that $\pi(\mathcal{U}(\mathcal{C}) \cup \mathcal{C}_{\mathcal{E}}) \subseteq \gtrsim$, $\pi(\mathcal{C}) \subseteq \gtrsim \cup >$, and $\pi(\mathcal{C}) \cap > \neq \varnothing$ then there are no $\mathcal{C}$-minimal rewrite sequences.*

Proof. Suppose to the contrary that there exists a $\mathcal{C}$-minimal rewrite sequence:

$$t_1 \to_{\mathcal{R}}^* u_1 \to_{\mathcal{C}} t_2 \to_{\mathcal{R}}^* u_2 \to_{\mathcal{C}} t_3 \to_{\mathcal{R}}^* \cdots \tag{3}$$

Let $\mathcal{G}$ be the set of defined symbols of $\mathcal{R} \setminus \mathcal{U}(\mathcal{C})$. We show that after applying the interpretation $I_{\mathcal{G}}$ we obtain an infinite rewrite sequence in $\mathcal{U}(\mathcal{C}) \cup \mathcal{C}_{\mathcal{E}} \cup \mathcal{C}$ in which every rule of $\mathcal{C}$ is used infinitely often. Since all terms in (3) belong to $\mathcal{T}_{\infty}^{\sharp}$, they are terminating with respect to $\mathcal{R}$ and hence we can indeed apply the interpretation $I_{\mathcal{G}}$. Let $i \geqslant 1$.

- First consider the dependency pair step $u_i \to_{\mathcal{C}} t_{i+1}$. There exist a dependency pair $l \to r \in \mathcal{C}$ and a substitution σ such that $u_i = l\sigma$ and $t_{i+1} = r\sigma$. We may assume that $\mathcal{D}\mathsf{om}(\sigma) \subseteq \mathcal{V}\mathsf{ar}(l)$. Since $u_i \in \mathcal{T}_{\infty}^{\sharp}$, $\sigma(x)$ is terminating for every variable $x \in \mathcal{V}\mathsf{ar}(l)$. Hence the substitution $\sigma_{I_{\mathcal{G}}}$ is well-defined. Since r lacks $\mathcal{G}$-symbols by Lemma 27, we have $I_{\mathcal{G}}(r\sigma) = r\sigma_{I_{\mathcal{G}}}$ by Lemma 25. Furthermore, $I_{\mathcal{G}}(l\sigma) \to_{\mathcal{C}_{\mathcal{E}}}^* l\sigma_{I_{\mathcal{G}}}$ by Lemma 25. Hence

$$I_{\mathcal{G}}(u_i) \to_{\mathcal{C}_{\mathcal{E}}}^* l\sigma_{I_{\mathcal{G}}} \to_{\mathcal{C}} r\sigma_{I_{\mathcal{G}}} = I_{\mathcal{G}}(t_{i+1})$$

- Next consider the rewrite sequence $t_i \to_{\mathcal{R}}^* u_i$. Because all terms in this sequence are terminating, we obtain $I_{\mathcal{G}}(t_i) \to_{\mathcal{U}(\mathcal{C}) \cup \mathcal{C}_{\mathcal{E}}}^* I_{\mathcal{G}}(u_i)$ by repeated applications of Lemma 28.

Next we apply the argument filtering π to all terms in the resulting infinite rewrite sequence in $\mathcal{U}(\mathcal{C}) \cup \mathcal{C}_{\mathcal{E}} \cup \mathcal{C}$. Because of the assumptions of this theorem, we can simply reuse the proof of Theorem 18 (where $\mathcal{U}(\mathcal{C}) \cup \mathcal{C}_{\mathcal{E}}$ takes the place of $\mathcal{R}$) and obtain the desired contradiction with the well-foundedness of $>$. □

Since $\mathcal{U}(\mathcal{C})$ in general is a proper subset of $\mathcal{R}$, the condition $\pi(\mathcal{U}(\mathcal{C})) \subseteq \gtrsim$ is easier to satisfy than the condition $\pi(\mathcal{R}) \subseteq \gtrsim$ of Theorem 18. What about the additional condition $\pi(\mathcal{C}_{\mathcal{E}}) \subseteq \gtrsim$? By choosing $\pi(\mathsf{cons}) = [1,2]$ the condition reduces to $\mathsf{cons}(x,y) \gtrsim x$ and $\mathsf{cons}(x,y) \gtrsim y$. Virtually all reduction pairs that are used in termination tools can be extended to satisfy this condition. For reduction pairs that are based on simplification orders, like $(\geq_{\mathrm{lpo}}, >_{\mathrm{lpo}})$, this is clear. A sufficient condition that makes the semantic construction described in Section 4 for generating reduction pairs work is that each pair of elements of the carrier has a least upper bound. For interpretations in the set $\mathbb{N}$ of natural numbers equipped with the standard order this is obviously satisfied. The necessity of the least upper bound condition follows by considering the term algebra associated with the famous rule $\mathsf{f}(\mathsf{a},\mathsf{b},x) \to \mathsf{f}(x,x,x)$ of Toyama [22] equipped with the well-founded order $\to^+$.

As a matter of fact, due to the condition $\pi(\mathcal{C}_{\mathcal{E}}) \subseteq \gtrsim$, Theorem 29 provides only a sufficient condition for the absence of $\mathcal{C}$-minimal rewrite sequences. A concrete example of a terminating TRS that cannot be proved terminating by the criterion of Theorem 29 will be presented at the end of this section.

Example 30. Let us take a final look at the SCC $\{11, 12, 13, 14\}$ in our leading example. There are no usable rules. By taking the linear polynomial interpretation $\wedge^\sharp_{\mathbb{N}}(x, y) = x + y$ and $\vee_{\mathbb{N}}(x, y) = x + y + 1$ the involved dependency pairs reduce the following inequalities:

$$\begin{aligned} 11: &\quad x + y + z + 1 > x + y \\ 12: &\quad x + y + z + 1 > x + z \\ 13: &\quad x + y + z + 1 > x + y \\ 14: &\quad x + y + z + 1 > x + z \end{aligned}$$

Hence there are no $\mathcal{C}$-minimal rewrite sequences for any nonempty subset $\mathcal{C} \subseteq \{11, 12, 13, 14\}$ and we conclude that the TRS is terminating.

The modularity result in Giesl *et al.* [10] can be expressed as the version of Theorem 29 where $\mathcal{U}(\mathcal{C})$ is replaced by

$$\mathcal{U}'(\mathcal{C}) = \bigcup_{l \to r \in \mathcal{C}} \mathcal{U}'(l^\flat)$$

The mapping $(\cdot)^\flat \colon \mathcal{T}^\sharp \to \mathcal{T}$ replaces the dependency pair symbol $f^\sharp$ at the root of its argument by the original defined function symbol f and $\mathcal{U}'(t)$ is computed like $\mathcal{U}(t)$ but with a different relation $\blacktriangleright'$ that relates more function symbols: $f \blacktriangleright' g$ if there exists a rewrite rule $l \to r \in \mathcal{R}$ such that $f = \mathrm{root}(l)$ and g is a defined function symbol in $\mathcal{F}\mathsf{un}(l) \cup \mathcal{F}\mathsf{un}(r)$.

Since $\mathcal{U}(r) \subseteq \mathcal{U}(r^\flat) \subseteq \mathcal{U}(l^\flat) \subseteq \mathcal{U}'(l^\flat)$ for every dependency pair $l \to r$, it is clear that $\mathcal{U}(\mathcal{C})$ is always a subset of $\mathcal{U}'(\mathcal{C})$. Very often it is a proper subset and that may affect the ability to prove termination. This will become clear from the experimental data in the next section.

Example 31. If we adopt the above definition of usable rules then for the SCC $\{7, 8, 9, 10\}$ in our leading example all five rewrite rules are usable whereas for the SCC $\{11, 12, 13, 14\}$ only rules 4 and 5 are usable. For the SCC $\{9\}$ in Example 13 all seven rewrite rules and for the SCC $\{10, 11, 12\}$ rules 3–7 are usable. Finally, for the SCC $\{7\}$ in Example 14 rules 1 and 2 are usable whereas for the SCC $\{8, 10\}$ all six rewrite rules are usable.

Combining the two main results of this paper, we arrive at the following corollary.

Corollary 32. *A TRS $\mathcal{R}$ is terminating if for every cycle $\mathcal{C}$ in $\mathsf{DG}(\mathcal{R})$ one of the following two conditions holds:*

- *there exists a simple projection π for $\mathcal{C}$ such that $\pi(\mathcal{C}) \subseteq \trianglerighteq$ and $\pi(\mathcal{C}) \cap \vartriangleright \neq \varnothing$,*
- *there exist an argument filtering π and a reduction pair $(\gtrsim, >)$ such that $\pi(\mathcal{U}(\mathcal{C}) \cup \mathcal{C}_{\mathcal{E}}) \subseteq \gtrsim$, $\pi(\mathcal{C}) \subseteq \gtrsim \cup >$, and $\pi(\mathcal{C}) \cap > \neq \varnothing$.* □

The final example in this paper shows that the reverse does not hold. This is in contrast to Theorem 18, which provides a sufficient and necessary condition for termination. The reason is that termination of a TRS $\mathcal{R}$ is equivalent to the termination of $\mathcal{R} \cup \mathsf{DP}(\mathcal{R})$, a result due to [1] (see [18] for a simple proof based on type introduction).

Example 33. Consider the terminating TRS $\mathcal{R}$ consisting of the following two rewrite rules:

$$\begin{aligned} 1:&\quad \mathsf{f}(\mathsf{s}(\mathsf{a}),\mathsf{s}(\mathsf{b}),x) \to \mathsf{f}(x,x,x) \\ 2:&\quad \mathsf{g}(\mathsf{f}(\mathsf{s}(x),\mathsf{s}(y),z)) \to \mathsf{g}(\mathsf{f}(x,y,z)) \end{aligned}$$

There are three dependency pairs:

$$\begin{aligned} 3:&\quad \mathsf{f}^\sharp(\mathsf{s}(\mathsf{a}),\mathsf{s}(\mathsf{b}),x) \to \mathsf{f}^\sharp(x,x,x) \\ 4:&\quad \mathsf{g}^\sharp(\mathsf{f}(\mathsf{s}(x),\mathsf{s}(y),z)) \to \mathsf{g}^\sharp(\mathsf{f}(x,y,z)) \\ 5:&\quad \mathsf{g}^\sharp(\mathsf{f}(\mathsf{s}(x),\mathsf{s}(y),z)) \to \mathsf{f}^\sharp(x,y,z) \end{aligned}$$

The dependency graph

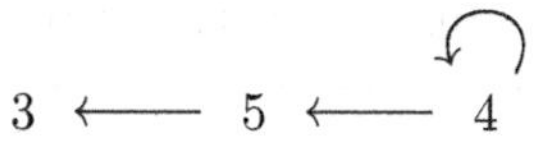

contains 1 cycle: $\mathcal{C} = \{4\}$. The only simple projection for $\mathsf{g}^\sharp$ transforms 4 into

$$4: \quad \mathsf{f}(\mathsf{s}(x),\mathsf{s}(y),z) \to \mathsf{f}(x,y,z)$$

and $\mathsf{f}(x,y,z)$ is not a proper subterm of $\mathsf{f}(\mathsf{s}(x),\mathsf{s}(y),z)$. We have $\mathcal{U}(\mathcal{C}) = \{1\}$. We claim that the inclusions $\pi(\mathcal{U}(\mathcal{C}) \cup \mathcal{C}_\mathcal{E}) \subseteq \gtrsim$ and $\pi(\mathcal{C}) \subseteq >$ are not satisfied for any argument filtering π and reduction pair $(\gtrsim, >)$. The reason is simply that the term $t = \mathsf{g}^\sharp(\mathsf{f}(u,u,u))$ with $u = \mathsf{s}(\mathsf{cons}(\mathsf{s}(\mathsf{a}),\mathsf{s}(\mathsf{b})))$ admits the following cyclic reduction in $\mathcal{U}(\mathcal{C}) \cup \mathcal{C}_\mathcal{E} \cup \mathcal{C}$:

$$\begin{aligned} t \to_{\mathcal{C}} &\quad \mathsf{g}^\sharp(\mathsf{f}(\mathsf{cons}(\mathsf{s}(\mathsf{a}),\mathsf{s}(\mathsf{b})),\mathsf{cons}(\mathsf{s}(\mathsf{a}),\mathsf{s}(\mathsf{b})),u)) \\ \to_{\mathcal{C}_\mathcal{E}} &\quad \mathsf{g}^\sharp(\mathsf{f}(\mathsf{s}(\mathsf{a}),\mathsf{cons}(\mathsf{s}(\mathsf{a}),\mathsf{s}(\mathsf{b})),u)) \\ \to_{\mathcal{C}_\mathcal{E}} &\quad \mathsf{g}^\sharp(\mathsf{f}(\mathsf{s}(\mathsf{a}),\mathsf{s}(\mathsf{b}),u)) \\ \to_{\mathcal{U}(\mathcal{C})} &\quad t \end{aligned}$$

6 Benchmarks

We implemented the new criteria presented in the preceding sections in the Tyrolean Termination Tool [16], the successor of the Tsukuba Termination Tool [15].

We tested the effect of the improvements described in the previous sections on 223 examples from three different sources:

- all 89 terminating TRSs from Arts and Giesl [2],
- all 23 TRSs from Dershowitz [7],
- all 119 terminating TRSs from Steinbach and Kühler [19, Sections 3 and 4].

Eight of these TRSs appear in more than one collection, so the total number is 223. In all experiments we used the EDG* approximation [17] of the dependency graph and, when the lexicographic path order is used, the divide and conquer algorithm described in the full version of [14] is used to search for suitable argument filterings. The experiments were performed on a PC equipped with a 2.20 GHz Mobile Intel Pentium 4 Processor - M and 512 MB of memory.

Table 1. Summary.

	`s`	`l`	`sl`	`ul`	`sul`
success	128	133	149	144 (138)	152 (151)
	0.01	0.07	0.01	0.01 (0.02)	0.01 (0.01)
failure	95	90	74	79 (85)	71 (72)
	0.01	0.01	0.01	0.01 (0.02)	0.01 (0.02)
timeout	0	0	0	0 (0)	0 (0)
total time	1.72	10.49	2.33	2.31 (4.46)	2.07 (2.80)
		`p`	`sp`	`up`	`sup`
success		139	180	179 (148)	189 (185)
		0.32	0.37	0.28 (0.33)	0.25 (0.40)
failure		77	39	44 (71)	34 (37)
		0.52	0.33	0.03 (0.66)	0.03 (0.59)
timeout		7	4	0 (4)	0 (1)
total time		294.13	198.83	51.30 (215.93)	47.79 (125.97)

The results are summarized in Table 1. The letters in the column headings have the following meaning:

- `s` the subterm criterion of Section 3,
- `u` the usable rules criterion of Section 5,
- `l` lexicographic path order in combination with the argument filtering heuristic that considers for an n-ary function symbol the full argument filtering $[1,\ldots,n]$ in addition to the n collapsing argument filterings $1,\ldots,n$,
- `p` polynomial interpretation restricted to linear polynomials with coefficients from $\{0,1\}$; the usefulness of the latter restriction has been first observed in [11].

We list the number of successful termination attempts, the number of failures (which means that no termination proof was found while fully exploring the search space implied by the options), and the number of timeouts, which we set to 30 seconds. The numbers in parentheses refer to the usable rules criterion of [10] which is described in the latter part of Section 5. The figures below the number of successes and failures indicate the average time in seconds. It is interesting to note that the subterm criterion could handle 279 of the 395 generated SCCs, resulting in termination proofs for 128 of the 223 TRSs.

References

1. T. Arts and J. Giesl. Termination of term rewriting using dependency pairs. *Theoretical Computer Science*, 236:133–178, 2000.
2. T. Arts and J. Giesl. A collection of examples for termination of term rewriting using dependency pairs. Technical Report AIB-2001-09, RWTH Aachen, 2001.
3. F. Baader and T. Nipkow. *Term Rewriting and All That.* Cambridge University Press, 1998.
4. L. Bachmair and N. Dershowitz. Commutation, transformation, and termination. In *Proceedings of the 8th International Conference on Automated Deduction*, volume 230 of *Lecture Notes in Computer Science*, pages 5–20, 1986.
5. C. Borralleras, M. Ferreira, and A. Rubio. Complete monotonic semantic path orderings. In *Proceedings of the 17th International Conference on Automated Deduction*, volume 1831 of *Lecture Notes in Artificial Intelligence*, pages 346–364, 2000.
6. E. Contejean, C. Marché, B. Monate, and X. Urbain. C*i*ME version 2, 2000. Available at `http://cime.lri.fr/`.
7. N. Dershowitz. 33 Examples of termination. In *French Spring School of Theoretical Computer Science*, volume 909 of *Lecture Notes in Computer Science*, pages 16–26, 1995.
8. N. Dershowitz. Termination dependencies. In *Proceedings of the 6th International Workshop on Termination*, Technical Report DSIC-II/15/03, Universidad Politécnica de Valencia, pages 27–30, 2003.
9. J. Giesl, T. Arts, and E. Ohlebusch. Modular termination proofs for rewriting using dependency pairs. *Journal of Symbolic Computation*, 34(1):21–58, 2002.
10. J. Giesl, R. Thiemann, P. Schneider-Kamp, and S. Falke. Improving dependency pairs. In *Proceedings of the 10th International Conference on Logic for Programming, Artificial Intelligence and Reasoning*, volume 2850 of *Lecture Notes in Artificial Intelligence*, pages 165–179, 2003.
11. J. Giesl, R. Thiemann, P. Schneider-Kamp, and S. Falke. Mechanizing dependency pairs. Technical Report AIB-2003-08, RWTH Aachen, Germany, 2003.
12. J. Giesl, R. Thiemann, P. Schneider-Kamp, and S. Falke. Automated termination proofs with AProVE. In *Proceedings of the 15th International Conference on Rewriting Techniques and Applications*, Lecture Notes in Computer Science, 2004. This volume.
13. B. Gramlich. Generalized sufficient conditions for modular termination of rewriting. *Applicable Algebra in Engineering, Communication and Computing*, 5:131–158, 1994.
14. N. Hirokawa and A. Middeldorp. Automating the dependency pair method. In *Proceedings of the 19th International Conference on Automated Deduction*, volume 2741 of *Lecture Notes in Artificial Intelligence*, pages 32–46, 2003. Full version submitted for publication.
15. N. Hirokawa and A. Middeldorp. Tsukuba termination tool. In *Proceedings of the 14th International Conference on Rewriting Techniques and Applications*, volume 2706 of *Lecture Notes in Computer Science*, pages 311–320, 2003.

16. N. Hirokawa and A. Middeldorp. Tyrolean termination tool, 2004. Available at `http://cl2-informatik.uibk.ac.at/ttt`.
17. A. Middeldorp. Approximations for strategies and termination. In *Proceedings of the 2nd International Workshop on Reduction Strategies in Rewriting and Programming*, volume 70(6) of *Electronic Notes in Theoretical Computer Science*, 2002.
18. A. Middeldorp and H. Ohsaki. Type introduction for equational rewriting. *Acta Informatica*, 36(12):1007–1029, 2000.
19. J. Steinbach and U. Kühler. Check your ordering – termination proofs and open problems. Technical Report SR-90-25, Universität Kaiserslautern, 1990.
20. Terese. *Term Rewriting Systems*, volume 55 of *Cambridge Tracts in Theoretical Computer Science*. Cambridge University Press, 2003.
21. R. Thiemann, J. Giesl, and P. Schneider-Kamp. Improved modular termination proofs using dependency pairs. In *Proceedings of the 2nd International Joint Conference on Automated Reasoning*, Lecture Notes in Artificial Intelligence, 2004. To appear.
22. Y. Toyama. Counterexamples to the termination for the direct sum of term rewriting systems. *Information Processing Letters*, 25:141–143, 1987.
23. X. Urbain. Modular & incremental automated termination proofs. *Journal of Automated Reasoning*, 2004. To appear.
24. H. Zantema. TORPA: Termination of rewriting proved automatically. In *Proceedings of the 15th International Conference on Rewriting Techniques and Applications*, Lecture Notes in Computer Science, 2004. This volume.

Inductive Theorems for Higher-Order Rewriting*

Takahito Aoto[1], Toshiyuki Yamada[2], and Yoshihito Toyama[1]

[1] Research Institute of Electrical Communication, Tohoku University, Japan
{aoto,toyama}@nue.riec.tohoku.ac.jp
[2] Faculty of Engineering, Mie University, Japan
toshi@cs.info.mie-u.ac.jp

Abstract. Based on the simply typed term rewriting framework, inductive reasoning in higher-order rewriting is studied. The notion of higher-order inductive theorems is introduced to reflect higher-order feature of simply typed term rewriting. Then the inductionless induction methods in first-order term rewriting are incorporated to verify higher-order inductive theorems. In order to ensure that higher-order inductive theorems are closed under contexts, the notion of higher-order sufficient completeness is introduced. Finally, the decidability of higher-order sufficient completeness is discussed.

1 Introduction

In a framework based on equational logic such as functional programming and algebraic specification, properties of a particular recursive data structure can be verified by induction. Once an appropriate translation from a specification to a set of equational axioms has been established, inductive properties of the specification can be proved using term rewriting. Automatically proving inductive theorems by term rewriting has been widely investigated since earlier works [5, 8, 17].

Higher-order functions are ubiquitous in functional programming. However, they can not be expressed directly in usual term rewriting, and thus various frameworks of higher-order extension of term rewriting have been proposed in the literature. In most of these frameworks, termination proof is difficult to cope with, and consequently only few attempts have been made to incorporate higher-order functions to inductive theorem proving.

Simply typed term rewriting proposed by Yamada [21] is a simple extension of first-order term rewriting which incorporates higher-order functions. Equational specification using higher-order functions, like functional programs, are naturally expressed in this framework. In contrast to the usual higher-order term rewriting frameworks [6, 9, 15], simply typed term rewriting dispenses with bound variables. In this respect, simply typed term rewriting reflects limited higher-order features. On the other hand, simply typed term rewriting framework is succinct and theoretically much easier to deal with.

* This work was partially supported by grants from Japan Society for the Promotion of Science, No. 14580357, No. 14780187 and a grant from Ministry of Education, Culture, Sports, Science and Technology, No. 15017203.

V. van Oostrom (Ed.): RTA 2004, LNCS 3091, pp. 269–284, 2004.

In this paper, we study inductive theorem proving in higher-order rewriting based on simply typed term rewriting framework. We show that there is an equation that is intuitively inductively valid, but is inductively invalid in terms of first-order inductive reasoning. We propose the notion of higher-order inductive theorem that reflects higher-order feature more reasonably. Then inductionless induction methods in first-order term rewriting are incorporated to simply typed term rewriting to verify higher-order inductive theorems. It also turns out that a higher-order inductive theorem is not always closed under contexts. To ensure that higher-order inductive theorems are closed under contexts, we introduce the notion of higher-order sufficient completeness. The decidability of higher-order sufficient completeness for a subclass of simply typed term rewriting systems is shown by introducing the notion of higher-order quasi-reducibility.

2 Preliminaries

In this section, we first fix our notation on abstract reduction systems, and recall a result about equivalence of abstract reduction systems. Based on this equivalence condition, we review notions and results on inductive theorem proving in first-order term rewriting. We assume the reader to be familiar with abstract reduction systems and (first-order) term rewriting [4, 19].

2.1 Equivalence of Abstract Reduction Systems

An *abstract reduction system* (*ARS*, for short) is a pair $\langle A, \rightarrow \rangle$ of a set A and a binary relation $\rightarrow$ on A. The reflexive transitive closure and the equivalence closure of $\rightarrow$ are denoted by $\rightarrow^*$ and $\leftrightarrow^*$, respectively. An element $a \in A$ is *reducible* when $a \rightarrow b$ for some $b \in A$; otherwise, it is *normal* or in *normal form*. The sets of reducible and normal elements of an ARS $\mathcal{A}$ are denoted by $\mathrm{RED}(\mathcal{A})$ and $\mathrm{NF}(\mathcal{A})$, respectively. Elements $a \in A$ and $b \in A$ are *joinable* if $a \rightarrow^* c \;{}^*\!\!\leftarrow b$ for some $c \in A$. An ARS $\mathcal{A}$ is *weakly normalizing* ($\mathrm{WN}(\mathcal{A})$) if for any $a \in A$ there exists $b \in \mathrm{NF}(A)$ such that $a \rightarrow^* b$, is *terminating* or *strongly normalizing* ($\mathrm{SN}(\mathcal{A})$) if there is no infinite reduction sequence, and is *confluent* ($\mathrm{CR}(\mathcal{A})$) if $a, b \in A$ are joinable whenever $a \;{}^*\!\!\leftarrow c \rightarrow^* b$ for some $c \in A$.

The following proposition provides a sufficient condition for the equivalence of two ARSs, which is the basis of all our results on inductive theorem proving (c.f. [10, 20]).

Proposition 1 (equivalence condition for ARSs). Let $\mathcal{A}_1 = \langle A, \rightarrow_1 \rangle$ and $\mathcal{A}_2 = \langle A, \rightarrow_2 \rangle$ be ARSs such that $\rightarrow_1 \subseteq \rightarrow_2$. Suppose that the following conditions are satisfied: (1) $\mathrm{WN}(\mathcal{A}_1)$, (2) $\mathrm{CR}(\mathcal{A}_2)$, (3) $\mathrm{NF}(\mathcal{A}_1) \subseteq \mathrm{NF}(\mathcal{A}_2)$. Then $\leftrightarrow_1^* = \leftrightarrow_2^*$.

The third condition above is equivalent to $\mathrm{RED}(\mathcal{A}_2) \subseteq \mathrm{RED}(\mathcal{A}_1)$. It is often useful to consider a set $X \subseteq \mathrm{RED}(\mathcal{A}_1)$ of intermediate elements reachable from $\mathrm{RED}(\mathcal{A}_2)$.

Proposition 2 (inclusion of normal elements). Let $\mathcal{A}_1 = \langle A, \rightarrow_1 \rangle$ and $\mathcal{A}_2 = \langle A, \rightarrow_2 \rangle$ be ARSs such that $\rightarrow_1 \subseteq \rightarrow_2$. Suppose there exists a subset X of A such that (1) for any $a \in \mathrm{RED}(\mathcal{A}_2)$ there exists $b \in X$ such that $a \rightarrow_1^* b$, (2) $X \subseteq \mathrm{RED}(\mathcal{A}_1)$. Then $\mathrm{NF}(\mathcal{A}_1) \subseteq \mathrm{NF}(\mathcal{A}_2)$.

In this paper, we consider two instances of the proposition above in order to give criteria for the inclusion between two sets of normal forms. One is an abstraction of the use of coverset and the other corresponds sufficient completeness.

Proposition 3 (coverset in ARS). Let $\mathcal{A}_1 = \langle A, \to_1 \rangle$ and $\mathcal{A}_2 = \langle A, \to_2 \rangle$ be ARSs such that $\to_1 \subseteq \to_2$. Suppose there exists a collection $\mathcal{C}$ of subsets of A such that (1) for any $a \in \mathrm{RED}(\mathcal{A}_2)$ there exist $C \in \mathcal{C}$ and $b \in C$ such that $a \to_1^* b$, (2) $C \subseteq \mathrm{RED}(\mathcal{A}_1)$ for any $C \in \mathcal{C}$. Then $\mathrm{NF}(\mathcal{A}_1) \subseteq \mathrm{NF}(\mathcal{A}_2)$.

Proof. Take $X = \bigcup \mathcal{C}$ in Proposition 2. □

Proposition 4 (sufficient completeness in ARS). Let $\mathcal{A}_1 = \langle A, \to_1 \rangle$ and $\mathcal{A}_2 = \langle A, \to_2 \rangle$ be ARSs such that $\to_1 \subseteq \to_2$. Suppose there exists a subset C of A such that (1) for any $a \in \mathrm{RED}(\mathcal{A}_2)$ there exists $b \in A \setminus C$ such that $a \to_1^* b$, (2) $A \setminus C \subseteq \mathrm{RED}(\mathcal{A}_1)$. Then $\mathrm{NF}(\mathcal{A}_1) \subseteq \mathrm{NF}(\mathcal{A}_2)$.

Proof. Take $X = A \setminus C$ in Proposition 2. □

2.2 First-Order Inductive Theorem Proving

We denote by $\mathrm{T}(F, V)$ the set of *many-sorted terms* over a set F of sorted function symbols and the set V of sorted variables. An equation is a pair of terms of the same sort, written as $l \approx r$. The sets of variables occurring in a term t and in an equation $l \approx r$ are written as $\mathrm{V}(t)$ and $\mathrm{V}(l \approx r)$. A *ground* term is a term without variables. A substitution θ is said to be ground when $\theta(x)$ is ground for every variable in its *domain* $\mathrm{Dom}(\theta) = \{x \mid x \neq \theta(x)\}$. We put subscript $_g$ to denote ground terms, substitutions, and contexts, e.g. t_g, θ_g, and C_g.

Let $\mathcal{R} = \langle S, F, R \rangle$ be a *many-sorted term rewriting system* (*many-sorted TRS*, for short) where S is a set of sorts, F is a set of function symbols, and R is a set rewrite rules. An equation $l \approx r$ is an *inductive theorem* of $\mathcal{R}$ if $l\theta_g \leftrightarrow_{\mathcal{R}}^* r\theta_g$ holds for all ground substitutions θ_g such that $V(l \approx r) \subseteq \mathrm{Dom}(\theta_g)$. Obviously, $\langle \mathrm{T}(F, \emptyset), \to_{\mathcal{R}} \rangle$ forms an ARS. We denote $\mathrm{NF}(\langle \mathrm{T}(F, \emptyset), \to_{\mathcal{R}} \rangle)$, $\mathrm{WN}(\langle \mathrm{T}(F, \emptyset), \to_{\mathcal{R}} \rangle)$, etc. by $\mathrm{NF}_g(\mathcal{R})$, $\mathrm{WN}_g(\mathcal{R})$, etc., respectively. Proposition 1 gives a sufficient condition for an equation to be an inductive theorem of a many-sorted TRS. Let $\mathcal{R} = \langle S, F, R \rangle$ be a many-sorted TRS and $l \approx r$ an equation on $\mathrm{T}(F, V)$. In order to prove the equation $l \approx r$ to be an inductive theorem of $\mathcal{R}$, we often consider the extended TRS $\mathcal{R}' = \langle S, F, R \cup \{l \to r\} \rangle$, which we denote by $\mathcal{R} \cup \{l \to r\}$.

Proposition 5 (inductive theorem proving). Let $\mathcal{R}$ be a many-sorted TRS and $\mathcal{R}' = \mathcal{R} \cup \{l \to r\}$. Suppose the following conditions are satisfied: (1) $\mathrm{WN}_g(\mathcal{R})$, (2) $\mathrm{CR}_g(\mathcal{R}')$, (3) $\mathrm{NF}_g(\mathcal{R}) \subseteq \mathrm{NF}_g(\mathcal{R}')$. Then $l \approx r$ is an inductive theorem of $\mathcal{R}$.

Proof. Let $\mathcal{R} = \langle S, F, R \rangle$. Take $\mathcal{A}_1 = \langle \mathrm{T}(F, \emptyset), \to_{\mathcal{R}} \rangle$ and $\mathcal{A}_2 = \langle \mathrm{T}(F, \emptyset), \to_{\mathcal{R}'} \rangle$ in Proposition 1. □

To check condition (3) of the proposition efficiently, the notion of *coverset* is used. For a many-sorted TRS $\mathcal{R}$ and a term t, a set Θ of substitutions is a *coverset of substitutions* for t w.r.t. $\mathcal{R}$ if for any ground substitution θ_g satisfying $\mathrm{V}(t) \subseteq \mathrm{Dom}(\theta_g)$, there exists a substitution $\sigma \in \Theta$ and a ground substitution θ'_g such that $t\theta_g \to^*_{\mathcal{R}} t\sigma\theta'_g$.

Proposition 6 (inductive theorem proving with coverset). Let $\mathcal{R}$ be a many-sorted TRS and $\mathcal{R}' = \mathcal{R} \cup \{l \to r\}$. Suppose that the following conditions are satisfied: (1) $\mathrm{WN}_g(\mathcal{R})$, (2) $\mathrm{CR}_g(\mathcal{R}')$, (3) $l\sigma \in \mathrm{RED}(\mathcal{R})$ for all $\sigma \in \Theta$ where Θ is a coverset of substitutions for l w.r.t. $\mathcal{R}$. Then $l \approx r$ is an inductive theorem of $\mathcal{R}$.

Proof. We use Proposition 5. The inclusion $\mathrm{NF}_g(\mathcal{R}) \subseteq \mathrm{NF}_g(\mathcal{R}')$ follows from condition (3) by taking $\mathcal{C} = \{\mathrm{RED}_\sigma \mid \sigma \in \Theta\}$ in Proposition 3 where $\mathrm{RED}_\sigma = \{C_g[l\sigma\theta_g] \mid \mathrm{V}(l\sigma) \subseteq \mathrm{Dom}(\theta_g)\}$. □

Another condition is obtained by strengthening condition (1). Let us divide the set of function symbols into two disjoint sets as follows. We say a function symbol f is a *defined symbol* of a many-sorted TRS $\langle S, F, R\rangle$ if $f = \mathrm{root}(l)$ for some $l \to r \in R$; otherwise, f is a *constructor symbol*. The sets of defined and constructor symbols are denoted by F_d and F_c, respectively. A many-sorted TRS $\mathcal{R}$ is *sufficiently complete* ($\mathrm{SC}(\mathcal{R})$) if for any ground term s_g there exists a ground constructor term t_g such that $s_g \to^*_{\mathcal{R}} t_g$.

Proposition 7 (inductive theorem proving with SC). Let $\mathcal{R}$ be a many-sorted TRS and $\mathcal{R}' = \mathcal{R} \cup \{l \to r\}$. Suppose the following conditions are satisfied: (1) $\mathrm{SC}(\mathcal{R})$, (2) $\mathrm{CR}_g(\mathcal{R}')$, (3) l contains a defined symbol. Then $l \approx r$ is an inductive theorem of $\mathcal{R}$.

Proof. Let $\mathcal{R} = \langle S, F, R\rangle$. We use Proposition 5. By definition, $\mathrm{SC}(\mathcal{R})$ implies $\mathrm{WN}_g(\mathcal{R})$. By $\mathrm{SC}(\mathcal{R})$, every ground term containing a defined symbol is reducible. Hence, the inclusion $\mathrm{NF}_g(\mathcal{R}) \subseteq \mathrm{NF}_g(\mathcal{R}')$ follows from condition (3) by taking $C = \mathrm{T}(F_\mathrm{c}, \emptyset)$ in Proposition 4. □

It is known that when $\mathrm{WN}_g(\mathcal{R})$ holds, it is decidable whether $\mathrm{SC}(\mathcal{R})$ for possibly non-left-linear $\mathcal{R}$ (see, e.g., [7, 18]).

3 Higher-Order Inductive Theorems

In this section, we first recall the basic notions and terminology of simply typed term rewriting, which were introduced in [21]. Then we introduce the notion of higher-order inductive theorems, which is suitable for STTRSs.

3.1 Simply Typed Term Rewriting Systems

For a set B of *basic types*, the set of *simple types* is the smallest set $\mathrm{ST}(B)$ such that (1) $B \subseteq \mathrm{ST}(B)$, and (2) $\tau_1 \times \cdots \times \tau_n \to \tau_0 \in \mathrm{ST}(B)$ whenever $\tau_0, \tau_1, \ldots, \tau_n \in \mathrm{ST}(B)$. A non-basic type is called a *higher-order* type. Note that our definition allows multiple basic types whereas the original one in [21] is based on a single basic type. When clear, simple type is abbreviated as *type*.

Each *constant* or *variable* is associated with its type; the sets of constants and variables of type τ are denoted by Σ^τ and V^τ, respectively. Σ and V stand for the sets of all constants and variables, respectively. We assume that V^τ is countably infinite for each simple type τ. The sets of variables of basic types and of higher-order types are denoted by V^{b} and V^{h}, respectively.

The set $\mathrm{T}(\Sigma, V)^\tau$ of *simply typed terms* of type τ over Σ and V is defined as follows: (1) $\Sigma^\tau \cup V^\tau \subseteq \mathrm{T}(\Sigma, V)^\tau$, and (2) if $s \in \mathrm{T}(\Sigma, V)^{\tau_1 \times \cdots \times \tau_n \to \tau}$ $(n \geq 1)$ and $t_i \in \mathrm{T}(\Sigma, V)^{\tau_i}$ for all $i \in \{1, \ldots, n\}$ then $(s\, t_1 \cdots t_n) \in \mathrm{T}(\Sigma, V)^\tau$. A simply typed term t has type τ (denoted by t^τ) when $t \in \mathrm{T}(\Sigma, V)^\tau$. It is clear that each simply typed term has a unique type; thus τ is also referred to as the type of t (denoted by $\mathrm{type}(t)$). The set of all simply typed terms is denoted by $\mathrm{T}(\Sigma, V)$. The set of simply typed terms of basic types and of higher-order types are denoted by $\mathrm{T}^{\mathrm{b}}(\Sigma, V)$ and $\mathrm{T}^{\mathrm{h}}(\Sigma, V)$, respectively.

The *head symbol* of a simply typed term is defined as follows: (1) $\mathrm{head}(t) = t$ if $t \in \Sigma \cup V$, and (2) $\mathrm{head}((s\, t_1 \cdots t_n)) = \mathrm{head}(s)$. A simply typed term is *linear* if no variable occurs more than once in it. A *simply typed equation* is a pair of simply typed terms of the same type, written as $l \approx r$. The sets of variables occurring in a term t and in an equations $l \approx r$ are written as $\mathrm{V}(t)$ and $\mathrm{V}(l \approx r)$, respectively.

A context is a simply typed term with a special symbol $\Box$ (hole). If a context C has n holes and $t_1, \ldots, t_n$ are simply typed terms then $C[t_1, \ldots, t_n]$ is the simply typed term obtained from C by replacing holes with $t_1, \ldots, t_n$ from left to right. A term s is a *subterm* of a term t (denoted by $s \trianglelefteq t$) if $t = C[s]$ for some context C with precisely one hole. A *substitution* is a mapping $\sigma : V \to \mathrm{T}(\Sigma, V)$ that satisfies the following conditions: (1) $\mathrm{Dom}(\sigma) = \{x \mid \sigma(x) \neq x\}$ is finite, and (2) for every variable x, x and $\sigma(x)$ have the same type. The homomorphic extension of σ to $\mathrm{T}(\Sigma, V)$ is also denoted by σ. As usual, $\sigma(t)$ is written as $t\sigma$.

A simply typed equation $l \approx r$ is called a *simply typed rewrite rule* if $\mathrm{head}(l) \in \Sigma$ and $\mathrm{V}(r) \subseteq \mathrm{V}(l)$. A simply typed rewrite rule $l \approx r$ will be often written as $l \to r$. A triple $\langle B, \Sigma, R \rangle$ consisting of a set B of basic types, a set Σ of constants, and a set R of simply typed rewrite rules is called a simply typed term rewriting system (STTRS, for short). The *rewrite relation* $\to_{\mathcal{R}}$ induced by a simply typed term rewriting system $\mathcal{R} = \langle B, \Sigma, R \rangle$ is the smallest relation over $\mathrm{T}(\Sigma, V)$ satisfying the following conditions: (1) $l\sigma \to_{\mathcal{R}} r\sigma$ for all $l \to r \in R$ and for all substitutions σ, and (2) if $s \to_{\mathcal{R}} t$ then $C[s] \to_{\mathcal{R}} C[t]$ for all contexts C. A constant symbol f is a *defined symbol* of $\mathcal{R}$ if $f = \mathrm{head}(l)$ for some $l \to r \in R$; otherwise, f is a *constructor symbol*. The sets of defined and constructor symbols are denoted by Σ_{d} and Σ_{c}, respectively.

Example 1 (simply typed term rewriting). Let $\mathcal{R} = \langle B, \Sigma, R \rangle$ be an STTRS where $B = \{\mathrm{N}, \mathrm{NList}\}$, $\Sigma = \{\ \mathsf{0}^{\mathrm{N}}, \mathsf{s}^{\mathrm{N} \to \mathrm{N}}, []^{\mathrm{NList}}, :^{\mathrm{N} \times \mathrm{NList} \to \mathrm{NList}}, \mathsf{map}^{(\mathrm{N} \to \mathrm{N}) \times \mathrm{NList} \to \mathrm{NList}}, \circ^{(\mathrm{N} \to \mathrm{N}) \times (\mathrm{N} \to \mathrm{N}) \to (\mathrm{N} \to \mathrm{N})}, \mathsf{twice}^{(\mathrm{N} \to \mathrm{N}) \to (\mathrm{N} \to \mathrm{N})}\ \}$, and

$$R = \left\{ \begin{array}{lcl} \mathsf{map}\ F\ [] & \to & [] \\ \mathsf{map}\ F\ (x : xs) & \to & (F\ x) : (\mathsf{map}\ F\ xs) \\ (F \circ G)\ x & \to & F\ (G\ x) \\ \mathsf{twice}\ F & \to & F \circ F \end{array} \right\}.$$

Here is an example of rewrite sequence of $\mathcal{R}$:

$$\begin{array}{lll} \mathsf{map}\ (\mathsf{twice}\ \mathsf{s})\ (0 : [\,]) & \to_{\mathcal{R}} & \mathsf{map}\ (\mathsf{s} \circ \mathsf{s})\ (0 : [\,]) \\ & \to_{\mathcal{R}} & ((\mathsf{s} \circ \mathsf{s})\ 0) : (\mathsf{map}\ (\mathsf{s} \circ \mathsf{s})\ [\,]) \\ & \to_{\mathcal{R}} & (\mathsf{s}\ (\mathsf{s}\ 0)) : (\mathsf{map}\ (\mathsf{s} \circ \mathsf{s})\ [\,]) \\ & \to_{\mathcal{R}} & (\mathsf{s}\ (\mathsf{s}\ 0)) : [\,]. \end{array}$$

3.2 Inductive Theorems of STTRSs

One may define the notion of inductive theorem of an STTRS exactly the same as that of a many-sorted STTRS, that is, an equation $l \approx r$ is an inductive theorem of an STTRS $\mathcal{R}$ if $l\theta_g \leftrightarrow^*_{\mathcal{R}} r\theta_g$ holds for any ground substitution θ_g such that $\mathrm{V}(l \approx r) \subseteq \mathrm{Dom}(\theta_g)$. Henceforth, we call this notion of inductive theorem *first-order*. Propositions 5 and 7 are extended to STTRSs in the obvious way.

Example 2 (first-order inductive theorem proving). Let $\mathcal{R} = \langle B, \Sigma, R\rangle$ be an STTRS where $B = \{\ \mathrm{N}, \mathrm{NList}\ \}$, $\Sigma = \{\ 0^{\mathrm{N}}, \mathsf{s}^{\mathrm{N}\to\mathrm{N}}, [\,]^{\mathrm{NList}}, :^{\mathrm{N}\times\mathrm{NList}\to\mathrm{NList}}, \mathsf{append}^{\mathrm{NList}\times\mathrm{NList}\to\mathrm{NList}}, \mathsf{map}^{(\mathrm{N}\to\mathrm{N})\times\mathrm{NList}\to\mathrm{NList}}\ \}$, and

$$R = \left\{ \begin{array}{lll} \mathsf{map}\ F\ [\,] & \to & [\,] \\ \mathsf{map}\ F\ (x : xs) & \to & (F\ x) : (\mathsf{map}\ F\ xs) \\ \mathsf{append}\ [\,]\ ys & \to & ys \\ \mathsf{append}\ (x : xs)\ ys & \to & x : (\mathsf{append}\ xs\ ys) \end{array} \right\}.$$

Let us now show that the equation

$$e \quad = \quad \mathsf{map}\ F\ (\mathsf{append}\ xs\ ys) \approx \mathsf{append}\ (\mathsf{map}\ F\ xs)\ (\mathsf{map}\ F\ ys)$$

is a first-order inductive theorem of $\mathcal{R}$. Let $\mathcal{R}' = \mathcal{R} \cup \{e\}$. Using the method in [1] or [2] one can show that $\mathrm{SN}(\mathcal{R}')$ holds, and thus $\mathrm{WN}_g(\mathcal{R})$ follows. All critical pairs of $\mathcal{R}'$ are joinable. By $\mathrm{SN}(\mathcal{R}')$ and the Critical Pair Lemma for STTRS [3], it follows that $\mathcal{R}'$ is confluent. Hence $\mathrm{CR}_g(\mathcal{R}')$. One can show that $\Theta = \{\{xs \mapsto [\,]\}, \{xs \mapsto y : ys\}\}$ is a coverset of substitutions for the left-hand side of e w.r.t. $\mathcal{R}$. It is easy to check $(\mathsf{map}\ F\ (\mathsf{append}\ xs\ ys))\sigma \in \mathrm{RED}(\mathcal{R})$ for any $\sigma \in \Theta$. Therefore, e is a first-order inductive theorem of $\mathcal{R}$.

As seen from the next example, the notion of first-order inductive theorem is inadequate for equational reasoning in STTRS because of the presence of higher-order terms.

Example 3 (inadequateness of first-order inductive theorem). Let $\mathcal{R} = \langle B, \Sigma, R\rangle$ be an STTRS where $B = \{\ \mathrm{N}\ \}$, $\Sigma = \{\ 0^{\mathrm{N}}, \mathsf{s}^{\mathrm{N}\to\mathrm{N}}, \mathsf{id}_0^{\mathrm{N}\to\mathrm{N}}, \mathsf{id}_1^{\mathrm{N}\to\mathrm{N}}\ \}$, and

$$R = \left\{ \begin{array}{lll} \mathsf{id}_0\ 0 & \to & 0 \\ \mathsf{id}_0\ (\mathsf{s}\ x) & \to & \mathsf{s}\ (\mathsf{id}_0\ x) \\ \mathsf{id}_1\ x & \to & x \end{array} \right\}.$$

Consider whether the equation $\mathsf{id}_0 \approx \mathsf{id}_1$ is an inductive theorem of $\mathcal{R}$. Since both id_0 and id_1 are ground, and $\mathsf{id}_0 \leftrightarrow^*_{\mathcal{R}_1} \mathsf{id}_1$ does not hold, the equation $\mathsf{id}_0 \approx \mathsf{id}_1$ is not a first-order inductive theorem of $\mathcal{R}$.

Since id_0 computes the same result as id_1 does for any argument of type N, it seems natural to regard the equation $\mathsf{id}_0 \approx \mathsf{id}_1$ inductively valid.

Inductive theorems are mostly useful to verify properties of functional programs, algebraic specification, etc. and from this point of view only "evaluation form" matters – since functional programs are evaluated in ground terms of basic types, it is reasonable to define inductive theorems in higher-order rewriting those equations that are valid on the set of ground terms of basic types.

Inductive theorems of a higher-order type are verified after turning them into an equation of basic type by appending fresh variables without changing its "inductive" meaning.

Definition 1 (extensional form).

1. Let t be a simply typed term of type $\tau_{11} \times \cdots \times \tau_{1n_1} \to (\tau_{21} \times \cdots \times \tau_{2n_2} \to \cdots \to (\tau_{m1} \times \cdots \times \tau_{mn_m} \to \rho) \cdots)$ such that ρ is a basic type. An *extensional form* of t is a term $((\cdots((t\, x_{11} \cdots x_{1n_1})\, x_{21} \cdots x_{2n_2}) \cdots)\, x_{m1} \cdots x_{mn_m})$ where $x_{11}, \ldots, x_{mn_m}$ are mutually distinct fresh variables. Clearly, any extensional form of a simply typed term has a basic type. The extensional form of t with fresh variables $x_{11}, \ldots, x_{mn_m}$ is written as $t{\uparrow}_{x_{11},\ldots,x_{mn_m}}$.
2. Let $\phi(s_1, \ldots, s_n)$ be an expression containing simply typed terms $s_1, \ldots, s_n$ of the same type. The extensional form $\phi(s_1{\uparrow}_{x_{11},\ldots,x_{mn_m}}, \ldots, s_n{\uparrow}_{x_{11},\ldots,x_{mn_m}})$ of $\phi(s_1, \ldots, s_n)$ is obtained by replacing each simply typed term s_i in the expression by $s_i{\uparrow}_{x_{11},\ldots,x_{mn_m}}$, where $x_{11}, \ldots, x_{mn_m}$ are mutually distinct fresh variables.
3. When the variables $x_{11}, \ldots, x_{mn_m}$ are not important, we omit them and simply write, e.g., $t{\uparrow}$ and $\phi(s_1{\uparrow}, \ldots, s_n{\uparrow})$.

Example 4 (extensional form). An extensional form of map is $\mathsf{map}\ F\ xs$, and an extensional form of $\circ$ is $(F \circ G)\ xs$. An extensional form of the equation $\mathsf{id}_0 \approx \mathsf{id}_1$ is $\mathsf{id}_0\ x \approx \mathsf{id}_1\ x$.

Definition 2 (higher-order inductive theorem). An equation $l \approx r$ is a *higher-order inductive theorem* of an STTRS $\mathcal{R}$ if $l{\uparrow}\theta_g \leftrightarrow^*_{\mathcal{R}} r{\uparrow}\theta_g$ holds for all ground substitutions θ_g such that $\mathrm{V}(l{\uparrow} \approx r{\uparrow}) \subseteq \mathrm{Dom}(\theta_g)$.

The next theorem follows immediately from the definition of higher-order inductive theorem and Proposition 1.

Theorem 1 (higher-order inductive theorem proving). Let $\mathcal{R}$ be an STTRS and $\mathcal{R}' = \mathcal{R} \cup \{l{\uparrow} \to r{\uparrow}\}$. Suppose the following conditions are satisfied: (1) $\mathrm{WN}^{\mathrm{b}}_g(\mathcal{R})$, (2) $\mathrm{CR}^{\mathrm{b}}_g(\mathcal{R}')$, (3) $\mathrm{NF}^{\mathrm{b}}_g(\mathcal{R}) \subseteq \mathrm{NF}^{\mathrm{b}}_g(\mathcal{R}')$. Then $l \approx r$ is a higher-order inductive theorem of $\mathcal{R}$.

The notion of coverset of substitutions is extended to STTRS in an obvious way as follows. For an STTRS $\mathcal{R}$ and a simply typed term t, a set Θ of substitutions is a *coverset of substitutions for t w.r.t. $\mathcal{R}$* if for any ground substitution θ_g satisfying $\mathrm{V}(t) \subseteq \mathrm{Dom}(\theta_g)$, there exists a substitution $\sigma \in \Theta$ and a ground substitution θ'_g such that $t\theta_g \to^*_{\mathcal{R}} t\sigma\theta'_g$.

Theorem 2 (higher-order inductive theorem proving with coverset). Let $\mathcal{R}$ be an STTRS and $\mathcal{R}' = \mathcal{R} \cup \{l\!\uparrow \to r\!\uparrow\}$. Suppose the following conditions are satisfied: (1) $\mathrm{WN}^{\mathrm{b}}_g(\mathcal{R})$, (2) $\mathrm{CR}^{\mathrm{b}}_g(\mathcal{R}')$, (3) $l\!\uparrow\!\sigma \in \mathrm{RED}(\mathcal{R})$ for all $\sigma \in \Theta$, where Θ is a coverset of substitutions for $l\!\uparrow$ w.r.t. $\mathcal{R}$. Then $l \approx r$ is a higher-order inductive theorem of $\mathcal{R}$.

Example 5 (higher-order inductive theorem proving). Let us consider Example 3 again. By taking an extensional form of the equation $\mathsf{id}_0 \approx \mathsf{id}_1$, we obtain the equation $\mathsf{id}_0\ x \approx \mathsf{id}_1\ x$. Let $\mathcal{R}' = \mathcal{R} \cup \{\mathsf{id}_0\ x \to \mathsf{id}_1\ x\}$. Then $\mathrm{WN}^{\mathrm{b}}_g(\mathcal{R})$ and $\mathrm{CR}^{\mathrm{b}}_g(\mathcal{R}')$ are proved in the same way as Example 2. By taking the coverset of substitutions $\Theta = \{\{x \mapsto \mathsf{0}\}, \{x \mapsto \mathsf{s}\ y\}\}$, condition (3) is easily verified. Therefore, the equation $\mathsf{id}_0 \approx \mathsf{id}_1$ is a higher-order inductive theorem of $\mathcal{R}$.

The notion of sufficient completeness is extended to STTRS as follows. An STTRS $\mathcal{R}$ is *sufficiently complete* ($\mathrm{SC}(\mathcal{R})$) if for any ground term s_g of basic type there exists a ground constructor term t_g of basic type such that $s_g \to^*_{\mathcal{R}} t_g$.

Theorem 3 (higher-order inductive theorem proving with SC). Let $\mathcal{R}$ be an STTRS and $\mathcal{R}' = \mathcal{R} \cup \{l\!\uparrow \to r\!\uparrow\}$. Suppose the following conditions are satisfied: (1) $\mathrm{SC}(\mathcal{R})$, (2) $\mathrm{CR}^{\mathrm{b}}_g(\mathcal{R}')$, (3) l contains a defined symbol. Then $l \approx r$ is a higher-order inductive theorem of $\mathcal{R}$.

4 Monotone Higher-Order Inductive Theorem

In contrast to the first-order inductive theorems, higher-order inductive theorems introduced in the previous section may not be monotone, that is, even if $l \approx r$ is a higher-order inductive theorem of an STTRS, $C[l] \approx C[r]$ may not be a higher-order inductive theorem depending on the choice of the context C – this observation is due to K. Kusakari. In this section, we study monotonicity of higher-order inductive theorems.

Definition 3 (monotone higher-order inductive theorem). Let $\mathcal{R}$ be an STTRS. A higher-order inductive theorem $l \approx r$ of $\mathcal{R}$ is *monotone* if $C[l] \approx C[r]$ is a higher-order inductive theorem of $\mathcal{R}$ for any context C.

First, let us examine examples of higher-order inductive theorems that are not monotone.

Example 6. Let $\mathcal{R} = \langle B, \Sigma, R\rangle$ be an STTRS such that $B = \{\ \mathrm{N}\ \}$, $\Sigma = \{\ \mathsf{0}^{\mathrm{N}}, \mathsf{s}^{\mathrm{N}\to\mathrm{N}}, +^{\mathrm{N}\to\mathrm{N}\to\mathrm{N}}, \mathsf{f}^{(\mathrm{N}\to\mathrm{N})\to\mathrm{N}}\ \}$, and

$$R = \left\{\begin{array}{lcl} (+\ x)\ \mathsf{0} & \to & x \\ (+\ x)\ (\mathsf{s}\ y) & \to & \mathsf{s}\ ((+\ x)\ y) \\ \mathsf{f}\ \mathsf{s} & \to & \mathsf{0} \\ \mathsf{f}\ (+\ x) & \to & \mathsf{s}\ \mathsf{0} \end{array}\right\}.$$

The equation $+\ (\mathsf{s}\ \mathsf{0}) \approx \mathsf{s}$ is an inductive theorem of $\mathcal{R}$. However, $\mathsf{f}\ (+\ (\mathsf{s}\ \mathsf{0})) \approx \mathsf{f}\ \mathsf{s}$ is not, since $\mathsf{f}\ (+\ (\mathsf{s}\ \mathsf{0})) \to_{\mathcal{R}} \mathsf{s}\ \mathsf{0} \neq \mathsf{0}\ {}_{\mathcal{R}}\!\!\leftarrow \mathsf{f}\ \mathsf{s}$. Thus the equation $+\ (\mathsf{s}\ \mathsf{0}) \approx \mathsf{s}$ is not monotone.

The key observation on what leads to non-monotonicity of higher-order inductive theorems is that $l \approx r$ is a higher-order inductive theorem implies only $C[l{\uparrow}\theta_g] \leftrightarrow^*_{\mathcal{R}} C[r{\uparrow}\theta_g]$ but not necessarily $(C[l]){\uparrow}\sigma_g \leftrightarrow^*_{\mathcal{R}} (C[r]){\uparrow}\sigma_g$. Let us examine an example where such a desired implication seems satisfied.

Example 7 (inductive theorem within a context). Consider the STTRS defined by $\mathcal{R} = \langle B, \Sigma, R\rangle$ where $B = \{\ \mathrm{N}, \mathrm{NList}\ \}$, $\Sigma = \{\ \mathsf{0}^{\mathrm{N}}, \mathsf{s}^{\mathrm{N}\to\mathrm{N}}, []^{\mathrm{NList}}, :^{\mathrm{N}\times\mathrm{NList}\to\mathrm{NList}}, +^{\mathrm{N}\to\mathrm{N}\to\mathrm{N}}, \mathsf{map}^{(\mathrm{N}\to\mathrm{N})\times\mathrm{NList}\to\mathrm{NList}}\ \}$, and

$$R = \left\{ \begin{array}{lcl} (+\ \mathsf{0})\ y & \to & y \\ (+\ (\mathsf{s}\ x))\ y & \to & \mathsf{s}\ ((+\ x)\ y) \\ \mathsf{map}\ F\ [] & \to & [] \\ \mathsf{map}\ F\ (x : xs) & \to & (F\ x) : (\mathsf{map}\ F\ xs) \end{array} \right\}.$$

The equation $+\ (\mathsf{s}\ \mathsf{0}) \approx \mathsf{s}$ is a higher-order inductive theorem of $\mathcal{R}$. Consider the context $C = \mathsf{map}\ \Box\ (\mathsf{0} : [])$. We are going to verify the equation $C[+\ (\mathsf{s}\ \mathsf{0})] \approx C[\mathsf{s}]$ also is a higher-order inductive theorem.

$$\begin{array}{rl} \mathsf{map}\ (+\ (\mathsf{s}\ \mathsf{0}))\ (\mathsf{0} : []) & \to_{\mathcal{R}} ((+\ (\mathsf{s}\ \mathsf{0}))\ \mathsf{0}) : (\mathsf{map}\ (+\ (\mathsf{s}\ \mathsf{0}))\ []) \\ & \to_{\mathcal{R}} (\mathsf{s}\ ((+\ \mathsf{0})\ \mathsf{0})) : (\mathsf{map}\ (+\ (\mathsf{s}\ \mathsf{0}))\ []) \\ & \to_{\mathcal{R}} (\mathsf{s}\ \mathsf{0}) : (\mathsf{map}\ (+\ (\mathsf{s}\ \mathsf{0}))\ []) \\ & \to_{\mathcal{R}} (\mathsf{s}\ \mathsf{0}) : [] \\ & {}_{\mathcal{R}}\!\leftarrow (\mathsf{s}\ \mathsf{0}) : \mathsf{map}\ \mathsf{s}\ [] \\ & {}_{\mathcal{R}}\!\leftarrow \mathsf{map}\ \mathsf{s}\ (\mathsf{0} : []) \end{array}$$

This rewrite sequence can be reconstructed from $(+\ (\mathsf{s}\ \mathsf{0}))\ \mathsf{0} \leftrightarrow^*_{\mathcal{R}} \mathsf{s}\ \mathsf{0}$ and

$$\mathsf{map}\ F\ (\mathsf{0} : []) \quad \to_{\mathcal{R}} \quad (F\ \mathsf{0}) : \mathsf{map}\ F\ [] \quad \to_{\mathcal{R}} \quad (F\ \mathsf{0}) : [].$$

The observation above motivates the notion of higher-order sufficient completeness.

Definition 4 (higher-order sufficient completeness).

1. Let t be a simply typed term such that $t = C[F]$ for some $F \in V^{\mathrm{h}}$. The occurrence of F is said to be *expanded* if the occurrence of F in t has the form $t = C'[F{\uparrow}\theta]$ for some context C' and a substitution θ.
2. An STTRS $\mathcal{R} = \langle B, \Sigma, R\rangle$ is said to be *higher-order sufficiently complete* (HSC($\mathcal{R}$)) if for any simply typed term $s \in \mathrm{T}^{\mathrm{b}}(\Sigma, V^{\mathrm{h}})$ there exists $t \in \mathrm{T}^{\mathrm{b}}(\Sigma, V^{\mathrm{h}})$ such that $s \to^*_{\mathcal{R}} t$ and that either t has an expanded variable occurrence or t is a ground constructor term.

Example 8 (higher-order sufficient completeness). Let $\mathcal{R}$ be the STTRS in Example 7, $s = \mathsf{map}\ F\ (\mathsf{0} : [])$, and $t = (F\ \mathsf{0}) : []$. The occurrence of variable F in s is non-expanded, while that of F in t is expanded. Using a result in Section 5, it will turn out that $\mathcal{R}$ is higher-order sufficiently complete. Indeed, we have $s \to^*_{\mathcal{R}} t \in \mathrm{T}^{\mathrm{b}}(\Sigma, V^{\mathrm{h}})$ and t has an expanded variable occurrence.

Note that HSC($\mathcal{R}$) implies SC($\mathcal{R}$) since reduction preserves groundness. Hence the notion of higher-order sufficient completeness extends the notion of (first-order) sufficient completeness.

Lemma 1 (sufficient condition for monotonicity). If an STTRS $\mathcal{R}$ satisfies $\mathrm{SN}_g^{\mathrm{b}}(\mathcal{R})$ and HSC($\mathcal{R}$), then every higher-order inductive theorem of $\mathcal{R}$ is monotone.

Proof. Suppose $l \approx r$ is a higher-order inductive theorem. We have to show that $C[l] \approx C[r]$ also is a higher-order inductive theorem. Let $s = C_g[l\theta_g, \ldots, l\theta_g]$ and $t = C_g[r\theta_g, \ldots, r\theta_g]$. We prove $s \leftrightarrow^*_{\mathcal{R}} t$ for all ground contexts C_g of basic type and ground substitutions θ_g, which implies the desired result as a special case when C_g contains precisely one hole. The proof proceeds by well-founded induction on s w.r.t. $\rightarrow_{\mathcal{R}}$. If l and r have a basic type, then $s \leftrightarrow^*_{\mathcal{R}} t$ immediately follows from the assumption that $l \approx r$ is an inductive theorem. Consider the case that l and r have a non-basic type. Let C' be a context obtained from C_g by replacing all (possibly zero) expanded occurrences of holes with an arbitrary variable F whose type is the same as that of l and r. Let σ and τ be substitutions satisfying $\sigma(F) = l\theta_g$ and $\tau(F) = r\theta_g$. We have $s = C'\sigma[l\theta_g, \ldots, l\theta_g]$ and $t = C'\tau[r\theta_g, \ldots, r\theta_g]$. Since all occurrences of F in C' are expanded, the assumption that $l \approx r$ is a higher-order inductive theorem implies $C'\sigma[F, \ldots, F] \leftrightarrow^*_{\mathcal{R}} C'\tau[F, \ldots, F]$. By HSC($\mathcal{R}$), there exists a term $u = D_g[F, \ldots, F]$ with some ground context D_g such that $C'\sigma[F, \ldots, F] \rightarrow^*_{\mathcal{R}} u$ and either u is a ground constructor term or there is an expanded occurrence of F in u. In the former case, D_g has no holes and hence both $s = C'\sigma[l\theta_g, \ldots, l\theta_g] \rightarrow^*_{\mathcal{R}} u\sigma = D_g$ and $t = C'\tau[r\theta_g, \ldots, r\theta_g] \leftrightarrow^*_{\mathcal{R}} C'\sigma[r\theta_g, \ldots, r\theta_g] \rightarrow^*_{\mathcal{R}} u\tau = D_g$ are satisfied. In the latter case, the reduction sequence from $C'\sigma[F, \ldots, F]$ to u consists of at least one step, which implies the non-emptiness of the corresponding rewrite sequences from s to $u\sigma$. By the induction hypothesis, we obtain $u\sigma = D_g[l\theta_g, \ldots, l\theta_g] \leftrightarrow^*_{\mathcal{R}} D_g[r\theta_g, \ldots, r\theta_g] = u\tau$. Hence, both $s \rightarrow^*_{\mathcal{R}} u\sigma \leftrightarrow^*_{\mathcal{R}} u\tau$ and $t = C'\tau[r\theta_g, \ldots, r\theta_g] \leftrightarrow^*_{\mathcal{R}} C'\sigma[r\theta_g, \ldots, r\theta_g] \rightarrow^*_{\mathcal{R}} u\tau$ are satisfied. In both cases, we obtained $s \leftrightarrow^*_{\mathcal{R}} t$. □

Combining Theorem 3 and Lemma 1, we obtain the following theorem.

Theorem 4 (monotone higher-order inductive theorem proving with HSC). Let $\mathcal{R}$ be an STTRS and $\mathcal{R}' = \mathcal{R} \cup \{l\!\uparrow\, \rightarrow r\!\uparrow\}$. Suppose the following conditions are satisfied: (1) $\mathrm{SN}_g^{\mathrm{b}}(\mathcal{R})$, (2) HSC($\mathcal{R}$), (3) $\mathrm{CR}_g^{\mathrm{b}}(\mathcal{R}')$, (4) l contains a defined symbol. Then $l \approx r$ is a monotone higher-order inductive theorem of $\mathcal{R}$.

Example 9 (monotone higher-order inductive theorem proving). Let $\mathcal{R} = \langle B, \Sigma, R\rangle$ be an STTRS where $B = \{\ \mathrm{N}, \mathrm{NList}\ \}$, $\Sigma_{\mathrm{d}} = \{\ \mathsf{map}^{(\mathrm{N}\to\mathrm{N})\to(\mathrm{NList}\to\mathrm{NList})}, \circ^{(\mathrm{N}\to\mathrm{N})\times(\mathrm{N}\to\mathrm{N})\to(\mathrm{N}\to\mathrm{N})}, \bullet^{(\mathrm{NList}\to\mathrm{NList})\times(\mathrm{NList}\to\mathrm{NList})\to(\mathrm{NList}\to\mathrm{NList})}\ \}$, $\Sigma_{\mathrm{c}} = \{\ 0^{\mathrm{N}}, \mathsf{s}^{\mathrm{N}\to\mathrm{N}}, [\,]^{\mathrm{NList}}, :^{\ \mathrm{N}\times\mathrm{NList}\to\mathrm{NList}}\ \}$, and

$$R = \left\{ \begin{array}{lcl} (\mathsf{map}\ F)\ [\,] & \rightarrow & [\,] \\ (\mathsf{map}\ F)\ (x : xs) & \rightarrow & (F\ x) : ((\mathsf{map}\ F)\ xs) \\ (F \circ G)\ x & \rightarrow & F\ (G\ x) \\ (X \bullet Y)\ xs & \rightarrow & X\ (Y\ xs) \end{array} \right\}.$$

HSC($\mathcal{R}$) will be shown in Example 11. So, for the moment, suppose HSC($\mathcal{R}$). Let us consider the equation $e = \mathsf{map}\ (F \circ G) \approx (\mathsf{map}\ F) \bullet (\mathsf{map}\ G)$. By taking an extensional form of this equation, we obtain $e' = (\mathsf{map}\ (F \circ G))\ xs \approx ((\mathsf{map}\ F) \bullet (\mathsf{map}\ G))\ xs$. Let $\mathcal{R}' = \mathcal{R} \cup \{e'\}$. In the same way as Example 2, one can show $\mathrm{SN}^{\mathrm{b}}_{g}(\mathcal{R})$ and $\mathrm{CR}^{\mathrm{b}}_{g}(\mathcal{R}')$. The term $\mathsf{map}\ (F \circ G)$ contains defined symbols map and $\circ$. Therefore, by Theorem 4, e is a monotone higher-order inductive theorem of $\mathcal{R}$.

5 Decidability of Higher-Order Sufficient Completeness

It is undecidable, in general, whether higher-order sufficient completeness holds for a given STTRS. In first-order term rewriting, the related property called quasi-reducibility plays an important role in proving the decidability of sufficient completeness for TRSs satisfying certain conditions [7, 18]. In order to present a class of STTRSs where the higher-order sufficient completeness is decidable, we need to extend the notion of quasi-reducibility.

Definition 5 (argument subterms). Every constant and variable has no *argument subterms.* The *argument subterms* of a simply typed term $(s\ t_1 \cdots t_n)$ are argument subterms of s together with the terms $t_1, \ldots, t_n$. The set of all argument subterms in t is denoted by $\mathrm{Arg}(t)$.

Example 10 (argument subterms). A simply typed term $\mathsf{map}\ F$ has a unique argument subterm F. The argument subterms of $(\mathsf{map}\ (\mathsf{s} \circ \mathsf{s}))\ (0 : [\,])$ are $\mathsf{s} \circ \mathsf{s}$ and $0 : [\,]$.

Definition 6 (higher-order quasi-reducibility). An STTRS $\mathcal{R} = \langle B, \Sigma, R\rangle$ is *higher-order quasi-reducible* (HQR($\mathcal{R}$)) when for any term t such that $\mathrm{head}(t) \in \Sigma_{\mathrm{d}}$, if every argument subterms of t of basic type is a ground constructor term and of higher-order type is a variable, then t is reducible.

We are going to show, under some conditions, that the equivalence of HQR($\mathcal{R}$) and HSC($\mathcal{R}$) holds and that HQR($\mathcal{R}$) is decidable. Consequently, HSC($\mathcal{R}$) is decidable for that class of STTRSs. We first present a condition by which HQR($\mathcal{R}$) implies HSC($\mathcal{R}$).

Definition 7 (elementary STTRS). The set $\mathrm{ST}_{\mathrm{el}}$ of *elementary types* is defined inductively as: (1) $B \subseteq \mathrm{ST}_{\mathrm{el}}$, (2) if $\tau_1, \ldots, \tau_n \in B$ $(n \geq 1)$ and $\tau_0 \in \mathrm{ST}_{\mathrm{el}}$ then $\tau_1 \times \cdots \times \tau_n \to \tau_0 \in \mathrm{ST}_{\mathrm{el}}$. Let $\Sigma_{\mathrm{el}} = \bigcup_{\tau \in \mathrm{ST}_{\mathrm{el}}} \Sigma^{\tau}$. An STTRS $\mathcal{R} = \langle B, \Sigma, R\rangle$ is said to be *elementary* (EL($\mathcal{R}$)) if $\Sigma_{\mathrm{c}} \subseteq \Sigma_{\mathrm{el}}$ is satisfied.

Observe that if a set Σ of function symbols satisfies $\Sigma_{\mathrm{c}} \subseteq \Sigma_{\mathrm{el}}$, then all argument subterms of a simply typed term t are of basic type whenever $\mathrm{head}(t)$ is a constructor symbol.

Lemma 2 (key property for ensuring HSC). Let $\mathcal{R}$ be an STTRS satisfying HQR($\mathcal{R}$) and EL($\mathcal{R}$). If t is a normal form of basic type and all variables in t are of higher-order type, then t either has some expanded variable occurrence or is a ground constructor term.

Proof. The proof proceeds by structural induction on t. Distinguish cases by head(t). □

The notions of higher-order quasi-reducibility and elementarity appear in [12] to give a sufficient condition of SC($\mathcal{R}$).

Theorem 5 (equivalence condition for HQR and HSC). Let $\mathcal{R}$ be an STTRS satisfying $\mathrm{WN}^{\mathrm{b}}(\mathcal{R})$ and $\mathrm{EL}(\mathcal{R})$. Then, $\mathrm{HQR}(\mathcal{R})$ iff $\mathrm{HSC}(\mathcal{R})$.

Proof. The "if" direction is obvious by definition. For a proof of the "only if" direction, let $\mathcal{R} = \langle B, \Sigma, R \rangle$ and $s \in \mathrm{T}^{\mathrm{b}}(\Sigma, V^{\mathrm{h}})$. By $\mathrm{WN}^{\mathrm{b}}(\mathcal{R})$, there exists a normal form t of s such that $t \in \mathrm{T}^{\mathrm{b}}(\Sigma, V^{\mathrm{h}})$. Hence the result follows from Lemma 2 using $\mathrm{EL}(\mathcal{R})$. □

Example 11 (proving higher-order sufficient completeness). We now show that STTRS in Example 9 is higher-order sufficiently complete. In that example, it has been shown $\mathrm{SN}^{\mathrm{b}}(\mathcal{R})$ holds and hence so does $\mathrm{WN}^{\mathrm{b}}(\mathcal{R})$. Clearly, $\mathrm{EL}(\mathcal{R})$ and $\mathrm{HQR}(\mathcal{R})$ hold. Thus by Theorem 5 it follows that $\mathcal{R}$ is higher-order sufficiently complete.

The condition $\mathrm{EL}(\mathcal{R})$ is essential, because $\mathrm{WN}^{\mathrm{b}}(\mathcal{R})$ and $\mathrm{HQR}(\mathcal{R})$ do not necessarily guarantee $\mathrm{HSC}(\mathcal{R})$ as the following example shows.

Example 12. Let $\mathcal{R} = \langle B, \Sigma, R \rangle$ be an STTRS where $B = \{\ \mathrm{N}, \mathrm{NPair}\ \}$, $\Sigma = \{\ \mathsf{0}^{\mathrm{N}}, \mathsf{s}^{\mathrm{N} \to \mathrm{N}}, +^{\mathrm{N} \to \mathrm{N} \to \mathrm{N}}, \mathsf{f}^{(\mathrm{N} \to \mathrm{N}) \to \mathrm{NPair}}, \langle -, - \rangle^{(\mathrm{N} \to \mathrm{N}) \times (\mathrm{N} \to \mathrm{N}) \to \mathrm{NPair}}\ \}$, and

$$R = \left\{ \begin{array}{lcl} \mathsf{f}\ F & \to & \langle F, F \rangle \\ (+\ x)\ \mathsf{0} & \to & x \\ (+\ x)\ (\mathsf{s}\ y) & \to & \mathsf{s}\ ((+\ x)\ y) \end{array} \right\}.$$

Then, both $\mathrm{WN}^{\mathrm{b}}(\mathcal{R})$ and $\mathrm{HQR}(\mathcal{R})$ are satisfied. The term $\langle (+\ \mathsf{0}), (+\ \mathsf{0}) \rangle$ is a normal form and contains defined symbols, while it has no expanded variables.

We next show that higher-order sufficient completeness is decidable for a class of elementary STTRSs. In our decidability proof, we need to distinguish occurences of argument subterms. The notions of argument depth and argument position are introduced for that purpose.

Definition 8 (argument depth). The *argument depth*, or simply the depth, $\mathrm{d}(t)$ of a simply typed term t is defined by $\mathrm{d}(t) = 1 + \max\{\mathrm{d}(s) \mid s \in \mathrm{Arg}(s)\}$, where $\max \emptyset = 0$. The argument depth $\mathrm{d}(\mathcal{R})$ of an STTRS $\mathcal{R} = \langle B, \Sigma, R \rangle$ is defined by $\mathrm{d}(\mathcal{R}) = \max\{\mathrm{d}(l) \mid l \to r \in R\}$.

Definition 9 (argument position). A finite sequence of pairs of positive integers is called an *argument position*. The set $\mathrm{APos}(t)$ of argument positions in a simply typed term t and the subterm $t_{|p}$ of t at $p \in \mathrm{APos}(t)$ are defined as follows:

$$\begin{array}{ll} \mathrm{APos}(t) & = \{\varepsilon\} \cup \bigcup_{ij} \{ij.p \mid p \in \mathrm{APos}(t_{ij})\} \\ t_{|\varepsilon} & = t \\ t_{|ij.p} & = {t_{ij}}_{|p} \end{array}$$

where $t = ((\cdots((a\ t_{11}\cdots t_{1n_1})\ t_{21}\cdots\ t_{2n_2})\cdots)\ t_{m1}\cdots t_{mn_m})$ with $a \in \Sigma \cup V$. The prefix order $\geq$ on argument positions is defined as follows: $p \geq q$ iff $p = q.q'$ for some q'. We write $p \parallel q$ when neither $p \geq q$ nor $p \leq q$.

Example 13 (argument depth/argument position). Let $s = (\mathsf{map}\ (\mathsf{s} \circ \mathsf{s}))\ (\mathsf{0} : [\,])$. Then the $\mathrm{d}(s) = 3$ and $\mathrm{APos}(s) = \{\varepsilon, 11, 21, 11.11, 11.12, 21.11, 21.12\}$. We have $s_{|p} = \mathsf{0}$ when $p = 21.11$.

Let $B^{\mathrm{f}}, B^{\mathrm{i}} \subseteq B$ be the sets of basic types τ such that $\mathrm{T}^{\tau}(\Sigma_{\mathrm{c}}, \emptyset)$ is finite and infinite, respectively. Without loss of generality, we assume that[1] there exists a constant of type τ for each $\tau \in B$ and that[2] for any $\tau \in B^{\mathrm{f}}$ all ground constructors of type τ are constants. Clearly, for each type in $\tau \in B^{\mathrm{i}}$ and each positive integer k, there exists a ground constructor of type τ whose depth is k; so, let δ_k^{τ} be a simply typed term in $\mathrm{T}^{\tau}(\Sigma_{\mathrm{c}}, \emptyset)$ such that $\mathrm{d}(\delta_k) = k$, for each $\tau \in B^{\mathrm{i}}$ and $k > 0$. We will omit superscript $^{\tau}$ of δ_k^{τ} when it is clear from its context.

Definition 10 (top of a term). Suppose $\Sigma_{\mathrm{c}} \subseteq \Sigma_{\mathrm{el}}$. Let k be a positive integer and t be a ground constructor term of basic type. Let $p_1, \ldots, p_n$ be the list of all argument positions p in t satisfying $|p| \geq k - 1$. Then the *top* of t of depth k is the term obtained from t by replacing each subterm $t_{|p_i}$ by x_i, which is denoted by $\mathrm{top}_k(t)$. Finally, let $\mathrm{Top}_k^{\tau} = \{\mathrm{top}_k(t) \mid t \in \mathrm{T}^{\tau}(\Sigma_{\mathrm{c}}, \emptyset)\}$ for all $\tau \in B$. We will omit superscript $^{\tau}$ of Top_k^{τ} whenever it is clear from the context.

Example 14 (top of a term). Let $B = \{\,\mathrm{N}, \mathrm{NList}\,\}$ and $\Sigma_{\mathrm{c}} = \{\,\mathsf{0}^{\mathrm{N}}, \mathsf{s}^{\mathrm{N}\to\mathrm{N}}, [\,]^{\mathrm{NList}}, :^{\mathrm{N}\times\mathrm{NList}\to\mathrm{NList}}\,\}$. Then $\mathrm{Top}_1^{\mathrm{N}} = \{x\}$, $\mathrm{Top}_2^{\mathrm{N}} = \{\mathsf{0},\ \mathsf{s}\ x\}$, $\mathrm{Top}_2^{\mathrm{NList}} = \{[\,],\ y : ys\}$, and $\mathrm{Top}_3^{\mathrm{NList}} = \{[\,],\ \mathsf{0} : [\,],\ \mathsf{0} : (y : ys),\ (\mathsf{s}\ x) : [\,],\ (\mathsf{s}\ z) : (w : ws)\}$.

Lemma 3 (finiteness of tops). Suppose $\Sigma_{\mathrm{c}} \subseteq \Sigma_{\mathrm{el}}$. If Σ_{c} is finite, then Top_k^{τ} is finite (modulo variable renaming) for any positive integer k and $\tau \in B$.

To present a class of STTRSs for which higher-order sufficient completeness is decidable, we need the notion of higher-order scheme.

Definition 11 (higher-order scheme). An STTRS $\mathcal{R} = \langle B, \Sigma, R\rangle$ is said to be a *higher-order scheme* ($\mathrm{HS}(\mathcal{R})$) if every rewrite rule $l \to r \in R$ satisfies the properties that no higher-order variable occurs more than once in l and that for each subterm $(s\ t_1 \cdots t_n)$ in l and for any $i \in \{1, \ldots, n\}$, t_i is a variable whenever it is of higher-order type.

Example 15 (higher-order scheme). STTRSs in Examples 1, 2, 3, 7, 9, and 12 are all higher-order schemes. The STTRS in Example 6 is not.

Lemma 4. Let $\mathcal{R} = \langle B, \Sigma, R\rangle$ be an STTRS satisfying $\mathrm{EL}(\mathcal{R})$ and $\mathrm{HS}(\mathcal{R})$. Let $k = \mathrm{d}(\mathcal{R})$ and $f \in \Sigma_{\mathrm{d}}$. Then the following two statements are equivalent:

---

[1] If there is no ground constructor term of type τ then just ignore that type; otherwise take a minimal ground constructor term of type τ and regard it as a constant.

[2] Regard each ground constructor term of type τ as a constant.

1. There exists a linear simply typed term t in normal form with $\mathrm{head}(t) = f$ such that for every argument subterm s of t, we have $s \in \mathrm{Top}_k$ if s has a basic type, otherwise s is a variable.
2. There exists a simply typed term t in normal form with $\mathrm{head}(t) = f$ such that every argument subterm s of t is a ground constructor term if s has a basic type and is a variable otherwise.

Combining Lemmata 3 and 4, we obtain the following theorem, which yields our main decidability result as a corollary with help of Theorem 5.

Theorem 6 (decidability of HQR). Let $\mathcal{R}$ be an STTRS which satisfies $\mathrm{HS}(\mathcal{R})$ and $\mathrm{EL}(\mathcal{R})$. Then it is decidable whether $\mathcal{R}$ is higher-order quasi-reducible.

Corollary 1 (decidability of HSC). Let $\mathcal{R}$ be an STTRS which satisfies $\mathrm{WN}^{\mathrm{b}}(\mathcal{R})$, $\mathrm{HS}(\mathcal{R})$ and $\mathrm{EL}(\mathcal{R})$. Then it is decidable whether $\mathcal{R}$ is higher-order sufficiently complete.

6 Related Works

Linestad et al. [14] extended Huet and Hullot's inductive theorem proving procedure [5] for Nipkow's higher-order rewriting framework [18]. They introduced the notion of first-order substitution – substitution that instantiates only first-order variables – and showed that when their inductive theorem proving procedure terminates with Proof, then the given equation $s \approx t$ is initially consistent with a convergent HRS $\mathcal{R}$, that is, the normal form of $s\sigma$ equals to that of $t\sigma$ for any first-order ground substitution σ. Clearly, the notion of first-order substitution is closely related to the way that we extend the notion of quasi-reducibility to higher-order. The notion of initial consistency is similar to the first-order inductive validity, but in our framework initial consistency is not easy to work with, since our notion of inductive validity starts from more abstract setting.

Kusakari et al. [13] introduced a higher-order equational logic and its extensional algebraic semantics based on higher-order term rewriting introduced by Kusakari [11] – regardless of difference of underlying formulation, Kusakari's higher-order term rewriting framework is included to simply typed term rewriting. They showed the quotient ground term algebra is an initial extensional model and introduced two kinds of inductive validity which correspond to first-order inductive theorem and monotone higher-order inductive theorem; they also presented inductive theorem proving framework similar to the one given in Example 2. In [12], Kusakari extends these results to S-expression rewriting systems (SRS, for short); there he also claims that any SRS $\mathcal{R}$ satisfies $\mathrm{SC}(\mathcal{R})$ whenever $\mathrm{WN}(\mathcal{R})$, $\mathrm{EL}(\mathcal{R})$, and $\mathrm{HQR}(\mathcal{R})$ hold.

Apart from the one mentioned above, there is a number of works on higher-order extensional algebraic structures and its equational proof theory. In a survey [16], higher-order extensional algebraic structures and its sound and complete equational calculus have been given based on a higher-order language with product and functional types. The equational calculus given there, however, seems difficult to relate with simply typed term rewriting.

7 Conclusion

We have proposed the notion of higher-order inductive theorem in simply typed term rewriting, which enables us to reason about equational specification in the presence of higher-order terms. It turned out that for the proof of an equation $l \approx r$ to be a higher-order inductive theorem of a given STTRS $\mathcal{R}$, it suffices to show the following three properties on ground terms: weak normalisation of $\mathcal{R}$, confluence of the extended system $\mathcal{R} \cup \{l\!\uparrow \to r\!\uparrow\}$, and the coincidence of reducibility in these two systems. There are two ways to guarantee the the third condition. One is to use the notion of coverset of substitutions, and the other is by sufficient completeness.

Since higher-order inductive theorems are not necessarily monotone, we have provided a method to guarantee this property. Higher-order sufficient completeness together with termination guarantees monotonicity. We have also proved that higher-order sufficient completeness is decidable for the class of weakly normalizing elementary higher-order schemes.

Acknowledgments

The authors would like to thank Keiichiro Kusakari for pointing out that higher-order inductive validity does not form a congruence. Thanks are due to anonymous referees for comments and suggestions on related pointers.

References

1. T. Aoto and T. Yamada. Proving termination of simply typed term rewriting systems automatically. *IPSJ Transactions on Programming*, 44(SIG 4 PRO 17):67–77, 2003. In Japanese.
2. T. Aoto and T. Yamada. Termination of simply typed term rewriting systems by translation and labelling. In *Proceedings of the 14th International Conference on Rewriting Techniques and Applications*, volume 2706 of *LNCS*, pages 380–394. Springer-Verlag, 2003.
3. T. Aoto, T. Yamada, and Y. Toyama. Proving inductive theorems of higher-order functional programs. In *Information Technology Letters*, volume 2, pages 21–22, 2003. In Japanese.
4. F. Baader and T. Nipkow. *Term Rewriting and All That.* Cambridge University Press, Cambridge, 1998.
5. G. Huet and J.-M. Hullot. Proof by induction in equational theories with constructors. *Journal of Computer and System Sciences*, 25(2):239–266, 1982.
6. J.-P. Jouannaud and M. Okada. Executable higher-order algebraic specification languages. In *Proceedings of the 6th IEEE Symposium on Logic in Computer Science*, pages 350–361. IEEE Press, 1991.
7. D. Kapur, P. Narendran, and H. Zhang. On sufficient-completeness and related properties of term rewriting systems. *Acta Informatica*, 24(4):395–415, 1987.
8. D. Kapur, P. Narendran, and H. Zhang. Automating inductionless induction using test sets. *Journal of Symbolic Computation*, 11(1–2):81–111, 1991.

9. J. W. Klop. *Combinatory Reduction Systems.* PhD thesis, Rijksuniversiteit, Utrecht, 1980.
10. H. Koike and Y. Toyama. Inductionless induction and rewriting induction. *Comptuter Software*, 17(6):1–12, 2000. In Japanese.
11. K. Kusakari. On proving termination of term rewriting systems with higher-order variables. *IPSJ Transactions on Programming*, 42(SIG 7 PRO 11):35–45, 2001.
12. K. Kusakari. Inductive theorems in SRS. Manuscript, 2003. In Japanese.
13. K. Kusakari, M. Sakai, and T. Sakabe. Characterizing inductive theorems by extensional initial models in a higher-order equational logic. Distributed at IPSJ seminar PRO–2003–3, 2003.
14. H. Linnestad, C. Prehofer, and O. Lysne. Higher-order proof by consistency. In *Proceedings of the 16th Annual Conference on Foundations of Software Technology and Theoretical Computer Science*, volume 1180 of *LNCS*, pages 274–285. Springer-Verlag, 1996.
15. R. Mayr and T. Nipkow. Higher-order rewrite systems and their confluence. *Theoretical Computer Science*, 192(1):3–29, 1998.
16. K. Meinke. Higher-order equational logic for specification, simulation and testing. In *Proceedings of the 2nd International Workshp on Higher-Order Algebra, Logic and Term Rewriting*, volume 1074 of *LNCS*, pages 124–143. Springer-Verlag, 1995.
17. D. R. Musser. On proving inductive properties of abstract data types. In *Proceedings of the 7th Annual ACM Symposium on Principles of Programming Languages*, pages 154–162. ACM Press, 1980.
18. T. Nipkow and G. Weikum. A decidability result about sufficient-completeness of axiomatically specified abstract data types. In *Proceedings of the 6th GI-Conference on Theoretical Computer Science*, volume 145 of *LNCS*, pages 257–267. Springer-Verlag, 1983.
19. Terese. *Term Rewriting Systems.* Cambridge University Press, 2003.
20. Y. Toyama. How to prove equivalence of term rewriting systems without induction. *Theoretical Computer Science*, 90(2):369–390, 1991.
21. T. Yamada. Confluence and termination of simply typed term rewriting systems. In *Proceedings of the 12th International Conference on Rewriting Techniques and Applications*, volume 2051 of *LNCS*, pages 338–352. Springer-Verlag, 2001.

The Joinability and Unification Problems for Confluent Semi-constructor TRSs

Ichiro Mitsuhashi, Michio Oyamaguchi,
Yoshikatsu Ohta, and Toshiyuki Yamada

Faculty of Engineering, Mie University,
1515 Kamihama-cho, Tsu-shi, 514-8507, Japan
{ichiro,mo,ohta,toshi}@cs.info.mie-u.ac.jp

Abstract. The unification problem for term rewriting systems (TRSs) is the problem of deciding, for a TRS R and two terms s and t, whether s and t are unifiable modulo R. Mitsuhashi et al. have shown that the problem is decidable for confluent simple TRSs. Here, a TRS is simple if the right-hand side of every rewrite rule is a ground term or a variable. In this paper, we extend this result and show that the unification problem for confluent semi-constructor TRSs is decidable. Here, a semi-constructor TRS is such a TRS that every subterm of the right-hand side of each rewrite rule is ground if its root is a defined symbol. We first show the decidability of joinability for confluent semi-constructor TRSs. Then, using the decision algorithm for joinability, we obtain a unification algorithm for confluent semi-constructor TRSs.

1 Introduction

The unification problem for term rewriting systems (TRSs) is the problem of deciding, for a TRS R and two terms s and t, whether s and t are unifiable modulo R. This problem is undecidable in general and even if we restrict to either right-ground TRSs [9] or terminating, confluent, monadic, and linear TRSs [7]. Here, a TRS is monadic if the height of the right-hand side of every rewrite rule is at most one [12]. On the other hand, it is known that unification is decidable for some subclasses of TRSs [2, 4, 5, 8, 11]. Recently, Mitsuhashi et al. have shown that the unification problem is decidable for confluent simple TRSs [7]. Here, a TRS is simple if the right-hand side of every rewrite rule is a ground term or a variable. In this paper, we extend the result of [7] and show that unification for confluent semi-constructor TRSs is decidable. Here, a semi-constructor TRS is such a TRS that every subterm of the right-hand side of each rewrite rule is ground if its root is a defined symbol.

In order to obtain this result, we first show the decidability of joinability for confluent semi-constructor TRSs. Joinability of several subclasses of TRSs has been shown to be decidable so far [13]. Many of these decidability results have been proved by reducing these problems to decidable ones for tree automata, so that these decidable subclasses are restricted to those of right-linear TRSs. In

V. van Oostrom (Ed.): RTA 2004, LNCS 3091, pp. 285–300, 2004.

this paper, we provide a decidability result of joinability for possibly non-right-linear TRSs. To our knowledge, such attempts were very few so far.

Next, we use a new unification algorithm obtained by refining those of [11, 7] to show the decidability of the unification problem for confluent semi-constructor TRSs. A main difference between the algorithms of the present paper and of the previous works [11, 7] is that the previous ones were constructed using decision algorithms for joinability and reachability, but the present using only a decision algorithm for joinability. Besides, complex typed pairs of terms used in the previous ones are changed to simplified typed pairs which are used in the present one.

Moreover, in this paper we show that confluence is necessary to show the decidability of joinability for semi-constructor TRSs, that is, joinability for (non-confluent) linear semi-constructor TRSs is undecidable.

2 Preliminaries

We assume that the reader is familiar with standard definitions of rewrite systems (see [1, 14]) and we just recall here the main notations used in this paper.

Let X be a set of variables, F a finite set of operation symbols graded by an arity function $\mathrm{ar}\colon F \to \mathbb{N}$, $F_n = \{f \in F \mid \mathrm{ar}(f) = n\}$, $Leaf = X \cup F_0$ the set of *leaf symbols*, and T the set of terms constructed from X and F. We use x, y, z as variables, b, c, d as constants, and r, s, t as terms. A term is *ground* if it has no variable. Let G be the set of ground terms and let $S = T \setminus (G \cup X)$. Let $\mathrm{V}(s)$ be the set of variables occurring in s. The *height* of s is defined as follows: $\mathrm{h}(a) = 0$ if a is a leaf symbol and $\mathrm{h}(f(t_1, \ldots, t_n)) = 1 + \max\{\mathrm{h}(t_1), \ldots, \mathrm{h}(t_n)\}$. The *root symbol* is defined as $\mathrm{root}(a) = a$ if a is a leaf symbol and $\mathrm{root}(f(t_1, \ldots, t_n)) = f$.

A position in a term is expressed by a sequence of positive integers, which are partially ordered by the prefix ordering $\leq$. To denote that positions u and v are disjoint, we use $u|v$. The subset of all minimal positions (w.r.t. $\leq$) of W is denoted by $\mathrm{Min}(W)$. Let $\mathcal{O}(s)$ be the set of positions of s.

Let $s_{|u}$ be the subterm of s at position u. We use $s[t]_u$ to denote the term obtained from s by replacing the subterm $s_{|u}$ by t. For a sequence $(u_1, \cdots, u_n)$ of pairwise disjoint positions and terms $r_1, \cdots, r_n$, we use $s[r_1, \cdots, r_n]_{(u_1, \ldots, u_n)}$ to denote the term obtained from s by replacing each subterm $s_{|u_i}$ by $r_i (1 \leq i \leq n)$.

A *rewrite rule* is defined as a directed equation $\alpha \to \beta$ that satisfies $\alpha \notin X$ and $\mathrm{V}(\alpha) \supseteq \mathrm{V}(\beta)$. Let $\leftarrow$ is the inverse of $\to$, $\leftrightarrow\ = \ \to \cup \leftarrow$ and $\downarrow\ = \ \to^* \cdot \leftarrow^*$. Let $\gamma\colon s_1 \overset{u_1}{\leftrightarrow} s_2 \cdots \overset{u_{n-1}}{\leftrightarrow} s_n$ be a *rewrite sequence*. This sequence is abbreviated to $\gamma\colon s_1 \leftrightarrow^* s_n$ and $\mathcal{R}(\gamma) = \{u_1, \cdots, u_{n-1}\}$ is the set of the redex positions of γ. If the root position ε is not a redex position of γ, then γ is called ε-invariant. For any sequence γ and position set W, $\mathcal{R}(\gamma) \geq W$ if for any $v \in \mathcal{R}(\gamma)$ there exists a $u \in W$ such that $v \geq u$. If $\mathcal{R}(\gamma) \geq W$, we write $\gamma\colon s_1 \overset{\geq W}{\leftrightarrow^*} s_n$.

Let $\mathcal{O}_G(s) = \{u \in \mathcal{O}(s) \mid s_{|u} \in G\}$. For any set $\Delta \subseteq X \cup F$, let $\mathcal{O}_\Delta(s) = \{u \in \mathcal{O}(s) \mid \mathrm{root}(s_{|u}) \in \Delta\}$. Let $\mathcal{O}_x(s) = \mathcal{O}_{\{x\}}(s)$. The set D of *defined symbols* for a TRS R is defined as $D = \{\mathrm{root}(\alpha) \mid \alpha \to \beta \in R\}$. A term s is *semi-constructor* if, for every subterm t of s such that $\mathrm{root}(t)$ is a defined symbol, t is ground.

Definition 1. A rule $\alpha \to \beta$ is *ground* if $\alpha, \beta \in G$, *right-ground* if $\beta \in G$, *semi-constructor* if β is semi-constructor, and *linear* if $|\mathcal{O}_x(\alpha)| \leq 1$ and $|\mathcal{O}_x(\beta)| \leq 1$ for every $x \in X$.

Example 1. Let $R_e = \{\mathsf{nand}(x, x) \to \mathsf{not}(x), \mathsf{nand}(\mathsf{not}(x), x) \to \mathsf{t}, \mathsf{t} \to \mathsf{nand}(\mathsf{f}, \mathsf{f}), \mathsf{f} \to \mathsf{nand}(\mathsf{t}, \mathsf{t})\}$.

R_e is semi-constructor, non-terminating and confluent [3]. We will use this R_e in examples given in Section 3.

Definition 2. [11] An equation is a pair of terms s and t denoted by $s \approx t$. An equation $s \approx t$ is *unifiable modulo* a TRS R (or simply R-*unifiable*) if there exists a substitution θ and a rewrite sequence γ such that γ: $s\theta \leftrightarrow^* t\theta$. Such θ and γ are called an R-*unifier* and a *proof* of $s \approx t$, respectively. This notion is extended to sets of term pairs: for $\Gamma \subseteq T \times T$, θ is an R-unifier of Γ if θ is an R-unifier of every pair in Γ. In this case, Γ is R-unifiable. As a special case of R-unifiability, $s \approx t$ is $\emptyset$-unifiable if there exists a substitution θ such that $s\theta = t\theta$, i.e., $\emptyset$-unifiability coincides with the usual unifiability. If $s \downarrow t$ then $s \approx t$ is *joinable*. If $s \to^* t$ then $s \approx t$ is *reachable*.

Definition 3. TRSs R and R' are *equivalent* if $\leftrightarrow^*_R = \leftrightarrow^*_{R'}$.

3 Joinability

First, we show that the joinability and reachability problems for (non-confluent) semi-constructor TRSs are undecidable.

Theorem 1. The joinability and reachability problems for linear semi-constructor term rewriting systems are undecidable.

Proof (sketch). The proof is by a reduction from the Post's correspondence problem (PCP). Let $P = \{\langle u_i, v_i \rangle \in \Sigma^* \times \Sigma^* \mid 1 \leq i \leq k\}$ be an instance of the PCP. The corresponding TRS R_P is constructed as follows: Let $F_0 = \{\mathsf{c}, \mathsf{d}, \$\}, F_1 = \Sigma \cup \{\mathsf{f}, \mathsf{h}\}, F_2 = \{\mathsf{g}\}$, $R_P = \{\mathsf{c} \to \mathsf{h}(\mathsf{c}), \mathsf{c} \to \mathsf{d}, \mathsf{d} \to \mathsf{f}(\mathsf{d})\} \cup \{\mathsf{d} \to \mathsf{g}(u_i(\$), v_i(\$)), \mathsf{f}(\mathsf{g}(x, y)) \to \mathsf{g}(u_i(x), v_i(y)) \mid 1 \leq i \leq k\} \cup \{\mathsf{h}(\mathsf{g}(a(x), a(y))) \to \mathsf{g}(x, y) \mid a \in \Sigma\}$. $u(x)$ is an abbreviation for $a_1(a_2(\cdots a_k(x)))$ where $u = a_1 a_2 \cdots a_k$ with $a_1, \cdots, a_k \in \Sigma$. R_P is linear and semi-constructor. For R_P, the following three propositions (1)–(3) are equivalent: (1) $\mathsf{c} \downarrow \mathsf{g}(\$, \$)$, (2) $\mathsf{c} \to^* \mathsf{g}(\$, \$)$, and (3) PCP P has a solution. □

3.1 Standard Semi-constructor TRSs

From now on, we consider only confluent semi-constructor TRSs, for which joinability is shown to be decidable. In order to facilitate the decidability proof, we transform a TRS into a simpler equivalent one.

Definition 4. For TRS R, we use R_{rg} and $\overline{R_{\mathrm{rg}}}$ to denote the sets of right-ground and non-right-ground rewrite rules in R, respectively.

If R is clear from the context, we write $\to_{\mathrm{rg}}$ instead of $\to_{R_{\mathrm{rg}}}$.

Definition 5. A TRS R is *standard* if the following condition holds: for every $\alpha \to \beta \in R$ either $\alpha \in F_0$ and $\mathrm{h}(\beta) \leq 1$ or $\alpha \notin F_0$ and for every $u \in \mathcal{O}(\beta)$ if $\beta_{|u} \in G$ then $\beta_{|u} \in F_0$.

Let R_0 be a confluent semi-constructor TRS. The corresponding standard TRS $R^{(\mathrm{i})}$ is constructed as follows. First, we choose $\alpha \to \beta \in R_k (k \geq 0)$ that does not satisfy the standardness condition. If $\alpha \in F_0$ then let $\{u_1, \cdots, u_m\} = \{1, \cdots, \mathrm{ar}(\mathrm{root}(\beta))\} \setminus \mathcal{O}_{F_0}(\beta)$, else let $\{u_1, \cdots, u_m\} = \mathrm{Min}(\mathcal{O}_G(\beta)) \setminus \mathcal{O}_{F_0}(\beta)$. Let $R_{k+1} = R_k \setminus \{\alpha \to \beta\} \cup \{\alpha \to \beta[d_1, \cdots, d_m]_{(u_1, \cdots, u_m)}\} \cup \{d_i \to \beta_{|u_i} \mid 1 \leq i \leq m\}$ where $d_1, \cdots, d_m$ are new pairwise distinct constants which do not appear in R_k. This procedure is applied repeatedly until the TRS satisfies the condition of standardness. The resulting TRS is denoted by $R^{(\mathrm{i})}$. For example, $\{\mathsf{f}_1(x) \to \mathsf{g}(x, \mathsf{g}(\mathsf{a}, \mathsf{b}))), \mathsf{f}_2(x) \to \mathsf{f}_2(\mathsf{g}(\mathsf{c}, \mathsf{d}))\}$ is transformed to $\{\mathsf{f}_1(x) \to \mathsf{g}(x, \mathsf{d}_1), \mathsf{d}_1 \to \mathsf{g}(\mathsf{a}, \mathsf{b}), \mathsf{f}_2(x) \to \mathsf{d}_2, \mathsf{d}_2 \to \mathsf{f}_2(\mathsf{d}_3), \mathsf{d}_3 \to \mathsf{g}(\mathsf{c}, \mathsf{d})\}$. This transformation preserves confluence, joinability and unifiability.

Lemma 1.

(1) $R^{(\mathrm{i})}$ is confluent.
(2) For any terms s, t which do not contain new constants, $s \downarrow_{R_0} t$ iff $s \downarrow_{R^{(\mathrm{i})}} t$.
(3) For any terms s, t which do not contain new constants, $s \approx t$ is R_0-unifiable iff $s \approx t$ is $R^{(\mathrm{i})}$-unifiable.

The proof is straightforward, since R_0 is confluent. By this lemma, we can assume that a given confluent semi-constructor TRS is standardized without loss of generality. By standardization, for any $\alpha \to \beta \in R_{\mathrm{rg}}$, $\alpha \in F_0$ or $\beta \in F_0$ holds and $\mathrm{h}(\beta) \leq 1$. However, by the transformation algorithm given in Section 3.2, the heights of the right-hand sides of ground rules (called R_C type rules later) may increase. This is the only exceptional case.

3.2 Adding Ground Rules

The joinability for right-ground TRSs is decidable [10]. In this paper, we show that the joinability for confluent semi-constructor TRSs is decidable, by reducing to the joinability for right-ground TRSs.

Let R_1 be a confluent TRS and R_2 be such a TRS that $\to_{R_2} \subseteq \downarrow_{R_1}$. Then, obviously $R_1 \cup R_2$ is equivalent to R_1 and confluent. Thus, even if we add pairs of joinable terms of R_1 to R_1 as new rewrite rules (called shortcuts), confluence, joinability and unifiability properties are preserved. Note that reachability is not necessarily preserved. Now, we show that the joinability of confluent semi-constructor TRSs reduces to that of right-ground TRSs by adding new finite ground rules. For this purpose, we need some definitions.

Definition 6. A rule $\alpha \to \beta$ has *type* C if $\alpha \in F_0, \beta \notin F_0$ and $\mathcal{O}_{D \setminus F_0}(\beta) = \emptyset$, and has *type* F_0 if $\alpha, \beta \in F_0$. Let $R_\tau = \{\alpha \to \beta \in R \mid \alpha \to \beta$ has type $\tau\}$.

That is, R_C is the subset of R_{rg} satisfying that for every rule $\alpha \to \beta \in R_C$, α is a constant, and β is non-constant and contains no defined non-constant symbol. Henceforth, we assume that $R \setminus R_C$ is standard.

Definition 7.

$$\mathrm{h_D}(s) = \begin{cases} w + \max\{\mathrm{h_D}(s_i) \mid 1 \le i \le n\} & (\text{if } s = f(s_1, \cdots, s_n), n > 0, f \in D) \\ 1 + \max\{\mathrm{h_D}(s_i) \mid 1 \le i \le n\} & (\text{if } s = f(s_1, \cdots, s_n), n > 0, f \notin D) \\ 0 & (\text{if } s \in \mathit{Leaf}) \end{cases}$$

where $w = 1 + 2\max\{\mathrm{h}(\beta) \mid \alpha \to \beta \in R\}$. Note that we give weight w to each defined non-constant symbol and 1 to each other non-constant symbol and define new heights derived from these weights. We define $\mathrm{H_D}(s) = \{\mathrm{h_D}(s_{|u}) \mid u \in \mathcal{O}(s)\}_{\mathrm{m}}$, which is a multiset of heights of all subterms of s. Here, we use $\{\cdots\}_{\mathrm{m}}$ to denote a multiset and $\sqcup$ to denote multiset union. For TRS R_{e} of Example 1, $w = 3$ and $\mathrm{H_D}(\mathsf{nand}(\mathsf{not}(x), x)) = \{0, 0, 1, 4\}_{\mathrm{m}}$.

Let $\ll$ be the multiset extension of the usual relation $<$ on $\mathbb{N}$ and let $\underline{\ll}$ be $\ll \cup =$. Let $\#(s) = (\mathrm{H_D}(s), \mathrm{g}(s))$. Here, function $\mathrm{g}(s)$ returns a natural number corresponding to s uniquely, and we assume that the ordering derived by this function is closed under context, i.e., for any terms r, s, t and any position $u \in \mathcal{O}(r)$, if $\mathrm{g}(s) < \mathrm{g}(t)$ then $\mathrm{g}(r[s]_u) < \mathrm{g}(r[t]_u)$. Such a function g is effectively computable [15]. In order to compare $\#(s)$ and $\#(t)$, we use lexicographic order $<_{\mathrm{lex}}$. Note that $<_{\mathrm{lex}}$ is a total order. *A term s_0 is minimum in set Δ* iff $s_0 \in \Delta$ and $\#(s_0) = \mathrm{Min}(\{\#(s') \mid s' \in \Delta\})$.

Definition 8.

(1) Function linearize(s) linearizes non-linear term s in the following manner. For each variable occurring more than once in s, the first occurrence is not renamed, and the other ones are replaced by new pairwise distinct variables. For example, linearize($\mathsf{nand}(x, x)$) $= \mathsf{nand}(x, x_1)$. If function linearize replaces x by x_1, then we use $x \equiv x_1$ to denote the replacement relation.
(2) For set $\Delta \subseteq T$, $\mathrm{Psub}(\Delta) = \{s_{|u} \mid s \in \Delta, u \in \mathcal{O}(s) \setminus \{\varepsilon\}\}$.
(3) For set $\Delta \subseteq T$, $\mathrm{Bud}(\Delta, R_C) = F_0 \cup \mathrm{Psub}(\Delta \cup \{\beta \mid \alpha \to \beta \in R_C\})$. Note that if $\Delta \subseteq F_0$ then $\mathrm{Bud}(\Delta, R_C) = \mathrm{Bud}(\emptyset, R_C)$.
(4) Substitution σ is *joinability preserving under relation $\equiv$ for TRS R_{rg}* if $x\sigma \downarrow_{R_{\text{rg}}} x'\sigma$ whenever $x \equiv x'$. In this case, we write $\sigma \in \downarrow (\equiv, R_{\text{rg}})$.
(5) For TRS R and term α, $R(\alpha) = \{\beta \mid \alpha \to \beta \in R\}$.
(6) Let $\{s_1, \cdots, s_m\} = R_C(d)$ and $\{u_1, \cdots, u_n\} = \mathrm{Min}(\cup_{1 \le i \le m} \mathcal{O}_{F_0}(s_i))$. Let d_j be the minimum term in $\{s_{i|u_j} \in F_0 \mid 1 \le i \le m\}$, $1 \le j \le n$. Then we define $\mathrm{Normalize}(d, R_C) = \{d \to s_1[d_1, \cdots, d_n]_{(u_1, \cdots, u_n)}\} \cup \{d_j \to s_{i|u_j} \mid 1 \le i \le m, 1 \le j \le n, d_j \ne s_{i|u_j}\}$. For example, Normalize(t, $\{\mathsf{t} \to \mathsf{not}(\mathsf{not}(\mathsf{t})), \mathsf{t} \to \mathsf{not}(\mathsf{f})\}$) $= \{\mathsf{t} \to \mathsf{not}(\mathsf{f}), \mathsf{f} \to \mathsf{not}(\mathsf{t})\}$.

The proofs of all the following lemmata of this paper are given in [15].

Lemma 2. Let $R \setminus R_C$ be standard. Let $\alpha \to \beta \in \overline{R_{\text{rg}}}$, $\theta : X \to T$ and $s \to^*_{\overline{R_{\text{rg}}}} \alpha\theta$. Let $\alpha' = \mathrm{linearize}(\alpha)$. Then, there exists a substitution $\sigma : \mathrm{V}(\alpha') \to \mathrm{Bud}(\{s\}, R_C)$ such that $s \to^*_{\overline{R_{\text{rg}}}} \alpha'\sigma \to^*_{\overline{R_{\text{rg}}}} \alpha\theta$, $\beta\sigma \to^*_{\overline{R_{\text{rg}}}} \beta\theta$ and $\sigma \in \downarrow (\equiv, R_{\text{rg}})$.

By Lemma 2, for a rewrite sequence $d \rightarrow^*_{R_{\mathrm{rg}}} \alpha\theta \rightarrow \beta\theta$, there exists $\alpha'\sigma$ such that $d \rightarrow^*_{R_{\mathrm{rg}}} \alpha'\sigma \rightarrow^*_{R_{\mathrm{rg}}} \alpha\theta$ and $\beta\sigma \rightarrow^*_{R_{\mathrm{rg}}} \beta\theta$. So, if we add a new ground rule $d \rightarrow \beta\sigma$ to R, then we have $d \rightarrow^*_{R'} \beta\theta$ for $R' = R_{\mathrm{rg}} \cup \{d \rightarrow \beta\sigma\}$. Thus, by adding shortcut rules such as $d \rightarrow \beta\sigma$, we can omit applications of $\alpha \rightarrow \beta$ which is a non-right-ground rule. Using this technique, the following algorithm takes as input a standard semi-constructor TRS $R^{(\mathrm{i})}$ and produces as output an equivalent semi-constructor TRS $R^{(\mathrm{f})}$ satisfying that if $d \rightarrow^*_{R^{(\mathrm{i})}} s$ then $d \rightarrow^*_{R^{(\mathrm{f})}_{\mathrm{rg}}} s$. We call $R^{(\mathrm{f})}$ a *quasi-right-ground TRS*, hereafter.

function MakeQuasiRightGround(R)
  $R :=$ Determinize(R);
  repeat
    $R' := R$;
    $R :=$ Determinize(AddShortcuts(R'))
  until $R = R'$;
  return R

function AddShortcuts(R)
  $R' := R$;
  for each $\alpha \rightarrow \beta \in \overline{R_{\mathrm{rg}}}$ **do**
    $\alpha' :=$ linearize(α);
    for each $d \in F_0, \sigma : \mathrm{V}(\alpha') \rightarrow \mathrm{Bud}(\emptyset, R_C)$ such that $\sigma \in\downarrow (\equiv, R_{\mathrm{rg}})$ **do**
      if $d \rightarrow^*_{R_{\mathrm{rg}}} \alpha'\sigma$ **then** $R' := R' \cup \{d \rightarrow \beta\sigma\}$
  return R'

function Determinize(R)
  while there exists d such that $|R_C(d)| > 1$ **do**
    $R := R \cup \mathrm{Normalize}(d, R_C) \setminus \{d \rightarrow s \mid d \rightarrow s \in R_C\}$
  return R

Example 2. For TRS R_{e} of Example 1, MakeQuasiRightGround(R_{e}) first computes Determinize(R_{e}). It returns the same R_{e} as output. Next, AddShortcuts(R_{e}) is called. Since $\mathsf{t} \rightarrow \mathsf{nand}(\mathsf{f},\mathsf{f}), \mathsf{nand}(x,x) \rightarrow \mathsf{not}(x) \in R_{\mathrm{e}}$, a new shortcut rule $\mathsf{t} \rightarrow \mathsf{not}(\mathsf{f})$ is added to R_{e}. Similarly, $\mathsf{f} \rightarrow \mathsf{not}(\mathsf{t})$ is added. Thus, AddShortcuts(R_{e}) $= R'$ where $R' = R_{\mathrm{e}} \cup \{\mathsf{t} \rightarrow \mathsf{not}(\mathsf{f}), \mathsf{f} \rightarrow \mathsf{not}(\mathsf{t})\}$. Next, Determinize($R'$) is called and returns the same R' as output. Then, AddShortcuts(R') is called. Note that $R'_C = \{\mathsf{t} \rightarrow \mathsf{not}(\mathsf{f}), \mathsf{f} \rightarrow \mathsf{not}(\mathsf{t})\}$. AddShortcuts($R'$) returns the same R' and also calls Determinize(R'). Then, the algorithm halts. Let $R^{(\mathrm{f})}_{\mathrm{e}}$ be this result: $R^{(\mathrm{f})}_{\mathrm{e}} = R_{\mathrm{e}} \cup \{\mathsf{t} \rightarrow \mathsf{not}(\mathsf{f}), \mathsf{f} \rightarrow \mathsf{not}(\mathsf{t})\}$, which will be used in later examples.

We apply this algorithm to standard TRS. But by an application of this algorithm, the heights of some right-hand side terms of type C rules may become greater than 1. This algorithm satisfies the following lemmata.

Lemma 3. MakeQuasiRightGround is terminating.

Lemma 4. Let $R^{(\mathrm{f})} = \mathrm{MakeQuasiRightGround}(R^{(\mathrm{i})})$.

(1) If $d \to^*_{R^{(\mathrm{i})}} s$ then $d \to^*_{R^{(\mathrm{f})}_{\mathrm{rg}}} s$.
(2) $\to_{R^{(\mathrm{f})}} \subseteq \downarrow_{R^{(\mathrm{i})}}$.

Corollary 1.

(1) $R^{(\mathrm{f})}$ is confluent (since $R^{(\mathrm{i})}$ is confluent).
(2) $c \downarrow_{R^{(\mathrm{i})}} d$ iff $c \downarrow_{R^{(\mathrm{f})}_{\mathrm{rg}}} d$.
(3) $s \approx t$ is $R^{(\mathrm{i})}$-unifiable iff $s \approx t$ is $R^{(\mathrm{f})}$-unifiable.

3.3 Auxiliary Terms

We have shown that all rewrite sequences from every constant in $R^{(\mathrm{i})}$ (i.e., $d \to^*_{R^{(\mathrm{i})}} s$) can be obtained by using only right-ground rules (i.e., $d \to^*_{R^{(\mathrm{f})}_{\mathrm{rg}}} s$). Now, we want to extend this result to that for rewrite sequences from any term. For this purpose, we need the notion of auxiliary terms. For $\Delta \subseteq G$

function Aux(Δ)
  repeat
    $\Delta' := \Delta$;
    $\Delta :=$ AddTerms(Δ')
  until $\Delta = \Delta'$;
  return Δ

function AddTerms(Δ)
  $\Delta' := \Delta$;
  for each $\alpha \to \beta \in \overline{R^{(\mathrm{f})}_{\mathrm{rg}}}$ **do**
    $\alpha' :=$ linearize(α);
    for each $s \in \Delta, p \in \mathcal{O}_{D \setminus F_0}(s)$,
        $\sigma : \mathrm{V}(\alpha') \to \mathrm{Bud}(\{s_{|p}\}, R^{(\mathrm{f})}_C)$ such that $\sigma \in\downarrow (\equiv, R^{(\mathrm{f})}_{\mathrm{rg}})$ **do**
      if $s_{|p} \to^*_{R^{(\mathrm{f})}_{\mathrm{rg}}} \alpha'\sigma$ **then** $\Delta' := \Delta' \cup \{s[\beta\sigma]_p\}$
  return Δ'

Example 3. In TRS $R^{(\mathrm{f})}_{\mathrm{e}}$ of Example 2,

$$\begin{aligned}\mathrm{Aux}(\{\mathsf{not}(\mathsf{nand}(\mathsf{t},\mathsf{t}))\}) &= \mathrm{AddTerms}(\{\mathsf{not}(\mathsf{nand}(\mathsf{t},\mathsf{t}))\}) \\ &= \{\mathsf{not}(\mathsf{nand}(\mathsf{t},\mathsf{t})), \mathsf{not}(\mathsf{not}(\mathsf{t}))\}.\end{aligned}$$

Lemma 5. For any ground term s,

(1) For any $s' \in \mathrm{Aux}(\{s\})$, $\mathrm{Aux}(\{s'\}) \subseteq \mathrm{Aux}(\{s\})$.
(2) $\mathrm{Aux}(\{s\})$ is finite and computable.
(3) For any $s' \in \mathrm{Aux}(\{s\})$, $s' \downarrow_{R^{(\mathrm{f})}} s$.
(4) If $s \to^*_{R^{(\mathrm{i})}} t$ then there exists $s' \in \mathrm{Aux}(\{s\})$ such that $s' \to^*_{R^{(\mathrm{f})}_{\mathrm{rg}}} t$.

We call s' in Lemma 5(4) an *auxiliary term* of (s,t). This will be used to transform non-right-ground rewrite sequence to right-ground rewrite sequence.

Example 4. For rewrite sequence $\mathsf{not}(\mathsf{nand}(\mathsf{t},\mathsf{t})) \to_{\mathrm{rg}}^{*} \mathsf{not}(\mathsf{nand}(\mathsf{not}(\mathsf{f}),\mathsf{not}(\mathsf{f}))) \to \mathsf{not}(\mathsf{not}(\mathsf{not}(\mathsf{f}))$, we can choose $\mathsf{not}(\mathsf{not}(\mathsf{t})) \in \mathrm{Aux}(\{\mathsf{not}(\mathsf{nand}(\mathsf{t},\mathsf{t}))\})$ and not $(\mathsf{not}(\mathsf{t})) \to_{\mathrm{rg}} \mathsf{not}(\mathsf{not}(\mathsf{not}(\mathsf{f})))$.

3.4 Joinability for Confluent Semi-constructor TRSs

Lemma 6. For any ground terms s and t, $s \downarrow_{R^{(\mathrm{i})}} t$ iff there exists $s' \in \mathrm{Aux}(\{s\})$, $t' \in \mathrm{Aux}(\{t\})$ such that $s' \downarrow_{R_{\mathrm{rg}}^{(\mathrm{f})}} t'$.

By Lemma 5(2) and decidablity of $s' \downarrow_{R_{\mathrm{rg}}^{(\mathrm{f})}} t'$ [10], $s \downarrow_{R^{(\mathrm{i})}} t$ is decidable for ground terms s and t. If s or t is non-ground, $s \downarrow_{R^{(\mathrm{i})}} t$ is equivalent to $s\sigma \downarrow_{R^{(\mathrm{i})}} t\sigma$ where $\sigma : \mathrm{V}(s) \cup \mathrm{V}(t) \to F_0'$ is a bijection and F_0' is a set of new pairwise distinct constants which do not appear in $R^{(\mathrm{i})}$. Thus, we have the following theorem.

Theorem 2. The joinability for confluent semi-constructor term rewriting systems is decidable.

By confluence, we have the following corollary too.

Corollary 2. The word problem for confluent semi-constructor term rewriting systems is decidable.

4 *R*-Unification Algorithm

In this section, We give an R-unification algorithm for confluent semi-constructor TRSs. Henceforth, we consider a fixed confluent semi-constructor TRS R. We assume that R is quasi-right-ground.

Definition 9. [11] Let $\mathcal{L}(s) = \{t \mid t \leftrightarrow^* s\}$.

It is decidable for any terms s and s', whether $s' \in \mathcal{L}(s)$ holds or not by Corollary 2. If term s_0 is minimum in $\mathcal{L}(s)$ then s_0 is obviously minimum in $\mathcal{L}(s_0)$. We say that s_0 is minimum, in the sense that s_0 is minimum in $\mathcal{L}(s_0)$.

Lemma 7. Let s_0 be minimum and let γ: $s_0 \to^* t$. Then, $\mathcal{R}(\gamma) \geq \mathcal{O}_{Leaf}(s_0)$. (That is, only leaf symbols of s_0 are rewritten in γ.)

Example 5. $\mathsf{nand}(\mathsf{t},x)$ is a minimum term. $\mathcal{O}_{Leaf}(\mathsf{nand}(\mathsf{t},x)) = \{1,2\}$. Only leaf symbols of $\mathsf{nand}(\mathsf{t},x)$ are rewritten in a rewrite sequence such as $\mathsf{nand}(\mathsf{t},x) \xrightarrow{1} \mathsf{nand}(\mathsf{not}(\mathsf{f}),x) \xrightarrow{11} \mathsf{nand}(\mathsf{not}(\mathsf{not}(\mathsf{t})),x) \xrightarrow{111} \cdots$.

Lemma 8. The minimum term in $\mathcal{L}(s)$ is computable.

4.1 Locally Minimum Unifiers and Typed Pairs of Terms

Definition 10. Let $\Gamma \subseteq T \times T$. A substitution θ is a *locally minimum* R-unifier of Γ if θ is an R-unifier of Γ and $x\theta$ is minimum for every $x \in Dom(\theta)$.

Our unification algorithm takes a pair $s \approx t$ as input and produces a locally minimum unifier θ of $s \approx t$ iff $s \approx t$ is R-unifiable. Different types of pairs are distinguished by using the notation $s \rhd t$ and $s \approx_{\mathrm{vf}} t$, which are said to be of type $\rhd$ and of type vf, respectively. These definitions are similar to those of [11]. Type $\rhd_U$ was used in [11], but parameter U is not essential, so omitted.

Definition 11. Let $E_0 = \{s \approx t, s \rhd t, s \approx_{\mathrm{vf}} t, \mathbf{fail} \mid s,t \in T\}$. Here, **fail** is introduced as a special symbol and we assume that there exists no R-unifier of **fail** [11]. For $\Gamma \subseteq E_0$ and substitution θ, let $\Gamma\theta = \{s\theta \approx t\theta \mid s \approx t \in \Gamma\} \cup \{s\theta \rhd t\theta \mid s \rhd t \in \Gamma\} \cup \{s\theta \approx_{\mathrm{vf}} t\theta \mid s \approx_{\mathrm{vf}} t \in \Gamma\} \cup \{\mathbf{fail} \mid \mathbf{fail} \in \Gamma\}$.

R-unifiers of these new pairs are required to satisfy additional conditions derived from these types.

Definition 12. A substitution θ is a (locally minimum) R-unifier of $s \rhd t$ if θ is a (locally minimum) R-unifier of $s \approx t$ and there exists a rewrite sequence $\gamma\colon s\theta \to^* r \overset{\geq \mathcal{O}_X(t)}{\longleftrightarrow^*} t\theta$ for some term r.

A substitution θ is a (locally minimum) R-unifier of $s \approx_{\mathrm{vf}} t$ if θ is a (locally minimum) R-unifier of $s \approx t$ and there exists $\gamma\colon s\theta \leftrightarrow^* t\theta$ such that $\mathcal{O}_X(t)$ is a frontier in γ, i.e., $u|v$ or $v \leq u$ holds for any $u \in \mathcal{R}(\gamma)$ and $v \in \mathcal{O}_X(t)$.

Note that if $t \notin S$,$s \rhd t$ can be replaced by $s \approx t$ and excluded from E_0.

Example 6. Let $R_{\mathrm{e}}^{(\mathrm{f})}$ be the TRS of Example 2.

1. $\mathsf{nand}(\mathsf{f},\mathsf{not}(\mathsf{not}(\mathsf{t}))) \rhd \mathsf{nand}(y,\mathsf{not}(y))$ is $R_{\mathrm{e}}^{(\mathrm{f})}$-unifiable, since any substitution θ satisfying $y\theta = \mathsf{f}$ is an $R_{\mathrm{e}}^{(\mathrm{f})}$-unifier: $\mathsf{nand}(\mathsf{f},\mathsf{not}(\mathsf{not}(\mathsf{t}))) \overset{21}{\leftarrow} \mathsf{nand}(\mathsf{f},\mathsf{not}(\mathsf{f}))$.
2. $\mathsf{nand}(\mathsf{t},\mathsf{not}(\mathsf{t})) \approx_{\mathrm{vf}} \mathsf{nand}(\mathsf{not}(\mathsf{f}),y)$ is $R_{\mathrm{e}}^{(\mathrm{f})}$-unifiable, since any substitution θ satisfying $y\theta = \mathsf{f}$ is an $R_{\mathrm{e}}^{(\mathrm{f})}$-unifier: $\mathsf{nand}(\mathsf{t},\mathsf{not}(\mathsf{t})) \overset{1}{\to} \mathsf{nand}(\mathsf{not}(\mathsf{f}),\mathsf{not}(\mathsf{t})) \overset{2}{\leftarrow} \mathsf{nand}(\mathsf{not}(\mathsf{f}),\mathsf{f})$.

To convert typed pairs into the untyped ones, we define the function Core.

Definition 13. [11] For $\Gamma \subseteq E_0$, let $\mathrm{Core}(\Gamma) = \{s \approx t \mid s \approx t \in \Gamma \text{ or } s \rhd t \in \Gamma \text{ or } s \approx_{\mathrm{vf}} t \in \Gamma\} \cup \{\mathbf{fail} \mid \mathbf{fail} \in \Gamma\}$.

Definition 14. [11] Substitutions θ and θ' are *consistent* if $x\theta = x\theta'$ for any $x \in Dom(\theta) \cap Dom(\theta')$.

4.2 Outline of R-Unification Algorithm

We now give our R-unification algorithm for confluent semi-constructor TRSs which is based on the unification algorithm in [11] applicable to confluent right-ground TRSs. The algorithm in [11] is constructed by using algorithms of deciding joinability and reachability for right-ground TRSs. But, the algorithm given in this section uses only the algorithm in Section 3 which decides joinability of semi-constructor TRSs. Thus, our unification algorithm can be considered as a refined version of that of [11] in the sense that no algorithm of deciding reachability of semi-constructor TRSs is needed, and some primitive operations are unified or simplified.

Each primitive operation Φ of our algorithm takes a finite set of pairs $\Gamma \subseteq E_0$ and produces some $\tilde{\Gamma} \subseteq E_0$, denoted by $\Gamma \Rightarrow_\Phi \tilde{\Gamma}$. This operation is called a transformation. Such a transformation is made nondeterministically: $\Gamma \Rightarrow_\Phi \Gamma_1, \Gamma \Rightarrow_\Phi \Gamma_2, \cdots, \Gamma \Rightarrow_\Phi \Gamma_k$ are allowed for some $\Gamma_1, \cdots, \Gamma_k \subseteq E_0$. In this case, we write $\Phi(\Gamma) = \{\Gamma_1, \cdots, \Gamma_k\}$ regarding Φ as a function. Let $\Rightarrow_\Phi^*$ be the reflexive transitive closure of $\Rightarrow_\Phi$. Our algorithm starts from $\Gamma_0 = \{s_0 \approx t_0\}$ and makes primitive transformations repeatedly. We will prove that there exists a sequence $\Gamma_0 \Rightarrow_\Phi^* \Gamma$ such that Γ is $\emptyset$-unifiable iff Γ_0 is R-unifiable.

Our algorithm is divided into three stages. Stage I repeatedly decomposes a set of term pairs Γ into another one $\tilde{\Gamma}$ by guessing a rewrite rule applied at the root position of a non-variable subterm of some term appearing in Γ. Finally, Stage I transforms Γ into a set of type vf pairs Γ_{f}, which becomes an input of the next Stage II. Stage II is similar to a usual $\emptyset$-unification algorithm and stops when a set of type vf pairs Γ is in solved form as explained later. The Final Stage only checks $\emptyset$-unifiability of Γ in solved form.

We give the definition related to validity of the algorithm.

Definition 15. [11] Let $\Phi\colon \mathcal{P}(E_0) \to \mathcal{P}(\mathcal{P}(E_0))$ be a transformation. Then, Φ is valid iff the following validity conditions (V1) and (V2) hold. For any $\Gamma \subseteq E_0$, let $\Phi(\Gamma) = \{\Gamma_1, \cdots, \Gamma_n\}$.

(V1) If θ is a locally minimum R-unifier of Γ, then there exists an i $(1 \leq i \leq n)$ and a substitution θ' such that θ' is consistent with θ and θ' is a locally minimum R-unifier of Γ_i.

(V2) If there exists an i $(1 \leq i \leq n)$ such that $\mathrm{Core}(\Gamma_i)$ is R-unifiable, then $\mathrm{Core}(\Gamma)$ is R-unifiable.

4.3 Stage I

The transformation Φ_1 of Stage I takes as input a finite subset of pairs $\Gamma \subseteq E_0$ and has a finite number of nondeterministic choices $\Gamma \Rightarrow_{\Phi_1} \Gamma_1, \cdots, \Gamma \Rightarrow_{\Phi_1} \Gamma_k$ for some $\Gamma_1, \cdots, \Gamma_k \subseteq E_0$. We consider all possibilities in order to ensure the correctness of the algorithm.

We begin with the initial $\Gamma = \{s_0 \approx t_0\}$ and repeatedly apply the transformation Φ_1 until the current Γ becomes $\emptyset$ or contains either **fail** or at least one type vf pair. This condition is called the **stop condition** of Stage I and defined as $\Gamma \cap \{\textbf{fail}, s \approx_{\mathrm{vf}} t \mid s,t \in T\} \neq \emptyset$ or $\Gamma = \emptyset$. If Γ satisfies this condition, then Γ becomes an input of the next stage.

To describe the transformations used in Stage I, we need the following auxiliary function:

$$\mathrm{Decompose}(s,t) = \{s_{|i} \rhd t_{|i} \mid 1 \leq i \leq k, t_{|i} \in S\} \cup \{s_{|i} \approx t_{|i} \mid 1 \leq i \leq k, t_{|i} \notin S\}$$

where $k = \mathrm{ar}(\mathrm{root}(s))$.

In Stage I, we nondeterministically apply *Conversion* or choose an element p in $\Gamma \setminus X \times X$ and apply one of the following transformations (TT, GG, VG, VT) to Γ according to form of the chosen $p := s \approx t$ or $= s \rhd t$.

$s \setminus t$	S	G	X
S	TT	TT	VT
G	TT	GG	VG
X	VT	VG	–

If no transformation is possible, $\Gamma \Rightarrow_{\Phi_1} \{\textbf{fail}\}$. We write $s \simeq t$ if $s \approx t$ or $t \approx s$. We say that $p = s \simeq t$ satisfies the TT condition if $s, t \notin X$ and either $s \notin G$ or $t \notin G$, the VT condition if $s \in X$ and $t \in S$, the VG condition if $s \in X$ and $t \in G$, and the GG condition if $s, t \in G$. Similarly, we say that $p = s \rhd t$ satisfies the TT condition if $s \notin X$ and $t \in S$, and the VT condition if $s \in X$ and $t \in S$. Note that if $p = s \rhd t$ then $t \in S$.

Let $\Gamma' = \Gamma \setminus \{p\}$. In the following explanations, we assume that θ is a locally minimum unifier of p and we list the conditions that are assumed on a proof γ of p. When applying the transformations we of course lack this information and so we just have to check that the conditions of the transformations are satisfied.

Conversion. If $\Gamma \subseteq \{x \approx r, r \approx x, x \rhd r \mid r \in T \setminus G\}$, then

$$\Gamma \Rightarrow_{\Phi_1} \mathrm{Conv}(\Gamma)$$

where $\mathrm{Conv}(\Gamma) = \{x \approx_{\mathrm{vf}} r \mid x \approx r \in \Gamma \text{ or } r \approx x \in \Gamma \text{ or } x \rhd r \in \Gamma\}$. Note that $\mathrm{Conv}(\Gamma)$ satisfies the **stop condition** of Stage I.

In the following examples, we use the TRS $R_{\mathrm{e}}^{(\mathrm{f})}$ of Example 2.

Example 7. $\{\mathsf{nand}(y, \mathsf{not}(y)) \approx y\} \Rightarrow_{\Phi_1} \{y \approx_{\mathrm{vf}} \mathsf{nand}(y, \mathsf{not}(y))\}$

TT Transformation.

1. If $p = s \simeq t$ satisfies the TT condition, we choose one of the following three cases. Let $k = \mathrm{ar}(\mathrm{root}(s))$. We guess that θ is a locally minimum R-unifier of p and that there exists a joinable sequence $\gamma\colon s\theta \downarrow t\theta$.
   (a) If $\mathrm{root}(s) = \mathrm{root}(t)$, then

$$\Gamma' \cup \{p\} \Rightarrow_{\Phi_1} \Gamma' \cup \{s_{|i} \approx t_{|i} \mid 1 \leq i \leq k\}$$

   In this transformation, we guess that $\gamma\colon s\theta \downarrow t\theta$ is ε-invariant.

(b) If $s \notin G$, then we choose a fresh variant of a rule $\alpha \to \beta \in R$ that satisfies $\mathrm{root}(s) = \mathrm{root}(\alpha)$ and

$$\Gamma' \cup \{p\} \Rightarrow_{\Phi_1} \Gamma' \cup \mathrm{Decompose}(s, \alpha) \cup \{\beta \approx t\}$$

In this transformation, we guess that $\alpha\sigma \to \beta\sigma$ is the leftmost ε-reduction step in $\gamma\colon s\theta \to^* \alpha\sigma \to \beta\sigma \downarrow t\theta$ for some substitution σ (where the subsequence $s\theta \to^* \alpha\sigma$ is ε-invariant).

(c) If $s \in G$, then we choose a term $s' \in \mathrm{Aux}(\{s\})$ and

i. If $\mathrm{root}(s') = \mathrm{root}(t)$,

$$\Gamma' \cup \{p\} \Rightarrow_{\Phi_1} \Gamma' \cup \{s'_{|i} \approx t_{|i} \mid 1 \leq i \leq k\}$$

ii. We choose a rule $\alpha \to \beta \in R_{\mathrm{rg}}$ that satisfies $s' \to^+_{\mathrm{rg}} \beta$ and

$$\Gamma' \cup \{p\} \Rightarrow_{\Phi_1} \Gamma' \cup \{\beta \approx t\}$$

and then do a single TT transformation on $t \approx \beta$ as in 1.a or 1.b. Note that it is decidable whether or not $s \to^+_{\mathrm{rg}} \beta$ [10]. In this transformation, we guess that $\alpha\sigma \to \beta$ is the rightmost ε-reduction step in $\gamma\colon s \to^*_{\mathrm{rg}} \alpha\sigma \to \beta \downarrow t\theta$ for some substitution σ.

In this transformation, we guess an auxiliary term s'. Then, we can assume $s \downarrow s' \to^*_{\mathrm{rg}} \leftarrow^* t\theta$.

2. If $p = s \rhd t$ satisfies the TT condition, we choose one of the following three cases. We guess that there exists a sequence $\gamma\colon s\theta \to^* r \overset{\geq \mathcal{O}_X(t)}{\leftrightarrow^*} t\theta$ for some term r.

(a) If $\mathrm{root}(s) = \mathrm{root}(t)$, then

$$\Gamma' \cup \{p\} \Rightarrow_{\Phi_1} \Gamma' \cup \mathrm{Decompose}(s, t)$$

and if $s \in G$, then apply the VG transformation described later to every $s' \approx x \in \mathrm{Decompose}(s, t) \cap (G \times X)$. In this transformation, we guess that $\gamma\colon s\theta \to^* r \overset{\geq \mathcal{O}_X(t)}{\leftrightarrow^*} t\theta$ is ε-invariant.

(b) If $s \notin G$, we choose a fresh variant of a rule $\alpha \to \beta \in R$ that satisfies $\mathrm{root}(s) = \mathrm{root}(\alpha)$ and

$$\Gamma' \cup \{p\} \Rightarrow_{\Phi_1} \Gamma' \cup \mathrm{Decompose}(s, \alpha) \cup \{\beta \rhd t\}$$

In this transformation, we guess that $\alpha\sigma \to \beta\sigma$ is the leftmost ε-reduction step in $\gamma\colon s\theta \to^* \alpha\sigma \to \beta\sigma \to^* r \overset{\geq \mathcal{O}_X(t)}{\leftrightarrow^*} t\theta$ for some substitution σ (where the subsequence $s\theta \to^* \alpha\sigma$ is ε-invariant).

(c) If $s \in G$, then we choose a term $s' \in \mathrm{Aux}(\{s\})$ and

i. If $\mathrm{root}(s') = \mathrm{root}(t)$,

$$\Gamma' \cup \{p\} \Rightarrow_{\Phi_1} \Gamma' \cup \mathrm{Decompose}(s', t)$$

ii. We choose a rule $\alpha \to \beta \in R_{\mathrm{rg}}$ that satisfies $s' \to^+_{\mathrm{rg}} \beta, \mathrm{root}(\beta) = \mathrm{root}(t)$ and

$$\Gamma' \cup \{p\} \Rightarrow_{\Phi_1} \Gamma' \cup \{\beta \rhd t\}$$

and then do a single TT transformation on $\beta \rhd t$ as in 2.a. Again it is decidable whether or not $s' \to^+_{\mathrm{rg}} \beta$ [10]. In this transformation, we guess that $\alpha\sigma \to \beta$ is the rightmost ε-reduction step in $\gamma\colon s' \to^* \alpha\sigma \to \beta \to^* r \overset{\geq \mathcal{O}_X(t)}{\leftrightarrow^*} t\theta$ for some substitution σ. Thus, the subsequence $\gamma'(\text{of}\gamma)\colon \beta \to^* r \overset{\geq \mathcal{O}_X(t)}{\leftrightarrow^*} t\theta$ is ε-invariant. This ensures that case 2.a of the TT transformation is applicable to $\beta \rhd t$.

Example 8. In case 1.a of the TT transformation,

$$\{\mathsf{nand}(\mathsf{not}(x), \mathsf{t}) \approx \mathsf{nand}(y, \mathsf{not}(y))\} \Rightarrow_{\Phi_1} \{\mathsf{not}(x) \approx y, \mathsf{t} \approx \mathsf{not}(y)\}$$

By choosing rule $\mathsf{nand}(x, x) \to \mathsf{not}(x)$ and applying case 1.b, we get

$$\begin{aligned}&\{\mathsf{nand}(\mathsf{not}(x), \mathsf{t}) \approx \mathsf{nand}(y, \mathsf{not}(y))\} \Rightarrow_{\Phi_1}\\ &\qquad \mathrm{Decompose}(\mathsf{nand}(\mathsf{not}(x), \mathsf{t}), \mathsf{nand}(x', x')) \cup \{\mathsf{not}(x') \approx \mathsf{nand}(y, \mathsf{not}(y))\}\\ &\qquad = \{\mathsf{not}(x) \approx x', \mathsf{t} \approx x', \mathsf{not}(x') \approx \mathsf{nand}(y, \mathsf{not}(y))\}\end{aligned}$$

By choosing auxiliary term t and rule $\mathsf{t} \to \mathsf{not}(\mathsf{f})$ and applying case 1.c, we get

$$\{\mathsf{t} \approx \mathsf{not}(y)\} \Rightarrow_{\Phi_1} \{\mathsf{not}(\mathsf{f}) \approx \mathsf{not}(y)\}$$

After that we apply case 1.a of the TT transformation to $\mathsf{not}(y) \approx \mathsf{not}(\mathsf{f})$ and get $\{y \approx \mathsf{f}\}$.

GG Transformation. If $p = s \approx t$ satisfies the GG condition and $s \downarrow t$ then

$$\Gamma' \cup \{p\} \Rightarrow_{\Phi_1} \Gamma'$$

Note that it is decidable whether or not $s \downarrow t$.

Example 9.

$$\{\mathsf{nand}(\mathsf{t}, \mathsf{not}(\mathsf{f})) \approx \mathsf{f}\} \Rightarrow_{\Phi_1} \emptyset$$

Note that $\mathsf{nand}(\mathsf{t}, \mathsf{not}(\mathsf{f})) \downarrow \mathsf{f}$ holds, e.g., $\mathsf{nand}(\mathsf{t}, \mathsf{not}(\mathsf{f})) \to \mathsf{nand}(\mathsf{not}(\mathsf{f}), \mathsf{not}(\mathsf{f})) \to \mathsf{not}(\mathsf{not}(\mathsf{f})) \leftarrow \mathsf{not}(\mathsf{t}) \leftarrow \mathsf{f}$.

VG Transformation. If $p = x \simeq s$ satisfies the VG condition,

$$\Gamma' \cup \{p\} \Rightarrow_{\Phi_1} \Gamma'\sigma$$

where, $\sigma = \{x \mapsto s'\}$ and s' is the minimum term in $\mathcal{L}(s)$.

Example 10. By choosing $p = y \approx \mathsf{f}$ and $s' = \mathsf{f}$ (note that f is a minimum term with respect to the ordering for the function #),

$$\{y \approx \mathsf{f}, \mathsf{nand}(y, \mathsf{not}(y)) \approx \mathsf{f}\} \Rightarrow_{\Phi_1} \{\mathsf{nand}(\mathsf{f}, \mathsf{not}(\mathsf{f})) \approx \mathsf{f}\}$$

VT Transformation.

1. If $p = x \simeq s$ satisfies the VT condition, we choose a position $v \in \mathcal{O}(s)$ such that $s_{|v} \in S$ and one of the following two cases.
   (a) We choose a rule $\alpha \to \beta \in R$ that satisfies $\mathrm{root}(s_{|v}) = \mathrm{root}(\alpha)$ and
   $$\Gamma' \cup \{p\} \Rightarrow_{\Phi_1} \Gamma' \cup \mathrm{Decompose}(s_{|v}, \alpha) \cup \{s[\beta]_v \approx x\}$$
   In this transformation, we guess the sequence γ: $s\theta \to^* s\theta[\alpha\sigma]_v \xrightarrow{v} s\theta[\beta\sigma]_v \downarrow x\theta$ for some σ and $v \in \mathrm{Min}(\mathcal{R}(\gamma))$.
   (b) We choose a constant c and
   $$\Gamma' \cup \{p\} \Rightarrow_{\Phi_1} \Gamma' \cup \{x \approx s[c]_v, c \approx s_{|v}\}$$
   and if $v = \varepsilon$, we apply the VG transformation to $x \approx c$. In this transformation, we guess that $x\theta_{|v} = c$ and $c \downarrow s\theta_{|v}$.
2. If $p = x \rhd s$ satisfies the VT condition, we choose a constant c and a position $v \in \mathcal{O}(s)$ such that $s|_v \in S$. Then
   $$\Gamma' \cup \{p\} \Rightarrow_{\Phi_1} \Gamma' \cup \{x \rhd s[c]_v, c \rhd s_{|v}\}$$
   If $s[c]_v \in G$, then $x \rhd s[c]_v$ is replaced by $x \approx s[c]_v$. If $v = \varepsilon$, then we apply the VG transformation to $x \approx c$.

Example 11. By choosing $v = \varepsilon$ and constant t and applying case 1.b, we get

$$\{y \approx \mathsf{nand}(y, \mathsf{not}(y))\} \Rightarrow_{\Phi_1} \{y \approx \mathsf{t}, \mathsf{t} \approx \mathsf{nand}(y, \mathsf{not}(y))\}$$

After that we apply the VG transformation to $y \approx \mathsf{t}$.

4.4 Stage II and Final Stage

Stage II and Final Stage are the same as those of [11]. Thus, we describe only definitions necessary to prove correctness of the algorithm. Let Φ_2 be the one step transformation of Stage II. We write $\Gamma \Rightarrow_{\Phi_2} \tilde{\Gamma}$ if $\Phi_2(\Gamma) \ni \tilde{\Gamma}$.

Definition 16. [11] Let $\Gamma_{\mathrm{X}} = \{x \approx y \mid x \approx_{\mathrm{vf}} y \in \Gamma\}$ and $\Gamma_{\mathrm{T}} = \{s \approx_{\mathrm{vf}} t \in \Gamma \mid s \notin X \text{ or } t \notin X\}$. Let $\sim_{\Gamma_{\mathrm{X}}}$ be the equivalence relation derived from Γ_{X}, i.e., the reflexive, transitive and symmetric closure of Γ_{X}. Let $[x]_{\sim_{\Gamma_{\mathrm{X}}}}$ be the equivalence class of x.

Definition 17. [6] Γ is in *solved form* if for any $x \approx_{\mathrm{vf}} s$ and $y \approx_{\mathrm{vf}} t$ in Γ_{T} with $x \sim_{\Gamma_{\mathrm{X}}} y$, $s = t$ holds.

The **stop condition** of Stage II is defined as either $\Gamma = \{\mathbf{fail}\}$ or for any $s \approx_{\mathrm{vf}} t \in \Gamma_{\mathrm{T}}$, $s \in X$ and Γ is in solved form.

Correctness condition of Φ:

(1) $\Rightarrow^*_{\Phi_1} \cdot \Rightarrow^*_{\Phi_2}$ is terminating, and
(2) $\Gamma_0 = \{s_0 \approx t_0\}$ is R-unifiable iff there exist Γ_1 and Γ_f such that $\Gamma_0 \Rightarrow^*_{\Phi_1} \Gamma_1 \Rightarrow^*_{\Phi_2} \Gamma_f$, Γ_1 satisfies the **stop condition** of Stage I, Γ_f satisfies the one of Stage II, and Γ_f is $\emptyset$-unifiable.

Now, we can deduce the main theorem.

Theorem 3. The unification problem for confluent semi-constructor term rewriting systems is decidable.

5 Conclusion

In this paper, we have shown that the joinability and unification problems for confluent semi-constructor TRSs are decidable. But, reachability remains open. Obviously, the class of semi-constructor TRSs is a subclass of strongly weight-preserving TRSs, for which some sufficient conditions to ensure confluence are given in [3].

Acknowledgements

We would like to thank all the anonymous referees of this paper for their helpful comments. This work was supported in part by Grand-in-Aid for Scientific Research 15500009 from Japan Society for the Promotion of Science.

References

1. F. Baader and T. Nipkow. *Term Rewriting and All That.* Cambridge University Press, 1998.
2. H. Comon, M. Haberstrau, and J.-P. Jouannaud. Syntacticness, cycle-syntacticness and shallow theories. *Information and Computation*, 111(1):154–191, 1994.
3. H. Gomi, M. Oyamaguchi, and Y. Ohta. On the Church-Rosser property of root-E-overlapping and strongly depth-preserving term rewriting systems. *Transactions of Information Processing Society of Japan*, 39(4):992–1005, 1998.
4. J.-M. Hullot. Canonical forms and unification. In *Proc. 5th Conf. on Automated Deduction*, pages 318–334. LNCS 87, 1980.
5. F. Jacquemard, C. Meyer, and C. Weidenbach. Unification in extensions of shallow equational theories. In *Proc. 9th Rewriting Techniques and Applications*, pages 76–90. LNCS 1379, 1998.
6. A. Martelli and G. Rossi. Efficient unification with infinite terms in logic programming. In *Proc. 5th Generation Computer Systems*, pages 202–209, 1984.
7. I. Mitsuhashi, M. Oyamaguchi, Y. Ohta, and T. Yamada. On the unification problem for confluent monadic term rewriting systems. *IPSJ Transactions on Programming*, 44(SIG 4(PRO 17)):54–66, 2003.
8. R. Nieuwenhuis. Basic paramodulation and decidable theories. In *Proc. 11th IEEE Symp. Logic in Computer Science*, pages 473–482, 1996.

9. M. Oyamaguchi. On the word problem for right-ground term-rewriting systems. *Trans. IEICE*, E73(5):718–723, 1990.
10. M. Oyamaguchi. The reachability and joinability problems for right-ground term-rewriting systems. *Journal of Information Processing*, 13(3):347–354, 1990.
11. M. Oyamaguchi and Y. Ohta. The unification problem for confluent right-ground term rewriting systems. *Information and Computation*, 183(2):187–211, 2003.
12. K. Salomaa. Deterministic tree pushdown automata and monadic tree rewriting systems. *Jornal of Computer and System Sciences*, 37:367–394, 1988.
13. T. Takai, Y. Kaji, and H. Seki. Right-linear finite path overlapping term rewriting systems effectively preserve recognizability. In *Proc. 11th Rewriting Techniques and Applications*, pages 246–260. LNCS 1833, 2000.
14. Terese. *Term Rewriting Systems.* Cambridge University Press, 2003.
15. http://www.cs.info.mie-u.ac.jp/~ichiro/rta04/.

A Visual Environment for Developing Context-Sensitive Term Rewriting Systems

Jacob Matthews[1], Robert Bruce Findler[1], Matthew Flatt[2], and Matthias Felleisen[3]

[1] University of Chicago
{jacobm,robby}@cs.uchicago.edu
[2] University of Utah
mflatt@cs.utah.edu
[3] Northeastern University
matthias@ccs.neu.edu

Abstract. Over the past decade, researchers have found context-sensitive term-rewriting semantics to be powerful and expressive tools for modeling programming languages, particularly in establishing type soundness proofs. Unfortunately, developing such semantics is an error-prone activity. To address that problem, we have designed PLT Redex, an embedded domain-specific language that helps users interactively create and debug context-sensitive term-rewriting systems. We introduce the tool with a series of examples and discuss our experience using it in courses and developing an operational semantics for R^5RS Scheme.

1 Introduction

Since the late 1980s researchers have used context-sensitive term-rewriting systems as one of their primary tools for specifying the semantics of programming languages. They do so with good reason. Syntactic rewriting systems have proved to be simple, flexible, composable, and expressive abstractions for describing programming language behaviors. In particular, the rewriting approach to operational semantics described by Felleisen and Hieb [1] and popularized by Wright and Felleisen [2] is widely referenced and used.

Unfortunately, designing context-sensitive rewriting systems is subtle and error-prone. People often make mistakes in their rewriting rules that can be difficult to detect, much less correct. Researchers have begun to acknowledge and respond to this difficulty. Xiao, Sabry, and Ariola [3], for instance, developed a tool that verifies that a given context-sensitive term-rewriting system satisfies the unique evaluation context lemma. In the same spirit, we present PLT Redex, a domain-specific language for context-sensitive reduction systems embedded in PLT Scheme [4]. It allows its users to express rewriting rules in a convenient and precise way, to visualize the chains of reductions that their rules produce for particular terms, and to test subject reduction theorems. In section 2 of this paper we briefly explain context-sensitive term rewriting, in section 3 we introduce the rewrite language through a series of examples, and in section 4 we discuss how the language helps with subject reduction proofs. We discuss our experience in section 5, related work in section 6, and conclude with section 7.

V. van Oostrom (Ed.): RTA 2004, LNCS 3091, pp. 301–311, 2004.

$$e = (e\ e) \mid x \mid v \qquad ((\lambda\ (x)\ e)\ v) \mapsto e[v/x] \qquad (\beta_v)$$
$$v = (\lambda\ (x)\ e) \mid f \qquad (f\ v) \mapsto \delta(f,v) \qquad (\delta_v)$$

Plotkin	**Felleisen/Hieb**
$\dfrac{e \mapsto e'}{e \rightarrow e'}$	$E = [] \mid (v\ E) \mid (E\ e)$
$\dfrac{e \rightarrow e'}{(e\ e'') \rightarrow (e'\ e'')}$	if $e \mapsto e'$, then $E[e] \rightarrow E[e']$
$\dfrac{e \rightarrow e'}{(v\ e) \rightarrow (v\ e')}$	

Fig. 1. Specifying an evaluator for λ_v

2 Context-Sensitive Rewriting: A Brief Overview

In his seminal paper [5] on the relationship among abstract machines, interpreters and the λ-calculus, Plotkin shows that an evaluator specified via an abstract machine defines the same function as an evaluator specified via a recursive interpreter. Furthermore, the standard reduction theorem for a λ-calculus generated from properly restricted reduction relations is also equivalent to this function. As Plotkin indicated, the latter definition is by far the most concise and the easiest to use for many proofs.

Figure 1 presents Plotkin's λ_v-calculus. The top portion defines expressions, values, and the two basic relations (β_v and δ_v). The rules below on the left are his specification of the strategy for applying those two basic rules in a leftmost-outermost manner.

In a 1989 paper, Felleisen and Hieb [1] develop an alternate presentation of Plotkin's λ-calculi. Like Plotkin, they use β_v and δ_v as primitive rewriting rules. Instead of inference rules, however, they specify a set of evaluation contexts. Roughly speaking, an *evaluation context* is a term with a hole at the point where the next rewriting step must take place. Placing a term in the hole is equivalent to the textual substitution of the hole by the term [6]. The right side of the bottom half of figure 1 shows how to specify Plotkin's evaluator function with evaluation contexts.

While the two specifications of a call-by-value evaluator are similar at first glance, Felleisen and Hieb's is more suitable for extensions with non-functional constructs (assignment, exceptions, control, threads, etc). Figure 2 shows how easy it is to extend the system of figure 1 (right) with assignable variables. Each program is now a pair of a store and an expression. The bindings in the store introduce the mutable variables and bind free variables in the expression. When a dereference expression for a store variable appears in the hole of an evaluation context, it is replaced with its value. When an assignment with a value on the right-hand side appears in the hole, the let-bindings are modified to capture the effect of the assignment statement. The entire extension consists of three rules, with the original two rules included verbatim. Felleisen and Hieb also showed that this system can be turned into a conventional context-free calculus like the λ-calculus.

$$p = ((store\ (x\ v)\ \ldots)\ e)$$
$$e = \cdots as\ before \cdots \mid (let\ ((x\ e))\ e) \mid (set!\ x\ e)$$
$$P = ((store\ (x\ v)\ \ldots)\ E)$$
$$E = \cdots as\ before \cdots \mid (let\ ((x\ E))\ e) \mid (set!\ x\ E)$$

$$((store\ (x_1\ v_1)\ldots\ (x_2\ v_2)\ (x_3\ v_3)\ldots)\ E[x_2]) \rightarrow ((store\ (x_1\ v_1)\ldots\ (x_2\ v_2)\ (x_3\ v_3)\ldots)\ E[v_2])$$

$$((store\ (x_1\ v_1)\ldots\ (x_2\ v_2)\ (x_3\ v_3)\ldots)\ E[(set!\ x_2\ v_4)]) \rightarrow ((store\ (x_1\ v_1)\ldots\ (x_2\ v_4)\ (x_3\ v_3)\ldots)\ E[v_4])$$

$$((store\ (x_1\ v_1)\ldots)\ E[(let\ ((x_2\ v_2))\ e)]) \rightarrow ((store\ (x_1\ v_1)\ldots\ (x_3\ v_2))\ E[e[x_3\ /\ x_2]])$$
where x_3 is fresh

if $e \mapsto e'$, then $P[e] \rightarrow P[e']$

Fig. 2. Specifying an evaluator for λ_S

Context-sensitive term-rewriting systems are ideally suited for proving the type soundness of programming languages. Wright and Felleisen [2] showed how this works for imperative extensions of the λ-calculus and a large number of people have adapted the technique to other languages since then.

Not surprisingly, though, as researchers have modelled more and more complex languages with these systems, they have found it more and more difficult to model them accurately, as two of the authors discovered when they used it to specify the kernel of Java [7].

3 A Language for Specifying Context-Sensitive Rewriting

To manage this complexity, we have developed PLT Redex, a declarative domain-specific language for specifying context-sensitive rewriting systems. The language is embedded in MzScheme [4], an extension of R^5RS Scheme. MzScheme is particularly suitable for our purposes for two reasons. First, as an extension of Scheme, its basic form of data includes S-expressions and primitives for manipulating S-expressions as patterns. Roughly speaking, an S-expression is an abstract syntax tree representation of a syntactic term, making it a natural choice for manipulating program text. Second, embedding PLT Redex in MzScheme gives PLT Redex programmers a program development environment and extensive libraries for free.

The three key forms PLT Redex introduces are **language**, **red**, and *traces* (we typeset syntactic forms in bold and functions in italics). The first,

(**language** (*<non-terminal-name>* *<rhs-pattern>* ...) ...)

specifies a BNF grammar for a regular tree language. Each right-hand side is written in PLT Redex's pattern language (consisting of a mixture of concrete syntax elements and non-terminals, roughly speaking). With a language definition in place, the **red** form is used to define the reduction relation:

(**red** *<language-name>* *<lhs-pattern>* *<consequence>*)

$$
\begin{aligned}
e &= v \mid (e\ e) \mid (+\ e\ e) \mid x \\
v &= (\lambda\ (x)\ e) \mid \mathit{number} \\
c &= [\,] \\
&\quad \mid (v\ c) \mid (c\ e) \\
&\quad \mid (+\ v\ c) \mid (+\ c\ e) \\
x &\in \mathit{Vars}
\end{aligned}
$$

```
(define λv
  (language (e v (e e) (+ e e) x)
            (v (lambda (x) e) number)
            (c hole
               (v c) (c e)
               (+ v c) (+ c e))
            (x (variable-except lambda +)))
```

$$
\begin{aligned}
&c[(+\ \ulcorner n_1 \urcorner\ \ulcorner n_2 \urcorner)] \\
&\rightarrow c[\ulcorner n_1 + n_2 \urcorner]
\end{aligned}
$$

```
(red λv (in-hole c_1 (+ number_1 number_2))
  (replace (term c_1) (term hole)
           (+ (term number_1) (term number_2))))
```

$$
\begin{aligned}
&c[(\lambda\ (x)\ e)\ v)] \\
&\rightarrow c[e[x/v]]
\end{aligned}
$$

```
(red λv (in-hole c_1 ((lambda (x_1) e_1) v_1))
  (replace (term c_1) (term hole)
           (substitute (term x_1)
                       (term v_1)
                       (term e_1)))))
```

Fig. 3. λ_v semantics

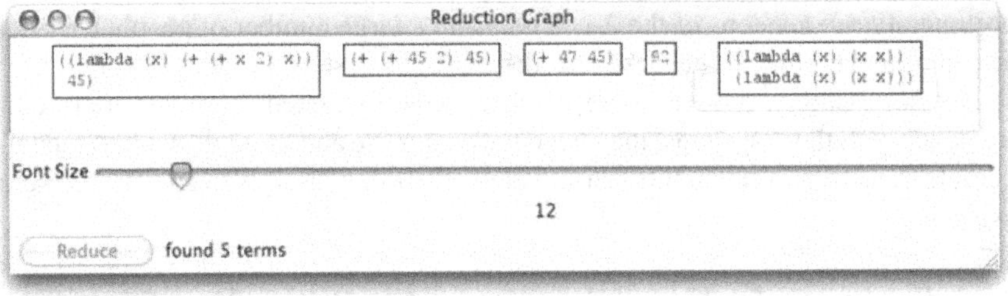

Fig. 4. Reduction of a simple λ_v term and of Ω

Syntactically, it consists of three sub-expressions: a language to which the reduction applies, a source pattern specifying which terms the rule matches, and Scheme code that, when evaluated, produces the resulting term as an S-expression. Finally, the function *traces* accepts a language, a list of reductions, and a term (in Scheme terms, an arbitrary S-expression). When invoked, it opens a window that shows the reduction graph of terms reachable from the initial term. All screenshots in this paper show the output of *traces*. The remainder of this section presents PLT Redex via a series of examples.

3.1 Example: λ_v

Our first example is Plotkin's call-by-value λ-calculus, extended with numbers and addition. Figure 3 shows its definition in Felleisen and Hieb's notation on the left, and in PLT Redex, on the right.

The λ_v language consists of abstractions, numbers, applications, sums, and variable references, and has only two rewriting rules. As figure 3 shows, the traditional mathematical notation translates directly into PLT Redex: each line in the BNF description of

λ_v's grammar becomes one line in **language**. The pattern (*variable-except lambda* +) matches any symbol except those listed (in this case, *lambda* and +).

The reduction rules also translate literally into uses of the **red** form. The first reduction rule defines the semantics of addition. The pattern in the second argument to **red** matches expressions where a syntactic term of the form (+ *number number*) is the next step to be performed. It also binds the pattern metavariables *c_1*, *number_1*, and *number_2* to the context and +'s operands, respectively. In general, pattern variables with underscores must match the non-terminal before the underscore.

The third subexpression of **red** constructs a new S-expression where the addition operation is replaced by the sum of the operands, using Scheme's + operator (numeric constants in S-expressions are identical to the numbers they represent). The *in-hole* pattern is dual to the *replace* function. The former decomposes an expression into a context and a hole, and the latter composes an expression from a context and its hole's new content. The **term** form is PLT Redex's general-purpose tool for building S-expressions. Here we use it only to dereference pattern variables. The second reduction rule, β_v, uses the function *substitute* to perform capture-avoiding variable substitution.[1] Figure 4 shows a term that reduces to 92 on the left and a term that diverges on the right. Arrows are drawn from each term to the terms it can directly reduce to; the circular arrow attached to the Ω term indicates that it reduces to itself. In general, the *traces* function generates a user-specified number of terms and then waits until the "Reduce" button is clicked.

3.2 Example: λ_S

Figure 5 contains PLT Redex definitions for λ_S in parallel to the definitions given in figure 2. The first rules uses an ellipses pattern to match a sequence of any length, including zero, whose elements match the pattern before the ellipses. In this case, the pattern used to match against the store is a common idiom, matching three instances of a pattern with ellipses after the first and the last. This idiom is used to select an interesting S-expression in a sequence; in this case it matches *x_i* to every variable in the store and *v_i* to the corresponding value. To restrict the scope of the match, we use the same pattern variable, *x_i*, in both the store and in the expression. This duplication constrains S-expressions matched in the two places to be structurally identical, and thus the variable in the store and the variable in the term must be the same.

In figure 5 we also see **term** used to construct construct large S-expressions rather than just to get the values of pattern metavariables. In addition to treating pattern variables specially, **term** also has special rules for commas and ellipses. The expression following a comma is evaluated as Scheme code and its result is placed into the S-expression at that point. Ellipses in a **term** expression are duals to the pattern ellipses. The pattern before an ellipsis in a pattern is matched against a sequence of S-expressions and the S-expression before an ellipsis in a **term** expression is expanded into a sequence

[1] Currently, *substitute* must be defined by the user using a more primitive built-in form called **subst** whose details we elide for space. We intend to eliminate this requirement in a future version of PLT Redex; see section 7.

```
(define λS
  (language
    (p ((store (x v) ...) e))
    (e ··· as before ···
      (let ((x e)) e)
      (set! x e))
    (x (variable-except
      lambda set! let))

    (PC ((store (x v) ...) EC))
    (EC ··· as before ···
      (set! x EC)
      (let ((x EC)) e)))

(red λS
  ((store (x_a v_a) ... (x_i v_i) (x_b v_b) ...)
   (in-hole EC_1 x_i))
  (term ((store (x_a v_a) ... (x_i v_i) (x_b v_b) ...)
    ,(replace (term EC_1) (term hole) (term v_i)))))
(red λS
  ((store (x_a v_a) ... (x_i v_old) (x_b v_b) ...)
   (in-hole EC_1 (set! x_i v_new)))
  (term ((store (x_a v_a) ... (x_i v_new) (x_b v_b) ...)
    ,(replace (term EC_1) (term hole) (term v_new)))))
(red λS
  ((store (x_a v_a) ...)
   (in-hole EC_1 (let ((x_i v_i)) e_1)))
  (let ((new-x (variable-not-in (term (x_a ...)) (term x_i))))
    (term ((store (x_a v_a) ... (,new-x v_i))
      ,(replace (term EC_1) (term hole)
                (substitute (term x_i) new-x (term e_1)))))))
```

Fig. 5. λ_S semantics

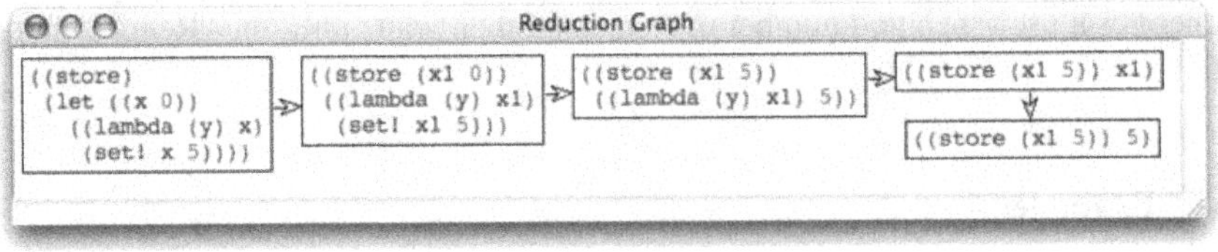

Fig. 6. Reduction of a simple λ_S term

of S-expressions and spliced into its context.[2] Accordingly, the first rule produces a term whose store is identical to the store in the term it consumed.

The final rule also introduces another PLT Redex function, *variable-not-in*, which takes an arbitrary syntactic term and a variable name and produces a new variable whose name is similar to the input variable's name and that does not occur in the given term.

Figure 6 shows a sample reduction sequence in λ_S using, in order, a *let* reduction, a *set!* reduction, a β_v reduction, and a dereference reduction.

3.3 Example: Threaded λ_S

We can add concurrency to λ_S with surprisingly few modifications. The language changes as shown in figure 7. A program still consists of a single store, but instead of just one expression it now contains one expression per thread. In addition, each reference to *EC* in the λ_S reductions becomes *TC*. No other changes need to be made, and in particular no reduction rules need modification.

[2] With the exception of ellipsis and pattern variables, **term** is identical to Scheme's **quasiquote**.

(**define** t-λ_S
  (**language**
    (*p* ((*store* (*x v*) ...) (*threads e* ...)))
    (*PC* ((*store* (*x v*) ...) *TC*))
    (*TC* (*threads e* ... *EC e* ...))
    $\cdots$ *as before* $\cdots$)

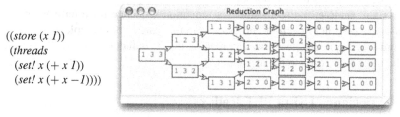

Fig. 7. Threaded λ_S

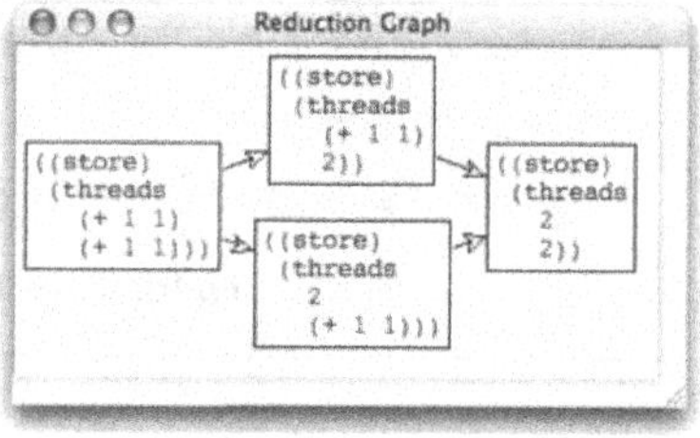

Fig. 8. Multiple reductions

((*store* (*x 1*))
 (*threads*
  (*set! x* (+ *x 1*))
  (*set! x* (+ *x* −*1*))))

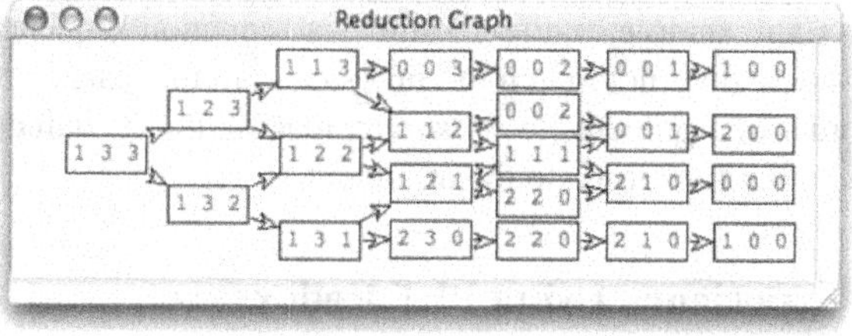

On the left, a threaded λ_S term and on the right, boxes containing *x*'s value and the number of subexpressions remaining in each thread

Fig. 9. Reduction summary using *traces*

To express non-determinism, PLT Redex's pattern language supports ambiguous patterns by finding *all possible* ways a pattern might match a term. Consider the *TC* evaluation context in figure 7, which uses the selection idiom described in section 3.2. Unlike that example, nothing restricts this selection to a particular thread, so PLT Redex produces multiple matches, one for each reducible thread. The *traces* window reflects this by displaying all of the reductions that apply to each term when constructing the reduction graph, as shown in figure 8.

Due to the possible interleaving of multiple threads, even simple expressions reduce many different ways and gaining insight from a thicket of terms can be difficult. Accordingly, *traces* has an optional extra argument that allows the user to provide an alternative view of the term that can express a summary or just the salient part of a term without affecting the underlying reduction sequence. Figure 9 shows an example summarized reduction sequence.

4 Subject Reduction

A widely used proof technique for establishing type soundness [2] is based on context-sensitive rewriting semantics. Each proof has a key subject-reduction lemma that guarantees that the type of a term before a reduction matches the type after the reduction.

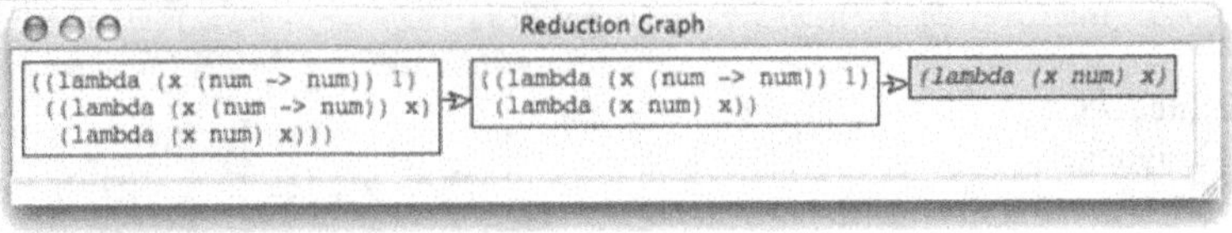
Reduction Graph
((lambda (x (num -> num)) 1)
((lambda (x (num -> num)) x)
(lambda (x num) x)))
((lambda (x (num -> num)) 1)
(lambda (x num) x))
(lambda (x num) x)

Fig. 10. Subject fails in second reduction

PLT Redex provides support for exploring and debugging such subject-reduction lemmas. The function *traces/predicate* allows the user to specify a predicate that implements the subject of the subject-reduction lemma. Then, *traces/predicate* highlights in red italics any terms that do not satisfy it.

As an example, figure 10 shows a reduction sequence for a term where the third step's type does not match the first step's. In this particular example, the user swapped the order of arguments to *substitute*, making the β_v reduction incorrectly substitute the body of the function into the argument.

5 Experience Using PLT Redex

As the examples in section 3 suggest, PLT Redex is suitable for modeling a wide variety of languages. That suggestion has been borne out in practice as we have developed a reduction semantics for R^{5}RS Scheme that captures the language in as much detail as the R^{5}RS formal semantics does [8, section 7.2]. Our experience developing one facet of the semantics highlights the strength of PLT Redex. In evaluating a procedure call, the R^{5}RS document deliberately leaves unspecified the order in which arguments are evaluated, but specifies that [8, section 4.1.3]

> *the effect of any concurrent evaluation of the operator and operand expressions is constrained to be consistent with* some *sequential order of evaluation. The order of evaluation may be chosen differently for each procedure call*

In the formal semantics section, the authors explain how they model this ambiguity:

> *[w]e mimic [the order of evaluation] by applying arbitrary permutations* permute *and* unpermute ... *to the arguments in a call before and after they are evaluated. This is not quite right since it suggests, incorrectly, that the order of evaluation is constant throughout a program* .... [8, Section 7.2].

We realized that our rules, in contrast, can capture the intended semantics using nondeterminism to select the argument to reduce. Our initial, incorrect idea of how to capture this was to change the definition of expression evaluation contexts (otherwise similar to those in λ_S) so that they could occur on either side of an application:

$$(\textbf{language}\ (EC\ (e\ EC)\ (EC\ e))\cdots \textit{as before}\cdots))$$

When we visualized a few reductions using that modification using *traces*, we quickly spotted an error in our approach. We had accidentally introduced non-deterministic concurrency to the language. The term

```
(define r5
  (language
   (inert (mark v)
          e)
   (EC ((mark EC) inert)
       (inert (mark EC))))
   ··· as before ···)

(red r5 (in-hole PC_1 (e_1 inert_1))
     (replace (term PC_1) (term hole)
              (term ((mark e_1) inert_1))))
(red r5 (in-hole PC_1 (inert_1 e_1))
     (replace (term PC_1) (term hole)
              (term (inert_1 (mark e_1)))))
(red r5 (in-hole c_1 ((mark (lambda (x_1) e_1)) (mark v_1)))
     (replace (term c_1)
              (term hole)
              (substitute (term x_1) (term v_1) (term e_1)))))
```

Fig. 11. Reduction rules for unspecified application order

```
(let ((x 1)) (((lambda (y) (lambda (z) x))
               (set! x (+ x 1)))
              (set! x (− x 1))))
```

should always reduce to *1*. Under our semantics, however, it could also reduce to *2* and *0*, just as the term in figure 9 does.

Experimenting with the faulty system gave us the insight into how to fix it. We realized that the choice of which term in an application should be ambiguous, but once the choice had been made, there should not be any further ambiguity. Accordingly, we introduced a mark in application expressions. The choice of where to place the mark is arbitrary, but once a mark is placed, evaluation under that mark must complete before the other subexpression of an application is evaluated.

Figure 11 shows the necessary revisions to λ_S to support R^5RS-style procedure applications. We introduce the non-terminal *inert* to stand for terms where evaluation does not occur, *i.e.*, unmarked expressions or marked values. The top two reductions on the right-hand side of the figure non-deterministically introduce marks into applications. The evaluation contexts change to ensure that evaluation only occurs inside marked expressions, and application changes to expect marked procedures and arguments.

In addition to providing a graphical interface to reduction graphs, PLT Redex also provides a programatic interface to the reduction graphs as a consequence of its being embedded in PLT Scheme. This interface let us build large automatic test suites for the R^5RS semantics system among others that we could run without having to call *traces* and produce visual output. We found these test cases to be invaluable during development, since changes to one section of our semantics often had effects on seemingly unrelated sections and inspecting visual output manually quickly became infeasible.

We gained additional experience by using PLT Redex as a pedagogical tool. The University of Utah's graduate-level course on programming languages introduces students to the formal specification of languages through context-sensitive rewriting. Students model a toy arithmetic language, the pure λ-calculus, the call-by-value λ-calculus (including extensions for state and exceptions), typed λ-calculi, and a model of Java.

In the most recent offering of the course, we implemented many of the course's reduction systems using PLT Redex, and students used PLT Redex to explore specific evaluations. Naturally, concepts such as confluence and determinism stood out particularly well in the graphical presentation of reduction sequences. In the part of the course where we derive an interpreter for the λ-calculus through a series of "machines," PLT Redex was helpful in exposing the usefulness of each machine change.

As a final project, students implemented context-sensitive rewriting models from recent conference papers as PLT Redex programs. This exercise provided students with a much deeper understanding of the models than they would have gained from merely reading the paper. For a typical paper, students had to fill significant gaps in the formal content (*e.g.*, the figures with grammars and reduction rules). This experience suggests that paper authors could benefit from creating a machine-checked version of a model, which would help to ensure that all relevant details are included in a paper's formalism.

6 Related Work

Many researchers have implemented programs similar to our reduction tool. For example, Elan [9], Maude [10], and Stratego [11] all allow users to implement term-rewriting systems (and more), but are focused more on context-free term-rewriting. The ASF+SDF compiler [12] has strong connections to PLT Redex but is geared towards language implementation rather than exploration and so makes tradeoffs that do not suit the needs of lightweight debugging (but that make it a better tool for building efficient large-scale language implementations).

Our reduction tool is focused on context-sensitive rewriting and aims to help its users visualize and understand rewriting systems rather than employ them for some other purpose. The $\mathbf{in}^2$ graphical interpreter for interaction nets [13] also helps its users visualize sequences of reductions, but is tailored to a single language.

7 Conclusion and Future Work

Our own efforts to develop novel rewriting systems and to teach operational semantics based on term-rewriting to students have been aided greatly by having an automatic way to visualize rewriting systems. We are confident that PLT Redex can be useful to others for the same purposes.

Our implementation is reasonably efficient. The test suite for our "beginner" language semantics, a system with 14 nonterminals with 55 total productions and 52 reduction rules that models a reasonable purely-functional subset of Scheme intended for beginning programmers, runs 90 reductions in just over 2 seconds on our test machine. This represents a huge slowdown over the speed one would expect from a dedicated interpreter, but in practice seems quick enough to be useful.

We plan to extend PLT Redex to allow simple ways to express the binding structure of a language, which will allow us to synthesize capture-avoiding substitution rules automatically. We also plan to add more support for the reduction rules commonly used in the literature, such as source patterns that match only if other patterns did not.

Our implementation of PLT Redex is available as an add-on package to DrScheme (http://www.drscheme.org). Choose DrScheme's File|Install .plt File menu item and supply this url: http://people.cs.uchicago.edu/%7Ejacobm/plt/pltredex.plt

Acknowledgements

We would like to thank Richard Cobbe, the students of the University of Utah's Spring 2003 CS 6520 class, and the anonymous reviewers of RTA 2004 for their helpful feedback on PLT Redex and this paper.

References

1. Felleisen, M., Hieb, R.: The revised report on the syntactic theories of sequential control and state. Theoretical Computer Science (1992) 235–271
2. Wright, A., Felleisen, M.: A syntactic approach to type soundness. Information and Computation (1994) 38–94 First appeared as Technical Report TR160, Rice University, 1991.
3. Xiao, Y., Sabry, A., Ariola, Z.M.: From syntactic theories to interpreters: Automating the proof of unique decomposition. In: Higher-Order and Symbolic Computation. Volume 4. (1999) 387–409
4. Flatt, M.: PLT MzScheme: Language manual. Technical Report TR97-280, Rice University (1997) http://www.mzscheme.org/.
5. Plotkin, G.D.: Call-by-name, call-by-value and the λ-calculus. Theoretical Computer Science **1** (1975) 125–159
6. Barendregt, H.: The Lambda Calculus, Its Syntax and Semantics. Volume 103 of Studies in Logics and the Foundations of Mathematics. North Holland, Amsterdam (1981)
7. Flatt, M., Krishnamurthi, S., Felleisen, M.: A programmer's reduction semantics for classes and mixins. Formal Syntax and Semantics of Java **1523** (1999) 241–269 Preliminary version appeared in proceedings of *Principles of Programming Languages*, 1998. Revised version is Rice University technical report TR 97-293, June 1999.
8. Kelsey, R., Clinger, W., (Editors), J.R.: Revised^5 report of the algorithmic language Scheme. ACM SIGPLAN Notices **33** (1998) 26–76
9. Borovansky, P., Kirchner, C., Kirchner, H., Moreau, P.E., Vittek, M.: ELAN: A logical framework based on computational systems. In: Proc. of the First Int. Workshop on Rewriting Logic. Volume 4., Elsevier (1996)
10. Clavel, M., Durán, F., Eker, S., Lincoln, P., Martí-Oliet, N., Meseguer, J., Quesada, J.F.: Maude: Specification and programming in rewriting logic. Theoretical Computer Science (2001)
11. Visser, E.: Stratego: A language for program transformation based on rewriting strategies. System description of Stratego 0.5. In Middeldorp, A., ed.: Rewriting Techniques and Applications (RTA'01). Volume 2051 of Lecture Notes in Computer Science., Springer-Verlag (2001) 357–361
12. van den Brand, M.G.J., Heering, J., Klint, P., Oliver, P.A.: Compiling language definitions: The ASF+SDF compiler. ACM Transactions on Programming Languages and Systems **24** (2002) 334–368
13. Lippi, S.: in2: A Graphical Interpreter for Interaction Nets (system description). In Tison, S., ed.: Rewriting Techniques and Applications, 13th International Conference, RTA-02. LNCS 2378, Copenhagen, Denmark, Springer (2002) 380–384

Author Index

Lecture Notes in Computer Science

For information about Vols. 1–2977

please contact your bookseller or Springer-Verlag

Vol. 3092: J. Eckstein, H. Baumeister (Eds.), Extreme Programming and Agile Processes in Software Engineering. XVI, 358 pages. 2004.

Vol. 3091: V. van Oostrom (Ed.), Rewriting Techniques and Applications. X, 313 pages. 2004.

Vol. 3084: A. Persson, J. Stirna (Eds.), Advanced Information Systems Engineering. XIV, 596 pages. 2004.

Vol. 3083: W. Emmerich, A.L. Wolf (Eds.), Component Deployment. X, 249 pages. 2004.

Vol. 3076: D. Buell (Ed.), Algorithmic Number Theory. XI, 451 pages. 2004.

Vol. 3074: B. Kuijpers, P. Revesz (Eds.), Constraint Databases and Applications. XII, 181 pages. 2004.

Vol. 3066: S. Tsumoto, R. Słowiński, J. Komorowski, J.W. Grzymala-Busse (Eds.), Rough Sets and Current Trends in Computing. XX, 853 pages. 2004. (Subseries LNAI).

Vol. 3065: A. Lomuscio, D. Nute (Eds.), Deontic Logic. X, 275 pages. 2004. (Subseries LNAI).

Vol. 3064: D. Bienstock, G. Nemhauser (Eds.), Integer Programming and Combinatorial Optimization. XI, 445 pages. 2004.

Vol. 3063: A. Llamosí, A. Strohmeier (Eds.), Reliable Software Technologies - Ada-Europe 2004. XIII, 333 pages. 2004.

Vol. 3062: J.L. Pfaltz, M. Nagl, B. Böhlen (Eds.), Applications of Graph Transformations with Industrial Relevance. XV, 500 pages. 2004.

Vol. 3060: A.Y. Tawfik, S.D. Goodwin (Eds.), Advances in Artificial Intelligence. XIII, 582 pages. 2004. (Subseries LNAI).

Vol. 3059: C.C. Ribeiro, S.L. Martins (Eds.), Experimental and Efficient Algorithms. X, 586 pages. 2004.

Vol. 3058: N. Sebe, M.S. Lew, T.S. Huang (Eds.), Computer Vision in Human-Computer Interaction. X, 233 pages. 2004.

Vol. 3056: H. Dai, R. Srikant, C. Zhang (Eds.), Advances in Knowledge Discovery and Data Mining. XIX, 713 pages. 2004. (Subseries LNAI).

Vol. 3054: I. Crnkovic, J.A. Stafford, H.W. Schmidt, K. Wallnau (Eds.), Component-Based Software Engineering. XI, 311 pages. 2004.

Vol. 3053: C. Bussler, J. Davies, D. Fensel, R. Studer (Eds.), The Semantic Web: Research and Applications. XIII, 490 pages. 2004.

Vol. 3052: W. Zimmermann, B. Thalheim (Eds.), Abstract State Machines 2004. Advances in Theory and Practice. XII, 235 pages. 2004.

Vol. 3051: R. Berghammer, B. Möller, G. Struth (Eds.), Relational and Kleene-Algebraic Methods in Computer Science. X, 279 pages. 2004.

Vol. 3047: F. Oquendo, B. Warboys, R. Morrison (Eds.), Software Architecture. X, 279 pages. 2004.

Vol. 3046: A. Laganà, M.L. Gavrilova, V. Kumar, Y. Mun, C.K. Tan, O. Gervasi (Eds.), Computational Science and Its Applications - ICCSA 2004. LIII, 1016 pages. 2004.

Vol. 3045: A. Laganà, M.L. Gavrilova, V. Kumar, Y. Mun, C.K. Tan, O. Gervasi (Eds.), Computational Science and Its Applications – ICCSA 2004. LIII, 1040 pages. 2004.

Vol. 3044: A. Laganà, M.L. Gavrilova, V. Kumar, Y. Mun, C.K. Tan, O. Gervasi (Eds.), Computational Science and Its Applications – ICCSA 2004. LIII, 1140 pages. 2004.

Vol. 3043: A. Laganà, M.L. Gavrilova, V. Kumar, Y. Mun, C.J.K. Tan, O. Gervasi (Eds.), Computational Science and Its Applications – ICCSA 2004. LIII, 1180 pages. 2004.

Vol. 3042: N. Mitrou, K. Kontovasilis, G.N. Rouskas, I. Iliadis, L. Merakos (Eds.), NETWORKING 2004, Networking Technologies, Services, and Protocols; Performance of Computer and Communication Networks; Mobile and Wireless Communications. XXXIII, 1519 pages. 2004.

Vol. 3039: M. Bubak, G.D.v. Albada, P.M. Sloot, J. Dongarra (Eds.), Computational Science - 2004. LXVI, 1271 pages. 2004.

Vol. 3038: M. Bubak, G.D.v. Albada, P.M. Sloot, J. Dongarra (Eds.), Computational Science - ICCS 2004. LXVI, 1311 pages. 2004.

Vol. 3037: M. Bubak, G.D.v. Albada, P.M. Sloot, J. Dongarra (Eds.), Computational Science - ICCS 2004. LXVI, 745 pages. 2004.

Vol. 3036: M. Bubak, G.D.v. Albada, P.M. Sloot, J. Dongarra (Eds.), Computational Science - ICCS 2004. LXVI, 713 pages. 2004.

Vol. 3035: M.A. Wimmer (Ed.), Knowledge Management in Electronic Government. XII, 326 pages. 2004. (Subseries LNAI).

Vol. 3034: J. Favela, E. Menasalvas, E. Chávez (Eds.), Advances in Web Intelligence. XIII, 227 pages. 2004. (Subseries LNAI).

Vol. 3033: M. Li, X.-H. Sun, Q. Deng, J. Ni (Eds.), Grid and Cooperative Computing. XXXVIII, 1076 pages. 2004.

Vol. 3032: M. Li, X.-H. Sun, Q. Deng, J. Ni (Eds.), Grid and Cooperative Computing. XXXVII, 1112 pages. 2004.

Vol. 3031: A. Butz, A. Krüger, P. Olivier (Eds.), Smart Graphics. X, 165 pages. 2004.

Vol. 3030: P. Giorgini, B. Henderson-Sellers, M. Winikoff (Eds.), Agent-Oriented Information Systems. XIV, 207 pages. 2004. (Subseries LNAI).

Vol. 3029: B. Orchard, C. Yang, M. Ali (Eds.), Innovations in Applied Artificial Intelligence. XXI, 1272 pages. 2004. (Subseries LNAI).

Vol. 3028: D. Neuenschwander, Probabilistic and Statistical Methods in Cryptology. X, 158 pages. 2004.

Vol. 3027: C. Cachin, J. Camenisch (Eds.), Advances in Cryptology - EUROCRYPT 2004. XI, 628 pages. 2004.

Vol. 3026: C. Ramamoorthy, R. Lee, K.W. Lee (Eds.), Software Engineering Research and Applications. XV, 377 pages. 2004.

Vol. 3025: G.A. Vouros, T. Panayiotopoulos (Eds.), Methods and Applications of Artificial Intelligence. XV, 546 pages. 2004. (Subseries LNAI).

Vol. 3024: T. Pajdla, J. Matas (Eds.), Computer Vision - ECCV 2004. XXVIII, 621 pages. 2004.

Vol. 3023: T. Pajdla, J. Matas (Eds.), Computer Vision - ECCV 2004. XXVIII, 611 pages. 2004.

Vol. 3022: T. Pajdla, J. Matas (Eds.), Computer Vision - ECCV 2004. XXVIII, 621 pages. 2004.

Vol. 3021: T. Pajdla, J. Matas (Eds.), Computer Vision - ECCV 2004. XXVIII, 633 pages. 2004.

Vol. 3019: R. Wyrzykowski, J. Dongarra, M. Paprzycki, J. Wasniewski (Eds.), Parallel Processing and Applied Mathematics. XIX, 1174 pages. 2004.

Vol. 3016: C. Lengauer, D. Batory, C. Consel, M. Odersky (Eds.), Domain-Specific Program Generation. XII, 325 pages. 2004.

Vol. 3015: C. Barakat, I. Pratt (Eds.), Passive and Active Network Measurement. XI, 300 pages. 2004.

Vol. 3014: F. van der Linden (Ed.), Software Product-Family Engineering. IX, 486 pages. 2004.

Vol. 3012: K. Kurumatani, S.-H. Chen, A. Ohuchi (Eds.), Multi-Agnets for Mass User Support. X, 217 pages. 2004. (Subseries LNAI).

Vol. 3011: J.-C. Régin, M. Rueher (Eds.), Integration of AI and OR Techniques in Constraint Programming for Combinatorial Optimization Problems. XI, 415 pages. 2004.

Vol. 3010: K.R. Apt, F. Fages, F. Rossi, P. Szeredi, J. Váncza (Eds.), Recent Advances in Constraints. VIII, 285 pages. 2004. (Subseries LNAI).

Vol. 3009: F. Bomarius, H. Iida (Eds.), Product Focused Software Process Improvement. XIV, 584 pages. 2004.

Vol. 3008: S. Heuel, Uncertain Projective Geometry. XVII, 205 pages. 2004.

Vol. 3007: J.X. Yu, X. Lin, H. Lu, Y. Zhang (Eds.), Advanced Web Technologies and Applications. XXII, 936 pages. 2004.

Vol. 3006: M. Matsui, R. Zuccherato (Eds.), Selected Areas in Cryptography. XI, 361 pages. 2004.

Vol. 3005: G.R. Raidl, S. Cagnoni, J. Branke, D.W. Corne, R. Drechsler, Y. Jin, C.G. Johnson, P. Machado, E. Marchiori, F. Rothlauf, G.D. Smith, G. Squillero (Eds.), Applications of Evolutionary Computing. XVII, 562 pages. 2004.

Vol. 3004: J. Gottlieb, G.R. Raidl (Eds.), Evolutionary Computation in Combinatorial Optimization. X, 241 pages. 2004.

Vol. 3003: M. Keijzer, U.-M. O'Reilly, S.M. Lucas, E. Costa, T. Soule (Eds.), Genetic Programming. XI, 410 pages. 2004.

Vol. 3002: D.L. Hicks (Ed.), Metainformatics. X, 213 pages. 2004.

Vol. 3001: A. Ferscha, F. Mattern (Eds.), Pervasive Computing. XVII, 358 pages. 2004.

Vol. 2999: E.A. Boiten, J. Derrick, G. Smith (Eds.), Integrated Formal Methods. XI, 541 pages. 2004.

Vol. 2998: Y. Kameyama, P.J. Stuckey (Eds.), Functional and Logic Programming. X, 307 pages. 2004.

Vol. 2997: S. McDonald, J. Tait (Eds.), Advances in Information Retrieval. XIII, 427 pages. 2004.

Vol. 2996: V. Diekert, M. Habib (Eds.), STACS 2004. XVI, 658 pages. 2004.

Vol. 2995: C. Jensen, S. Poslad, T. Dimitrakos (Eds.), Trust Management. XIII, 377 pages. 2004.

Vol. 2994: E. Rahm (Ed.), Data Integration in the Life Sciences. X, 221 pages. 2004. (Subseries LNBI).

Vol. 2993: R. Alur, G.J. Pappas (Eds.), Hybrid Systems: Computation and Control. XII, 674 pages. 2004.

Vol. 2992: E. Bertino, S. Christodoulakis, D. Plexousakis, V. Christophides, M. Koubarakis, K. Böhm, E. Ferrari (Eds.), Advances in Database Technology - EDBT 2004. XVIII, 877 pages. 2004.

Vol. 2991: R. Alt, A. Frommer, R.B. Kearfott, W. Luther (Eds.), Numerical Software with Result Verification. X, 315 pages. 2004.

Vol. 2990: J. Leite, A. Omicini, L. Sterling, P. Torroni (Eds.), Declarative Agent Languages and Techniques. XII, 281 pages. 2004. (Subseries LNAI).

Vol. 2989: S. Graf, L. Mounier (Eds.), Model Checking Software. X, 309 pages. 2004.

Vol. 2988: K. Jensen, A. Podelski (Eds.), Tools and Algorithms for the Construction and Analysis of Systems. XIV, 608 pages. 2004.

Vol. 2987: I. Walukiewicz (Ed.), Foundations of Software Science and Computation Structures. XIII, 529 pages. 2004.

Vol. 2986: D. Schmidt (Ed.), Programming Languages and Systems. XII, 417 pages. 2004.

Vol. 2985: E. Duesterwald (Ed.), Compiler Construction. X, 313 pages. 2004.

Vol. 2984: M. Wermelinger, T. Margaria-Steffen (Eds.), Fundamental Approaches to Software Engineering. XII, 389 pages. 2004.

Vol. 2983: S. Istrail, M.S. Waterman, A. Clark (Eds.), Computational Methods for SNPs and Haplotype Inference. IX, 153 pages. 2004. (Subseries LNBI).

Vol. 2982: N. Wakamiya, M. Solarski, J. Sterbenz (Eds.), Active Networks. XI, 308 pages. 2004.

Vol. 2981: C. Müller-Schloer, T. Ungerer, B. Bauer (Eds.), Organic and Pervasive Computing – ARCS 2004. XI, 339 pages. 2004.

Vol. 2980: A. Blackwell, K. Marriott, A. Shimojima (Eds.), Diagrammatic Representation and Inference. XV, 448 pages. 2004. (Subseries LNAI).

Vol. 2979: I. Stoica, Stateless Core: A Scalable Approach for Quality of Service in the Internet. XVI, 219 pages. 2004.

Vol. 2978: R. Groz, R.M. Hierons (Eds.), Testing of Communicating Systems. XII, 225 pages. 2004.